This book provides an extensive introduction to the nu class of integral equations. The initial chapters provide a general name for the numerical analysis of Fredholm integral equations of the second kind, covering degenerate kernel, projection, and Nyström methods. Additional discussions of multivariable integral equations and iteration methods update the reader on the present state of the art in this area.

The final chapters focus on the numerical solution of boundary integral equation (BIE) reformulations of Laplace's equation, in both two and three dimensions. Two chapters are devoted to planar BIE problems, which include both existing methods and remaining questions. Practial problems for BIE such as the set up and solution of the discretized BIE are also discussed.

Each chapter concludes with a discussion of the literature, and a large bibliography serves as an extended resource for students and researchers needing more information on solving particular integral equations.

CAMBRIDGE MONOGRAPHS ON
APPLIED AND COMPUTATIONAL
MATHEMATICS

Series Editors
P. G. CIARLET, A. ISERLES, R. V. KOHN, M. H. WRIGHT

4 The Numerical Solution of Integral Equations of the Second Kind

The *Cambridge Monographs on Applied and Computational Mathematics* reflects the crucial role of mathematical and computational techniques in contemporary science. The series publishes expositions on all aspects of applicable and numerical mathematics, with an emphasis on new developments in this fast-moving area of research.

State-of-the-art methods and algorithms as well as modern mathematical descriptions of physical and mechanical ideas are presented in a manner suited to graduate research students and professionals alike. Sound pedagogical presentation is a prerequisite. It is intended that books in the series will serve to inform a new generation of researchers.

Also in this series:

A Practical Guide to Pseudospectral Methods, *Bengt Fornberg*

Level Set Methods, *J.A. Sethian*

Dynamical Systems and Numerical Analysis, *A.M. Stuart and A.R. Humphries*

The Numerical Solution of Integral Equations of the Second Kind

KENDALL E. ATKINSON
University of Iowa

CAMBRIDGE
UNIVERSITY PRESS

CAMBRIDGE UNIVERSITY PRESS
Cambridge, New York, Melbourne, Madrid, Cape Town, Singapore, São Paulo, Delhi

Cambridge University Press
The Edinburgh Building, Cambridge CB2 8RU, UK

Published in the United States of America by Cambridge University Press, New York

www.cambridge.org
Information on this title: www.cambridge.org/9780521102834

First published 1997
This digitally printed version 2009

A catalogue record for this publication is available from the British Library

Library of Congress Cataloguing in Publication data
Atkinson, Kendall E.
The numerical solution of integral equations of the second kind /
Kendall E. Atkinson.
p.   cm.
Includes bibliographical references (p.   –   ) and index.
ISBN 0-521-58391-8 (hc)
1. Integral equations – Numerical solutions.   I. Title.
QA431.A837   1997
551'.46 – dc20   96-45961
                                                                CIP

ISBN 978-0-521-58391-6 hardback
ISBN 978-0-521-10283-4 paperback

To Alice

Contents

vii

Preface

In this book, numerical methods are presented and analyzed for the solution of integral equations of the second kind, especially Fredholm integral equations. Major additions have been made to this subject in the twenty years since the publication of my survey [39], and I present here an up-to-date account of the subject. In addition, I am interested in methods that are suitable for the solution of boundary integral equation reformulations of Laplace's equation, and three chapters are devoted to the numerical solution of such boundary integral equations. Boundary integral equations of the first kind that have a numerical theory closely related to that for boundary integral equations of the second kind are also discussed.

This book is directed toward several audiences. It is first directed to numerical analysts working on the numerical solution of integral equations. Second, it is directed toward applied mathematicians, including both those interested directly in integral equations and those interested in solving elliptic boundary value problems by use of boundary integral equation reformulations. Finally, it is directed toward that very large group of engineers needing to solve problems involving integral equations. In all of these cases, I hope the book is also readable and useful to well-prepared graduate students, as I had them in mind when writing the book.

During the period of 1960–1990, there has been much work on developing and analyzing numerical methods for solving linear Fredholm integral equations of the second kind, with the integral operator being compact on a suitable space of functions. I believe this work is nearing a stage in which there will be few major additions to the theory, especially as regards equations for functions of a single variable; but I hope to be proven wrong. In Chapters 2 through 6, the main aspects of the theory of numerical methods for such integral equations is presented, including recent work on solving integral equations on surfaces in $\mathbf{R}^3$. Chapters 7 through 9 contain a presentation of numerical methods for solving some boundary integral equation reformulations of Laplace's equation, for problems in both two and three dimensions. By restricting the presentation to Laplace's equation, a simpler development can be given than is needed when dealing with the large variety of boundary integral equations that have been studied during the past twenty years. For a more complete development of the numerical solution of all forms of boundary integral equations for planar problems, see Prößdorf and Silbermann [438].

In Chapter 1, a brief introduction/review is given of the classical theory of Fredholm integral equations of the second kind in which the integral operator is compact. In presenting the theory of this and the following chapters, a functional analysis framework is used, which is generally set in the space of continuous functions $C(D)$ or the space of square integrable functions $L^2(D)$. Much recent work has been in the framework of Sobolev spaces $H^r(D)$, which is used in portions of Chapter 7, but I believe the simpler framework given here is accessible to a wider audience in the applications community. Therefore, I have chosen this simpler sframework in preference to regarding boundary integral equations as pseudodifferential operator equations on Sobolev spaces. The reader still will need to have some knowledge of functional analysis, although not a great deal, and a summary of some of the needed results from functional analysis is presented in the appendix.

I would like to thank Mihai Anitescu, Paul Martin, Matthew Schuette, and Jaehoon Seol, who found many typographical and other errors in the book. It is much appreciated. I thank the production staff at TechBooks, Inc. for their fine work in turning my manuscript into the book you see here. Finally, I also thank my wife Alice, who as always has been very supportive of me during the writing of this book.

1

A brief discussion of integral equations

The theory and application of integral equations is an important subject within applied mathematics. Integral equations are used as mathematical models for many and varied physical situations, and integral equations also occur as reformulations of other mathematical problems. We begin with a brief classification of integral equations, and then in later sections, we give some of the classical theory for one of the most popular types of integral equations, those that are called *Fredholm integral equations of the second kind*, which are the principal subject of this book. There are many well-written texts on the theory and application of integral equations, and we note particularly those of Hackbusch [249] Hochstadt [272], Kress [325], Mikhlin [380], Pogorzelski [426], Schmeidler [492], Widom [568], and Zabreyko, et al. [586].

1.1. Types of integral equations

This book is concerned primarily with the numerical solution of what are called Fredholm integral equations, but we begin by discussing the broader category of integral equations in general. In classifying integral equations, we say, very roughly, that those integral equations in which the integration domain varies with the independent variable in the equation are *Volterra integral equations*; and those in which the integration domain is fixed are Fredholm integral equations. We first consider these two types of equations, and the section concludes with some other important integral equations.

1.1.1. Volterra integral equations of the second kind

The general form that is studied is

$$x(t) + \int_a^t K(t, s, x(s)) \, ds = y(t), \quad t \geq a \qquad (1.1.1)$$

1

The functions $K(t, s, u)$ and $y(t)$ are given, and $x(t)$ is sought. This is a *nonlinear* integral equation, and it is in this form that the equation is most commonly applied and solved. Such equations can be thought of as generalizations of

$$x'(t) = f(t, x(t)), \quad t \geq a, \quad x(a) = x_0 \tag{1.1.2}$$

the *initial value problem for ordinary differential equations*. This equation is equivalent to the integral equation

$$x(t) = x_0 + \int_a^t f(s, x(s)) \, ds, \quad t \geq a$$

which is a special case of (1.1.1).

For an introduction to the theory of Volterra integral equations, see R. Miller [384]. The numerical methods for solving (1.1.1) are closely related to those for solving the initial value problem (1.1.2). These integral equations are not studied in this book, and the reader is referred to Brunner and de Riele [96] and Linz [345]. Volterra integral equations are most commonly studied for functions x of one variable, as above, but there are examples of Volterra integral equations for functions of more than one variable.

1.1.2. Volterra integral equations of the first kind

The general nonlinear Volterra integral equation of the first kind has the form

$$\int_a^t K(t, s, x(s)) \, ds = y(t), \quad t \geq a \tag{1.1.3}$$

The functions $K(t, s, u)$ and $y(t)$ are given functions, and the unknown is $x(s)$. The general linear Volterra integral equation of the first kind is of the form

$$\int_a^t K(t, s)x(s) \, ds = y(t), \quad t \geq a \tag{1.1.4}$$

For Volterra equations of the first kind, the linear equation is the more commonly studied case. The difficulty with these equations, linear or nonlinear, is that they are "ill-conditioned" to some extent, and that makes their numerical solution more difficult. (Loosely speaking, an *ill-conditioned problem* is one in which small changes in the data y can lead to much larger changes in the solution x.)

A very simple but important example of (1.1.4) is

$$\int_a^t x(s) \, ds = y(t), \quad t \geq a \tag{1.1.5}$$

This is equivalent to $y(a) = 0$ and $x(t) = y'(t)$, $t \geq a$. Thus the numerical solution of (1.1.5) is equivalent to the numerical differentiation of $y(t)$. For a discussion of the numerical differentiation problem from this perspective, see Cullum [149] and Anderssen and Bloomfield [11], and for the numerical solution of the more general equation (1.1.4), see Linz [345].

1.1.3. Abel integral equations of the first kind

An important case of (1.1.4) is the *Abel integral equation*

$$\int_0^t \frac{H(t,s)x(s)}{(t^p - s^p)^\alpha} \, ds = y(t), \quad t > 0 \tag{1.1.6}$$

Here $0 < \alpha < 1$ and $p > 0$, and particularly important cases are $p = 1$ and $p = 2$ (both with $\alpha = \frac{1}{2}$). The function $H(t,s)$ is assumed to be smooth (that is, several times continuously differentiable). Special numerical methods have been developed for these equations, as they occur in a wide variety of applications. For a general solvability theory for (1.1.6), see Ref. [35], and for a discussion of numerical methods for their solution, see Linz [345], Brunner and de Riele [96], and Anderssen and de Hoog [12].

1.1.4. Fredholm integral equations of the second kind

The general form of such an integral equation is

$$\lambda x(t) - \int_D K(t,s)x(s) \, ds = y(t), \quad t \in D, \ \lambda \neq 0 \tag{1.1.7}$$

with D a closed bounded set in $\mathbf{R}^m$, some $m \geq 1$. The *kernel function* $K(t,s)$ is assumed to be absolutely integrable, and it is assumed to satisfy other properties that are sufficient to imply the Fredholm Alternative Theorem (see Theorem 1.3.1 in §1.3). For $y \neq 0$, we have λ and y given, and we seek x; this is the *nonhomogeneous problem*. For $y = 0$, equation (1.1.7) becomes an *eigenvalue problem*, and we seek both the *eigenvalue* λ and the *eigenfunction* x. The principal focus of the numerical methods presented in the following chapters is the numerical solution of (1.1.7) with $y \neq 0$. In the next two sections we present some theory for the integral operator in (1.1.7).

1.1.5. Fredholm integral equations of the first kind

These equations take the form

$$\int_D K(t,s)x(s) \, ds = y(t), \quad t \in D \tag{1.1.8}$$

with the assumptions on K and D the same as in (1.1.7). Such equations are usually classified as ill-conditioned, because their solution x is sensitive to small changes in the data function y. For practical purposes, however, these problems need to be subdivided into two categories. First, if $K(t, s)$ is a smooth function, then the solution $x(s)$ of (1.1.8) is extremely sensitive to small changes in $y(t)$, and special methods of solution are needed. For excellent introductions to this topic, see Groetsch [241], [242], Kress [325, Chaps. 15–17] and Wing [572]. If however, $K(t, s)$ is a *singular* function, then the ill-conditioning of (1.1.8) is quite manageable; and indeed, much of the theory for such equations is quite similar to that for the second-kind equation (1.1.7). Examples of this type of first-kind equation occur quite frequently in the subject of *potential theory*, and a well-studied example is

$$\int_\Gamma \log |t - s| x(s)\, ds = y(t), \quad t \in \Gamma \tag{1.1.9}$$

with Γ a curve in $\mathbf{R}^2$. This and other similarly behaved first-kind equations will be discusssed in Chapters 7 and 8.

1.1.6. Boundary integral equations

These equations are integral equation reformulations of partial differential equations. They are widely studied and applied in connection with solving boundary value problems for elliptic partial differential equations, but they are also used in connection with other types of partial differential equations.

As an example, consider solving the problem

$$\Delta u(P) = 0, \quad P \in D \tag{1.1.10}$$

$$u(P) = g(P), \quad P \in \Gamma \tag{1.1.11}$$

where D is a bounded region in $\mathbf{R}^3$ with nonempty interior, and Γ is the boundary of D. From the physical setting for (1.1.10)–(1.1.11), there is reason to believe that u can be written as a *single layer potential:*

$$u(P) = \int_\Gamma \frac{\rho(Q)}{|P - Q|} dQ, \quad P \in D \tag{1.1.12}$$

In this, $|P - Q|$ denotes the ordinary Euclidean distance between P and Q. The function $\rho(Q)$ is called a *single layer density function*, and it is the unknown in the equation. Using the boundary condition (1.1.11), it is straightforward to show that

$$\int_\Gamma \frac{\rho(Q)}{|P - Q|} dQ = g(P), \quad P \in \Gamma \tag{1.1.13}$$

This equation is solved for ρ, and then (1.1.12) is used to obtain the solution of (1.1.10)–(1.1.11).

Boundary integral equations can be Fredholm integral equations of the first or second kind, Cauchy singular integral equations (see 1.1.8 below), or modifications of them. In the literature, boundary integral equations are often referred to as *BIE*, and methods for solving partial differential equations via the boundary integral equation reformulation are called *BIE methods*. There are many books and papers written on BIE methods; for example, see Refs. [50], [55], Jaswon and Symm [286], Kress [325, Chaps. 6, 8, 9], Sloan [509] and the references contained therein.

1.1.7. Wiener-Hopf integral equations

These have the form

$$\lambda x(t) - \int_0^\infty k(t - s) x(s) \, ds = y(t), \quad 0 \le t < \infty \tag{1.1.14}$$

Originally, such equations were studied in connection with problems in radiative transfer, and more recently, they have been related to the solution of boundary integral equations for planar problems in which the boundary is only piecewise smooth. A very extensive theory for such equations is given in Krein [322], and more recently, a simpler introduction to some of the more important parts of the theory has been given by Anselone and Sloan [20] and deHoog and Sloan [163].

1.1.8. Cauchy singular integral equations

Let Γ be an open or closed contour in the complex plane. The general form of a Cauchy singular integral equation is given by

$$a(z)\phi(z) + \frac{b(z)}{\pi i} \int_\Gamma \frac{\phi(\zeta)}{\zeta - z} \, d\zeta + \int_\Gamma K(z, \zeta)\phi(\zeta) \, d\zeta = \psi(z), \quad z \in \Gamma \tag{1.1.15}$$

The functions a, b, ψ, and K are given complex-valued functions, and ϕ is the unknown function. The function K is to be absolutely integrable; and in addition, it is to be such that the associated integral operator is a Fredholm

integral operator in the sense of 1.1.4 above. The first integral in (1.1.15) is
interpreted as a *Cauchy principal value integral*:

$$\int_\Gamma \frac{\phi(\zeta)}{\zeta - z}\, d\zeta = \lim_{\epsilon \to 0^+} \int_{\Gamma_\epsilon} \frac{\phi(\zeta)}{\zeta - z}\, d\zeta \qquad (1.1.16)$$

with $\Gamma_\epsilon = \{\zeta \in \Gamma \mid |\zeta - z| \geq \epsilon\}$. Cauchy singular integral equations occur
in a variety of physical problems, especially in connection with the solution
of partial differential equations in $\mathbf{R}^2$. Among the best known books on the
theory and application of Cauchy singular integral equations are Muskhelishvili
[390] and Gakhov [208]; and an important more recent work is that of Mikhlin
and Prößdorf [381]. For an introduction to the numerical solution of Cauchy
singular integral equations, see Elliott [177], [178] and Prößdorf and Silbermann
[438].

1.2. Compact integral operators

The framework we present in this and the following sections is fairly abstract,
and it might seem far removed from the numerical solution of actual integral
equations. In fact, our framework is needed to understand the behavior of
most numerical methods for solving Fredholm integral equations, including
the answering of questions regarding convergence, numerical stability, and
asymptotic error estimates. The language of functional analysis has become
more standard in the past few decades, and so in contrast to our earlier book [39,
Part I], we do not develop here any results from functional analysis, but rather
state them in the appendix and refer the reader to other sources for proofs.

When $\mathcal{X}$ is a finite dimensional vector space and $\mathcal{A} : \mathcal{X} \to \mathcal{X}$ is linear, the
equation $\mathcal{A}x = y$ has a well-developed solvability theory. To extend these
results to infinite dimensional spaces, we introduce the concept of a *compact
operator* $\mathcal{K}$; and then in the following section, we give a theory for operator
equations $\mathcal{A}x = y$ in which $\mathcal{A} = I - \mathcal{K}$.

Definition. *Let $\mathcal{X}$ and $\mathcal{Y}$ be normed vector spaces, and let $\mathcal{K} : \mathcal{X} \to \mathcal{Y}$ be
linear. Then $\mathcal{K}$ is* compact *if the set*

$$\{\mathcal{K}x \mid \|x\|_\mathcal{X} \leq 1\}$$

*has compact closure in $\mathcal{Y}$. This is equivalent to saying that for every bounded
sequence $\{x_n\} \subset \mathcal{X}$, the sequence $\{\mathcal{K}x_n\}$ has a subsequence that is convergent
to some point in $\mathcal{Y}$. Compact operators are also called* completely continuous
operators. (By a set S having compact closure *in $\mathcal{Y}$, we mean its closure $\bar{S}$ is a
compact set in $\mathcal{Y}$.)*

There are other definitions for a compact operator, but the above is the one used most commonly. In the definition, the spaces $\mathcal{X}$ and $\mathcal{Y}$ need not be complete; but in virtually all applications, they are complete. With completeness, some of the proofs of the properties of compact operators become simpler, and we will always assume $\mathcal{X}$ and $\mathcal{Y}$ are complete (that is, Banach spaces) when dealing with compact operators.

1.2.1. Compact integral operators on C(D)

Let D be a closed bounded set in $\mathbf{R}^m$, some $m \geq 1$, and define

$$\mathcal{K}x(t) = \int_D K(t, s)x(s)\, ds, \quad t \in D, \quad x \in C(D) \qquad (1.2.17)$$

Using $C(D)$ with $\|\cdot\|_\infty$, we want to show that $\mathcal{K}: C(D) \to C(D)$ is both bounded and compact. We assume $K(t, s)$ is Riemann-integrable as a function of s, for all $t \in D$, and further we assume the following.

K1. $\lim_{h \to 0} \omega(h) = 0$, with

$$\omega(h) \equiv \max_{t, \tau \in D} \max_{|t - \tau| \leq h} \int_D |K(t, s) - K(\tau, s)|\, ds \qquad (1.2.18)$$

K2.

$$\max_{t \in D} \int_D |K(t, s)|\, ds < \infty \qquad (1.2.19)$$

Using K1, if $x(s)$ is bounded and integrable, then $\mathcal{K}x(t)$ is continuous, with

$$|\mathcal{K}x(t) - \mathcal{K}x(\tau)| \leq \omega(|t - \tau|)\|x\|_\infty \qquad (1.2.20)$$

Using K2, we have boundedness of $\mathcal{K}$, with

$$\|\mathcal{K}\| = \max_{t \in D} \int_D |K(t, s)|\, ds \qquad (1.2.21)$$

To discuss compactness of $\mathcal{K}$, we first need to identify the compact sets in $C(D)$. To do this, we use the Arzela-Ascoli theorem from advanced calculus. It states that a subset $S \subset C(D)$ has compact closure if and only if (1) S is a uniformly bounded set of functions, and (2) S is an equicontinuous family. Now consider the set $S = \{\mathcal{K}x \mid x \in C(D) \text{ and } \|x\|_\infty \leq 1\}$. This is uniformly bounded, since $\|\mathcal{K}x\|_\infty \leq \|\mathcal{K}\|\|x\|_\infty \leq \|\mathcal{K}\|$. In addition, S is equicontinuous from (1.2.20). Thus S has compact closure in $C(D)$, and $\mathcal{K}$ is a compact operator on $C(D)$ to $C(D)$.

What are the kernel functions K that satisfy K1–K2? Easily, these assumptions are satisfied if $K(t, s)$ is a continuous function of $(t, s) \in D$. In addition, let $D = [a, b]$ and consider

$$\mathcal{K}x(t) = \int_a^b \log|t - s| \, x(s) \, ds \qquad (1.2.22)$$

and

$$\mathcal{K}x(t) = \int_a^b \frac{1}{|s - t|^\beta} x(s) \, ds \qquad (1.2.23)$$

with $\beta < 1$. These definitions of $\mathcal{K}$ can be shown to satisfy K1–K2, although we will not do so here. Later we will show by other means that these are compact operators.

Another way to show that $K(t, s)$ satisfies K1 and K2 is to rewrite K in the form

$$K(t, s) = \sum_{i=0}^p H_i(t, s) L_i(t, s) \qquad (1.2.24)$$

for some $p > 0$, with each $L_i(t, s)$ continuous for $a \leq t, s \leq b$ and each $H_i(t, s)$ satisfying K1–K2. It is left to the reader to show that in this case, K also satisfies K1–K2. The utility of this approach is that it is sometimes difficult to show directly that K satisfies K1–K2, whereas showing (1.2.24) may be easier.

Example. Let $[a, b] = [0, \pi]$ and $K(t, s) = \log|\cos t - \cos s|$. Then rewrite this as

$$K(t, s) = \underbrace{|s - t|^{-\frac{1}{2}}}_{H(t,s)} \underbrace{|s - t|^{\frac{1}{2}} \log|\cos t - \cos s|}_{L(t,s)} \qquad (1.2.25)$$

Easily, L is continuous; and from the discussion following (1.2.23), H satisfies K1–K2. Thus K is the kernel of a compact integral operator on $C[0, \pi]$ to $C[0, \pi]$.

1.2.2. Properties of compact operators

Another way of obtaining compact operators is to look at limits of simpler "finite-dimensional operators" in $L[\mathcal{X}, \mathcal{Y}]$, the Banach space of bounded linear operators from $\mathcal{X}$ to $\mathcal{Y}$. This gives another perspective on compact operators, one that leads to improved intuition by emphasizing their relationship to operators on finite dimensional spaces.

Definition. *Let $\mathcal{X}$ and $\mathcal{Y}$ be vector spaces. The linear operator $\mathcal{K}: \mathcal{X} \to \mathcal{Y}$ is a* finite rank operator *if $Range(\mathcal{K})$ is finite dimensional.*

Lemma 1.2.1. Let $\mathcal{X}$ and $\mathcal{Y}$ be normed linear spaces, and let $\mathcal{K}: \mathcal{X} \to \mathcal{Y}$ be a bounded finite rank operator. Then $\mathcal{K}$ is a compact operator.

Proof. Let $\mathcal{R} = Range(\mathcal{K})$. Then $\mathcal{R}$ is a normed finite-dimensional space, and therefore it is complete. Consider the set

$$S = \{\mathcal{K}x \mid \|x\| \le 1\}$$

The set S is bounded by $\|\mathcal{K}\|$. Also $S \subset \mathcal{R}$. Then S has compact closure, since all bounded closed sets in a finite dimensional space are compact. This shows $\mathcal{K}$ is compact. $\qquad\Box$

Example. Let $\mathcal{X} = \mathcal{Y} = C[a, b]$ with $\|\cdot\|_\infty$. Consider the kernel function

$$K(t, s) = \sum_{i=1}^{n} \beta_i(t)\gamma_i(s) \tag{1.2.26}$$

with each β_i continuous on $[a, b]$ and each $\gamma_i(s)$ absolutely integrable on $[a, b]$. Then the associated integral operator $\mathcal{K}$ is a bounded, finite rank operator on $C[a, b]$ to $C[a, b]$:

$$\mathcal{K}x(t) = \sum_{i=1}^{n} \beta_i(t) \int_a^b \gamma_i(s)x(s)\,ds, \quad x \in C[a, b] \tag{1.2.27}$$

$$\|\mathcal{K}\| \le \sum_{i=1}^{n} \|\beta_i\|_\infty \int_a^b |\gamma_i(s)|\,ds$$

From (1.2.27), $\mathcal{K}x \in C[a, b]$ and $Range(\mathcal{K}) \subset Span\{\beta_1, \ldots, \beta_n\}$, a finite dimensional space. Kernel functions of the form (1.2.26) are called *degenerate*.

In the following section we will see that the associated integral equation $(\lambda - \mathcal{K})x = y$, $\lambda \ne 0$ is essentially a finite dimensional equation.

Lemma 1.2.2. Let $\mathcal{K} \in L[\mathcal{X}, \mathcal{Y}]$ and $\mathcal{L} \in L[\mathcal{Y}, \mathcal{Z}]$, and let $\mathcal{K}$ or $\mathcal{L}$ (or both) be compact. Then $\mathcal{L}\mathcal{K}$ is compact on $\mathcal{X}$ to $\mathcal{Z}$.

Proof. This is left as an exercise for the reader. $\qquad\Box$

The following lemma gives the framework for using finite rank operators to obtain similar, but more general compact operators.

Lemma 1.2.3. Let $\mathcal{X}$ and $\mathcal{Y}$ be normed linear spaces, with $\mathcal{Y}$ complete. Let $\mathcal{K} \in L[\mathcal{X}, \mathcal{Y}]$, let $\{\mathcal{K}_n\}$ be a sequence of compact operators in $L[\mathcal{X}, \mathcal{Y}]$, and assume $\mathcal{K}_n \to \mathcal{K}$ in $L[\mathcal{X}, \mathcal{Y}]$, i.e., $\|\mathcal{K}_n - \mathcal{K}\| \to 0$. Then $\mathcal{K}$ is compact.

Proof. Let $\{x_n\}$ be a sequence in $\mathcal{X}$ satisfying $\|x_n\| \leq 1, n \geq 1$. We must show that $\{\mathcal{K}x_n\}$ contains a convergent subsequence.

Since $\mathcal{K}_1$ is compact, the sequence $\{\mathcal{K}_1 x_n\}$ contains a convergent subsequence. Denote the convergent subsequence by $\{\mathcal{K}_1 x_n^{(1)} \mid n \geq 1\}$, and let its limit be denoted by $y_1 \in \mathcal{Y}$. For $k \geq 2$, inductively pick a subsequence $\{x_n^{(k)} \mid n \geq 1\} \subset \{x_n^{(k-1)}\}$ such that $\{\mathcal{K}_k x_n^{(k)}\}$ converges to a point $y_k \in \mathcal{Y}$. Thus,

$$\lim_{n \to \infty} \mathcal{K}_k x_n^{(k)} = y_k \quad \text{and} \quad \{x_n^{(k)}\} \subset \{x_n^{(k-1)}\}, \quad k \geq 1 \qquad (1.2.28)$$

We will now choose a special subsequence $\{z_k\} \subset \{x_n\}$ for which $\{\mathcal{K}z_k\}$ is convergent in $\mathcal{Y}$. Let $z_1 = x_j^{(1)}$ for some j, such that $\|\mathcal{K}_1 x_n^{(1)} - y_1\| \leq 1$ for all $n \geq j$. Inductively, for $k \geq 2$, pick $z_k = x_j^{(k)}$ for some j, such that z_k is further along in the sequence $\{x_n\}$ than is z_{k-1} and such that

$$\left\| \mathcal{K}_k x_n^{(k)} - y_k \right\| \leq \frac{1}{k}, \quad n \geq j \qquad (1.2.29)$$

The sequence $\{\mathcal{K}z_k\}$ is a Cauchy sequence in $\mathcal{Y}$. To show this, consider

$$\|\mathcal{K}z_{k+p} - \mathcal{K}z_k\| \leq \|\mathcal{K}z_{k+p} - \mathcal{K}_k z_{k+p}\| + \|\mathcal{K}_k z_{k+p} - \mathcal{K}_k z_k\|$$

$$+ \|\mathcal{K}_k z_k - \mathcal{K}z_k\|$$

$$\leq 2\|\mathcal{K} - \mathcal{K}_k\| + \|\mathcal{K}_k z_{k+p} - y_k\| + \|y_k - \mathcal{K}_k z_k\|$$

$$\leq 2\|\mathcal{K} - \mathcal{K}_k\| + \frac{2}{k}, \quad p \geq 1$$

The last statement uses (1.2.28)–(1.2.29), noting that $z_{k+p} \in \{x_n^{(k)}\}$ for all $p \geq 1$. Use the assumption that $\|\mathcal{K} - \mathcal{K}_k\| \to 0$ to conclude the proof that $\{\mathcal{K}z_k\}$ is a Cauchy sequence in $\mathcal{Y}$. Since $\mathcal{Y}$ is complete, $\{\mathcal{K}z_k\}$ is convergent in $\mathcal{Y}$, and this shows that $\mathcal{K}$ is compact. $\square$

For almost all function spaces $\mathcal{X}$ of interest, the compact operators can be characterized as being the limit of a sequence of bounded finite-rank operators. This gives a further justification for the presentation of Lemma 1.2.3.

Example. Let D be a closed and bounded set in $\mathbf{R}^m$, some $m \geq 1$. For example, D could be a region with nonempty interior, a piecewise smooth surface, or a piecewise smooth curve. Let $K(t, s)$ be a continuous function of $t, s \in D$. Then suppose we can define a sequence of continuous degenerate kernel functions $K_n(t, s)$ for which

$$\max_{t \in D} \int_D |K(t, s) - K_n(t, s)| \, ds \to 0 \quad \text{as} \quad n \to \infty \quad (1.2.30)$$

Then for the associated integral operators, it easily follows that $\mathcal{K}_n \to \mathcal{K}$; and by Lemma 1.2.3, $\mathcal{K}$ is compact. The result (1.2.30) is true for general continuous functions $K(t, s)$, and we leave to the problems the proof for various choices of D. Of course, we already knew that $\mathcal{K}$ was compact in this case, from the discussion following (1.2.23). But the present approach shows the close relationship of compact operators and finite dimensional operators.

Example. Let $\mathcal{X} = \mathcal{Y} = C[a, b]$ with $\|\cdot\|_\infty$. Consider the kernel function

$$K(t, s) = \frac{1}{|t - s|^\gamma} \quad (1.2.31)$$

for some $0 < \gamma < 1$. Define a sequence of continuous kernel functions to approximate it:

$$K_n(t, s) = \begin{cases} \dfrac{1}{|t - s|^\gamma}, & |t - s| \geq \dfrac{1}{n} \\[2ex] n^\gamma, & |t - s| \leq \dfrac{1}{n} \end{cases} \quad (1.2.32)$$

For the associated integral operators,

$$\|\mathcal{K} - \mathcal{K}_n\| = \frac{2\gamma}{1 - \gamma} n^{\gamma - 1}$$

which converges to zero as $n \to \infty$. By Lemma 1.2.3, $\mathcal{K}$ is a compact operator on $C[a, b]$.

1.2.3. Integral operators on $L^2(a, b)$

Let $\mathcal{X} = \mathcal{Y} = L^2(a, b)$, and let $\mathcal{K}$ be the integral operator associated with $K(t, s)$. We first show that under suitable assumptions on K, $\mathcal{K} : L^2(a, b) \to L^2(a, b)$. Let

$$M = \left[\int_a^b \int_a^b |K(t, s)|^2 \, ds \, dt \right]^{1/2} \quad (1.2.33)$$

and assume $M < \infty$. For $x \in L^2(a, b)$, use the Cauchy-Schwarz inequality to obtain

$$\|\mathcal{K}x\|_2^2 = \int_a^b \left| \int_a^b K(t,s)x(s)\,ds \right|^2 dt$$

$$\leq \int_a^b \left[\int_a^b |K(t,s)|^2\,ds \right] \left[\int_a^b |x(s)|^2\,ds \right] dt$$

$$= M^2 \|x\|_2^2$$

This proves that $\mathcal{K}x \in L^2(a,b)$, and it shows

$$\|\mathcal{K}\| \leq M \qquad (1.2.34)$$

As a side note, the *Cauchy-Schwarz inequality* says the following: If $\mathcal{X}$ is an inner product space with inner product $(\cdot,\cdot)$ and norm $\|\cdot\|$, then

$$|(x,y)| \leq \|x\|\,\|y\|, \quad x,y \in \mathcal{X}$$

For example, see [48, p. 208].

This bound is comparable to the use of the Frobenius matrix norm to bound the operator norm of a matrix $A \colon \mathbf{R}^n \to \mathbf{R}^m$, when the vector norm $\|\cdot\|_2$ is being used. Kernel functions K for which $M < \infty$ are called *Hilbert-Schmidt kernel functions*, and the quantity M in (1.2.33) is called the *Hilbert-Schmidt norm* of $\mathcal{K}$.

For integral operators $\mathcal{K}$ with a degenerate kernel function as in (1.2.26), the operator $\mathcal{K}$ is bounded if all $\beta_i, \gamma_i \in L^2(a,b)$. This is a straightforward result, which we leave as a problem for the reader. From Lemma 1.2.1, the integral operator is then compact.

To examine the compactness of $\mathcal{K}$ for more general kernel functions, we assume there is a sequence of kernel functions $K_n(t,s)$ for which (*i*) $\mathcal{K}_n \colon L^2(a,b) \to L^2(a,b)$ is compact, and (*ii*)

$$M_n \equiv \left[\int_a^b \int_a^b |K(t,s) - K_n(t,s)|^2\,ds\,dt \right]^{\frac{1}{2}} \to 0 \quad \text{as} \quad n \to \infty$$

$$(1.2.35)$$

For example, if K is continuous, then this follows from (1.2.30). The operator $\mathcal{K} - \mathcal{K}_n$ is an integral operator, and we apply (1.2.33)–(1.2.34) to it to obtain

$$\|\mathcal{K} - \mathcal{K}_n\| \leq M_n \to 0 \quad \text{as} \quad n \to \infty$$

From Lemma 1.2.3, this shows $\mathcal{K}$ is compact. For any Hilbert-Schmidt kernel function, (1.2.35) can be shown to hold for a suitable choice of degenerate kernel functions K_n.

We leave it to the problems to show that $\log|t-s|$ and $|t-s|^{-\gamma}$, $\gamma < \frac{1}{2}$ are Hilbert-Schmidt kernel functions. For $\frac{1}{2} \leq \gamma < 1$, the kernel function

$|t - s|^{-\gamma}$ still defines a compact integral operator $\mathcal{K}$ on $L^2(a, b)$, but the above theory for Hilbert-Schmidt kernel functions does not apply. For a proof of the compactness of $\mathcal{K}$ in this case, see Mikhlin [380, p. 160].

1.3. The Fredholm alternative theorem

Integral equations were studied in the nineteenth century as one means of investigating boundary value problems for Laplace's equation (1.1.10)–(1.1.11) and other elliptic partial differential equations. In the early 1900s, Ivar Fredholm gave necessary and sufficient conditions for the solvability of a large class of Fredholm integral equations of the second kind; and with these results, he then was able to give much more general existence theorems for the solution of boundary value problems such as (1.1.10)–(1.1.11). In this section we state and prove the most important result of Fredholm; and in the following section, we give additional results without proof.

The theory of integral equations has been due to many people, with David Hilbert being among the most important popularizers of the area. Integral equations continue as an important area of study in applied mathematics; for an introduction that includes a review of much recent literature, see Kress [325]. For an interesting historical account of the development of functional analysis as it was affected by the theory of integral equations, see Bernkopf [78].

Theorem 1.3.1 (Fredholm alternative). *Let $\mathcal{X}$ be a Banach space, and let $\mathcal{K}: \mathcal{X} \to \mathcal{X}$ be compact. Then the equation $(\lambda - \mathcal{K})x = y$, $\lambda \neq 0$ has a unique solution $x \in \mathcal{X}$ if and only if the homogeneous equation $(\lambda - \mathcal{K})z = 0$ has only the trivial solution $z = 0$. In such a case, the operator $\lambda - \mathcal{K}: \mathcal{X} \overset{1-1}{\underset{onto}{\to}} \mathcal{X}$ has a bounded inverse $(\lambda - \mathcal{K})^{-1}$.*

Proof. The theorem is true for any compact operator $\mathcal{K}$, but our proof is only for those compact operators that are the limit of a sequence of bounded finite-rank operators. For a more general proof, see Kress [325, Chap. 3]. We remark that the theorem is a generalization of the following standard result for finite dimensional vector spaces $\mathcal{X}$: For A a matrix of order n, with $\mathcal{X} = \mathbf{R}^n$ or $\mathbf{C}^n$ (with A having real entries for the former case), the linear system $Ax = y$ has a unique solution $x \in \mathcal{X}$ for all $y \in \mathcal{X}$ if and only if the homogeneous linear system $Ax = 0$ has only the zero solution $x = 0$.

(a) We begin with the case where $\mathcal{K}$ is finite-rank and bounded. Let $\{u_1, \ldots, u_n\}$ be a basis for $Range(\mathcal{K})$. Rewrite the equation $(\lambda - \mathcal{K})x = y$ as

$$x = \frac{1}{\lambda}[y + \mathcal{K}x] \tag{1.3.36}$$

If this equation has a unique solution $x \in \mathcal{X}$, then

$$x = \frac{1}{\lambda}[y + c_1 u_1 + \cdots + c_n u_n] \tag{1.3.37}$$

for some uniquely determined set of constants $c_1, \ldots, c_n$.

By substituting (1.3.37) into the equation, we have

$$\lambda \left\{ \frac{1}{\lambda} y + \frac{1}{\lambda} \sum_{i=1}^{n} c_i u_i \right\} - \frac{1}{\lambda} \mathcal{K} y - \frac{1}{\lambda} \sum_{j=1}^{n} c_j \mathcal{K} u_j = y$$

Multiply by λ, and then simplify to obtain

$$\lambda \sum_{i=1}^{n} c_i u_i - \sum_{j=1}^{n} c_j \mathcal{K} u_j = \mathcal{K} y \tag{1.3.38}$$

Using the basis $\{u_i\}$ for $Range(\mathcal{K})$, write

$$\mathcal{K} y = \sum_{i=1}^{n} \gamma_i u_i, \qquad \mathcal{K} u_j = \sum_{i=1}^{n} a_{ij} u_i, \quad 1 \le j \le n$$

The coefficients $\{\gamma_i\}$ and $\{a_{ij}\}$ are uniquely determined. Substituting into (1.3.38) and rearranging,

$$\sum_{i=1}^{n} \left\{ \lambda c_i - \sum_{j=1}^{n} a_{ij} c_j \right\} u_i = \sum_{i=1}^{n} \gamma_i u_i$$

By the independence of the basis elements u_i, we obtain the linear system

$$\lambda c_i - \sum_{j=1}^{n} a_{ij} c_j = \gamma_i, \quad 1 \le i \le n \tag{1.3.39}$$

Claim. This linear system and the equation $(\lambda - \mathcal{K})x = y$ are completely equivalent in their solvability, with (1.3.37) furnishing a one-to-one correspondence between the solutions of the two of them.

We have shown above that if x is a solution of $(\lambda - \mathcal{K})x = y$, then $(c_1, \ldots, c_n)$ is a solution of (1.3.39). In addition, suppose x_1 and x_2 are distinct solutions of $(\lambda - \mathcal{K})x = y$. Then

$$\mathcal{K} x_1 = \lambda x_1 - y \quad \text{and} \quad \mathcal{K} x_2 = \lambda x_2 - y, \quad \lambda \ne 0,$$

are also distinct vectors in *Range*($\mathcal{K}$), and thus the associated vectors of coordinates $(c_1^{(1)}, \ldots, c_n^{(1)})$ and $(c_1^{(2)}, \ldots, c_n^{(2)})$,

$$\mathcal{K}x_i = \sum_{k=1}^{n} c_k^{(i)} u_k, \quad i = 1, 2$$

must also be distinct.

For the converse statement, suppose $(c_1, \ldots, c_n)$ is a solution of (1.3.39). Define a vector $x \in \mathcal{X}$ by using (1.3.37), and then check whether this x satisfies the integral equation (1.3.36):

$$
\begin{aligned}
(\lambda - \mathcal{K})x &= \lambda \left\{ \frac{1}{\lambda} y + \frac{1}{\lambda} \sum_{i=1}^{n} c_i u_i \right\} - \frac{1}{\lambda} \mathcal{K} y - \frac{1}{\lambda} \sum_{j=1}^{n} c_j \mathcal{K} u_j \\
&= y + \frac{1}{\lambda} \left\{ \lambda \sum_{i=1}^{n} c_i u_i - \mathcal{K} y - \sum_{j=1}^{n} c_j \mathcal{K} u_j \right\} \\
&= y + \frac{1}{\lambda} \left\{ \sum_{i=1}^{n} \lambda c_i u_i - \sum_{i=1}^{n} \gamma_i u_i - \sum_{j=1}^{n} c_j \sum_{i=1}^{n} a_{ij} u_i \right\} \\
&= y + \frac{1}{\lambda} \sum_{i=1}^{n} \underbrace{\left\{ \lambda c_i - \gamma_i - \sum_{j=1}^{n} a_{ij} c_j \right\}}_{=0,\ i=1,\ldots,n} u_i \\
&= y
\end{aligned}
$$

Also, distinct coordinate vectors $(c_1, \ldots, c_n)$ lead to distinct solution vectors x in (1.3.37), because of the linear independence of the basis vectors $\{u_1, \ldots, u_n\}$. This completes the proof of the claim given above.

Now consider the Fredholm alternative theorem for $(\lambda - \mathcal{K})x = y$ with this finite rank $\mathcal{K}$. Suppose $\lambda - \mathcal{K} : \mathcal{X} \overset{1-1}{\underset{onto}{\to}} \mathcal{X}$. Then trivially, $Null(\lambda - \mathcal{K}) = \{0\}$. For the converse, assume $(\lambda - \mathcal{K})z = 0$ has only the solution $z = 0$, and note that we want to show that $(\lambda - \mathcal{K})x = y$ has a unique solution for every $y \in \mathcal{X}$.

Consider the associated linear system (1.3.39). It can be shown to have a unique solution for all right-hand sides $(\gamma_1, \ldots, \gamma_n)$ by showing that the homogeneous linear system has only the zero solution. The latter is done by means of the equivalence of the homogeneous linear system to the homogeneous equation $(\lambda - \mathcal{K})z = 0$, which implies $z = 0$. But since (1.3.39) has a unique solution, so must $(\lambda - \mathcal{K})x = y$, and it is given by (1.3.37).

We must also show that $(\lambda - \mathcal{K})^{-1}$ is bounded. This can be done directly by a further examination of the consequences of $\mathcal{K}$ being a bounded and finite rank operator; but it is simpler to just cite the Open Mapping Theorem (cf. Theorem A.2 in the Appendix).

(b) Assume now that $\|\mathcal{K} - \mathcal{K}_n\| \to 0$, with $\mathcal{K}_n$ finite rank and bounded. Rewrite $(\lambda - \mathcal{K})x = y$ as

$$[\lambda - (\mathcal{K} - \mathcal{K}_n)]x = y + \mathcal{K}_n x, \quad n \geq 1 \tag{1.3.40}$$

Pick an index $m > 0$ for which

$$\|\mathcal{K} - \mathcal{K}_m\| < |\lambda| \tag{1.3.41}$$

and fix it. By the Geometric Series Theorem (cf. Theorem A.1 in the Appendix),

$$Q_m \equiv [\lambda - (\mathcal{K} - \mathcal{K}_m)]^{-1}$$

exists and is bounded, with

$$\|Q_m\| \leq \frac{1}{|\lambda| - \|\mathcal{K} - \mathcal{K}_m\|}$$

The equation (1.3.40) can now be written in the equivalent form

$$x - Q_m \mathcal{K}_m x = Q_m y \tag{1.3.42}$$

The operator $Q_m \mathcal{K}_m$ is bounded and finite rank. The boundedness follows from that of Q_m and $\mathcal{K}_m$. To show it is finite rank, let $Range(\mathcal{K}_m) = Span\{u_1, \ldots, u_m\}$. Then

$$Range(Q_m \mathcal{K}_m) = Span\{Q_m u_1, \ldots, Q_m u_m\}$$

a finite-dimensional space.

The equation (1.3.42) is one to which we can apply part (a) of this proof. Assume $(\lambda - \mathcal{K})z = 0$ implies $z = 0$. By the above equivalence, this yields

$$(I - Q_m \mathcal{K}_m)z = 0 \Rightarrow z = 0$$

But from part (a), this says $(I - Q_m \mathcal{K}_m)x = w$ has a unique solution x for every $w \in \mathcal{X}$, and in particular, for $w = Q_m y$ as in (1.3.42). By the equivalence of (1.3.42) and $(\lambda - \mathcal{K}) = y$, we have that the latter is uniquely solvable for every $y \in \mathcal{X}$. The boundedness of $(\lambda - \mathcal{K})^{-1}$ follows from part (a) and the boundedness of Q_m; or the Open Mapping Theorem can be cited, as earlier in part (a). □

For many practical problems in which $\mathcal{K}$ is not compact, it is important to note what makes this proof work. It is *not* necessary that a sequence of bounded and finite rank operators $\{\mathcal{K}_n\}$ exists for which $\|\mathcal{K} - \mathcal{K}_n\| \to 0$. Rather, it is necessary to have the inequality (1.3.41) be satisfied for one finite rank operator $\mathcal{K}_m$; and in applying the proof to other operators $\mathcal{K}$, it is necessary only that $\mathcal{K}_m$ be compact. In such a case, the proof following (1.3.41) remains valid, and the Fredholm Alternative still applies to such an equation $(\lambda - \mathcal{K})x = y$.

1.4. Additional results on Fredholm integral equations

In this section we give additional results on the solvability of compact equations of the second kind, $(\lambda - \mathcal{K})x = y$, with $\lambda \neq 0$. No proofs are given, and the reader is referred to the books cited at the beginning of the chapter or to one of the many other books on such equations.

Definition. *Let* $\mathcal{K} : \mathcal{X} \to \mathcal{X}$. *If there is a scalar* λ *and an associated vector* $x \neq 0$ *for which* $\mathcal{K}x = \lambda x$, *then* λ *is called an* eigenvalue *and* x *an associated* eigenvector *of the operator* $\mathcal{K}$. *(When dealing with compact operators* $\mathcal{K}$, *we generally are interested in only the nonzero eigenvalues of* $\mathcal{K}$.*)*

Theorem 1.4.1. *Let* $\mathcal{K} : \mathcal{X} \to \mathcal{X}$ *be compact, and let* $\mathcal{X}$ *be a Banach space. Then:*

(1) The eigenvalues of $\mathcal{K}$ *form a discrete set in the complex plane* **C**, *with 0 as the only possible limit point.*

(2) For each nonzero eigenvalue λ *of* $\mathcal{K}$, *there is only a finite number of linearly independent eigenvectors.*

(3) Each nonzero eigenvalue λ *of* $\mathcal{K}$ *has finite index* $\nu(\lambda) \geq 1$. *This means*

$$Null(\lambda - \mathcal{K}) \underset{\neq}{\subset} Null((\lambda - \mathcal{K})^2) \underset{\neq}{\subset} \cdots \underset{\neq}{\subset} Null\big((\lambda - \mathcal{K})^{\nu(\lambda)}\big)$$

$$= Null\big((\lambda - \mathcal{K})^{\nu(\lambda)+1}\big) \tag{1.4.43}$$

In addition, $Null((\lambda - \mathcal{K})^{\nu(\lambda)})$ *is finite dimensional. The elements of* $Null((\lambda - \mathcal{K})^{\nu(\lambda)})$ *are called generalized eigenvectors of* $\mathcal{K}$.

(4) For all $\lambda \neq 0$, $Range(\lambda - \mathcal{K})$ *is closed in* $\mathcal{X}$.

(5) For each nonzero eigenvalue λ *of* $\mathcal{K}$,

$$\mathcal{X} = Null\big((\lambda - \mathcal{K})^{\nu(\lambda)}\big) \oplus Range\big((\lambda - \mathcal{K})^{\nu(\lambda)}\big) \tag{1.4.44}$$

is a decomposition of $\mathcal{X}$ into invariant subspaces. This implies that every $x \in \mathcal{X}$ can be written as $x = x_1 + x_2$ with unique choices

$$x_1 \in Null\big((\lambda - \mathcal{K})^{\nu(\lambda)}\big) \quad and \quad x_2 \in Range\big((\lambda - \mathcal{K})^{\nu(\lambda)}\big).$$

Being invariant means that

$$\mathcal{K}: Null\big((\lambda - \mathcal{K})^{\nu(\lambda)}\big) \to Null\big((\lambda - \mathcal{K})^{\nu(\lambda)}\big)$$
$$\mathcal{K}: Range\big((\lambda - \mathcal{K})^{\nu(\lambda)}\big) \to Range\big((\lambda - \mathcal{K})^{\nu(\lambda)}\big)$$

(6) The Fredholm Alternative Theorem and the above results (1)–(5) remain true if $\mathcal{K}^m$ is compact for some $m > 1$.

For results on the speed with which the eigenvalues $\{\lambda_n\}$ of compact integral operators $\mathcal{K}$ converge to zero, see Hille and Tamarkin [271] and Fenyö and Stolle [198, §8.9]. Generally, as the differentiability of the kernel function $K(t, s)$ increases, the speed of convergence to zero of the eigenvalues also increases.

For the following theorem, we need to introduce the concept of an adjoint operator. Let $\mathcal{A}: \mathcal{X} \to \mathcal{Y}$ be a bounded operator, and let both $\mathcal{X}$ and $\mathcal{Y}$ be complete inner product spaces (that is, Hilbert spaces), with associated inner products $(\cdot, \cdot)_\mathcal{X}$ and $(\cdot, \cdot)_\mathcal{Y}$. Then there is a unique *adjoint* operator $\mathcal{A}^*: \mathcal{Y} \to \mathcal{X}$ satisfying

$$(\mathcal{A}x, y)_\mathcal{Y} = (x, \mathcal{A}^*y)_\mathcal{X}, \quad x \in \mathcal{X}, \ y \in \mathcal{Y}$$

This can be proven using the Riesz representation theorem (cf. Theorem A.4 in the Appendix).

Theorem 1.4.2. *Let $\mathcal{X}$ be a Hilbert space with scalars the complex numbers $\mathbb{C}$, let $\mathcal{K}: \mathcal{X} \to \mathcal{X}$ be a compact operator, and let λ be a nonzero eigenvalue of $\mathcal{K}$. Then:*

(1) $\bar{\lambda}$ is an eigenvalue of the adjoint operator $\mathcal{K}^*$. In addition, $Null(\lambda - \mathcal{K})$ and $Null(\bar{\lambda} - \mathcal{K}^*)$ have the same dimension.

(2) The equation $(\lambda - \mathcal{K})x = y$ is solvable if and only if

$$(y, z) = 0, \quad z \in Null(\bar{\lambda} - \mathcal{K}^*) \tag{1.4.45}$$

An equivalent way of writing this is

$$Range(\lambda - \mathcal{K}) = Null(\bar{\lambda} - \mathcal{K}^*)^\perp$$

the subspace orthogonal to $Null(\bar{\lambda} - \mathcal{K}^*)$. With this, we can write the decomposition

$$\mathcal{X} = Null(\bar{\lambda} - \mathcal{K}^*) \oplus Range(\lambda - \mathcal{K}) \tag{1.4.46}$$

Theorem 1.4.3. *Let $\mathcal{X}$ be a Hilbert space with scalars the complex numbers $\mathbf{C}$, and let $\mathcal{K} : \mathcal{X} \to \mathcal{X}$ be a self-adjoint compact operator. Then all eigenvalues of $\mathcal{K}$ are real and of index $\nu(\lambda) = 1$. In addition, the corresponding eigenvectors can be written as an orthonormal set. Order and index the nonzero eigenvalues as follows:*

$$|\lambda_1| \geq |\lambda_2| \geq \cdots \geq |\lambda_n| \geq \cdots > 0 \tag{1.4.47}$$

with each eigenvalue repeated according to its multiplicity [that is, the dimension of $Null(\lambda - \mathcal{K})$]. Then we write

$$\mathcal{K}x_i = \lambda_i x_i, \quad i \geq 1 \tag{1.4.48}$$

with

$$(x_i, x_j) = \delta_{ij}$$

Also, the eigenvectors $\{x_i\}$ form an orthonormal basis for $Range(\bar{\lambda} - \mathcal{K})$.

Much of the theory of self-adjoint boundary value problems for ordinary and partial differential equations is based on Theorems 1.4.2 and 1.4.3. Moreover, the completeness in $L^2(D)$ of many families of functions is proven by showing they are the eigenfunctions to a self-adjoint differential equation or integral equation problem.

Example. Let $D = \{P \in \mathbf{R}^3 \mid |P| = 1\}$, the unit sphere in $\mathbf{R}^3$, and let $\mathcal{X} = L^2(D)$. In this, $|P|$ denotes the Euclidean length of P. Define

$$\mathcal{K}x(P) = \int_D \frac{x(Q)}{|P - Q|} dS_Q, \quad P \in D \tag{1.4.49}$$

This is a compact operator, a proof of which is given in Mikhlin [380, p. 160]. The eigenfunctions of $\mathcal{K}$ are called *spherical harmonics*, a much-studied set of functions; for example, see MacRobert [360]. For each integer $k \geq 0$, there are $2k + 1$ independent spherical harmonics of *degree* k, and for each such spherical harmonic x_k, we have

$$\mathcal{K}x_k = \frac{4\pi}{2k + 1}x_k, \quad k = 0, 1, \ldots \tag{1.4.50}$$

Letting $\mu_k = 4\pi/(2k+1)$, we have $Null(\mu_k - \mathcal{K})$ has dimension $2k+1, k \geq 0$. It is well-known that the set of all spherical harmonics form a basis for $L^2(D)$, in agreement with Theorem 1.4.3.

1.5. Noncompact integral operators

We give a few examples of important integral operators that are not compact.

1.5.1. An Abel integral equation

Let $\mathcal{X} = C[0, 1]$, and define

$$\mathcal{A}x(t) = \int_0^t \frac{x(s)}{\sqrt{t^2 - s^2}} \, ds \quad t > 0$$

$$\mathcal{A}x(0) = \frac{\pi}{2}x(0)$$

(1.5.51)

Then $\mathcal{A} : C[0, 1] \to C[0, 1]$ with $\|\mathcal{A}\| = \frac{\pi}{2}$.

However, $\mathcal{A}$ is not compact. To see this, first introduce

$$x_\alpha(t) = t^\alpha, \quad \alpha \geq 0.$$

Then for $t > 0$, use the change of variable $s = ut, 0 \leq u \leq 1$, to obtain

$$\mathcal{A}x_\alpha(t) = \int_0^t \frac{s^\alpha ds}{\sqrt{t^2 - s^2}} = t^\alpha \int_0^1 \frac{u^\alpha du}{\sqrt{1 - u^2}} \equiv \lambda_\alpha t^\alpha$$

(1.5.52)

The number λ_α is a continuous function of α; and as α varies in $[0, \infty)$, λ_α varies in $(0, \frac{\pi}{2}]$. Every such number λ_α is an eigenvalue of $\mathcal{A}$. Thus $\mathcal{A}^m$ is not compact, $m \geq 1$, as that would contradict Theorem 1.4.1 (Parts 1 and 6) above, that the set of eigenvalues is a discrete set.

1.5.2. Cauchy singular integral operators

Let Γ be a smooth simple closed curve in the complex plane $\mathbf{C}$, and define the Cauchy principal value integral operator

$$\mathcal{T}\phi(z) = \frac{1}{\pi i} \int_\Gamma \frac{\phi(\zeta)}{\zeta - z} \, d\zeta, \quad z \in \Gamma, \quad \phi \in L^2(\Gamma)$$

(1.5.53)

It can be shown that $\mathcal{T}\phi(z)$ exists for almost all $x \in \Gamma$ and that $\mathcal{T}\phi \in L^2(\Gamma)$. If ϕ is the limit of a function that is analytic inside Γ, then $\mathcal{T}\phi = \phi$. Thus, 1 is an eigenvalue for $\mathcal{T}$ of infinite multiplicity; and therefore, $\mathcal{T}$ cannot be compact. In fact, it can be shown that $\mathcal{T}^2 = I$ with respect to $\mathcal{X} = L^2(\Gamma)$. Since I is not compact on an infinite dimensional space, $\mathcal{T}$ cannot be compact, as otherwise it would imply I was compact.

1.5.3. Wiener-Hopf integral operators

Let $\mathcal{X} = C_0[0, \infty)$, the bounded continuous functions $x(t)$ on $[0, \infty)$ with $\lim_{t \to \infty} x(t) = 0$. The norm is $\|\cdot\|_\infty$. Define

$$\mathcal{K}x(t) = \int_0^\infty e^{-|t-s|} x(s)\, ds, \quad t \geq 0 \qquad (1.5.54)$$

The Wiener-Hopf integral equation

$$\lambda x(t) - \int_0^\infty e^{-|t-s|} x(s)\, ds = y(t), \quad 0 \leq t < \infty \qquad (1.5.55)$$

can be converted to an equivalent boundary-value problem for a second-order differential equation. From the differential equation, it can be shown that for every $\lambda \in [0, 2]$, the operator $(\lambda - \mathcal{K})^{-1}$ does *not* have a bounded inverse on $C_0[0, \infty)$; but none of the numbers λ are eigenvalues of $\mathcal{K}$. Again, $\mathcal{K}$ cannot be compact, as that would contradict Theorem 1.3.1.

Discussion of the literature

The area of integral equations is quite old, going back almost 300 years, but most of the theory of integral equations dates from the twentieth century. An excellent presentation of the history of Fredholm integral equations can be found in Bernkopf [78], who traces the historical development of both functional analysis and integral equations and shows how they are related.

There are many texts on the theory of integral equations. Among such, we note Fenyö and Stolle [197]–[200], Green [236], Hackbusch [251], Hochstadt [272], Jörgens [297], Kanwal [306], Krasnoselskii [319], Kress [325], Mikhlin [377]–[380], Mikhlin and Prößdorf [381], Muskhelishvili [390], Pogorzelski [426], Porter and Stirling [427], Schmeidler [492], Smirnov [518], Smithies [520], and Tricomi [544]. The monograph of Ingham and Kelmanson [283] gives a general introduction to boundary integral equation reformulations of some important elliptic partial differential equations, and that of Ramm [445] covers a wide variety of integral equations used in significant, but more specialized areas of applications. For the study of boundary integral equation methods for solving the important Helmholtz equation, see Colton and Kress [127]. For an extensive bibliography on integral equations up to 1970, including their application and numerical solution, see Noble [404].

The state of the art before 1960 for the numerical solution of integral equations is well described in the book of Kantorovich and Krylov [305]. From 1960 to the present day many new numerical methods have been developed for the solution of many types of integral equations, and this includes methods for the integral equations discussed in §1.1. For reasons of personal interest, we consider only the numerical solution of Fredholm integral equations of the second kind and of some closely related Fredholm integral equations of the first kind. It is also necessary for practical reasons to limit ourselves in some way, as a book considering all of the types of integral equations would be far too lengthy, as some would already consider the present text.

There are a number of texts on the numerical solution of the remaining types of integral equations, some of which also contain material on the numerical solution Fredholm integral equations of the second kind. Among such texts on the numerical solution of other types of integral equations, we note especially Albrecht and Collatz [3]; Anderssen, de Hoog, and Lukas [13]; Baker [73]; Baker and Miller [75]; Bückner [97]; Brunner and deRiele [96]; Delves and Walsh [165]; Delves and Mohamed [166]; Fenyö and Stolle [200, Part 5]; Golberg [218], [219]; Hackbusch [251]; Ivanov [284]; Kantorovich and Akilov [304]; Krasnoselskii, Vainikko, et al. [320]; Kress [325]; Linz [344], [345]; Prößdorf and Silbermann [437]; Reinhardt [458]; and Vainikko [549], [550].

2
Degenerate kernel methods

Integral equations with a degenerate kernel function were introduced earlier, in the proof of Theorem 1.3.1 in Chapter 1, and now we consider the solution of such equations as a numerical method for solving general Fredholm integral equations of the second kind. The degenerate kernel method is a well-known classical method for solving Fredholm integral equations of the second kind, and it is one of the easiest numerical methods to define and analyze.

In this chapter we first consider again the reduction of a degenerate kernel integral equation to an equivalent linear system, and we reduce the assumptions made in the earlier presentation in the proof of Theorem 1.3.1 in Chapter 1. Following this, we consider various ways of producing degenerate kernel approximations of more general Fredholm integral equations of the second kind.

2.1. General theory

Consider the integral equation

$$\lambda x(t) - \int_D K(t,s)x(s)\,ds = y(t), \quad t \in D \qquad (2.1.1)$$

with $\lambda \neq 0$ and $D \subset \mathbf{R}^m$, some $m \geq 1$. We assume throughout this and the following chapters that D is a closed bounded set. Generally, it is an m-dimensional set with a piecewise smooth boundary; or it can be the piecewise smooth boundary itself. We usually work in the space $\mathcal{X} = C(D)$ with $\|\cdot\|_\infty$, and occasionally in $\mathcal{X} = L^2(D)$. The integral operator $\mathcal{K}$ of (2.1.1) is assumed to be a compact operator on $\mathcal{X}$ into $\mathcal{X}$.

The kernel function K is to be approximated by a sequence of degenerate kernel functions,

$$K_n(t, s) = \sum_{i=1}^{n} \alpha_{i,n}(t)\beta_{i,n}(s), \quad n \geq 1 \qquad (2.1.2)$$

in such a way that the associated integral operators $\mathcal{K}_n$ satisfy

$$\lim_{n \to \infty} \|\mathcal{K} - \mathcal{K}_n\| = 0 \qquad (2.1.3)$$

Generally, we want this convergence to be rapid to obtain rapid convergence of x_n to x, where x_n is the solution of the approximating equation

$$\lambda x_n(t) - \int_D K_n(t, s)x_n(s)\, ds = y(t), \quad t \in D \qquad (2.1.4)$$

Among the questions to be addressed are the following: (1) How do we show $x_n \to x$ as $n \to \infty$? (2) How do we obtain x_n in the equation (2.1.4)? (3) How do we obtain $K_n(t, s)$ and show $\mathcal{K}_n \to \mathcal{K}$? We begin with the first of these questions.

Theorem 2.1.1. *Assume $\lambda - \mathcal{K}: \mathcal{X} \overset{1-1}{\underset{onto}{\to}} \mathcal{X}$, with $\mathcal{X}$ a Banach space and $\mathcal{K}$ bounded. Further, assume $\{\mathcal{K}_n\}$ is a sequence of bounded linear operators with*

$$\lim_{n \to \infty} \|\mathcal{K} - \mathcal{K}_n\| = 0$$

Then the operators $(\lambda - \mathcal{K}_n)^{-1}$ exist from $\mathcal{X}$ onto $\mathcal{X}$ for all sufficiently large n, say $n \geq N$, and

$$\|(\lambda - \mathcal{K}_n)^{-1}\| \leq \frac{\|(\lambda - \mathcal{K})^{-1}\|}{1 - \|(\lambda - \mathcal{K})^{-1}\|\|\mathcal{K} - \mathcal{K}_n\|}, \quad n \geq N \qquad (2.1.5)$$

For the equations $(\lambda - \mathcal{K})x = y$ and $(\lambda - \mathcal{K}_n)x_n = y$, $n \geq N$, we have

$$\|x - x_n\| \leq \|(\lambda - \mathcal{K}_n)^{-1}\|\|\mathcal{K}x - \mathcal{K}_n x\|, \quad n \geq N \qquad (2.1.6)$$

Proof. Use the identity

$$\lambda - \mathcal{K}_n = \lambda - \mathcal{K} + (\mathcal{K} - \mathcal{K}_n)$$

$$= (\lambda - \mathcal{K})[I + (\lambda - \mathcal{K})^{-1}(\mathcal{K} - \mathcal{K}_n)] \qquad (2.1.7)$$

Choose N so that

$$\|\mathcal{K} - \mathcal{K}_n\| < \frac{1}{\|(\lambda - \mathcal{K})^{-1}\|}, \quad n \geq N \qquad (2.1.8)$$

By the geometric series theorem (cf. Theorem A.1 in the Appendix), the quantity $I + (\lambda - \mathcal{K})^{-1}(\mathcal{K} - \mathcal{K}_n)$ has a bounded inverse, with

$$\|[I + (\lambda - \mathcal{K})^{-1}(\mathcal{K} - \mathcal{K}_n)]^{-1}\| \leq \frac{1}{1 - \|(\lambda - \mathcal{K})^{-1}\| \, \|\mathcal{K} - \mathcal{K}_n\|}$$

Using (2.1.7), this yields the existence of $(\lambda - \mathcal{K}_n)^{-1}$ and its bound (2.1.5).

For the error bound (2.1.6), use the identity

$$\begin{aligned}
x - x_n &= (\lambda - \mathcal{K})^{-1}y - (\lambda - \mathcal{K}_n)^{-1}y \\
&= (\lambda - \mathcal{K}_n)^{-1}[\mathcal{K} - \mathcal{K}_n](\lambda - \mathcal{K})^{-1}y \\
&= (\lambda - \mathcal{K}_n)^{-1}[\mathcal{K}x - \mathcal{K}_n x]
\end{aligned}$$

The error bound follows immediately.

A modification of the above also yields

$$\|(\lambda - \mathcal{K})^{-1} - (\lambda - \mathcal{K}_n)^{-1}\| \leq \|(\lambda - \mathcal{K}_n)^{-1}\| \|(\lambda - \mathcal{K})^{-1}\| \|\mathcal{K} - \mathcal{K}_n\| \tag{2.1.9}$$

From (2.1.3) and (2.1.5), this shows $(\lambda - \mathcal{K}_n)^{-1} \to (\lambda - \mathcal{K})^{-1}$ in $L(\mathcal{X}, \mathcal{X})$. Also, $\|(\lambda - \mathcal{K}_n)^{-1}\| \to \|(\lambda - \mathcal{K})^{-1}\|$. This completes the proof. $\square$

An important consequence of the above convergence theorem is that the speed of convergence need not depend on the differentiability of the unknown x, since (2.1.6) implies

$$\|x - x_n\| \leq \|(\lambda - \mathcal{K}_n)^{-1}\| \|\mathcal{K} - \mathcal{K}_n\| \|x\| \tag{2.1.10}$$

If $\|\mathcal{K} - \mathcal{K}_n\|$ converges rapidly to zero, then the same is true of $\|x - x_n\|$, independent of the differentiability of x. This will not be true of most other types of numerical methods for solving (2.1.1), as will be seen in the following two chapters.

With $\mathcal{X} = C(D)$, we choose the degenerate kernel (2.1.2) so that the functions $\alpha_i(t)$ are all continuous and the functions $\beta_i(s)$ are all absolutely integrable. To apply the above convergence theorem, note that

$$\|\mathcal{K} - \mathcal{K}_n\| = \max_{t \in D} \int_D |K(t, s) - K_n(t, s)| \, ds \tag{2.1.11}$$

With $\mathcal{X} = L^2(D)$, we require that all $\alpha_i, \beta_i \in L^2(D)$. To apply the convergence theorem, we can use

$$\|\mathcal{K} - \mathcal{K}_n\| \leq \left[\int_D \int_D |K(t, s) - K_n(t, s)|^2 \, ds \, dt \right]^{\frac{1}{2}} \tag{2.1.12}$$

If this is not sufficient, then other bounds are often possible. The kernels $K_n(t, s)$ should be chosen to make $\|\mathcal{K} - \mathcal{K}_n\|$ converge to zero as rapidly as practicable.

2.1.1. Solution of degenerate kernel integral equation

Using the formula (2.1.2) for $K_n(t, s)$, the integral equation $(\lambda - \mathcal{K}_n)x_n = y$ becomes

$$\lambda x_n(t) - \sum_{j=1}^{n} \alpha_j(t) \int_D \beta_j(s) x_n(s)\, ds = y(t), \quad t \in D \qquad (2.1.13)$$

Then the solution x_n is given by

$$x_n(t) = \frac{1}{\lambda}\left[y(t) + \sum_{j=1}^{n} c_j \alpha_j(t) \right] \qquad (2.1.14)$$

To determine $\{c_j\}$, multiply (2.1.13) by $\beta_i(t)$ and integrate over D. This yields the system

$$\lambda c_i - \sum_{j=1}^{n} c_j(\alpha_j, \beta_i) = (y, \beta_i), \quad i = 1, \dots, n \qquad (2.1.15)$$

with

$$(\alpha_j, \beta_i) = \int_D \beta_i(t) \alpha_j(t)\, dt \qquad (2.1.16)$$

The system (2.1.15) is solved, and x_n is obtained from (2.1.14).

There are many practical aspects to the use of (2.1.14)–(2.1.15). In accordance with Theorem 2.1.1, we can assume $(\lambda - \mathcal{K}_n)x_n = y$ is uniquely solvable for all $y \in \mathcal{X}$. Then we would like the associated linear system (2.1.15) to be nonsingular. In addition, this system requires the evaluation of the integrals (α_j, β_i) and (y, β_i); and since this must often be done numerically, we should choose the functions $\alpha_j(t)$ and $\beta_i(s)$ with this in mind. Finally, we would like to have the linear system (2.1.15) be as well-conditioned as the original problem $(\lambda - \mathcal{K})x = y$. The following theorem addresses the first of these concerns, and the other concerns are taken up later.

Theorem 2.1.2. *Assume* $\lambda - \mathcal{K}_n : \mathcal{X} \overset{1-1}{\underset{onto}{\rightarrow}} \mathcal{X}$, *with* $\lambda \neq 0$ *and with* $\mathcal{X} = C(D)$ *or* $L^2(D)$; *and let* $\mathcal{K}_n$ *have the degenerate kernel* (2.1.2). *Then the linear system* (2.1.15) *is nonsingular.*

Proof. We prove the result for $\mathcal{X} = L^2(D)$, and for definiteness, we assume that $\mathbf{R}$ is the field of scalars for $\mathcal{X}$. The case for $\mathcal{X} = C(D)$ can be proven similarly, or it can be considered as a corollary of the case $\mathcal{X} = L^2(D)$. The following proof is divided into two cases, depending on whether $\{\beta_1, \ldots, \beta_n\}$ is an independent or dependent set of functions.

Case (1). Assume $\{\beta_1, \ldots, \beta_n\}$ is an independent set of functions.

From the assumption that $(\lambda - \mathcal{K}_n)^{-1}$ exists, we know that the linear system (2.1.15) is solvable for all right-hand sides of the form

$$\gamma \equiv \begin{bmatrix} \gamma_1 \\ \vdots \\ \gamma_n \end{bmatrix} = \begin{bmatrix} (y, \beta_1) \\ \vdots \\ (y, \beta_n) \end{bmatrix} \quad \text{for some} \quad y \in L^2(D) \tag{2.1.17}$$

A solution of (2.1.15) in this case is furnished by the coefficients in the expansion (2.1.14), with $x_n = (\lambda - \mathcal{K}_n)^{-1}y$. Given any $\gamma \in \mathbf{R}^n$, we will show there exists $y \in L^2(D)$ for which (2.1.17) is valid.

Given $\gamma \in \mathbf{R}^n$, define

$$y(t) = \sum_{j=1}^{n} \delta_j \beta_j(t) \tag{2.1.18}$$

for some $(\delta_1, \ldots, \delta_n)$. These coefficients are to be chosen so as to have (2.1.17) be true, i.e., $\gamma_i = (y, \beta_i)$, $i = 1, \ldots, n$. This requires $(\delta_1, \ldots, \delta_n)$ to be the solution of the linear system

$$\sum_{j=1}^{n} \delta_j (\beta_j, \beta_i) = \gamma_i, \quad i = 1, \ldots, n \tag{2.1.19}$$

The matrix for this system is called a *Gram matrix*, and we show below that the linear independence of $\{\beta_i\}$ implies that (2.1.19) is nonsingular, thus making it uniquely solvable for $(\delta_1, \ldots, \delta_n)$.

To show (2.1.19) is nonsingular, consider the homogeneous system

$$\sum_{j=1}^{n} \delta_j (\beta_j, \beta_i) = 0, \quad i = 1, \ldots, n \tag{2.1.20}$$

Multiply equation i by δ_i, for each i, and then sum over i. This yields

$$\sum_{i=1}^{n} \sum_{j=1}^{n} \delta_i \delta_j (\beta_j, \beta_i) = 0$$

or equivalently,

$$(g, g) = 0, \qquad g = \sum_{i=1}^{n} \delta_i \beta_i$$

This implies $g = 0$, and by the linear independence of $\{\beta_i\}$, $g = 0$ implies $\delta_i = 0$, $i = 1, \ldots, n$. Since the homogeneous system (2.1.20) has only the zero solution, the nonhomogeneous system (2.1.19) has a unique solution for every right-hand side γ.

This shows that for each $\gamma \in \mathbf{R}^n$, there is $y \in L^2(D)$ for which (2.1.17) is valid. Consequently, the linear system (2.1.15) has a solution for every right-hand side γ. By standard results of linear algebra, this proves the system (2.1.15) is nonsingular.

Case (2). Assume $\{\beta_1, \ldots, \beta_n\}$ is a dependent set of functions.

We reduce this case to that of case (1), and then we cite its conclusions to obtain a similar conclusion for the present case. Rather than proving the nonsingularity of (2.1.15) for general $n \geq 1$, we do a simple case to illustrate the general idea of the proof.

Let $n = 3$, let β_1 and β_2 be independent, and let $\beta_3 = c_1 \beta_1 + c_2 \beta_2$. The matrix of coefficients for (2.1.15) is

$$A = \begin{bmatrix} \lambda - (\alpha_1, \beta_1) & -(\alpha_2, \beta_1) & -(\alpha_3, \beta_1) \\ -(\alpha_1, \beta_2) & \lambda - (\alpha_2, \beta_2) & -(\alpha_3, \beta_2) \\ -(\alpha_1, \beta_3) & -(\alpha_2, \beta_3) & \lambda - (\alpha_3, \beta_3) \end{bmatrix}$$

Based on the dependence of β_3 on β_1 and β_2, introduce the matrices

$$P = \begin{bmatrix} 1 & 0 & 0 \\ 0 & 1 & 0 \\ c_1 & c_2 & 1 \end{bmatrix} \qquad P^{-1} = \begin{bmatrix} 1 & 0 & 0 \\ 0 & 1 & 0 \\ -c_1 & -c_2 & 1 \end{bmatrix}$$

to carry out elementary row and column operations on A. Doing this, we obtain the following.

Add c_1 times column 3 to column 1; and add c_2 times column 3 to column 2:

$$AP = \begin{bmatrix} \lambda - (\alpha_1 + c_1\alpha_3, \beta_1) & -(\alpha_2 + c_2\alpha_3, \beta_1) & -(\alpha_3, \beta_1) \\ -(\alpha_1 + c_1\alpha_3, \beta_2) & \lambda - (\alpha_2 + c_2\alpha_3, \beta_2) & -(\alpha_3, \beta_2) \\ c_1\lambda - (\alpha_1 + c_1\alpha_3, \beta_3) & c_2\lambda - (\alpha_2 + c_2\alpha_3, \beta_3) & \lambda - (\alpha_3, \beta_3) \end{bmatrix}$$

Then subtract c_1 times row 1 and c_2 times row 2 from row 3, and use $\beta_3 = c_1\beta_1 + c_2\beta_2$ to obtain:

$$P^{-1}AP = \begin{bmatrix} \lambda - (\alpha_1 + c_1\alpha_3, \beta_1) & -(\alpha_2 + c_2\alpha_3, \beta_1) & -(\alpha_3, \beta_1) \\ -(\alpha_1 + c_1\alpha_3, \beta_2) & \lambda - (\alpha_2 + c_2\alpha_3, \beta_2) & -(\alpha_3, \beta_2) \\ 0 & 0 & \lambda \end{bmatrix}$$

$$\equiv \begin{bmatrix} \hat{A} & * \\ 0 & \lambda \end{bmatrix}$$

with $\hat{A}$ of order 2×2. Thus $\det(A) = \lambda \, \det(\hat{A})$. The matrix $\hat{A}$ is the matrix of coefficients associated with the degenerate kernel

$$\hat{K}_3(t, s) = [\alpha_1(t) + c_1\alpha_3(t)]\beta_1(s) + [\alpha_2(t) + c_2\alpha_3(t)]\beta_2(s)$$

and this equals the original degenerate kernel K_3 when β_3 has been replaced by $c_1\beta_1 + c_2\beta_2$. By case (1), $\hat{A}$ is nonsingular, and thus A is nonsingular. This proof can be generalized to handle any dependent set $\{\beta_i\}$, by reducing the matrix of coefficients A for (2.1.2) to a lower order matrix $\hat{A}$ for a rewritten form $\hat{K}_n$ of K_n, one that fits under case (1).   □

This theorem assures us that the system (2.1.15) is nonsingular for the cases in which we are interested, namely those for which $(\lambda - \mathcal{K}_n)^{-1}$ exists. The second case in the proof is especially important, as it is often not possible to know in advance whether or not $\{\beta_i\}$ is an independent or dependent set. Most texts appear to imply that $\{\beta_i\}$ must be independent, but this is not necessary.

In setting up the degenerate kernel $K_n(t, s)$ that approximates $K(t, s)$, it is necessary to consider the setup costs for the system (2.1.15). In particular, the coefficients (y, β_i) and (α_j, β_i) must be calculated, thus, it makes sense to choose the functions α_i, β_i with this in mind. Often the coefficients must be calculated using numerical integration, and we would like to minimize the cost of such numerical quadratures. In the following sections we consider three methods for constructing degenerate kernel approximations, and we consider the calculation of their coefficients.

2.2. Taylor series approximations

Consider the one-dimensional integral equation

$$\lambda x(t) - \int_a^b K(t, s)x(s)\,ds = y(t), \quad a \le t \le b \qquad (2.2.21)$$

Often we can write K as a power series in s,

$$K(t, s) = \sum_{i=0}^{\infty} k_i(t)(s - a)^i \qquad (2.2.22)$$

or in t,

$$K(t, s) = \sum_{i=0}^{\infty} \kappa_i(s)(t - a)^i \qquad (2.2.23)$$

Let K_n denote the partial sum of the first n terms on the right side of (2.2.22),

$$K_n(t, s) = \sum_{i=0}^{n-1} k_i(t)(s - a)^i \qquad (2.2.24)$$

Using the notation of (2.1.2), K_n is a degenerate kernel with

$$\alpha_i(t) = k_{i-1}(t), \qquad \beta_i(s) = (s - a)^{i-1}, \quad i = 1, \ldots, n \qquad (2.2.25)$$

The linear system (2.1.15) becomes

$$\lambda c_i - \sum_{j=1}^{n} c_j \int_a^b (s - a)^{i-1} k_{j-1}(s) \, ds = \int_a^b y(s)(s - a)^{i-1} \, ds,$$

$$i = 1, \ldots, n \qquad (2.2.26)$$

and the solution x_n is given by

$$x_n(t) = \frac{1}{\lambda} \left[y(t) + \sum_{i=0}^{n-1} c_{i+1} k_i(t) \right] \qquad (2.2.27)$$

The integrals in (2.2.26) must often be calculated numerically. However, there is not much that can be said for integrals of this generality. First, they involve the entire interval $[a, b]$, as contrasted with some later methods we consider. In addition, most of the integrands will be zero or quite small, in the neighborhood of $s = a$, the left end of the interval. The latter may aid in choosing a more efficient method of numerical integration. The following example avoids the numerical calculation of most of these integrals.

Example. Consider the integral equation

$$\lambda x(t) - \int_0^b e^{st} x(s) \, ds = y(t), \quad 0 \le t \le b \qquad (2.2.28)$$

Write

$$e^{st} = \sum_{i=0}^{\infty} \frac{s^i t^i}{i!}$$

and define K_n as in (2.2.24). The linear system (2.2.26) becomes

$$\lambda c_i - \sum_{j=1}^{n} c_j \frac{b^{i+j-1}}{(j-1)!(i+j-1)} = \int_0^b y(s)s^{i-1}\,ds, \quad i = 1, \dots, n$$

$$(2.2.29)$$

and the solution x_n of the degenerate kernel equation $(\lambda - \mathcal{K}_n)x_n = y$ is given by

$$x_n(t) = \frac{1}{\lambda}\left[y(t) + \sum_{i=0}^{n-1} c_{i+1} \frac{t^i}{i!} \right]$$

For the error analysis, let $\mathcal{X} = C[0, b]$ with $\|\cdot\|_\infty$. Then

$$\|\mathcal{K} - \mathcal{K}_n\| = \max_{0 \le t \le b} \int_0^b |e^{st} - K_n(t, s)|\,ds$$

$$= \max_{0 \le t \le b} \int_0^b \frac{(st)^n}{n!} e^{\zeta}\,ds, \quad \zeta = \zeta(t, s) \in [0, b]$$

$$\le \frac{b^{2n+1}}{(n+1)!} e^{b^2}$$

This converges to zero as $n \to \infty$. By Theorem 2.1.1, we obtain convergence of x_n to x, along with the error bound (2.1.6) [or (2.1.10)] whenever $(\lambda - \mathcal{K})^{-1}$ exists.

It is straightforward to calculate

$$\|\mathcal{K}\| = \frac{e^{b^2} - 1}{b}$$

If $|\lambda| > \|\mathcal{K}\|$, then $(\lambda - \mathcal{K})^{-1}$ exists by the geometric series theorem, and

$$\|(\lambda - \mathcal{K})^{-1}\| \le \frac{1}{|\lambda| - \|\mathcal{K}\|} \qquad (2.2.30)$$

For values of n for which

$$\|\mathcal{K} - \mathcal{K}_n\| < |\lambda| - \|\mathcal{K}\| \qquad (2.2.31)$$

Table 2.1. *Degenerate kernel example:* $\lambda = 5$,
$b = 1$

n	E_1	B_1	E_2	B_2
1	6.37E−2	9.57E−1	1.31E−1	8.07E−1
2	1.74E−2	2.17E−1	4.33E−2	1.83E−1
3	3.72E−3	4.84E−2	1.05E−2	4.08E−2
4	6.59E−4	9.41E−3	2.03E−3	7.93E−3
5	1.00E−4	1.56E−3	3.30E−4	1.31E−3
6	1.33E−5	2.23E−4	4.61E−5	1.88E−4
7	1.57E−6	2.78E−5	5.67E−6	2.35E−5
8	1.67E−7	3.09E−6	6.22E−7	2.60E−6
9	1.61E−8	3.09E−7	6.15E−8	2.60E−7
10	1.42E−9	2.81E−8	5.54E−9	2.37E−8

Table 2.2. *Degenerate kernel example:* $\lambda = 75$,
$b = 2$

n	E_1	B_1	E_2	B_2
4	1.75E−3		1.92E−1	
8	2.93E−4	1.07E+0	1.18E−2	1.11E+0
12	4.59E−6	9.48E−3	1.50E−4	9.87E−3
16	2.05E−8	4.22E−5	6.15E−7	4.40E−5
20	3.61E−11	7.53E−8	1.04E−9	7.84E−8

we can use the arguments of Theorem 2.1.1 to obtain the bound

$$\|(\lambda - \mathcal{K}_n)^{-1}\| \leq \frac{1}{|\lambda| - \|\mathcal{K}\| - \|\mathcal{K} - \mathcal{K}_n\|} \qquad (2.2.32)$$

Use $x = (\lambda - \mathcal{K})^{-1}y$ in (2.1.10) to obtain a computable error bound, with the needed norms of inverses obtained from (2.2.30) and (2.2.32).

We give numerical examples for several values of λ, b, and y in Tables 2.1 and 2.2. For true solutions, we use

$$x^{(1)}(t) = e^{-t}\cos(t), \quad x^{(2)}(t) = \sqrt{t} \qquad (2.2.33)$$

with the right-hand sides y defined accordingly. The tables contain the error

$$E_i = \left\|x^{(i)} - x_n^{(i)}\right\|_\infty, \quad i = 1, 2$$

and the corresponding error bound (2.1.10), labeled as B_i [provided (2.2.31) is satisfied]. The value of λ is so chosen that $\|\mathcal{K}\| \,/\, |\lambda|$ is approximately constant.

In Table 2.1, with $b = 1$, the error bounds are not too much larger than the actual errors; and in Table 2.2, with $b = 2$, there is a greater disparity between the true error and the error bound, but the bounds are still reasonable. In Figure 2.1, we give the graph of the errors $x^{(i)} - x_n^{(i)}$ for $b = 1$ and $n = 10$; and in Figure 2.2, we graph the errors as n varies, again with $b = 1$. This second graph shows there is a regularity in the behavior of the error as n increases. The

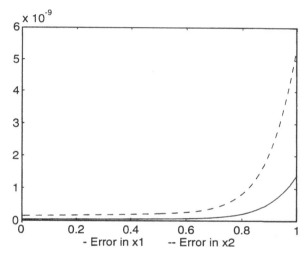

Figure 2.1. Errors in solving (2.2.28) by (2.2.29) for $n = 10$ and $b = 1$.

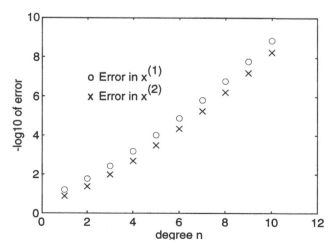

Figure 2.2. n vs. $-\log_{10} \|x - x_n^{(i)}\|_\infty$.

Table 2.3. *Condition numbers of (2.2.29)*

$b = 1$ n	$\lambda = 5$ $cond(A)$	$b = 2$ n	$\lambda = 75$ $cond(A)$
4	1.47	4	1.32
6	1.51	8	6.14
8	1.54	12	3.47E+2
10	1.56	16	4.68E+4
12	1.58	20	7.68E+6
14	1.59	24	1.37E+9

rate of change in the error is also independent of the differentiability of the unknown x.

In general, however, bounds such as (2.1.10) tend to be quite conservative and inaccurate. To obtain accurate bounds requires accurate bounds for the inverses $(\lambda - \mathcal{K})^{-1}$ and $(\lambda - \mathcal{K}_n)^{-1}$. Obtaining realistic bounds for these inverses is discussed in some detail in Linz [346].

2.2.1. Conditioning of the linear system

What is the condition number of the linear system $A_n c = r$ of (2.2.26), and is it well-conditioned? Before discussing what is known about this question, we use an example to show that the question is worthy of being raised.

Example. Consider again the example (2.2.28) and the linear system $A_n c = r$ of (2.2.29). In Table 2.3 we give condition numbers for $b = 1$ and $b = 2$, for varying values of n. The condition number is defined by

$$cond(A) = \|A\| \, \|A^{-1}\|$$

with the matrix norm the *spectral norm*

$$\|A\| = \sqrt{r_\sigma(A^T A)}$$

The notation $r_\sigma(B)$ denotes the maximum of the absolute values of the eigenvalues of the matrix B, and it is called the *spectral radius* of B. This indicates a problem for some systems, and further numerical examples show that as b increases, the condition numbers become larger and they increase more rapidly with increasing n.

The only general theory of a priori condition numbers for the linear system (2.1.15) associated with degenerate kernel methods was given in Whitley [567] (or this is the only reference of which we have knowledge). But his results do not apply directly to the methods we have considered, with the Taylor expansion based on expanding about the left endpoint of the integration interval. Instead, he gives results for the case of the Taylor approximation

$$K(t, s) = \sum_{i=0}^{\infty} b_i(s) (t - 1)^i, \quad -1 \le t, s \le 1$$

as a basis for defining the degenerate kernel approximations. Define

$$K_n(t, s) = \sum_{i=1}^{n} \alpha_i(t) \beta_i(s)$$

with

$$\alpha_i(t) = \frac{1}{d_i} (t - 1)^i, \qquad \beta_i(s) = d_i b_i(s)$$

$$d_i = \min_{j \in I_i} \left[\max_{-1 \le t \le 1} |t^{i-1} - t^{j-1}| \right], \qquad I_i = \{j \mid 0 \le j \le n - 1, \ j \ne i\}$$

The numbers d_i are chosen so as to try to minimize the condition number $cond(A)$ of (2.1.15), with respect to the matrix norms induced by $\| \cdot \|_\infty$ on $\mathbf{R}^n$. Then

$$cond(A) = (1 + \sqrt{2})^{n-2}[1 + 1/\sqrt{2}] + (1 - \sqrt{2})^{n-2}[1 - 1/\sqrt{2}]$$

See Whitley [567, p. 131].

This is a very negative result, as it seems to imply that the condition number will become quite large as n increases. From our earlier example in Table 2.3, we see partial agreement with this conclusion, but we also see that the Taylor series method works well if the interval length b is sufficiently small. More work needs to be done to understand further the conditioning of the linear system $A_n c = r$ for Taylor series methods.

For other types of degenerate kernel methods, the results of Whitley are much more encouraging of obtaining well-conditioned systems. In Chapter 3 we will be able to say more about the conditioning of some of the matrices arising in degenerate kernel methods.

2.3. Interpolatory degenerate kernel approximations

Interpolation is a simple way to obtain degenerate kernel approximations.
There are many kinds of interpolation, but we consider interpolation using only
the values of $K(t, s)$. There are many candidates for interpolation functions,
including polynomials, trigonometric polynomials, piecewise polynomial functions (including spline functions), and others. We give a general framework for
all of these, and then we illustrate these ideas with particular cases.

Let $\phi_1(t), \ldots, \phi_n(t)$ be a basis for the space of interpolation functions we
are using. For example, with polynomial interpolation of degree $<n$, we would
use

$$\phi_i(t) = t^{i-1}, \quad 1 \le i \le n \tag{2.3.34}$$

Let $t_1, \ldots, t_n$ be interpolation nodes in the integration region D. The *interpolation problem* is as follows: Given data $y_1, \ldots, y_n$, find

$$z(t) = \sum_{j=1}^{n} c_j \phi_j(t) \tag{2.3.35}$$

with

$$z(t_i) = y_i, \quad i = 1, \ldots, n \tag{2.3.36}$$

Thus, we want to find the coefficients $c_1, \ldots, c_n$ solving the linear system

$$\sum_{j=1}^{n} c_j \phi_j(t_i) = y_i, \quad i = 1, \ldots, n \tag{2.3.37}$$

In order for the interpolation problem to have a unique solution for all possible data $\{y_i\}$, it is necessary and sufficient that

$$\det(\Gamma_n) \ne 0, \quad \Gamma_n = [\phi_j(t_i)] \tag{2.3.38}$$

With polynomial interpolation and the basis of (2.3.34),

$$\Gamma_n = \left[t_i^{j-1} \right]_{i,j=1}^{n}$$

This is called a *Vandermonde* matrix, and it is known that $\det(\Gamma_n) \ne 0$ for all
distinct choices of $t_1, \ldots, t_n$; for example, see Ref. [48, p. 185, prob. 1].

To give an explicit formula for $K_n(t, s)$, we introduce a special basis for the
interpolation method. Define $\ell_k(t)$ to be the interpolation function for which

$$\ell_k(t_i) = \delta_{ik}, \quad i = 1, \ldots, n$$

Then the solution to the interpolation problem is given by

$$z(t) = \sum_{j=1}^{n} y_j \ell_j(t) \tag{2.3.39}$$

For polynomial interpolation, this is called *Lagrange's form* of the interpolation polynomial. We often use this name when dealing with other types of interpolation, and the functions $\ell_k(t)$ are usually called *Lagrange basis functions*. With polynomial interpolation,

$$\ell_k(t) = \prod_{\substack{i=1 \\ i \neq k}}^{n} \left(\frac{t - t_i}{t_k - t_i} \right)$$

2.3.1. Interpolation with respect to the variable t

Define

$$K_n(t, s) = \sum_{j=1}^{n} \underbrace{\ell_j(t)}_{\alpha_j(t)} \underbrace{K(t_j, s)}_{\beta_j(s)} \tag{2.3.40}$$

Then $K_n(t_i, s) = K(t_i, s)$, $i = 1, \ldots, n$, all $s \in D$. For the case $D = [a, b]$, with $K(t, s)$ being considered on the domain $[a, b] \times [a, b]$, we have that $K_n(t, s)$ equals $K(t, s)$ along all lines $t = t_i$.

The linear system $A_n c = r$ associated with the degenerate kernel method $(\lambda - \mathcal{K}_n) x_n = y$ is

$$\lambda c_i - \sum_{j=1}^{n} c_j \int_D \ell_j(s) K(t_i, s)\, ds = \int_D K(t_i, s) y(s)\, ds, \quad i = 1, \ldots, n \tag{2.3.41}$$

The solution x_n is given by

$$x_n(t) = \frac{1}{\lambda} \left\{ y(t) + \sum_{j=1}^{n} c_j \ell_j(t) \right\} \tag{2.3.42}$$

Note the integrals in (2.3.41) must generally be evaluated numerically.

When analyzing this degenerate kernel method within the context of the space $C(D)$, the error depends on

$$\|\mathcal{K} - \mathcal{K}_n\| = \max_{t \in D} \int_D |K(t, s) - K_n(t, s)|\, ds \tag{2.3.43}$$

which in turn depends on the interpolation error $K(t, s) - K_n(t, s)$. Some special cases are considered below.

2.3.2. Interpolation with respect to the variable s

Define

$$K_n(t,s) = \sum_{j=1}^{n} \underbrace{K(t,t_j)}_{\alpha_j(t)} \underbrace{\ell_j(s)}_{\beta_j(s)} \qquad (2.3.44)$$

Then $K_n(t,t_i) = K(t,t_i)$, all $t \in D$, $i = 1,\ldots,n$. The degenerate kernel method is given by

$$x_n(t) = \frac{1}{\lambda}\left\{y(t) + \sum_{j=1}^{n} c_j K(t,t_j)\right\} \qquad (2.3.45)$$

with $\{c_j\}$ satisfying

$$\lambda c_i - \sum_{j=1}^{n} c_j \int_D \ell_i(t)K(t,t_j)\,dt = \int_D \ell_i(t)y(t)\,dt, \quad i = 1,\ldots,n$$
$$(2.3.46)$$

The integrals in the coefficient matrices of systems (2.3.41) and (2.3.46) are probably of equal complexity when being evaluated. The right-hand side of (2.3.46) may be easier to evaluate, because often the functions $\ell_i(s)$ are identically zero over most of D, and this usually decreases the cost of numerically evaluating the integrals (y, ℓ_i). On the other hand, the formula (2.3.42) for $x_n(t)$ is usually easier and faster to evaluate than the corresponding formula (2.3.45). From the perspective of the computational cost of the degenerate kernel method, there seems to be no general reason to prefer either of the degenerate kernels (2.3.40) and (2.3.44) over the other one.

We should also consider the condition numbers of the linear systems (2.3.41) and (2.3.46) with these degenerate kernel methods. However, we defer this until Chapter 3, where we show that these systems are usually as well-conditioned as the integral equations with which they are associated.

2.3.3. Piecewise linear interpolation

Let $D = [a,b]$, $n > 0$, $h = (b-a)/n$, and $t_i = a + ih$, $i = 0,\ldots,n$. Given a function $g \in C[a,b]$, we interpolate it at the node points $\{t_i\}$ using piecewise linear interpolation. For $i = 1,\ldots,n$, define

$$\mathcal{P}_n g(t) = \frac{(t_i-t)g(t_{i-1}) + (t-t_{i-1})g(t_i)}{h}, \quad t_{i-1} \le t \le t_i \qquad (2.3.47)$$

2.3. *Interpolatory degenerate kernel approximations*   39

This defines a projection operator $\mathcal{P}_n : C[a, b] \to C[a, b]$, with $\|\mathcal{P}_n\| = 1$.
Note that we have $n + 1$ interpolation node points, thus necessitating a minor
change in the earlier notation for (2.3.35)–(2.3.46).
  Introduce

$$\ell_i(t) = \begin{cases} 1 - \dfrac{|t - t_i|}{h}, & t_{i-1} \leq t \leq t_{i+1} \\ 0, & \text{otherwise} \end{cases}$$

These are sometimes called "hat functions," from the shape of their graph.
Using them,

$$\mathcal{P}_n g(t) = \sum_{i=0}^{n} g(t_i)\ell_i(t)$$

It is well-known that

$$\|g - \mathcal{P}_n g\|_\infty \leq \begin{cases} \omega(g, h), & g \in C[a, b] \\ \dfrac{h^2}{8}\|g''\|_\infty, & g \in C^2[a, b] \end{cases} \tag{2.3.48}$$

The function ω is defined by

$$\omega(g, h) = \sup_{\substack{a \leq t, s \leq b \\ |t-s| \leq h}} |g(t) - g(s)|$$

and it is called the *modulus of continuity* of the function g. A function g is
continous on $[a, b]$ if and only if $\lim_{h \to 0} \omega(g, h) = 0$ as $h \to 0$.
  When interpolating $K(t, s)$ with respect to t, we have

$$K_n(t, s) = \frac{(t_i - t)K(t_{i-1}, s) + (t - t_{i-1})K(t_i, s)}{h}, \quad t_{i-1} \leq t \leq t_i \tag{2.3.49}$$

for $i = 1, \ldots, n, a \leq s \leq b$. For the elements of the coefficient matrix (2.3.41),

$$(\alpha_j, \beta_i) = \int_{t_{i-1}}^{t_{i+1}} \ell_j(s)K(t_i, s)\, ds \tag{2.3.50}$$

for $i = 1, \ldots, n - 1$. For the cases $i = 0$ and $i = n$, the interval of integration
will be $[a, a + h]$ and $[b - h, b]$, respectively. Such integrals can usually be
evaluated more simply and efficiently than an integral over the entire interval
$[a, b]$. We try to choose a numerical integration method that will require only
a few quadrature points on the interval of integration, with the error in the

integration being of the same order as the error in the approximate solution
$x_n(t)$. In the present case, the two point Gauss-Legendre formula

$$\int_0^h f(s)\, ds \approx \frac{h}{2}\left[f\left(\left[1 - \frac{1}{\sqrt{3}}\right]\frac{h}{2}\right) + f\left(\left[1 + \frac{1}{\sqrt{3}}\right]\frac{h}{2}\right)\right] \qquad (2.3.51)$$

is quite sufficient, when applied on each subinterval $[t_{i-1}, t_i]$. A framework for
showing the adequacy of this formula is given in the following subsection.

The right-hand sides of (2.3.41) are given by

$$(y, \beta_i) = \int_a^b y(s) K(t_i, s)\, ds \qquad (2.3.52)$$

These must be evaluated over the entire interval $[a, b]$. If our interpolation of
$K(t, s)$ had instead been based on (2.3.44), then the right sides (y, β_i) would
also be defined on only a smaller interval. But then the formula for $x_n(t)$ would
use (2.3.45), and it would not be as simple as in (2.3.42).

For the error in this degenerate kernel method, we apply (5.3.14) to obtain

$$\|\mathcal{K} - \mathcal{K}_n\| \le \frac{h^2(b-a)}{8}\left[\int_a^b \max_{a \le t \le b}\left|\frac{\partial^2 K(t, s)}{\partial t^2}\right| ds\right] \qquad (2.3.53)$$

Thus the error in convergence of $x_n = (\lambda - \mathcal{K}_n)^{-1}y$ to $x = (\lambda - \mathcal{K})^{-1}y$ is
bounded by ch^2 for some $c > 0$. The above bound assumes $K(t, s)$ is twice
continuously differentiable with respect to t.

Example. Consider again the integral equation (2.2.21) with the unknown
functions $x^{(i)}(t)$ of (2.2.33). Numerical results for varying b and n are given in
Tables 2.4 and 2.5. The columns labeled *"Ratio"* give the ratio of successive
maximum errors. Recall the notation

$$E_i = \left\|x^{(i)} - x_n^{(i)}\right\|_\infty, \qquad i = 1, 2$$

We give error bounds B_i that are based on (2.1.10), in the same manner as was
done for the earlier degenerate kernel example in Tables 2.1 and 2.2.

In addition, we give the condition numbers for the coefficient matrix A of
(2.3.41), and the reader should compare these results with those in Table 2.3.
Note that the condition numbers for $b = 2$ are comparable to those for $b = 1$,
in contrast to the results in Table 2.3. Later, in Chapter 3, we give a rigorous
justification that the size of b does not affect the size of cond(A) for the linear
system (2.3.41) when using piecewise polynomial interpolation.

Note that the errors $\|x^{(i)} - x_n^{(i)}\|_\infty$ are decreasing by a factor of approximately
4 whenever n is doubled, which is consistent with (2.3.53). Piecewise linear

Table 2.4. *Piecewise linear degenerate kernel:* $b = 1, \lambda = 5$

n	E_1	Ratio	B_1	E_2	Ratio	B_2	$cond(A_n)$
4	4.24E−4		2.32E−3	1.10E−3		1.95E−3	1.65
8	1.11E−4	3.8	5.79E−4	2.89E−4	3.8	4.88E−4	1.72
16	2.83E−5	3.9	1.45E−4	7.40E−5	3.9	1.22E−4	1.77
32	7.15E−6	4.0	3.62E−5	1.87E−5	4.0	3.05E−5	1.80
64	1.80E−6	4.0	9.05E−6	4.71E−6	4.0	7.62E−6	1.81

Table 2.5. *Piecewise linear degenerate kernel:* $b = 2, \lambda = 75$

n	E_1	Ratio	B_1	E_2	Ratio	B_2	$cond(A_n)$
4	1.45E−4		8.41E−1	2.84E−2		8.76E−1	1.34
8	5.59E−5	2.6	1.49E−1	8.69E−3	3.3	1.55E−1	1.36
16	1.73E−5	3.2	3.48E−2	2.41E−3	3.6	3.62E−2	1.42
32	4.79E−6	3.6	8.55E−3	6.36E−4	3.8	8.91E−3	1.48
64	1.26E−6	3.8	2.13E−3	1.63E−4	3.9	2.22E−3	1.52

interpolation does not converge rapidly, but it illustrates the general idea of the use of piecewise polynomial interpolation of any degree. For a more rapidly convergent method, use piecewise polynomial interpolation with polynomials of degree $p \geq 0$. With such interpolation, it is straightforward to show that the error $\|x - x_n\|_\infty$ is $O(h^{p+1})$, provided $K(t, s)$ and $x(s)$ are sufficiently differentiable.

Example. Consider the two-dimensional integral equation

$$\lambda x(s, t) - \int_0^b \int_0^b K(s, t, \sigma, \tau)x(\sigma, \tau)\, d\sigma\, d\tau = y(s, t), \quad 0 \leq s, t \leq b \tag{2.3.54}$$

The Banach space is $C(D)$ with the uniform norm, $D = [0, b] \times [0, b]$. We approximate K with a degenerate kernel based on piecewise linear interpolation with respect to both the variables s and t. Use the piecewise linear interpolation of the previous example, and define

$$K_n(s, t, \sigma, \tau) = \sum_{i=0}^n \sum_{j=0}^n \ell_i(s)\ell_j(t)K(s_i, t_j, \sigma, \tau) \tag{2.3.55}$$

Table 2.6. *Degenerate kernel
solution of (2.3.58)*

n	N	$\|x - x_n\|_\infty$	Ratio
2	9	1.61E$-$2	
4	25	4.19E$-$3	3.84
8	81	1.06E$-$3	3.95
16	289	2.65E$-$4	4.00

with $t_i = s_i = ih$. For the error, assume that $K(s, t, \sigma, \tau)$ is twice continuously differentiable with respect to both s and t, for all $(s, t), (\sigma, \tau) \in D$. Then we can show

$$\|\mathcal{K} - \mathcal{K}_n\| \le \gamma\, h^2 \qquad (2.3.56)$$

for a suitable value of $\gamma > 0$. The proof is left as a problem for the reader. Using (2.3.56), we obtain the standard error results, including

$$\|x - x_n\|_\infty = O(h^2) \qquad (2.3.57)$$

when $x \in C(D)$.

As a specific example, consider the integral equation

$$\lambda x(s, t) - \int_0^{\sqrt{\pi}} \int_0^{\sqrt{\pi}} \cos(s\sigma) \cos(t\tau) x(\sigma, \tau)\, d\sigma\, d\tau = y(s, t) \qquad (2.3.58)$$

with $\lambda = 5$. Easily $\|\mathcal{K}\| = \pi$, and (2.3.58) is uniquely solvable by the geometric series theorem. For illustrative purposes we choose $x(s, t) \equiv 1$ and define $y(s, t)$ accordingly. Numerical results are given in Table 2.6, and the empirical rate of convergence agrees with (2.3.57). In the table, $N = (n+1)^2$ is the order of the linear system associated with the degenerate kernel method. This also illustrates a difficulty with problems in more than one variable: the order of the linear system increases quite rapidly, and often the system must be solved iteratively, a point to which we return in a later chapter.

2.3.4. *Approximate calculation of the linear system*

The coefficients in the linear systems (2.3.41) and (2.3.46) must usually be calculated numerically, as was discussed above for (2.3.50). The use of approximations in the matrix coefficients or in the right-hand constants will lead to errors in the solution values $\{c_i\}$. Denote the original system by $A_n c = r$, and denote the linear system obtained by numerical integration by $\tilde{A}_n \tilde{c} = \tilde{r}$. For definiteness, we consider only (2.3.41) here, with the solution formula (2.3.42) for x_n.

Let $\tilde{x}_n$ denote the solution obtained by replacing $\{c_i\}$ by $\{\tilde{c}_i\}$. For the error, we have

$$x_n(t) - \tilde{x}_n(t) = \frac{1}{\lambda} \sum_{i=1}^n (c_i - \tilde{c}_i)\ell_i(t)$$

$$\|x_n - \tilde{x}_n\|_\infty \leq \frac{1}{|\lambda|}\|\mathcal{P}_n\|\|c - \tilde{c}\|_\infty \tag{2.3.59}$$

We have used

$$\mathcal{P}_n z(t) \equiv \sum_{i=1}^n z(t_i)\ell_i(t), \quad z \in C(D) \tag{2.3.60}$$

to write the function interpolating $z(t)$ at $t = t_1, \ldots, t_n$. Then $\mathcal{P}_n : C(D) \to C(D)$ is a bounded projection, and

$$\|\mathcal{P}_n\| = \max_{t \in D} \sum_{i=1}^n |\ell_i(t)| \tag{2.3.61}$$

For the particular case of piecewise linear interpolation, defined above in (2.3.47), we have $\|\mathcal{P}_n\| = 1$, and in this case,

$$\|x_n - \tilde{x}_n\|_\infty \leq \frac{1}{|\lambda|}\|c - \tilde{c}\|_\infty \tag{2.3.62}$$

Since $\|x - x_n\|_\infty \leq ch^2$, some $c > 0$, we should choose the numerical integration method used in obtaining $\tilde{A}_n \tilde{c} = \tilde{r}$ so as to ensure that $\|x_n - \tilde{x}_n\|_\infty$ is also $O(h^2)$. Moreover, we may also want to ensure that

$$\|x_n - \tilde{x}_n\|_\infty < \|x - x_n\| \tag{2.3.63}$$

perhaps much less.

To do the above, we need a bound on $\|c - \tilde{c}\|_\infty$ that is based on bounds for the integration errors $\|A_n - \tilde{A}_n\|$ and $\|r - \tilde{r}\|_\infty$. The bounds on $\|c - \tilde{c}\|_\infty$ are obtained by standard perturbation arguments for linear systems. To proceed further, we need a bound for $\|A_n^{-1}\|$, the matrix norm induced by $\|\cdot\|_\infty$ on $\mathbf{R}^n$ (see Ref. [48, p. 485]). In (3.6.151) of Chapter 3 we show

$$\|A_n^{-1}\| \leq \|\mathcal{P}_n\|\|(\lambda - \mathcal{P}_n\mathcal{K})^{-1}\| \tag{2.3.64}$$

and from (2.1.5),

$$\|(\lambda - \mathcal{P}_n\mathcal{K})^{-1}\| \leq \gamma \|(\lambda - \mathcal{K})^{-1}\| \tag{2.3.65}$$

for some $\gamma > 0$ and for all sufficiently large n.

Assume

$$\|A_n - \tilde{A}_n\| \le \frac{1}{\|A_n^{-1}\|} \tag{2.3.66}$$

Then by the geometric series theorem (cf. Theorem A.1 in the Appendix), we can show $\tilde{A}_n^{-1}$ exists, and that it satisfies

$$\|\tilde{A}_n^{-1}\| \le \frac{\|A_n^{-1}\|}{1 - \|A_n^{-1}\|} \|A_n - \tilde{A}_n\| \tag{2.3.67}$$

$$\|A_n^{-1} - \tilde{A}_n^{-1}\| \le \|\tilde{A}_n^{-1}\| \|A_n - \tilde{A}_n\| \|A_n^{-1}\| \tag{2.3.68}$$

The proof is very similar to that of the existence of $(\lambda - \mathcal{P}_n\mathcal{K})^{-1}$ in Theorem 2.1.1.

For the error $c - \tilde{c}$,

$$\begin{aligned} c - \tilde{c} &= A_n^{-1}r - \tilde{A}_n^{-1}r \\ &= [A_n^{-1} - \tilde{A}_n^{-1}]r + \tilde{A}_n^{-1}[r - \tilde{r}] \end{aligned}$$

$$\|c - \tilde{c}\|_\infty \le \|A_n^{-1} - \tilde{A}_n^{-1}\| \|r\|_\infty + \|\tilde{A}_n^{-1}\| \|r - \tilde{r}\|_\infty$$

From (2.3.41), it is straightforward to show

$$\|r\|_\infty \le \|\mathcal{K}\| \|y\|_\infty$$

Thus

$$\|c - \tilde{c}\|_\infty \le \|A_n^{-1} - \tilde{A}_n^{-1}\| \|\mathcal{K}\| \|y\|_\infty + \|\tilde{A}_n^{-1}\| \|r - \tilde{r}\|_\infty \tag{2.3.69}$$

Combining (2.3.64)–(2.3.69) with (2.3.62), we have bounds for the error $\|x_n - \tilde{x}_n\|_\infty$.

To make these ideas more precise, consider the case (2.3.49), in which $K_n(t, s)$ is defined using piecewise linear interpolation. The inequalities (2.3.64)–(2.3.65) become

$$\|A_n^{-1}\| \le \|(\lambda - \mathcal{P}_n\mathcal{K})^{-1}\| \le \gamma \|(\lambda - \mathcal{K})^{-1}\|$$

for all sufficiently large n. The inequality (2.3.66) can be satisfied uniformly in n if

$$\|A_n - \tilde{A}_n\| < \frac{1}{\gamma \|(\lambda - \mathcal{K})^{-1}\|}$$

Then

$$\left\| \tilde{A}_n^{-1} \right\| \le \frac{\left\| A_n^{-1} \right\|}{1 - \gamma \left\| (\lambda - \mathcal{K})^{-1} \right\|} \left\| A_n - \tilde{A}_n \right\|$$

Using (2.3.62) and (2.3.69), we have

$$\| x_n - \tilde{x}_n \|_\infty \le \frac{1}{\lambda} \left[\left\| A_n^{-1} - \tilde{A}_n^{-1} \right\| \| \mathcal{K} \| \| y \|_\infty + \left\| \tilde{A}_n^{-1} \right\| \| r - \tilde{r} \|_\infty \right] \quad (2.3.70)$$

To have $\| x_n - \tilde{x}_n \|_\infty = O(h^2)$, to agree with $\| x - x_n \|_\infty = O(h^2)$, (2.3.70) implies we should choose $\tilde{A}_n$ and $\tilde{r}$ so that

$$\| A_n - \tilde{A}_n \|, \| r - \tilde{r} \|_\infty = O(h^2)$$

This suggests that each element of A_n should be integrated with an accuracy of $O(h^3)$, and each element of r should be integrated with an accuracy of $O(h^2)$. This gives an indication of the type of quadrature rule that should be used in numerically evaluating A_n and r. It is left as a problem to consider the specific numerical integration method (2.3.51).

2.4. Orthonormal expansions

Let $\mathcal{X} = L^2(D)$, a Hilbert space over $\mathbf{R}$, and let $\mathcal{K} : \mathcal{X} \to \mathcal{X}$ be a compact integral operator. In textbooks on integral equations, a very popular degenerate kernel method for approximating $\mathcal{K}$ is based on using orthonormal expansions of the kernel function $K(t, s)$.

Let $L^2(D)$ have the inner product

$$(f, g) = \int_D w(t) f(t) g(t) \, dt$$

The weight function $w(t)$ is assumed to satisfy the usual properties

- For almost all $t \in D$, $w(t) \ge 0$.
- For all $n \ge 0$,

$$\int_D w(t) |t|^n \, dt < \infty$$

- If $f \in C(D)$ and is nonnegative on D, then

$$\int_D w(t) f(t) \, dt = 0 \implies f(t) \equiv 0$$

Let $\{\phi_1, \ldots, \phi_n, \ldots\}$ be a *complete orthonormal set* in $L^2(D)$ with respect to a weight function $w(t)$. This means that

1. $(\phi_n, \phi_m) = \delta_{nm}$, for $1 \le m, n < \infty$.
2. If $x \in L^2(D)$ and if $(x, \phi_n) = 0$ for all $n \ge 1$, then $x = 0$.

Then given $x \in L^2(D)$, we can write

$$x(t) = \sum_{i=1}^{\infty} (x, \phi_i)\phi_i(t)$$

This is the Fourier series of x with respect to $\{\phi_n\}$, and it converges in $L^2(D)$. We can apply this construction to the approximation of $K(t, s)$, with respect to either variable.

Expanding with respect to t, we have

$$K(t, s) = \sum_{i=1}^{\infty} \phi_i(t)\beta_i(s) \qquad (2.4.71)$$

with

$$\beta_i(s) = (K(\cdot, s), \phi_i) = \int_D w(t)K(t, s)\phi_i(t)\,dt \qquad (2.4.72)$$

Define the approximating degenerate kernel by

$$K_n(t, s) = \sum_{i=1}^{n} \phi_i(t)\beta_i(s) \qquad (2.4.73)$$

The equation $(\lambda - \mathcal{K}_n)x_n = y$ is solved by

$$x_n(t) = \frac{1}{\lambda}\left[y(t) + \sum_{i=1}^{n} c_i\phi_i(t)\right] \qquad (2.4.74)$$

From (2.1.15), the coefficients $\{c_i\}$ satisfy the linear system

$$\lambda c_i - \sum_{j=1}^{n} c_j \int_D \phi_j(s)\beta_i(s)\,ds = \int_D \beta_i(s)y(s)\,ds, \quad i = 1, \ldots, n \qquad (2.4.75)$$

If we write out the needed integrals more completely, we have

$$\int_D \phi_j(s)\beta_i(s)\,ds = \int_D \phi_j(s) \int_D w(t)K(t, s)\phi_i(t)\,dt\,ds$$

$$\int_D \beta_i(s)y(s)\,ds = \int_D y(s) \int_D w(t)K(t, s)\phi_i(t)\,dt\,ds$$

Thus all coefficients in the linear system (2.4.75) are double integrals over D, and these must usually be evaluated numerically. Compare this with the use of degenerate kernels based on interpolation, in which all integrals are single integrals over D. For this reason, we generally do not use degenerate kernels based on orthonormal expansions, although if the integrals can be evaluated at a reasonable cost, then the methods may be considered more seriously.

For the error in using $\mathcal{K}_n$, we have

$$\|\mathcal{K} - \mathcal{K}_n\| \le \left[\sum_{n+1}^{\infty} \|\beta_i\|_2^2\right]^{\frac{1}{2}}$$

We leave the proof as a problem for the reader. We also leave as a problem the derivation of the corresponding degenerate kernel approximations when the expansion of $K(t, s)$ is done with respect to the variable s.

Discussion of the literature

The degenerate kernel method is of theoretical use in the development of the theory of compact integral operators, and as such it can be found in a large number of books on integral operators. For example, see Fenyö and Stolle [198, p. 22], Hochstadt [272, p. 37], Mikhlin [378, p. 19], Pogorzelski [426, p. 71], Porter and Stirling [427, p. 57], and Zabreyko et al. [586, §2.3]. For general treatments of the use of degenerate kernel approximations to solve integral equations of the second kind, see Ref. [39, pp. 37–50], Baker [73, p. 385], Bückner [97], Fenyö and Stolle [200, Chap. 17], Hackbusch [251, §4.2], Kress [325, Chap. 11], Kantorovich and Krylov [305, §2.4], Linz [344, §6.2], Lonseth [347], and Mikhlin and Smolitskiy [382].

There have been a number of research papers written on degenerate kernel methods. Hämmerlin and Schumaker [256] show how to produce accurate degenerate kernel approximations and associated error bounds, and Arthur [27] discusses the use of degenerate kernel approximations using spline functions. In [501], [502], [513], I. Sloan and his colleagues developed a new type of degenerate kernel approximation, one leading to much more rapid convergence than earlier methods. This led subsequently to the development in [503] of the *Sloan iterate* for projection methods, which is examined in §3.4 of the next chapter. The degenerate kernel method is used in Bückner [98, p. 459] to carry out an error analysis for quadrature methods for solving Fredholm integral equations of the second kind, by converting such a discretization to an equivalent one based on a degenerate kernel approximation.

In §3.5 of Chapter 3, we introduce the idea of *regularizing* the equation $(\lambda - \mathcal{K})x = y$. It is then shown that if a projection method is applied to the regularized problem, the resulting numerical method is essentially a degenerate kernel method. Most of the important degenerate kernel methods of this chapter can be obtained in this manner, and thus their stability and convergence can be studied by using the corresponding results for the associated projection methods.

Degenerate kernel methods for solving $(\lambda - \mathcal{K})x = y$ have never been as popular as the methods presented in the following chapters. The degenerate kernel method often requires a significant amount of preliminary mathematical analysis if you want an efficient algorithm. But the method is conceptually simple, and it is completely general with respect to λ, $\mathcal{K}$, and y. The smoothness of x is unlikely to affect the rate of convergence of the method, since the rate is generally that of $\|\mathcal{K} - \mathcal{K}_n\|$.

3

Projection methods

To solve approximately the integral equation

$$\lambda x(t) - \int_D K(t,s)x(s)\,ds = y(t), \quad t \in D \tag{3.0.1}$$

choose a finite dimensional family of functions that is believed to contain a function $\bar{x}(s)$ close to the true solution $x(s)$. The desired numerical solution $\bar{x}(s)$ is selected by having it satisfy (3.0.1) approximately. There are various senses in which $\bar{x}(s)$ can be said to "satisfy (3.0.1) approximately," and these lead to different types of methods. The most popular of these are *collocation methods* and *Galerkin methods*, and they are defined below. When these methods are formulated in an abstract framework using functional analysis, they all make essential use of *projection operators*. Since the error analysis is most easily carried out within such a functional analysis framework, we refer collectively to all such methods as *projection methods*.

3.1. General theory

We write the integral equation (3.0.1) in the operator form

$$(\lambda - \mathcal{K})x = y$$

and the operator $\mathcal{K}$ is assumed to be compact on a Banach space $\mathcal{X}$ to $\mathcal{X}$. The most popular choices are $C(D)$ and $L^2(D)$. For Galerkin's method and its generalizations, Sobolev spaces $H^r(D)$ are also used commonly, with $H^0(D) \equiv L^2(D)$.

In practice, we choose a sequence of finite dimensional subspaces $\mathcal{X}_n \subset \mathcal{X}$, $n \geq 1$, with $\mathcal{X}_n$ having dimension d_n. Let $\mathcal{X}_n$ have a basis $\{\phi_1, \ldots, \phi_d\}$, with

$d \equiv d_n$ for notational simplicity. We seek a function $x_n \in \mathcal{X}_n$, and it can be written as

$$x_n(t) = \sum_{j=1}^{d} c_j \phi_j(t), \quad t \in D \tag{3.1.2}$$

This is substituted into (3.0.1), and the coefficients $\{c_1, \ldots, c_d\}$ are determined by forcing the equation to be almost exact in some sense. For later use, introduce

$$
\begin{aligned}
r_n(t) &= \lambda x_n(t) - \int_D K(t,s) x_n(s)\, ds - y(t) \\
&= \sum_{j=1}^{d} c_j \left\{ \lambda \phi_j(t) - \int_D K(t,s)\phi_j(s)\, ds \right\} - y(t), \quad t \in D \tag{3.1.3}
\end{aligned}
$$

This is called the *residual* in the approximation of the equation when using $x \approx x_n$. Symbolically,

$$r_n = (\lambda - \mathcal{K}) x_n - y$$

The coefficients $\{c_1, \ldots, c_d\}$ are chosen by forcing $r_n(t)$ to be approximately zero in some sense. The hope and expectation are that the resulting function $x_n(t)$ will be a good approximation of the true solution $x(t)$.

3.1.1. Collocation methods

Pick distinct node points $t_1, \ldots, t_d \in D$, and require

$$r_n(t_i) = 0, \quad i = 1, \ldots, d_n \tag{3.1.4}$$

This leads to determining $\{c_1, \ldots, c_d\}$ as the solution of the linear system

$$\sum_{j=1}^{d} c_j \left\{ \lambda \phi_j(t_i) - \int_D K(t_i,s)\phi_j(s)\, ds \right\} = y(t_i), \quad i = 1, \ldots, d_n \tag{3.1.5}$$

An immediate question is whether this system has a solution, and if so, whether it is unique. If so, does x_n converge to x? Note also that the linear system contains integrals that must usually be evaluated numerically, a point we return to later. We should have written the node points as $\{t_{1,n}, \ldots, t_{d,n}\}$; but for notational simplicity, the explicit dependence on n has been suppressed, to be understood only implicitly.

As a part of writing (3.1.5) in a more abstract form, we introduce a projection operator $\mathcal{P}_n$ that maps $\mathcal{X} = C(D)$ onto $\mathcal{X}_n$. Given $x \in C(D)$, define $\mathcal{P}_n x$ to be that element of $\mathcal{X}_n$ that interpolates x at the nodes $\{t_1, \ldots, t_d\}$. This means writing

$$\mathcal{P}_n x(t) = \sum_{j=1}^{d} \alpha_j \phi_j(t)$$

with the coefficients $\{\alpha_j\}$ determined by solving the linear system

$$\sum_{j=1}^{d} \alpha_j \phi_j(t_i) = x(t_i), \quad i = 1, \ldots, d_n$$

This linear system has a unique solution if

$$\det[\phi_j(t_i)] \neq 0 \qquad (3.1.6)$$

Henceforth in this chapter, we assume this is true whenever the collocation method is being discussed. By a simple argument, this condition also implies that the functions $\{\phi_1, \ldots, \phi_d\}$ are an independent set over D. In the case of polynomial interpolation for functions of one variable, the determinant in (3.1.6) is referred to as the *Vandermonde* determinant (for example, see Ref. [48, Chap. 3, p. 185]).

To see more clearly that $\mathcal{P}_n$ is linear, and to give a more explicit formula, we introduce a new set of basis functions. For each i, $1 \leq i \leq d_n$, let $\ell_i \in \mathcal{X}_n$ be that element that satisfies the interpolation conditions

$$\ell_i(t_j) = \delta_{ij}, \quad j = 1, \ldots, d_n$$

By (3.1.6), there is a unique such ℓ_i, and the set $\{\ell_1, \ldots, \ell_d\}$ is a new basis for $\mathcal{X}_n$. With polynomial interpolation, such functions ℓ_i are called *Lagrange basis functions*, and we will use this name with all types of approximating subspaces $\mathcal{X}_n$. With this new basis, we can write

$$\mathcal{P}_n x(t) = \sum_{j=1}^{d} x(t_j) \ell_j(t), \quad t \in D \qquad (3.1.7)$$

Clearly, $\mathcal{P}_n$ is linear and finite rank. In addition, as an operator on $C(D)$ to $C(D)$,

$$\|\mathcal{P}_n\| = \max_{t \in D} \sum_{j=1}^{d} |\ell_j(t)| \qquad (3.1.8)$$

Example. Let $\mathcal{X}_n = \mathrm{Span}\{1, t, \ldots, t^n\}$. Then for $i = 0, 1, \ldots, n$,

$$\ell_i(t) = \prod_{\substack{j=0 \\ j \neq i}}^{n} \left(\frac{t - t_j}{t_i - t_j} \right) \tag{3.1.9}$$

and formula (3.1.7) is called *Lagrange's form of the polynomial interpolation polynomial.*

We note that

$$\mathcal{P}_n z = 0 \quad \text{if and only if} \quad z(t_i) = 0, \ i = 1, \ldots, d_n \tag{3.1.10}$$

The condition (3.1.4) can now be rewritten as

$$\mathcal{P}_n r_n = 0$$

or equivalently,

$$\mathcal{P}_n (\lambda - \mathcal{K}) x_n = \mathcal{P}_n y, \quad x_n \in \mathcal{X}_n \tag{3.1.11}$$

We return to this later.

3.1.2. Galerkin's method

Let $\mathcal{X} = L^2(D)$ or some other Hilbert space, and let $(\cdot, \cdot)$ denote the inner product for $\mathcal{X}$. Require r_n to satisfy

$$(r_n, \phi_i) = 0, \quad i = 1, \ldots, d_n \tag{3.1.12}$$

The left side is the Fourier coefficient of r_n associated with ϕ_i. If $\{\phi_1 \ldots, \phi_d\}$ are the leading members of an orthonormal family $\Phi \equiv \{\phi_1, \ldots, \phi_d, \ldots\}$ that is complete in $\mathcal{X}$, then (3.1.12) requires the leading terms to be zero in the Fourier expansion of r_n with respect to Φ.

To find x_n, apply (3.1.12) to (3.1.3). This yields the linear system

$$\sum_{j=1}^{d} c_j \{\lambda(\phi_j, \phi_i) - (\mathcal{K}\phi_j, \phi_i)\} = (y, \phi_i), \quad i = 1, \ldots, d_n \tag{3.1.13}$$

This is Galerkin's method for obtaining an approximate solution to (3.0.1). Does the system have a solution? If so, is it unique? Does the resulting sequence of approximate solutions x_n converge to x in $\mathcal{X}$? Does the sequence converge in $C(D)$? Note also that the above formulation contains double integrals $(\mathcal{K}\phi_j, \phi_i)$. These must often be computed numerically. We return to a consideration of this later.

As a part of writing (3.1.13) in a more abstract form, we introduce a projection operator $\mathcal{P}_n$ that maps $\mathcal{X}$ onto $\mathcal{X}_n$. For general $x \in \mathcal{X}$, define $\mathcal{P}_n x$ to be the solution of the following minimization problem.

$$\|x - \mathcal{P}_n x\| = \min_{z \in \mathcal{X}_n} \|x - z\| \qquad (3.1.14)$$

Since $\mathcal{X}_n$ is finite dimensional, it can be shown that this problem has a solution; and by $\mathcal{X}_n$ being an inner product space, the solution can be shown to be unique.

To obtain a better understanding of $\mathcal{P}_n$, we give an explicit formula for $\mathcal{P}_n x$. Introduce a new basis $\{\psi_1, \ldots, \psi_d\}$ for $\mathcal{X}_n$ by using the *Gram-Schmidt* process to create an orthonormal basis from $\{\phi_1, \ldots, \phi_d\}$. The element ψ_i is a linear combination of $\{\phi_1, \ldots, \phi_i\}$, and moreover

$$(\psi_i, \psi_j) = \delta_{ij}, \quad i, j = 1, \ldots, d_n$$

With this new basis, it is straightforward to show that

$$\mathcal{P}_n x = \sum_{i=1}^{d} (x, \psi_i) \, \psi_i \qquad (3.1.15)$$

This shows immediately that $\mathcal{P}_n$ is a linear operator.

With this formula, we can show the following results.

$$\|x\|^2 = \|\mathcal{P}_n x\|^2 + \|x - \mathcal{P}_n x\|^2 \qquad (3.1.16)$$

$$\|\mathcal{P}_n x\|^2 = \sum_{i=1}^{d} |(x, \psi_i)|^2$$

$$(\mathcal{P}_n x, y) = (x, \mathcal{P}_n y), \quad x, y \in \mathcal{X} \qquad (3.1.17)$$

$$((I - \mathcal{P}_n)x, \mathcal{P}_n y) = 0, \quad x, y \in \mathcal{X} \qquad (3.1.18)$$

Because of the latter, $\mathcal{P}_n x$ is called the *orthogonal projection of x onto* $\mathcal{X}_n$. The operator $\mathcal{P}_n$ is called an *orthogonal projection* operator. The result (3.1.16) leads to

$$\|\mathcal{P}_n\| = 1 \qquad (3.1.19)$$

Using (3.1.18), we can show

$$\|x - z\|^2 = \|x - \mathcal{P}_n x\|^2 + \|\mathcal{P}_n x - z\|^2, \quad z \in \mathcal{X}_n \qquad (3.1.20)$$

This shows $\mathcal{P}_n x$ is the unique solution to (3.1.14).

We note that

$$\mathcal{P}_n z = 0 \quad \text{if and only if} \quad (z, \phi_i) = 0, \quad i = 1, \ldots, d_n \qquad (3.1.21)$$

With $\mathcal{P}_n$, we can rewrite (3.1.12) as

$$\mathcal{P}_n r_n = 0$$

or equivalently,

$$\mathcal{P}_n (\lambda - \mathcal{K}) x_n = \mathcal{P}_n y, \quad x_n \in \mathcal{X}_n \qquad (3.1.22)$$

Note the similarity to (3.1.11).

There is a variant on Galerkin's method, known as the *Petrov-Galerkin method*. With it, we still choose $x_n \in \mathcal{X}_n$, but now we require

$$(r_n, w) = 0, \quad \text{all } w \in \mathcal{W}_n$$

with $\mathcal{W}_n$ another finite dimensional subspace, also of dimension d_n. This method is not considered further in this chapter, but it is an important method when looking at the numerical solution of boundary integral equations. Another approach to Galerkin's method is to set it within a variational framework, and we do this in Chapter 7.

3.1.3. The general framework

Let $\mathcal{X}$ be a Banach space, and let $\{\mathcal{X}_n \mid n \geq 1\}$ be a sequence of finite dimensional subspaces, say of dimension d_n. Let $\mathcal{P}_n : \mathcal{X} \to \mathcal{X}_n$ be a bounded projection operator. This means that $\mathcal{P}_n$ is a bounded linear operator with

$$\mathcal{P}_n x = x, \quad x \in \mathcal{X}_n$$

Note that this implies $\mathcal{P}_n^2 = \mathcal{P}_n$, and thus

$$\|\mathcal{P}_n\| = \|\mathcal{P}_n^2\| \leq \|\mathcal{P}_n\|^2$$
$$\|\mathcal{P}_n\| \geq 1 \qquad (3.1.23)$$

We already have examples of $\mathcal{P}_n$ in the interpolatory projection operator of (3.1.7) and the orthogonal projection operator (3.1.15). Also, the interpolatory projection operator associated with piecewise linear interpolation on an interval $[a, b]$ was introduced in (2.3.47) in Chapter 2.

Motivated by (3.1.11) and (3.1.22), we approximate (3.0.1) by attempting to solve the problem

$$\mathcal{P}_n (\lambda - \mathcal{K}) x_n = \mathcal{P}_n y, \quad x_n \in \mathcal{X}_n \qquad (3.1.24)$$

This is the form in which the method is implemented, as it leads directly to equivalent finite linear systems such as (3.1.5) and (3.1.13). For the error analysis, however, we write (3.1.24) in an equivalent but more convenient form.

If x_n is a solution of (3.1.24), then by using $\mathcal{P}_n x_n = x_n$, the equation can be written as

$$(\lambda - \mathcal{P}_n \mathcal{K})x_n = \mathcal{P}_n y, \quad x_n \in \mathcal{X} \tag{3.1.25}$$

To see that a solution of this is also a solution of (3.1.24), note that if (3.1.25) has a solution $x_n \in \mathcal{X}$, then

$$x_n = \frac{1}{\lambda}[\mathcal{P}_n y + \mathcal{P}_n \mathcal{K} x_n] \in \mathcal{X}_n$$

Thus $\mathcal{P}_n x_n = x_n$,

$$(\lambda - \mathcal{P}_n \mathcal{K})x_n = \mathcal{P}_n(\lambda - \mathcal{K})x_n$$

and this shows that (3.1.25) implies (3.1.24).

For the error analysis, we compare (3.1.25) with the original equation

$$(\lambda - \mathcal{K})x = y \tag{3.1.26}$$

since both equations are defined on the original space $\mathcal{X}$. The theoretical analysis is based on the approximation of $\lambda - \mathcal{P}_n \mathcal{K}$ by $\lambda - \mathcal{K}$:

$$\lambda - \mathcal{P}_n \mathcal{K} = (\lambda - \mathcal{K}) + (\mathcal{K} - \mathcal{P}_n \mathcal{K})$$
$$= (\lambda - \mathcal{K})[I + (\lambda - \mathcal{K})^{-1}(\mathcal{K} - \mathcal{P}_n \mathcal{K})] \tag{3.1.27}$$

We use this in the following theorem.

Theorem 3.1.1. *Assume* $\mathcal{K} : \mathcal{X} \to \mathcal{X}$ *is bounded, with* $\mathcal{X}$ *a Banach space, and assume* $\lambda - \mathcal{K} : \mathcal{X} \overset{1-1}{\underset{onto}{\to}} \mathcal{X}$. *Further assume*

$$\|\mathcal{K} - \mathcal{P}_n \mathcal{K}\| \to 0 \quad as \ n \to \infty \tag{3.1.28}$$

Then for all sufficiently large n, say $n \geq N$, *the operator* $(\lambda - \mathcal{P}_n \mathcal{K})^{-1}$ *exists as a bounded operator from* $\mathcal{X}$ *to* $\mathcal{X}$. *Moreover, it is uniformly bounded:*

$$\sup_{n \geq N} \|(\lambda - \mathcal{P}_n \mathcal{K})^{-1}\| < \infty \tag{3.1.29}$$

For the solutions of (3.1.25) and (3.1.26),

$$x - x_n = \lambda(\lambda - \mathcal{P}_n \mathcal{K})^{-1}(x - \mathcal{P}_n x) \tag{3.1.30}$$

$$\frac{|\lambda|}{\|\lambda - \mathcal{P}_n \mathcal{K}\|}\|x - \mathcal{P}_n x\| \leq \|x - x_n\| \leq |\lambda|\|(\lambda - \mathcal{P}_n \mathcal{K})^{-1}\|\|x - \mathcal{P}_n xt\|$$

$$\tag{3.1.31}$$

This leads to $\|x - x_n\|$ *converging to zero at exactly the same speed as* $\|x - \mathcal{P}_n x\|$.

Proof.

(a) Pick N such that

$$\epsilon_N \equiv \sup_{n \geq N} \|\mathcal{K} - \mathcal{P}_n \mathcal{K}\| < \frac{1}{\|(\lambda - \mathcal{K})^{-1}\|}$$

Then the inverse $[I + (\lambda - \mathcal{K})^{-1}(\mathcal{K} - \mathcal{P}_n \mathcal{K})]^{-1}$ exists and is uniformly bounded by the geometric series theorem (cf. Theorem A.1 in the Appendix):

$$\|[I + (\lambda - \mathcal{K})^{-1}(\mathcal{K} - \mathcal{P}_n \mathcal{K})]^{-1}\| \leq \frac{1}{1 - \epsilon_N \|(\lambda - \mathcal{K})^{-1}\|}$$

Using (3.1.27), $(\lambda - \mathcal{P}_n \mathcal{K})^{-1}$ exists,

$$(\lambda - \mathcal{P}_n \mathcal{K})^{-1} = [I + (\lambda - \mathcal{K})^{-1}(\mathcal{K} - \mathcal{P}_n \mathcal{K})]^{-1}(\lambda - \mathcal{K})^{-1}$$

$$\|(\lambda - \mathcal{P}_n \mathcal{K})^{-1}\| \leq \frac{\|(\lambda - \mathcal{K})^{-1}\|}{1 - \epsilon_N \|(\lambda - \mathcal{K})^{-1}\|} \equiv M \qquad (3.1.32)$$

This shows (3.1.29).

(b) For the error formula (3.1.30), multiply $(\lambda - \mathcal{K})x = y$ by $\mathcal{P}_n$, and then rearrange to obtain

$$(\lambda - \mathcal{P}_n \mathcal{K})x = \mathcal{P}_n y + \lambda(x - \mathcal{P}_n x)$$

Subtract $(\lambda - \mathcal{P}_n \mathcal{K})x_n = \mathcal{P}_n y$ to get

$$(\lambda - \mathcal{P}_n \mathcal{K})(x - x_n) = \lambda(x - \mathcal{P}_n x) \qquad (3.1.33)$$

$$x - x_n = \lambda(\lambda - \mathcal{P}_n \mathcal{K})^{-1}(x - \mathcal{P}_n x)$$

which is (3.1.30). Taking norms and using (3.1.32),

$$\|x - x_n\| \leq |\lambda| M \|x - \mathcal{P}_n x\| \qquad (3.1.34)$$

Thus if $\mathcal{P}_n x \to x$, then $x_n \to x$ as $n \to \infty$.

(c) The upper bound in (3.1.31) follows directly from (3.1.30), as we have just seen. The lower bound follows by taking bounds in (3.1.33), to obtain

$$|\lambda| \|x - \mathcal{P}_n x\| \leq \|\lambda - \mathcal{P}_n \mathcal{K}\| \|x - x_n\|$$

This is equivalent to the lower bound in (3.1.31).

To obtain a lower bound that is uniform in n, note that for $n \geq N$,

$$\|\lambda - \mathcal{P}_n \mathcal{K}\| \leq \|\lambda - \mathcal{K}\| + \|\mathcal{K} - \mathcal{P}_n \mathcal{K}\|$$
$$\leq \|\lambda - \mathcal{K}\| + \epsilon_N$$

The lower bound in (3.1.31) can now be replaced by

$$\frac{|\lambda|}{\|\lambda - \mathcal{K}\| + \epsilon_N} \|x - \mathcal{P}_n x\| \leq \|x - x_n\|$$

Combining this and (3.1.34), we have

$$\frac{|\lambda|}{\|\lambda - \mathcal{K}\| + \epsilon_N} \|x - \mathcal{P}_n x\| \leq \|x - x_n\| \leq |\lambda| M \|x - \mathcal{P}_n x\| \qquad (3.1.35)$$

This shows that x_n converges to x if and only if $\mathcal{P}_n x$ converges to x. Moreover, if convergence does occur, then $\|x - \mathcal{P}_n x\|$ and $\|x - x_n\|$ tend to zero with exactly the same speed. □

To apply the above theorem, we need to know whether $\|\mathcal{K} - \mathcal{P}_n \mathcal{K}\| \to 0$ as $n \to \infty$. The following two lemmas address this question

Lemma 3.1.1. Let $\mathcal{X}$, $\mathcal{Y}$ be Banach spaces, and let $\mathcal{A}_n : \mathcal{X} \to \mathcal{Y}, n \geq 1$ be a sequence of bounded linear operators. Assume $\{\mathcal{A}_n x\}$ converges for all $x \in \mathcal{X}$. Then the convergence is uniform on compact subsets of $\mathcal{X}$.

Proof. By the principle of uniform boundedness (cf. Theorem A.3 in the Appendix), the operators $\mathcal{A}_n$ are uniformly bounded:

$$M \equiv \sup_{n \geq 1} \|\mathcal{A}_n\| < \infty$$

The functions $\mathcal{A}_n$ are also equicontinuous:

$$\|\mathcal{A}_n x - \mathcal{A}_n y\| \leq M \|x - y\|$$

Let S be a compact subset of $\mathcal{X}$. Then $\{\mathcal{A}_n\}$ is a uniformly bounded and equicontinuous family of functions on the compact set S; it is then a standard result of analysis that $\{\mathcal{A}_n x\}$ is uniformly convergent for $x \in S$. □

Lemma 3.1.2. Let $\mathcal{X}$ be a Banach space, and let $\{\mathcal{P}_n\}$ be a family of bounded projections on $\mathcal{X}$ with

$$\mathcal{P}_n x \to x \quad \text{as } n \to \infty, \quad x \in \mathcal{X} \qquad (3.1.36)$$

Let $\mathcal{K} : \mathcal{X} \rightarrow \mathcal{X}$ be compact. Then

$$\|\mathcal{K} - \mathcal{P}_n \mathcal{K}\| \rightarrow 0 \quad \text{as } n \rightarrow \infty$$

Proof. From the definition of operator norm,

$$\|\mathcal{K} - \mathcal{P}_n \mathcal{K}\| = \sup_{\|x\| \leq 1} \|\mathcal{K}x - \mathcal{P}_n \mathcal{K}x\| = \sup_{z \in \mathcal{K}(U)} \|z - \mathcal{P}_n z\|$$

with $\mathcal{K}(U) = \{\mathcal{K}x \mid \|x\| \leq 1\}$. The set $\overline{\mathcal{K}(U)}$ is compact. Therefore, by the preceding Lemma 3.1.1 and the assumption (3.1.36),

$$\sup_{z \in \mathcal{K}(U)} \|z - \mathcal{P}_n z\| \rightarrow 0 \quad \text{as } n \rightarrow \infty$$

This proves the lemma. $\square$

This last lemma includes most cases of interest, but not all. There are situations where $\mathcal{P}_n x \rightarrow x$ for most $x \in \mathcal{X}$, but not all x. In such cases, it is necessary to show directly that $\|\mathcal{K} - \mathcal{P}_n \mathcal{K}\| \rightarrow 0$. In such cases, of course, we see from (3.1.35) that $x_n \rightarrow x$ if and only if $\mathcal{P}_n x \rightarrow x$; and thus the method is not convergent for some solutions x. This would occur, for example, if $\mathcal{X}_n$ is the set of polynomials of degree $\leq n$ and $\mathcal{X} = C[a, b]$.

3.2. Examples of the collocation method

We give examples that illustrate the two main types of collocation methods:

- Decompose the integration region D into elements $\Delta_1, \ldots, \Delta_m$ and then approximate a function $x \in C(D)$ by a low degree polynomial over each of the elements Δ_i. These are often referred to as *piecewise polynomial collocation methods*, and when D is the boundary of a region, such methods are called *boundary element methods*.

- Approximate an $x \in C(D)$ by using a family of functions that are defined globally over all of D, for example, polynomials, trigonometric polynomials, or spherical polynomials. Generally, these approximating functions are also infinitely differentiable. Sometimes these types of collocation methods are referred to as *spectral methods*, especially when trigonometric polynomials are used.

In later sections we return to the examples presented below to illustrate newly presented material.

3.2.1. Piecewise linear interpolation

Recall the definition of piecewise linear interpolation given in and following (2.3.47) of Chapter 2. For convenience, we repeat those results here. Let $D = [a, b]$ and $n \geq 1$, and define $h = (b - a)/n$,

$$t_j = a + jh, \quad j = 0, 1, \ldots, n$$

The subspace $\mathcal{X}_n$ is the set of all functions that are piecewise linear on $[a, b]$, with breakpoints $\{t_0, \ldots, t_n\}$. Its dimension is $n + 1$.

Introduce the Lagrange basis functions for piecewise linear interpolation:

$$\ell_i(t) = \begin{cases} 1 - \dfrac{|t - t_i|}{h}, & t_{i-1} \leq t \leq t_{i+1} \\ 0, & \text{otherwise} \end{cases} \tag{3.2.37}$$

with the obvious adjustment of the definition for $\ell_0(t)$ and $\ell_n(t)$. The projection operator is defined by

$$\mathcal{P}_n x(t) = \sum_{i=0}^{n} x(t_i)\ell_i(t) \tag{3.2.38}$$

For convergence of $\mathcal{P}_n x$,

$$\|x - \mathcal{P}_n x\|_\infty \leq \begin{cases} \omega(x, h), & x \in C[a, b] \\ \dfrac{h^2}{8}\|x''\|_\infty, & x \in C^2[a, b] \end{cases} \tag{3.2.39}$$

This shows that $\mathcal{P}_n x \to x$ for all $x \in C[a, b]$. For any compact operator $\mathcal{K}: C[a, b] \to C[a, b]$, Lemma 3.1.2 implies $\|\mathcal{K} - \mathcal{P}_n\mathcal{K}\| \to 0$ as $n \to \infty$. Therefore the results of Theorem 3.1.1 can be applied directly to the numerical solution of the integral equation $(\lambda - \mathcal{K})x = y$. For sufficiently large n, say $n \geq N$, the equation $(\lambda - \mathcal{P}_n\mathcal{K})x_n = \mathcal{P}_n y$ has a unique solution x_n for each $y \in C[a, b]$; and by (3.1.34),

$$\|x - x_n\|_\infty \leq |\lambda| M \|x - \mathcal{P}_n x\|_\infty$$

For $x \in C^2[a, b]$,

$$\|x - x_n\|_\infty \leq |\lambda| M \frac{h^2}{8}\|x''\|_\infty \tag{3.2.40}$$

The linear system (3.1.5) takes the simpler form

$$\lambda x_n(t_i) - \sum_{j=0}^{n} x_n(t_j) \int_a^b K(t_i, s)\ell_j(s)\, ds = y(t_i), \quad i = 0, \ldots, n \tag{3.2.41}$$

The integrals can be simplified. For $j = 1, \ldots, n-1$,

$$\int_a^b K(t_i, s)\ell_j(s)\, ds = \frac{1}{h}\int_{t_{j-1}}^{t_j} K(t_i, s)(s - t_{j-1})\, ds$$

$$+ \frac{1}{h}\int_{t_j}^{t_{j+1}} K(t_i, s)(t_j - s)\, ds \qquad (3.2.42)$$

The integrals for $j = 0$ and $j = n$ are modified accordingly. These must usually be calculated numerically, and we want to use a quadrature method that will retain the order of convergence in (3.2.40) at a minimum cost in calculation time. We return to this in Chapter 4, under the heading of "Discrete Collocation Methods."

There is some interest in looking more carefully at the operator $\mathcal{P}_n\mathcal{K}$. Using (3.2.38), we write

$$\mathcal{P}_n\mathcal{K}x(t) = \int_a^b K_n(t, s)x(s)\, ds \qquad (3.2.43)$$

$$K_n(t, s) = \sum_{i=0}^n K(t_i, s)\ell_i(t) \qquad (3.2.44)$$

This shows $\mathcal{P}_n\mathcal{K}$ is a degenerate kernel integral operator, and in fact, it is the degenerate kernel introduced in (2.3.40) in Chapter 2. Using (3.2.43),

$$\|\mathcal{K} - \mathcal{P}_n\mathcal{K}\| = \max_{a \le t \le b} \int_a^b |K(t, s) - K_n(t, s)|\, ds \qquad (3.2.45)$$

If $K(t, s)$ is twice continuously differentiable with respect to t, uniformly for $a \le s \le b$, then

$$\|\mathcal{K} - \mathcal{P}_n\mathcal{K}\| \le \frac{h^2}{8}\int_a^b \max_{a \le t \le b}\left|\frac{\partial^2 K(t, s)}{\partial t^2}\right| ds \qquad (3.2.46)$$

Example. Recall the integral equation

$$\lambda x(t) - \int_0^b e^{st}x(s)ds = y(t), \quad 0 \le t \le b \qquad (3.2.47)$$

which was used as a numerical example in §§2.2 and 2.3 of Chapter 2. We use the same two unknowns as were used previously,

$$x^{(1)}(t) = e^{-t}\cos(t), \quad x^{(2)}(t) = \sqrt{t}, \quad 0 \le t \le b \qquad (3.2.48)$$

Table 3.1. *Example of piecewise linear
collocation for solving (3.2.47)*

n	$E_n^{(1)}$	Ratio	$E_n^{(2)}$	Ratio
2	5.25E−3		2.32E−2	
4	1.31E−3	4.01	7.91E−3	2.93
8	3.27E−4	4.01	2.75E−3	2.88
16	8.18E−5	4.00	9.65E−4	2.85
32	2.04E−5	4.00	3.40E−4	2.84
64	5.11E−6	4.00	1.20E−4	2.83
128	1.28E−6	4.00	4.24E−5	2.83

The results of the use of piecewise linear collocation are given in Table 3.1. The parameters are $b = 1$, $\lambda = 5$, as in Table 2.4 in Chapter 2. The errors given in the table are the maximum errors on the collocation node points,

$$E_n^{(k)} = \max_{0 \le i \le n} \left| x^{(k)}(t_i) - x_n^{(k)}(t_i) \right|$$

The column labeled *Ratio* is the ratio of the successive values of $E_n^{(k)}$ as n is doubled.

The function $x^{(2)}(t)$ is not continuously differentiable on $[0, b]$, and we have no reason to expect a rate of convergence of $O(h^2)$. Empirically, the errors $E_n^{(2)}$ appear to be $O(h^{1.5})$. From Theorem 3.1.1 we know that $\|x^{(2)} - x_n^{(2)}\|_\infty$ converges at exactly the same speed as $\|x^{(2)} - \mathcal{P}_n x^{(2)}\|_\infty$, and it can be shown that the latter is only $O(h^{0.5})$. This apparent contradiction between the empirical and theoretical rates is due to $x_n(t)$ being *superconvergent* at the collocation node points: for the numerical solution $x_n^{(2)}$,

$$\lim_{n \to \infty} \frac{E_n^{(2)}}{\left\| x^{(2)} - x_n^{(2)} \right\|_\infty} = 0 \qquad (3.2.49)$$

This is examined in much greater detail in §3.4 of this chapter.

We could have used interpolation with higher degree polynomials, and that would result in a more rapid speed of convergence for the collocation solution. However, the essential ideas would remain the same. When using approximants that are piecewise polynomial functions of degree r, it can be shown that the speed of convergence of $x_n(t)$ to $x(t)$ is $O(h^{r+1})$, provided $x \in C^{r+1}[a, b]$, the space of $(r + 1)$-times continuously differentiable functions on $[a, b]$. The proof is based on showing $\|x - \mathcal{P}_n x\|_\infty = (h^{r+1})$ for $x \in C^{r+1}[a, b]$, using the standard error formula for polynomial interpolation (cf. [48, §3.5]).

3.2.2. Collocation with trigonometric polynomials

We solve the integral equation

$$\lambda x(t) - \int_0^{2\pi} K(t,s)x(s)\,ds = y(t), \quad 0 \le t \le 2\pi \tag{3.2.50}$$

in which the kernel function is assumed to be 2π-periodic in both s and t:

$$K(t+2\pi,s) \equiv K(t,s+2\pi) \equiv K(t,s)$$

Define $\mathcal{X} = C_p(2\pi)$ to be the set of all 2π-periodic and continuous functions $x(t)$ on $-\infty < t < \infty$:

$$x(t+2\pi) \equiv x(t)$$

With the uniform norm $\|\cdot\|_\infty$, $C_p(2\pi)$ is a Banach space. We consider the solution of (3.2.50) for $y \in C_p(2\pi)$, which then implies $x \in C_p(2\pi)$.

Since the solution $x(t)$ is 2π-periodic, we approximate it with *trigonometric polynomials*. For $n \ge 1$, define the trigonometric polynomials of degree $\le n$ by

$$\mathcal{X}_n = \mathrm{Span}\{1, \cos(t), \sin(t), \ldots, \cos(nt), \sin(nt)\} \tag{3.2.51}$$

To define interpolation using $\mathcal{X}_n$, let $h = 2\pi/(2n+1)$ and $t_j = jh$, $j = 0, \pm 1, \pm 2, \ldots$. We use the points $\{t_0, \ldots, t_{2n}\}$ as our interpolation node points; but any $2n+1$ consecutive nodes t_j could have been used because of the 2π-periodicity of the functions being considered. The condition (3.1.6) can be shown to be satisfied, thus showing the interpolation projection operator to be well-defined.

Introduce the *Dirichlet kernel*,

$$D_n(t) = \frac{\sin\left(n+\frac{1}{2}\right)t}{2\sin\left(\frac{1}{2}t\right)} = \frac{1}{2} + \sum_{j=1}^n \cos(jt) \tag{3.2.52}$$

This is an even trigonometric polynomial of degree n, and it satisfies

$$D_n(t_j) = \begin{cases} \dfrac{2n+1}{2}, & j = 0, \pm(2n+1), \pm 2(2n+1), \ldots \\ 0, & \text{all other } j \end{cases}$$

Using it, we define the Lagrange basis functions by

$$\ell_j(t) = \frac{2}{2n+1} D_n(t-t_j), \quad j = 0, \ldots, 2n \tag{3.2.53}$$

The interpolatory projection of $C_p(2\pi)$ onto $\mathcal{X}_n$ is given by

$$\mathcal{P}_n x(t) = \sum_{j=0}^{2n} x(t_j)\ell_j(t) \tag{3.2.54}$$

It satisfies

$$\|\mathcal{P}_n\| = O(\log n) \tag{3.2.55}$$

which is proven in Zygmund [591, Chap. 10, pp. 19, 37]. Since $\|\mathcal{P}_n\| \to \infty$ as $n \to \infty$, it follows from the principle of uniform boundedness that there exist $x \in C_p(2\pi)$ for which $\mathcal{P}_n x$ does not converge to x in $C_p(2\pi)$ (cf. Theorem A.3 in the Appendix). For a general introduction to trigonometric interpolation, see Ref. [48, Section 3.8] and Zygmund [591, Chap. 10].

An important concept in studying the convergence of both Fourier series and trigonometric interpolation polynomials is the *minimax approximation*. Given $x \in C_p(2\pi)$, define the *minimax error* in approximating x by elements of $\mathcal{X}_n$ as follows:

$$\rho_n(x) = \inf_{z \in \mathcal{X}_n} \|x - z\|_\infty$$

The minimax approximation to x by elements of $\mathcal{X}_n$ is that element $q_n \in \mathcal{X}_n$ for which

$$\|x - q_n\|_\infty = \rho_n(x)$$

It can be shown that there is a unique such q_n, and for any $x \in C_p(2\pi)$, it can be shown that $\rho_n(x) \to 0$ as $n \to \infty$ (see Meinardus [376, p. 47]). To obtain a more precise estimate of the speed of convergence to zero of $\rho_n(x)$, let $x \in C_p(2\pi)$ be m-times continuously differentiable, with $x^{(m)}(t)$ satisfying the Hölder condition

$$\left| x^{(m)}(t) - x^{(m)}(s) \right| \le d_{m,\alpha}(x) \, |t - s|^\alpha \tag{3.2.56}$$

for some $0 < \alpha \le 1$. Then

$$\rho_n(x) \le \frac{c_{m,\alpha}(x)}{n^{m+\alpha}} \tag{3.2.57}$$

with

$$c_{m,\alpha}(x) = d_{m,\alpha}(x) \left(1 + \frac{\pi^2}{2} \right)^{m+1}$$

This is called *Jackson's theorem*, and a proof is given in Meinardus [376, p. 55]. For use later, we note that the dependence of $c_{m,\alpha}(x)$ on x is only through the Hölder constant $d_{m,\alpha}(x)$ in (3.2.56).

We now give a more precise bound for the error in trigonometric interpolation. For $x \in C_p(2\pi)$, let $q_n \in \mathcal{X}_n$ be its minimax approximation. Then

$$x - \mathcal{P}_n x = (x - q_n) - \mathcal{P}_n(x - q_n)$$

since $\mathcal{P}_n q_n = q_n$. Taking norms,

$$\|x - \mathcal{P}_n x\|_\infty \leq (1 + \|\mathcal{P}_n\|) \, \rho_n(x) = O(\log n)\rho_n(x) \qquad (3.2.58)$$

This can be combined with (3.2.57) to obtain convergence of $\mathcal{P}_n x$ to x, provided x is Hölder continuous (that is, it satisfies (3.2.56) with $m = 0$). As noted above, however, there are continuous 2π-periodic functions $x(t)$ for which $\mathcal{P}_n x$ does not converge uniformly to x.

Consider the use of the above trigonometric interpolation in solving (3.2.50) by collocation. To simplify the notation in writing the linear system associated with the collocation methods, let the basis in (3.2.51) be denoted by $\{\phi_1(t), \ldots, \phi_d(t)\}$ with $d = d_n = 2n + 1$. The linear system (3.1.5) i

$$\sum_{j=1}^{d} c_j \left\{ \lambda\phi_j(t_i) - \int_0^{2\pi} K(t_i, s)\phi_j(s) \, ds \right\} = y(t_i), \quad i = 1, \ldots, d_n \qquad (3.2.59)$$

and the solution is

$$x_n(t) = \sum_{j=1}^{d} c_j \phi_j(t)$$

The integrals in (3.2.59) are usually evaluated numerically, and for that we recommend using the trapezoidal rule. With periodic integrands, the trapezoidal rule is very effective; for example, see [48, Section 5.4]. We return to the evaluation of these coefficients in the next chapter, under the heading of "Discrete Collocation Methods."

To discuss the convergence of this collocation method, we must show $\|\mathcal{K} - \mathcal{P}_n\mathcal{K}\| \to 0$ as $n \to \infty$. Since Lemma 3.1.2 cannot be used, we must examine $\|\mathcal{K} - \mathcal{P}_n\mathcal{K}\|$ directly. We note that as in (3.2.43)–(3.2.44), we can write

$$\mathcal{P}_n\mathcal{K}x(t) = \int_0^{2\pi} K_n(t, s)x(s) \, ds \qquad (3.2.60)$$

$$K_n(t, s) = \sum_{i=1}^{d} K(t_i, s)\ell_i(t)$$

and

$$\|\mathcal{K} - \mathcal{P}_n\mathcal{K}\| = \max_{a \leq t \leq b} \int_0^{2\pi} |K(t, s) - K_n(t, s)| \, ds \qquad (3.2.61)$$

Table 3.2. *Errors in trigonometric
interpolation of (3.2.64)*

n	$\|x - \mathcal{P}_n x\|_\infty$	n	$\|x - \mathcal{P}_n x\|_\infty$
1	5.39E−1	6	6.46E−6
2	9.31E−2	7	4.01E−7
3	1.10E−2	8	2.22E−8
4	1.11E−3	9	1.10E−9
5	9.11E−5	10	5.00E−11

This can be shown to converge to zero by applying (3.2.57)–(3.2.58). We
assume that $K(t, s)$ satisfies

$$|K(t, s) - K(\tau, s)| \le c(K)\,|t - \tau|^\alpha, \tag{3.2.62}$$

for all s, t, τ. Then

$$\|\mathcal{K} - \mathcal{P}_n\mathcal{K}\| \le \frac{c \log n}{n^\alpha} \tag{3.2.63}$$

Since this converges to zero, we can apply Theorem 3.1.1 to the error analysis
of the collocation method with trigonometric interpolation.

Assuming (3.2.50) is uniquely solvable, the collocation $(\lambda - \mathcal{P}_n\mathcal{K})x_n = \mathcal{P}_n y$ has a unique solution x_n for all sufficiently large n, and $\|x - x_n\|_\infty \to 0$
if and only if $\|x - \mathcal{P}_n x\|_\infty \to 0$. We know there are cases for which the latter
is not true; but from (3.2.57)–(3.2.58), this is only for functions x with very
little smoothness. For functions x that are infinitely differentiable, the bound
(3.2.58) shows the rate of convergence will be very rapid, faster than $O(n^{-k})$
for any k.

There are kernel functions $K(t, s)$ that do not satisfy (3.2.62), but to which
the above collocation method can still be applied. Their error analysis requires
a more detailed knowledge of the smoothing properties of the operator $\mathcal{K}$. Such
cases occur when solving boundary integral equations with singular kernel
functions, and we consider such an equation in a later chapter.

Example. Rather than solve an integral equation, we show the rapid speed
of convergence of trigonometric interpolation when it is applied to a smooth
function,

$$x(t) = e^{\sin t}, \quad -\infty < t < \infty \tag{3.2.64}$$

The results are given in Table 3.2.

3.3. Examples of Galerkin's method

The applications of Galerkin's method can be classified into the same two
general types as described at the beginning of §3.2 for the collocation method.
The error analysis of Galerkin's method is usually carried out in a Hilbert space,
generally $L^2(D)$ or some Sobolev space $H^r(D)$. Following this, an analysis
within $C(D)$ is often also given to obtain results on convergence of the solution
in the uniform norm $\|\cdot\|_\infty$.

3.3.1. Piecewise linear approximations

Let $\mathcal{X} = L^2(a, b)$, and let its norm and inner product be denoted by simply $\|\cdot\|$
and $(\cdot, \cdot)$, respectively. Let $\mathcal{X}_n$ be the subspace of piecewise linear functions as
described in §3.2, following (3.2.37). The dimension of $\mathcal{X}_n$ is $n + 1$, and the
Lagrange functions of (3.2.37) are a basis for $\mathcal{X}_n$. However, now $\mathcal{P}_n$ denotes the
orthogonal projection of $L^2(a, b)$ onto $\mathcal{X}_n$. We begin by showing that $\mathcal{P}_n x \to x$
for all $x \in L^2(a, b)$.

Begin by assuming $x(t)$ is continuous on $[a, b]$. Let $\mathcal{I}_n x(t)$ denote the piecewise linear function in $\mathcal{X}_n$ that interpolates $x(t)$ at $t = t_0, \ldots, t_n$; see (3.2.38).
Recall that $\mathcal{P}_n x$ minimizes $\|x - z\|$ as z ranges over $\mathcal{X}_n$, a fact that is embodied
in the identity (3.1.20). Therefore,

$$
\begin{aligned}
\|x - \mathcal{P}_n x\| &\leq \|x - \mathcal{I}_n x\| \\
&\leq \sqrt{b - a}\, \|x - \mathcal{I}_n x\|_\infty \\
&\leq \sqrt{b - a}\, \omega(x; h)
\end{aligned}
\tag{3.3.65}
$$

The last inequality uses the error bound (3.2.39) with $\omega(x; h)$ the modulus of
continuity for x. This shows $\mathcal{P}_n x \to x$ for all continuous functions x on $[a, b]$.

It is well known that the set of all continuous functions on $[a, b]$ is dense in
$L^2(a, b)$. Also, the orthogonal projection $\mathcal{P}_n$ satisfies $\|\mathcal{P}_n\| = 1$; cf. (3.1.19).
For a given $x \in L^2(a, b)$, let $\{x_m\}$ be a sequence of continuous functions that
converge to x in $L^2(a, b)$. Then

$$
\begin{aligned}
\|x - \mathcal{P}_n x\| &\leq \|x - x_m\| + \|x_m - \mathcal{P}_n x_m\| + \|\mathcal{P}_n(x - x_m)\| \\
&\leq 2\|x - x_m\| + \|x_m - \mathcal{P}_n x_m\|
\end{aligned}
$$

Given an $\epsilon > 0$, pick m such that $\|x - x_m\| < \epsilon/4$ and fix m. This then implies
that for all n,

$$
\|x - \mathcal{P}_n x\| \leq \frac{\epsilon}{2} + \|x_m - \mathcal{P}_n x_m\|
$$

Letting $n \to \infty$, we have that

$$\|x - \mathcal{P}_n x\| \leq \epsilon$$

for all sufficiently large values of n. Since ϵ was arbitrary, this shows that $\mathcal{P}_n x \to x$ for general $x \in L^2(a, b)$.

For the integral equation $(\lambda - \mathcal{K})x = y$, we can use Lemma 3.1.2 to obtain $\|\mathcal{K} - \mathcal{P}_n \mathcal{K}\| \to 0$. This justifies the use of Theorem 3.1.1 to carry out the error analysis for the Galerkin equation $(\lambda - \mathcal{P}_n \mathcal{K})x_n = \mathcal{P}_n y$. As before, $\|x - x_n\|$ converges to zero with the same speed as $\|x - \mathcal{P}_n x\|$. For the latter, combine (3.3.65), (3.1.34), and (3.2.39), to obtain

$$\begin{aligned}
\|x - x_n\| &\leq |\lambda|\, M\, \|x - \mathcal{P}_n x\| \\
&\leq |\lambda|\, M\sqrt{b - a}\, \|x - \mathcal{I}_n x\|_\infty \\
&\leq |\lambda|\, M\sqrt{b - a}\, \frac{h^2}{8} \|x''\|_\infty
\end{aligned} \tag{3.3.66}$$

for $x \in C^2[a, b]$.

For the linear system we use the Lagrange basis functions of (3.2.37). These are not orthogonal, but they are still a very convenient basis from which to work. Moreover, producing an orthogonal basis for $\mathcal{X}_n$ is a nontrivial task. The solution x_n of $(\lambda - \mathcal{P}_n \mathcal{K})x_n = \mathcal{P}_n y$ is given by

$$x_n(t) = \sum_{j=0}^{n} c_j \ell_j(t)$$

The coefficients $\{c_j\}$ are obtained by solving the linear system

$$\sum_{j=0}^{n} c_j \left\{ \lambda(\ell_i, \ell_j) - \int_a^b \int_a^b K(t, s)\ell_i(t)\ell_j(s)\, ds\, dt = \int_a^b y(t)\ell_i(t)\, dt \right\},$$

$$i = 0, \dots, n \tag{3.3.67}$$

For the coefficients (ℓ_i, ℓ_j),

$$(\ell_i, \ell_j) = \begin{cases} 0, & |i - j| > 1 \\[2mm] \dfrac{2h}{3}, & 0 < i = j < n \\[2mm] \dfrac{h}{3}, & i = j = 0 \text{ or } n \\[2mm] \dfrac{h}{6}, & |i - j| = 1 \end{cases}$$

The double integrals in (3.3.67) will reduce to integrals over much smaller subintervals, because the basis functions $\ell_i(t)$ are zero over most of $[a, b]$. In Chapter 4 we return to the numerical evaluation of the integrals in (3.3.67) under the heading of "Discrete Galerkin Methods."

Just as was true with collocation methods, we can easily generalize the above presentation to include the use of piecewise polynomial functions of any fixed degree. Since the theory is entirely analogous to that presented above, we omit it here.

We defer to the following case a consideration of the uniform convergence of $x_n(t)$ to $x(t)$.

3.3.2. Galerkin's method with trigonometric polynomials

We consider again the use of trigonometric polynomials as approximations in solving the integral equation (3.2.50), with $K(t, s)$ and $y(t)$ being 2π-periodic functions as before. Initially, we use the space $\mathcal{X} = L^2(0, 2\pi)$, the space of all complex-valued and square integrable Lebesgue measurable functions on $(0, 2\pi)$. The inner product is defined by

$$(x, y) = \int_0^{2\pi} x(s)\overline{y(s)}\, dt$$

Later we will consider the space $C_p(2\pi)$, the set of all complex-valued 2π-periodic continuous functions, with the uniform norm.

The approximating subspace $\mathcal{X}_n$ is again the set of all trigonometric polynomials of degree $\leq n$. As a basis, however, we now use the complex exponentials,

$$\phi_j(s) = e^{ijs}, \quad j = 0, \pm 1, \ldots, \pm n \tag{3.3.68}$$

It is very well known that each x in $L^2(0, 2\pi)$ can be written as the Fourier series

$$x(s) = \frac{1}{2\pi} \sum_{j=-\infty}^{\infty} (x, \phi_j)\phi_j(s) \tag{3.3.69}$$

which is convergent in $L^2(0, 2\pi)$. The orthogonal projection of $L^2(0, 2\pi)$ onto $\mathcal{X}_n$ is just the nth partial sum of this series:

$$\mathcal{P}_n x(s) = \frac{1}{2\pi} \sum_{j=-n}^{n} (x, \phi_j)\phi_j(s) \tag{3.3.70}$$

From the convergence of (3.3.69), it follows that $\mathcal{P}_n x \to x$ for all $x \in L^2(0, 2\pi)$.

The speed of convergence of $\mathcal{P}_n x$ to x has been well-examined from many perspectives. In $L^2(0, 2\pi)$, we have

$$\|x - \mathcal{P}_n x\| = \left[\frac{1}{2\pi} \sum_{|j|>n} |(x, \phi_j)|^2\right]^{\frac{1}{2}} \tag{3.3.71}$$

and this can be used to derive bounds for the case that $x \in H^r(0, 2\pi)$, $r > 0$, the Sobolev space of 2π-periodic functions with smoothness r. The norm on $H^r(0, 2\pi)$ is given by

$$\|x\|_r = \left[\frac{1}{2\pi} \sum_{j=-\infty}^{\infty} [\max\{1, |j|\}]^{2r} |(x, \phi_j)|^2\right]^{\frac{1}{2}}, \quad x \in H^r(0, 2\pi) \tag{3.3.72}$$

for $r \geq 0$. For r an integer, this is equivalent to the norm

$$\|x\|_{r,*} = \left[\sum_{j=0}^{r} \left\|x^{(j)}\right\|^2\right]^{\frac{1}{2}} \tag{3.3.73}$$

In this, $\|\cdot\|$ is the standard norm on $L^2(0, 2\pi)$, and $x^{(j)}(s)$ is the distributional derivative of order j of $x(s)$. For such functions, we can show

$$\|x - \mathcal{P}_n x\| \leq \frac{c}{n^r} \left[\frac{1}{2\pi} \sum_{|j|>n} |j|^{2r} |(x, \phi_j)|^2\right]^{\frac{1}{2}} \leq \frac{c}{n^r} \|x\|_{H^r} \tag{3.3.74}$$

The constant c depends on r, but is independent of n and x. The greater the differentiability (or smoothness) of $x(t)$, the faster the speed of convergence of $\mathcal{P}_n x$ to x. Later in this section we give another way of examining the speed of convergence, one based on using the minimax error result (3.2.57).

Using Lemma 3.1.2, we have that $\|\mathcal{K} - \mathcal{P}_n\mathcal{K}\| \to 0$ as $n \to \infty$. Thus Theorem 3.1.1 can be applied to the error analysis of the approximating equation $(\lambda - \mathcal{P}_n\mathcal{K})x_n = \mathcal{P}_n y$. For all sufficiently large n, say $n \geq N$, the inverses $(\lambda - \mathcal{P}_n\mathcal{K})^{-1}$ are uniformly bounded, and $\|x - x_n\|$ is proportional to $\|x - \mathcal{P}_n x\|$, and thus x_n converges to x. One result on the rate of convergence is obtained by applying (3.3.74) to (3.1.35):

$$\|x - x_n\| \leq \frac{c|\lambda|M}{n^r} \|x\|_{H^r}, \quad n \geq N, \quad x \in H^r(0, 2\pi) \tag{3.3.75}$$

with c the same as in (3.1.35).

The linear system (3.1.13) for $(\lambda - \mathcal{P}_n \mathcal{K})x_n = \mathcal{P}_n y$ is given by

$$2\pi\lambda c_k - \sum_{j=-n}^{n} c_j \int_0^{2\pi} \int_0^{2\pi} e^{i(js-kt)} K(t,s)\, ds\, dt = \int_0^{2\pi} e^{-ikt} y(t)\, dt,$$

$$k = -n, \ldots, n \qquad (3.3.76)$$

with the solution x_n given by

$$x_n(t) = \sum_{j=-n}^{n} c_j e^{ijt}$$

The integrals in this system are usually evaluated numerically, and in Chapter 4 we examine this in some detail.

3.3.3. *Uniform convergence*

We often are interested in obtaining uniform convergence of x_n to x. For this we regard the operator $\mathcal{P}_n$ of (3.3.70) as an operator on $C_p(2\pi)$ to $\mathcal{X}_n$, and we take $\mathcal{X} = C_p(2\pi)$. Unfortunately, it is no longer true that $\mathcal{P}_n x$ converges to x for all $x \in \mathcal{X}$, and therefore Lemma 3.1.2 cannot be applied. In fact,

$$\|\mathcal{P}_n\| = O(\log n) \qquad (3.3.77)$$

For a proof of this, see Zygmund [591, Chap. 1].

To obtain convergence results for $\mathcal{P}_n x$, we use Jackson's Theorem (3.2.57). Let q_n be the minimax approximation to x from $\mathcal{X}_n$. Then

$$x - \mathcal{P}_n x = (x - q_n) - \mathcal{P}_n(x - q_n)$$

$$\|x - \mathcal{P}_n x\|_\infty \leq (1 + \|\mathcal{P}_n\|)\|x - q_n\|_\infty$$

$$\leq O(\log n)\, \rho_n(x) \qquad (3.3.78)$$

with $\rho_n(x)$ the minimax error. Let x be Hölder continuous with exponent α, as in (3.2.56). Apply (3.2.57) to (3.3.78), obtaining

$$\|x - \mathcal{P}_n x\|_\infty \leq \frac{c \log n}{n^\alpha} \qquad (3.3.79)$$

for some c. This converges to zero as $n \to \infty$. Higher speeds of convergence can be obtained by assuming greater smoothness for $x(t)$ and applying the complete form of (3.2.57).

To show convergence of $\|\mathcal{K} - \mathcal{P}_n \mathcal{K}\|$ to zero, we must show directly that

$$\|\mathcal{K} - \mathcal{P}_n \mathcal{K}\| = \max_{a \leq t \leq b} \int_0^{2\pi} |K(t,s) - K_n(t,s)|\, ds$$

converges to zero, much as was done earlier in (3.2.61) when using trigonometric interpolation. In this instance,

$$K_n(t, s) = \mathcal{P}_n K_s(t), \quad K_s(t) \equiv K(t, s)$$

Also as before in (3.2.62)–(3.2.63) and (3.3.78)–(3.3.79), if $K(t, s)$ satisfies the Hölder condition

$$|K(t, s) - K(\tau, s)| \le c(K)\,|t - \tau|^\alpha,$$

for all s, t, τ, then

$$\|\mathcal{K} - \mathcal{P}_n \mathcal{K}\| \le \frac{c \log n}{n^\alpha}$$

With this we can apply Theorem 3.1.1 and obtain a complete convergence analysis within $C_p(2\pi)$, thus obtaining results on the uniform convergence of x_n to x. In the next section we will consider another way of obtaining such uniform convergence results.

Another important example of the use of globally defined and smooth approximations is the use of spherical polynomials as approximations to functions defined on the unit sphere in $\mathbf{R}^3$. In Chapter 9 we use a Galerkin method with spherical polynomial approximations to solve a boundary integral equation on smooth closed surfaces in $\mathbf{R}^3$.

3.4. Iterated projection methods

For the integral equation $(\lambda - \mathcal{K})x = y$, consider the iteration

$$x^{(k+1)} = \frac{1}{\lambda}\left[y + \mathcal{K}x^{(k)}\right], \quad k = 0, 1, \dots$$

From the geometric series theorem (cf. Theorem A.1 in the Appendix), it can be shown that this iteration converges to the solution x if $\|\mathcal{K}\| < |\lambda|$, and in that case

$$\left\|x - x^{(k+1)}\right\| \le \frac{\|\mathcal{K}\|}{|\lambda|}\left\|x - x^{(k)}\right\|$$

Sloan [503] showed that one such iteration is always a good idea if the initial guess is the solution obtained by the Galerkin method, regardless of the size of $\|\mathcal{K}\|$. We examine this idea and its consequences, beginning with general projection methods.

Let x_n be the solution of the projection equation $(\lambda - \mathcal{P}_n \mathcal{K})x_n = \mathcal{P}_n y$. Define the *iterated projection solution* by

$$\hat{x}_n = \frac{1}{\lambda}[y + \mathcal{K}x_n] \qquad (3.4.80)$$

This new approximation is often an improvement on x_n. Moreover, it is used to better understand the behavior of the original projection solution.

Applying $\mathcal{P}_n$ to both sides of (3.4.80), we have

$$\mathcal{P}_n \hat{x}_n = \frac{1}{\lambda}[\mathcal{P}_n y + \mathcal{P}_n \mathcal{K}x_n] = x_n$$

$$\mathcal{P}_n \hat{x}_n = x_n \qquad (3.4.81)$$

Thus, x_n is the projection of $\hat{x}_n$ into $\mathcal{X}_n$. Substituting into (3.4.80) and rearranging terms, we have $\hat{x}_n$, which satisfies the equation

$$(\lambda - \mathcal{K}\mathcal{P}_n)\hat{x}_n = y \qquad (3.4.82)$$

Often we can directly analyze this equation, and then information can be obtained on x_n by applying (3.4.81). In particular,

$$x - \hat{x}_n = \frac{1}{\lambda}[y + \mathcal{K}x] - \frac{1}{\lambda}[y + \mathcal{K}x_n] = \frac{1}{\lambda}\mathcal{K}(x - x_n)$$

$$\|x - \hat{x}_n\| \le \frac{1}{|\lambda|}\|\mathcal{K}\|\|x - x_n\| \qquad (3.4.83)$$

This proves the convergence of $\hat{x}_n$ to x is at least as rapid as that of x_n to x. Often it will be more rapid, because operating on $x - x_n$ with $\mathcal{K}$ sometimes causes cancellation due to the smoothing behavior of integration.

From the above, we see that if $(\lambda - \mathcal{P}_n\mathcal{K})^{-1}$ exists, then so does $(\lambda - \mathcal{K}\mathcal{P}_n)^{-1}$. Moreover, from the definition of the solution x_n and (3.4.80),

$$\hat{x}_n = \frac{1}{\lambda}[y + \mathcal{K}x_n] = \frac{1}{\lambda}[y + \mathcal{K}(\lambda - \mathcal{P}_n\mathcal{K})^{-1}\mathcal{P}_n y]$$

$$(\lambda - \mathcal{K}\mathcal{P}_n)^{-1} = \frac{1}{\lambda}[I + \mathcal{K}(\lambda - \mathcal{P}_n\mathcal{K})^{-1}\mathcal{P}_n] \qquad (3.4.84)$$

Conversely, if $(\lambda - \mathcal{K}\mathcal{P}_n)^{-1}$ exists, then so does $(\lambda - \mathcal{P}_n\mathcal{K})^{-1}$. This follows from the following general Lemma 3.4.1, which also shows that

$$(\lambda - \mathcal{P}_n\mathcal{K})^{-1} = \frac{1}{\lambda}[I + \mathcal{P}_n(\lambda - \mathcal{K}\mathcal{P}_n)^{-1}\mathcal{K}] \qquad (3.4.85)$$

By combining (3.4.84) and (3.4.85), or by returning to the definitions of x_n and $\hat{x}_n$, we also have

$$(\lambda - \mathcal{P}_n \mathcal{K})^{-1} \mathcal{P}_n = \mathcal{P}_n (\lambda - \mathcal{K} \mathcal{P}_n)^{-1} \qquad (3.4.86)$$

We can choose to show the existence of either $(\lambda - \mathcal{P}_n \mathcal{K})^{-1}$ or $(\lambda - \mathcal{K} \mathcal{P}_n)^{-1}$, whichever is the more convenient, and the existence of the other inverse will follow immediately. Bounds on one inverse in terms of the other can also be given by using (3.4.84) and (3.4.85).

Lemma 3.4.1. Let $\mathcal{X}$ be a Banach space, and let $\mathcal{A}$, $\mathcal{B}$ be bounded linear operators on $\mathcal{X}$ to $\mathcal{X}$. Assume $(\lambda - \mathcal{A}\mathcal{B})^{-1}$ exists from $\mathcal{X}$ onto $\mathcal{X}$. Then $(\lambda - \mathcal{B}\mathcal{A})^{-1}$ also exists, and

$$(\lambda - \mathcal{B}\mathcal{A})^{-1} = \frac{1}{\lambda}[I + \mathcal{B}(\lambda - \mathcal{A}\mathcal{B})^{-1}\mathcal{A}] \qquad (3.4.87)$$

Proof. Calculate

$$(\lambda - \mathcal{B}\mathcal{A})\frac{1}{\lambda}[I + \mathcal{B}(\lambda - \mathcal{A}\mathcal{B})^{-1}\mathcal{A}]$$

$$= \frac{1}{\lambda}\{\lambda - \mathcal{B}\mathcal{A} + (\lambda - \mathcal{B}\mathcal{A})\mathcal{B}(\lambda - \mathcal{A}\mathcal{B})^{-1}\mathcal{A}\}$$

$$= \frac{1}{\lambda}\{\lambda - \mathcal{B}\mathcal{A} + \mathcal{B}(\lambda - \mathcal{A}\mathcal{B})(\lambda - \mathcal{A}\mathcal{B})^{-1}\mathcal{A}\}$$

$$= \frac{1}{\lambda}\{\lambda - \mathcal{B}\mathcal{A} + \mathcal{B}\mathcal{A}\}$$

$$= I$$

A similar proof works to show

$$\frac{1}{\lambda}[I + \mathcal{B}(\lambda - \mathcal{A}\mathcal{B})^{-1}\mathcal{A}](\lambda - \mathcal{B}\mathcal{A}) = I$$

This proves (3.4.87). □

For the error in $\hat{x}_n$, first rewrite $(\lambda - \mathcal{K})x = y$ as

$$(\lambda - \mathcal{K}\mathcal{P}_n)x = y + \mathcal{K}x - \mathcal{K}\mathcal{P}_n x$$

Subtract (3.4.82) to obtain

$$(\lambda - \mathcal{K}\mathcal{P}_n)(x - \hat{x}_n) = \mathcal{K}(I - \mathcal{P}_n)x \qquad (3.4.88)$$

Below, we will examine this apparently simple equation in much greater detail.

3.4.1. The iterated Galerkin solution

Assume that $\mathcal{X}$ is a Hilbert space and that x_n is the Galerkin solution of $(\lambda - \mathcal{K})x = y$. Then

$$(I - \mathcal{P}_n)^2 = I - \mathcal{P}_n$$

and

$$\|\mathcal{K}(I - \mathcal{P}_n)x\| = \|\mathcal{K}(I - \mathcal{P}_n)(I - \mathcal{P}_n)x\|$$
$$\leq \|\mathcal{K}(I - \mathcal{P}_n)\| \, \|(I - \mathcal{P}_n)x\| \qquad (3.4.89)$$

Using the fact that we are in a Hilbert space and that $\mathcal{P}_n$ is a self-adjoint projection (cf. (3.1.17)), we have

$$\|\mathcal{K}(I - \mathcal{P}_n)\| = \| [\mathcal{K}(I - \mathcal{P}_n)]^* \|$$
$$= \|(I - \mathcal{P}_n)\mathcal{K}^*\| \qquad (3.4.90)$$

The first line follows from the general principle that the norm of an operator is equal to the norm of its adjoint operator. The second line follows from (3.1.17) and properties of the adjoint operation.

With Galerkin methods it is generally the case that when $\mathcal{P}_n$ is regarded as an operator on the Hilbert space $\mathcal{X}$, then $\mathcal{P}_n x \to x$ for all $x \in \mathcal{X}$. This follows if the sequence of spaces $\{\mathcal{X}_n \mid n \geq 1\}$ has the *approximating property* on $\mathcal{X}$: for each $x \in \mathcal{X}$, there is a sequence $\{z_n\}$ with $z_n \in \mathcal{X}_n$ and

$$\lim_{n \to \infty} \|x - z_n\| = 0 \qquad (3.4.91)$$

When this is combined with the definition of $\mathcal{P}_n$, implicit in (3.1.14), we have $\mathcal{P}_n x \to x$ for all $x \in \mathcal{X}$.

Another result from functional analysis is that if $\mathcal{K}$ is a compact operator, then so is its adjoint $\mathcal{K}^*$. Combining this with Lemma 3.1.2 and the above assumption of the pointwise convergence of $\mathcal{P}_n$ to I on $\mathcal{X}$, we have that

$$\lim_{n \to \infty} \|(I - \mathcal{P}_n)\mathcal{K}^*\| = 0 \qquad (3.4.92)$$

We can also apply Theorem 3.1.1 to obtain the existence and uniform boundedness of $(\lambda - \mathcal{P}_n \mathcal{K})^{-1}$ for all sufficiently large n, say $n \geq N$. From (3.4.84) we also know that $(\lambda - \mathcal{P}_n \mathcal{K})^{-1}$ exists and is uniformly bounded for $n \geq N$. Apply this and (3.4.89) to (3.4.88), to obtain

$$\|x - \hat{x}_n\| \leq \|(\lambda - \mathcal{K}\mathcal{P}_n)^{-1}\| \|\mathcal{K}(I - \mathcal{P}_n)x\|$$
$$\leq c\|(I - \mathcal{P}_n)\mathcal{K}^*\| \|(I - \mathcal{P}_n)x\| \qquad (3.4.93)$$

Combining this with (3.4.92), we see that $\|x - \hat{x}_n\|$ converges to zero more rapidly than does $\|(I - \mathcal{P}_n)x\|$, or equivalently, $\|x - x_n\|$. Thus

$$\lim_{n \to \infty} \frac{\|x - \hat{x}_n\|}{\|x - x_n\|} = 0$$

The quantity $\|(I - \mathcal{P}_n)\mathcal{K}^*\|$ can generally be estimated in the same manner as is done for $\|(I - \mathcal{P}_n)\mathcal{K}\|$. Taking $\mathcal{K}$ to be an integral operator on $L^2(D)$, the operator $\mathcal{K}^*$ is an integral operator, with

$$\mathcal{K}^* x(t) = \int_D K(s,t)x(s)\,ds, \quad x \in L^2(D) \tag{3.4.94}$$

Example. Consider the integral equation

$$\lambda x(t) - \int_0^1 e^{st}x(s)\,ds = y(t), \quad 0 \le t \le 1 \tag{3.4.95}$$

with $\lambda = 50$ and $x(t) = e^t$. For $n \ge 1$, define the mesh size $h = 1/n$ and the mesh $t_j = jh$, $j = 0, 1, \ldots, n$. Let $\mathcal{X}_n$ be the set of functions that are piecewise linear on $[0, 1]$ with breakpoints $t_1, \ldots, t_{n-1}$. In contrast with the example given at the beginning of §3.3, we do not insist that the elements of $\mathcal{X}_n$ be continuous at the breakpoints. The dimension of $\mathcal{X}_n$ is $d_n = 2n$, and this is also the order of the linear system associated with solving $(\lambda - \mathcal{P}_n\mathcal{K})x_n = \mathcal{P}_n y$.

It is straightforward to show $\|(I - \mathcal{P}_n)\mathcal{K}^*\| = O(h^2)$ in this case. Also, if $y \in C^2[0, 1]$, then the solution x of (3.4.95) also belongs to $C^2[0, 1]$; consequently, we have $\|x - \mathcal{P}_n x\| = O(h^2)$. These results lead to

$$\|x - x_n\| = O(h^2) \tag{3.4.96}$$

$$\|x - \hat{x}_n\| = O(h^4) \tag{3.4.97}$$

This is confirmed empirically in the numerical calculations given in Table 3.3. The error columns give the maximum error at the node points, not the norm of the error in $L^2(0, 1)$. But we show below that (3.4.96)–(3.4.97) generalize to $C[0, 1]$ with the uniform norm.

3.4.2. *Uniform convergence of iterated Galerkin approximations*

We can examine the uniform convergence of $\hat{x}_n$ by analyzing the equation $(\lambda - \mathcal{K}\mathcal{P}_n)\hat{x}_n = y$, derived in (3.4.82), within the space $\mathcal{X} = C(D)$. Such an analysis is possible if we can show $\|\mathcal{K} - \mathcal{K}\mathcal{P}_n\| \to 0$, and this is often possible by using a variety of stratagems depending on the type of subspace $\mathcal{X}_n$ being used. We

Table 3.3. *A Galerkin and iterated Galerkin method for (3.4.95)*

n	$\|x - x_n\|$	Ratio	$\|x - \hat{x}_n\|$	Ratio
2	4.66E−2		5.45E−6	
4	1.28E−2	3.6	3.48E−7	15.7
8	3.37E−3	3.8	2.19E−8	15.9

will proceed in a simpler way, taking advantage of the convergence of $\hat{x}_n$ in the Hilbert space $\mathcal{X}$, which we assume to be $\mathcal{X} = L^2(D)$.

Subtract $(\lambda - \mathcal{K}\mathcal{P}_n)\hat{x}_n = y$ from $(\lambda - \mathcal{K})x = y$, and then rearrange the terms to get

$$\lambda(x - \hat{x}_n) = \mathcal{K}x - \mathcal{K}\mathcal{P}_n\hat{x}_n$$
$$= \mathcal{K}(I - \mathcal{P}_n)x + \mathcal{K}\mathcal{P}_n(x - \hat{x}_n) \qquad (3.4.98)$$

Introduce

$$K_t(s) = K(t, s), \quad s \in D$$

Write the integral

$$\mathcal{K}x(t) = \int_D K(t, s)x(s)\,ds, \quad t \in D$$

as

$$\mathcal{K}x(t) = (K_t, x), \quad t \in D$$

using the inner product of $L^2(D)$. Then (3.4.98) can be written

$$\lambda[x(t) - \hat{x}_n(t)] = (K_t, (I - \mathcal{P}_n)x) + (K_t, \mathcal{P}_n(x - \hat{x}_n))$$
$$= ((I - \mathcal{P}_n)K_t, (I - \mathcal{P}_n)x) + (K_t, \mathcal{P}_n(x - \hat{x}_n))$$

Using the Cauchy-Schwarz inequality,

$$|\lambda|\,|x(t) - \hat{x}_n(t)| \leq \|(I - \mathcal{P}_n)K_t\|\,\|(I - \mathcal{P}_n)x\| + \|K_t\|\,\|\mathcal{P}_n(x - \hat{x}_n)\|$$

The norm is that of $L^2(D)$. Recall from (3.1.19) that $\|\mathcal{P}_n\| = 1$ when $\mathcal{P}_n$ is regarded as an orthogonal operator on $L^2(D)$. Taking bounds as t ranges over D, we have

$$\|x - \hat{x}_n\|_\infty \leq \frac{1}{|\lambda|}\Big\{\|(I - \mathcal{P}_n)x\|\max_{t \in D}\|(I - \mathcal{P}_n)K_t\|$$
$$+ \|x - \hat{x}_n\|\max_{t \in D}\|K_t\|\Big\} \qquad (3.4.99)$$

Example. Consider the above example for the integral equation (3.4.95) in which piecewise linear functions were used to approximate $x(t)$. Recall the interpolation operator $\mathcal{I}_n$:

$$\mathcal{I}_n z(t) = \frac{(t_j - t)z(t_{j-1}) + (t - t_{j-1})z(t_j)}{h},$$

$$t_{j-1} \leq t \leq t_j, \quad j = 1, \dots, n$$

for $z \in C[0, 1]$. Then

$$\|(I - \mathcal{P}_n)K_t\| \leq \|(I - \mathcal{I}_n)K_t\| \leq \|(I - \mathcal{I}_n)K_t\|_\infty$$

$$\leq \frac{h^2}{8} \left[\max_{0 \leq s \leq 1} \left| \frac{\partial^2 K_t(s)}{\partial s^2} \right| \right]$$

$$\max_{t \in D} \|(I - \mathcal{P}_n)K_t\| \leq \frac{h^2}{8} \left[\max_{0 \leq t, s \leq 1} \left| \frac{\partial^2 K_t(s)}{\partial s^2} \right| \right]$$

Also,

$$\max_{0 \leq t \leq 1} \|K_t\| = \max_{0 \leq t \leq 1} \int_0^1 |K(t, s)|^2 \, ds$$

Combining these with (3.4.99) and with earlier error results, we have

$$\|x - \hat{x}_n\|_\infty = O(h^4)$$

for the error in the iterated Galerkin solution. The numbers in Table 3.3 are a numerical illustration of this result.

With higher degree piecewise polynomial functions on $[0, 1]$, we obtain higher order analogs of the above results. With $\mathcal{X}_n$ the set of piecewise polynomial functions on the grid $\{t_0, \dots, t_n\}$ with mesh size h, and with sufficient differentiability of the unknown solution $x(t)$, we obtain the results

$$\|x - x_n\|_\infty = O(h^{r+1})$$

$$\|x - \hat{x}_n\|_\infty = O(h^{2r+2}) \tag{3.4.100}$$

The first result assumes $x \in C^{r+1}[0, 1]$, and the second result further assumes that $K_t \in C^{r+1}[0, 1]$ for $0 \leq t \leq 1$.

3.4.3. The iterated collocation solution

With collocation, the iterated solution $\hat{x}_n$ is not always an improvement on the original collocation solution x_n, but it is for many cases of interest. The abstract theory is still applicable, and the error equation (3.4.88) is still the focus for the error analysis:

$$x - \hat{x}_n = (\lambda - \mathcal{K}\mathcal{P}_n)^{-1}\mathcal{K}(I - \mathcal{P}_n)x \qquad (3.4.101)$$

Recall that the projection $\mathcal{P}_n$ is now the interpolatory operator of (3.1.7). In contrast to the iterated Galerkin method, we do not have that $\|\mathcal{K} - \mathcal{K}\mathcal{P}_n\|$ converges to zero. In fact, it can be shown that

$$\|\mathcal{K}(I - \mathcal{P}_n)\| \geq \|\mathcal{K}\| \qquad (3.4.102)$$

To show the possibly faster convergence of $\hat{x}_n$, we must examine collocation methods on a case-by-case basis. With some, there will be an improvement. We begin with a simple example to show one of the main tools used in proving higher orders of convergence.

Consider using collocation with piecewise quadratic interpolation to solve the integral equation

$$\lambda x(t) - \int_a^b K(t, s)x(s)\, ds = y(t), \quad a \leq t \leq b \qquad (3.4.103)$$

Let $n \geq 2$ be an even integer. Define $h = (b - a)/n$ and $t_j = a + jh$, $j = 0, 1, \ldots, n$. Let $\mathcal{X}_n$ be the set of all continuous functions that are quadratic polynomials when restricted to each of the subintervals $[t_0, t_2], \ldots, [t_{n-2}, t_n]$. Easily, the dimension of $\mathcal{X}_n$ is $d_n = n + 1$, based on each element of $\mathcal{X}_n$ being completely determined by its values at the $n + 1$ nodes $\{t_0, \ldots, t_n\}$. Let $\mathcal{P}_n$ be the interpolatory projection operator from $\mathcal{X} = C[a, b]$ to $\mathcal{X}_n$.

We can write $\mathcal{P}_n x$ in its Lagrange form:

$$\mathcal{P}_n x(t) = \sum_{j=0}^n x(t_j)\ell_j(t) \qquad (3.4.104)$$

For the Lagrange basis functions $\ell_j(t)$ we must distinguish the cases of even and odd indices j. For j odd,

$$\ell_j(t) = \begin{cases} -\dfrac{1}{h^2}(t - t_{j-1})(t - t_{j+1}), & t_{j-1} \leq t \leq t_{j+1} \\ 0, & \text{otherwise} \end{cases}$$

For j even, $2 \leq j \leq n - 2$,

$$\ell_j(t) = \begin{cases} \dfrac{1}{2h^2}(t - t_{j-1})(t - t_{j-2}), & t_{j-2} \leq t \leq t_j \\ \dfrac{1}{2h^2}(t - t_{j+1})(t - t_{j+2}), & t_j \leq t \leq t_{j+2} \\ 0, & \text{otherwise} \end{cases}$$

The functions $\ell_0(t)$ and $\ell_n(t)$ are appropriate modifications of this last case.

For the interpolation error on $[t_{j-2}, t_j]$, for j even, we have two formulas:

$$x(t) - \mathcal{P}_n x(t) = (t - t_{j-2})(t - t_{j-1})(t - t_j)x[t_{j-2}, t_{j-1}, t_j, t] \tag{3.4.105}$$

and

$$x(t) - \mathcal{P}_n x(t) = \frac{(t - t_{j-2})(t - t_{j-1})(t - t_j)}{6} x'''(c_t), \quad t_{j-2} \le t \le t_j \tag{3.4.106}$$

for some $c_t \in [t_{j-2}, t_j]$, with $x \in C^3[a, b]$. The quantity $x[t_{j-2}, t_{j-1}, t_j, t]$ is a *Newton divided difference* of order three for the function $x(t)$. From the above formulas,

$$\|x - \mathcal{P}_n x\|_\infty \le \frac{\sqrt{3}}{27} h^3 \|x'''\|_\infty, \quad x \in C^3[a, b] \tag{3.4.107}$$

See [48, pp. 143,156] for details.

In using piecewise quadratic functions to define the collocation method to solve (3.4.103), the result (3.4.107) implies

$$\|x - x_n\|_\infty = O(h^3) \tag{3.4.108}$$

To examine the error in $\hat{x}_n$, we make a detailed examination of $\mathcal{K}(I - \mathcal{P}_n)x$.
Using (3.4.104),

$$\mathcal{K}(I - \mathcal{P}_n)x(t) = \int_a^b K(t, s)\left\{ x(s) - \sum_{j=0}^n x(t_j)\ell_j(s) \right\} ds$$

From (3.4.105),

$$\mathcal{K}(I - \mathcal{P}_n)x(t) = \sum_{k=1}^{n/2} \int_{t_{2k-2}}^{t_{2k}} K(t, s)(s - t_{2k-2})(s - t_{2k-1})$$
$$\times (s - t_{2k})\, x[t_{2k-2}, t_{2k-1}, t_{2k}, s]\, ds \tag{3.4.109}$$

Examining the integral in more detail, write it as

$$\int_{t_{2k-2}}^{t_{2k}} g_t(s)\omega(s)\, ds \tag{3.4.110}$$

with

$$\omega(s) = (s - t_{2k-2})(s - t_{2k-1})(s - t_{2k})$$
$$g_t(s) = K(t, s)\, x[t_{2k-2}, t_{2k-1}, t_{2k}, s]$$

Introduce

$$v(s) = \int_{t_{2k-2}}^{s} (\sigma - t_{2k-2})(\sigma - t_{2k-1})(\sigma - t_{2k}) \, d\sigma, \quad t_{2k-2} \le s \le t_{2k}$$

Then $v'(s) = \omega(s)$, $v(s) \ge 0$ on $[t_{2k-2}, t_{2k}]$, and $v(t_{2k-2}) = v(t_{2k}) = 0$. The integral (3.4.110) becomes

$$\int_{t_{2k-2}}^{t_{2k}} g_t(s) v'(s) \, ds = \underbrace{v(s) g(s)]_{t_{2k-2}}^{t_{2k}}}_{=0} - \int_{t_{2k-2}}^{t_{2k}} g_t'(s) v(s) \, ds$$

$$\left| \int_{t_{2k-2}}^{t_{2k}} g_t'(s) v(s) \, ds \right| \le \|g'\|_\infty \int_{t_{2k-2}}^{t_{2k}} v(s) \, ds = \frac{4h^5}{15} \|g'\|_\infty$$

In this,

$$g'(s) = \frac{\partial}{\partial s} \{ K(t, s) \, x[t_{2k-2}, t_{2k-1}, t_{2k}, s] \}$$

$$= \frac{\partial K(t, s)}{\partial s} x[t_{2k-2}, t_{2k-1}, t_{2k}, s] + K(t, s) \, x[t_{2k-2}, t_{2k-1}, t_{2k}, s, s]$$

The last formula uses a standard result for the differentiation of Newton divided differences: see [48, p. 147]. To have this derivation be valid, we must have $g \in C^1[a, b]$, and this is true if $x \in C^4[a, b]$ and $K_t \in C^1[a, b]$.

Combining these results, we have

$$\mathcal{K}(I - \mathcal{P}_n) x(t) = O(h^4) \tag{3.4.111}$$

With this, we have the following theorem.

Theorem 3.4.1. *Assume that the integral equation (3.4.103) is uniquely solvable for all $y \in C[a, b]$. Further assume that the solution $x \in C^4[a, b]$ and that the kernel function $K(t, s)$ is at least once continuously differentiable with respect to s. Let $\mathcal{P}_n$ be the interpolatory projection (3.4.104) defined by piecewise quadratic interpolation. Then the collocation equation $(\lambda - \mathcal{P}_n \mathcal{K}) x_n = \mathcal{P}_n y$ is uniquely solvable for all sufficiently large n, say $n \ge N$, and the inverses $(\lambda - \mathcal{P}_n \mathcal{K})^{-1}$ are uniformly bounded, say by $M > 0$. Moreover,*

$$\|x - x_n\|_\infty \le |\lambda| M \|x - \mathcal{P}_n x\|_\infty \le \frac{\sqrt{3} \, |\lambda| \, M}{27} h^3 \|x'''\|_\infty, \quad n \ge N \tag{3.4.112}$$

For the iterated collocation method,

$$\|x - \hat{x}_n\|_\infty \le c h^4 \tag{3.4.113}$$

for a suitable constant c > 0. Consequently,

$$\max_{j=0,\ldots,n} |x(t_j) - x_n(t_j)| = O(h^4) \qquad (3.4.114)$$

Proof. Formula (3.4.112) and the remarks preceding it are just a restatement of results from Theorem 3.1.1 applied to the particular $\mathcal{P}_n$ being considered here. The final bound in (3.4.112) comes from (3.4.107). The bound (3.4.113) comes from (3.4.101) and (3.4.111). The final result (3.4.114) comes from noting first that the property $\mathcal{P}_n \hat{x}_n = x_n$ [cf. (3.4.81)] implies

$$x_n(t_j) = \hat{x}_n(t_j), \quad j = 0, \ldots, n$$

and second from applying (3.4.113).   □

This theorem, and (3.4.113) in particular, shows that the iterated collocation method converges more rapidly when using piecewise quadratic interpolation as described preceding the theorem. However, when using piecewise linear interpolation to define $\mathcal{P}_n$, the iterated collocation solution $\hat{x}_n$ does not converge any more rapidly than the original solution x_n. Generally, if piecewise polynomial functions of degree r are used to define $\mathcal{X}_n$, and if r is an even integer, then the iterated solution will gain one power of h in its error bound; but this will not be true if r is an odd integer.

The result (3.4.114) is an example of *superconvergence*. The rate of convergence of $x_n(t)$ at the node points $\{t_0, \ldots, t_n\}$ is greater than it is over the interval $[a, b]$ as a whole. We first encountered it with the example (3.2.47)–(3.2.48), when using piecewise linear interpolation; see Table 3.1 and the case of $x^{(2)}(t) = \sqrt{t}$. The proof in that case that the error at the node points should be $O(h^{1.5})$ can be given along lines similar to what was done above, by showing that $\|x - \hat{x}_n\|_\infty = O(h^{1.5})$, although the argument is a bit more subtle.

3.4.4. Piecewise polynomial collocation at Gauss-Legendre nodes

We now consider an important example of collocation methods for which there is a significant increase in the rate of convergence, one which duplicates the orders of convergence seen in (3.4.100) for the Galerkin and iterated Galerkin solutions with an approximating subspace of piecewise polynomial functions. The error analysis appears to be due first to Chandler [105], and later a generalized and extended result was given by Chatelin and Lebbar [114].

We consider the integral equation

$$\lambda x(t) - \int_a^b K(t, s)x(s)\, ds = y(t), \quad a \le t \le b$$

Let $n > 0$ and $h = (b - a)/n$. Define $\tau_j = a + jh$, $j = 0, 1, \ldots, n$. Recall the Gauss-Legendre quadrature formula with $r + 1$ node points, written relative to $[0, 1]$:

$$\int_0^1 g(\sigma)\, d\sigma \approx \sum_{k=0}^r \omega_k g(\sigma_k) \qquad (3.4.115)$$

This formula has degree of precision $2r + 1$. For a review of Gaussian quadrature, see [48, Section 5.3]. Introduce the collocation node points

$$t_{i,k} = \tau_{i-1} + \sigma_k h, \quad k = 0, \ldots, r, \quad i = 1, \ldots, n \qquad (3.4.116)$$

There are $n(r + 1)$ such node points.

Let the subspace $\mathcal{X}_n$ be the set of all functions $z \in L^\infty(a, b)$ that are polynomials of degree $\le r$ on each of subintervals $[\tau_{i-1}, \tau_i]$, $i = 1, \ldots, n$. There are no continuity restrictions at the breakpoints $\tau_1, \ldots, \tau_{n-1}$, and the dimension of $\mathcal{X}_n$ is $d_n = n(r + 1)$. We can define a Lagrange basis for $\mathcal{X}_n$, as follows. Define

$$\alpha_k(\sigma) = \prod_{\substack{i=0 \\ i \ne k}}^r \left(\frac{\sigma - \sigma_i}{\sigma_k - \sigma_i} \right), \quad k = 0, \ldots, r$$

$$\ell_{i,k}(t) = \begin{cases} \alpha_k((t - \tau_{i-1})/h), & \tau_{i-1} \le t \le \tau_i \\ 0, & \text{otherwise} \end{cases}$$

for $k = 0, \ldots, r$, $i = 1, \ldots, n$. Using these, we can write the interpolation of $x \in C[a, b]$ by an element of $\mathcal{X}_n$ by

$$\mathcal{P}_n x(t) = \sum_{i=1}^n \sum_{k=0}^r x(t_{i,k}) \ell_{i,k}(t) \qquad (3.4.117)$$

For the error in this interpolation, we can show

$$\|x - \mathcal{P}_n x\|_\infty = O(h^{r+1}), \quad x \in C^{r+1}[a, b] \qquad (3.4.118)$$

The set $\mathcal{X}_n$ is not a subspace of $C[a, b]$, and therefore $\mathcal{P}_n$ is not a projection operator on $C[a, b]$. It can, however, be extended to a bounded projection on $L^\infty(a, b)$ into $\mathcal{X}_n$, with no increase in the size of $\|\mathcal{P}_n\|$. For a discussion of this and of other ways to deal with collocation methods that use approximations that are not continuous functions, see Atkinson, Graham, and Sloan [62].

The operator $\mathcal{P}_n\mathcal{K}$ can be written as a degenerate kernel integral operator and examined directly:

$$\mathcal{P}_n\mathcal{K}x(t) = \int_a^b K_n(t,s)x(s)\,ds, \quad a \le t \le b, \ x \in C[a,b]$$

$$K_n(t,s) = \sum_{i=1}^n \sum_{k=0}^r K(t_{i,k},s)\ell_{i,k}(t)$$

It is well-defined as an operator from $L^\infty(a,b)$ and $C[a,b]$. If $K(t,s)$ is $(r+1)$-times continuously differentiable with respect to t, for all $a \le s \le b$, then (3.4.118) can be used to show

$$\|\mathcal{K} - \mathcal{P}_n\mathcal{K}\| = O(h^{r+1})$$

Therefore, if $(\lambda - \mathcal{K})x = y$ is uniquely solvable, then $(\lambda - \mathcal{P}_n\mathcal{K})^{-1}$ exists and is uniformly bounded for all sufficiently large n, say $n \ge N$. The remaining standard results of Theorem 3.1.1 follow, and we have the usual error bound

$$\|x - x_n\|_\infty \le |\lambda|\|(\lambda - \mathcal{P}_n\mathcal{K})^{-1}\|\|x - \mathcal{P}_n x\|_\infty, \quad n \ge N$$

$$\|x - x_n\|_\infty \le ch^{r+1}\|x^{(r+1)}\|_\infty, \quad x \in C^{r+1}[a,b] \qquad (3.4.119)$$

The iterated collocation solution $\hat{x}_n$ is defined as before, and the inverses $(\lambda - \mathcal{K}\mathcal{P}_n)^{-1}$ are uniformly bounded for $n \ge N$. For the error in $\hat{x}_n$, we use (3.4.88) in the form

$$x - \hat{x}_n = (\lambda - \mathcal{K}\mathcal{P}_n)^{-1}\mathcal{K}(I - \mathcal{P}_n)x \qquad (3.4.120)$$

and we analyze $\mathcal{K}(I - \mathcal{P}_n)x$. Writing it out,

$$\mathcal{K}x(t) - \mathcal{K}\mathcal{P}_n x(t)$$

$$= \sum_{i=1}^n \int_{\tau_{i-1}}^{\tau_i} K(t,s)\left[x(s) - \sum_{k=0}^r x(t_{i,k})\ell_{i,k}(s)\right]ds$$

$$= \sum_{i=1}^n \int_{\tau_{i-1}}^{\tau_i} K(t,s)(s - t_{i,0})\cdots(s - t_{i,r})x[t_{i,0},\ldots,t_{i,r},s]\,ds$$

$$\qquad (3.4.121)$$

To simplify the notation, introduce

$$w_i(s) = (s - t_{i,0})\cdots(s - t_{i,r})$$

Expand $g_i(s) \equiv K(t, s)x[t_{i,0}, \ldots, t_{i,r}, s]$ in a Taylor series about $s = \tau_{i-1}$:

$$g_i(s) = \sum_{j=0}^{r} g_i^{(j)}(\tau_{i-1})(s - \tau_{i-1})^j + g_i^{(r+1)}(\xi)(s - \tau_{i-1})^{r+1} \qquad (3.4.122)$$

for some $\xi \in [\tau_{i-1}, s]$. To justify the needed differentiability of $g_i(s)$, note first that if $x \in C^{2r+2}[a, b]$, then $x[t_{i,0}, \ldots, t_{i,r}, s] \in C^{r+1}[\tau_{i-1}, \tau_i]$. A proof of this can be based on the Hermite-Gennochi formula (see [48, p. 144]), and it can be shown that

$$\max_{i=1,\ldots,n} \max_{\tau_{i-1} \leq s \leq \tau_i} \left| \frac{d^{r+1}}{ds^{r+1}} x[t_{i,0}, \ldots, t_{i,r}, s] \right| \leq \left\| x^{(2r+2)} \right\|_\infty$$

Thus if $x \in C^{2r+2}[a, b]$ and if $K(t, s)$ is $(r + 1)$-times continuously differentiable with respect to s, for $a \leq t \leq b$, then $g_i \in C^{r+1}[\tau_{i-1}, \tau_i]$, with

$$\max_{i=1,\ldots,n} \max_{\tau_{i-1} \leq s \leq \tau_i} \left| g_i^{(r+1)}(s) \right| \leq c \max_{r+1 \leq j \leq 2r+2} \left\| x^{(j)} \right\|_\infty \qquad (3.4.123)$$

The constant c involves the maximum values of $\partial^j K(t, s)/\partial s^j$, $j = 0, \ldots,$ $r + 1$. This justifies the above Taylor series expansion of $g_i(s)$, and it allows us to bound the remainder term in the expansion.

We next use another result for the Gauss-Legendre formula (3.4.115). Define

$$\Omega_r(\sigma) = (\sigma - \sigma_0) \cdots (\sigma - \sigma_r)$$

This is the Legendre polynomial of degree $r + 1$ relative to the interval $[0, 1]$, scaled so that its leading coefficient equals 1. Consequently, it is orthogonal to all polynomials of degree less than $r + 1$, and in particular,

$$\int_0^1 \Omega_r(\sigma)\sigma^j \, d\sigma = 0, \qquad j = 0, 1, \ldots, r \qquad (3.4.124)$$

Note that

$$w_i(s) = \Omega\left(\frac{s - \tau_{i-1}}{h}\right), \qquad i = 1, \ldots, n \qquad (3.4.125)$$

Use the change of variable $s = \tau_{i-1} + \sigma h$ in the inner integral of (3.4.121). Then apply (3.4.122)–(3.4.125) to obtain

$$\int_{\tau_{i-1}}^{\tau_i} K(t, s)w_i(s)x[t_{i,0}, \ldots, t_{i,r}, s] \, ds$$

$$= \int_{\tau_{i-1}}^{\tau_i} g_i^{(r+1)}(\xi)w_i(s)(s - \tau_{i-1})^{r+1} \, ds$$

When this last integral is bounded, using (3.4.123) and some straightforward bounds, we have

$$\left| \int_{\tau_{i-1}}^{\tau_i} g_i^{(r+1)}(\xi) w_i(s)(s - \tau_{i-1})^{r+1} ds \right| \le ch^{2r+3} \max_{r+1 \le j \le 2r+2} \left\| x^{(j)} \right\|_\infty$$

When combined with (3.4.121), we have

$$\| \mathcal{K}(I - \mathcal{P}_n)x \|_\infty = O(h^{2r+2}), \quad x \in C^{2r+2}[a, b]$$

Using (3.4.120),

$$\| x - \hat{x}_n \|_\infty = O(h^{2r+2}), \quad n \ge N, \ x \in C^{2r+2}[a, b] \tag{3.4.126}$$

When this is compared to (3.4.119), we see that the order of convergence of $\hat{x}_n$ is double that of the Galerkin solution x_n. Also, from $\mathcal{P}_n \hat{x}_n = x_n$, we have the following superconvergence result for $x_n(t)$ at the collocation node points:

$$\max_{\substack{1 \le i \le n \\ 0 \le k \le r}} |x(t_{i,j}) - x_n(t_{i,j})| = O(h^{2r+2}), \quad x \in C^{2r+2}[a, b] \tag{3.4.127}$$

This could be stated as a theorem, but it would be much the same in form as that of Theorem 3.4.3, and therefore we omit such a formal statement.

For a general discussion of superconvergence in solving integral equations, see Sloan [507].

3.4.5. The linear system for the iterated collocation solution

Let the interpolatory projection operator be

$$\mathcal{P}_n x(t) = \sum_{j=1}^n x(t_j) \ell_j(t), \quad x \in C(D) \tag{3.4.128}$$

When written out, the approximating equation $\lambda \hat{x}_n - \mathcal{K} \mathcal{P}_n \hat{x}_n = y$ becomes

$$\lambda \hat{x}_n(t) - \sum_{j=1}^n \hat{x}_n(t_j) \int_D K(t, s) \ell_j(s) \, ds = y(t), \quad t \in D \tag{3.4.129}$$

Evaluating at each node point t_i, we obtain the linear system

$$\lambda \hat{x}_n(t_i) - \sum_{j=1}^n \hat{x}_n(t_j) \int_D K(t_i, s) \ell_j(s) \, ds = y(t_i), \quad i = 1, \ldots, n \tag{3.4.130}$$

This is also the linear system for the collocation solution at the node points, as given, for example, in (3.2.41). This is not surprising since x_n and $\hat{x}_n$ agree at the node points.

The two solutions differ, however, at the remaining points in D. For general $t \in D$, $x_n(t)$ is based on the interpolation formula (3.4.128). However, the iterated collocation solution $\hat{x}_n(t)$ is given by using (3.4.129) in the form

$$\hat{x}_n(t) = \frac{1}{\lambda} \left\{ y(t) + \sum_{j=1}^{n} \hat{x}_n(t_j) \int_D K(t, s) \ell_j(s) \, ds \right\}, \quad t \in D \qquad (3.4.131)$$

In the following chapter, we will see that (3.4.129)–(3.4.131) is a special case of the Nyström method for solving $(\lambda - \mathcal{K})x = y$, and (3.4.131) is called the Nyström interpolation function. Generally, it is more accurate than ordinary polynomial interpolation.

3.5. Regularization of the solution

Consider the regularity (that is, the differentiability) of solutions of the integral equation

$$\lambda x(t) - \int_D K(t, s) x(s) \, ds = y(t), \quad t \in D \qquad (3.5.132)$$

Let $K(t, s)$ be m-times continuously differentiable with respect to t, for all $s \in D$. Then for $1 \le j \le m$, the solution $x \in C^j(D)$ if and only if $y \in C^j(D)$. As we have seen in the preceding sections, higher rates of convergence are associated with smoother solution functions x, and thus we want $x(t)$ to be a smooth function. In accordance with this, what should be done if $K(t, s)$ is smooth, as described, but $y(t)$ is not? In Kantorovich [303] a method of *regularization* was described that leads to a new equation in which the solution function is smoother.

From the equation

$$x = \frac{1}{\lambda} [y + \mathcal{K}x] \qquad (3.5.133)$$

we consider $z = \mathcal{K}x$ as a new and smoother unknown function. If

$$x = \frac{1}{\lambda} [y + z] \qquad (3.5.134)$$

is substituted into the original equation $(\lambda - \mathcal{K})x = y$, then after simplication we have

$$(\lambda - \mathcal{K})z = \mathcal{K}y \qquad (3.5.135)$$

This is the same basic equation as before, but now the right-hand side is at least as smooth as the kernel function $K(t, s)$ when considered as a function of t. This new equation can be solved numerically by any method, and then the formula (3.5.134) can be used to obtain an approximate solution of the original equation.

Apply a projection method to (3.5.135),

$$(\lambda - \mathcal{P}_n \mathcal{K})z_n = \mathcal{P}_n \mathcal{K}y \qquad (3.5.136)$$

and then define

$$\check{x}_n = \frac{1}{\lambda}[y + z_n] \qquad (3.5.137)$$

How well does $\check{x}_n$ converge to x? We begin by showing that

$$(\lambda - \mathcal{P}_n \mathcal{K})\check{x}_n = y \qquad (3.5.138)$$

Substituting (3.5.137) into the left side of (3.5.138),

$$
\begin{aligned}
(\lambda - \mathcal{P}_n \mathcal{K})\check{x}_n &= (\lambda - \mathcal{P}_n \mathcal{K})\frac{1}{\lambda}[y + z_n] \\
&= \frac{1}{\lambda}[\lambda y + \lambda z_n - \mathcal{P}_n \mathcal{K}y - \mathcal{P}_n \mathcal{K}z_n] \\
&= \frac{1}{\lambda}[\lambda y - \mathcal{P}_n \mathcal{K}y + \underbrace{\lambda z_n - \mathcal{P}_n \mathcal{K}z_n}_{=\mathcal{P}_n \mathcal{K}y}] \\
&= y
\end{aligned}
$$

Thus to analyze the convergence of $\check{x}_n$, we can work directly with (3.5.138). We use the same hypotheses as are assumed for the projection method, namely that $\|\mathcal{K} - \mathcal{P}_n \mathcal{K}\| \to 0$ as $n \to \infty$. As in Theorem 3.1.1, we can then conclude the existence and uniform boundedness of $(\lambda - \mathcal{P}_n \mathcal{K})^{-1}$. For the error,

$$
\begin{aligned}
x - \check{x}_n &= (\lambda - \mathcal{K})^{-1}y - (\lambda - \mathcal{P}_n \mathcal{K})^{-1}y \\
&= (\lambda - \mathcal{P}_n \mathcal{K})^{-1}[(\lambda - \mathcal{P}_n \mathcal{K}) - (\lambda - \mathcal{K})](\lambda - \mathcal{K})^{-1}y \\
&= (\lambda - \mathcal{P}_n \mathcal{K})^{-1}[\mathcal{K} - \mathcal{P}_n \mathcal{K}]x
\end{aligned}
$$

and

$$\|x - \check{x}_n\| \le \|(\lambda - \mathcal{P}_n \mathcal{K})^{-1}\| \|\mathcal{K} - \mathcal{P}_n \mathcal{K}\| \|x\|$$

In this, the speed of convergence is apparently independent of the smoothness of x.

This error analysis should seem familiar, as it is exactly the same analysis as was given in Chapter 2 for degenerate kernel methods. In fact, equation (3.5.138) is a degenerate kernel integral equation. When the collocation method is applied to the regularized equation (3.5.135), the resulting method (3.5.138) is the method of §2.3 of Chapter 2, with the degenerate kernel based on interpolation of $K(t, s)$ with respect to t, as in (2.3.40). When the Galerkin method is used to solve (3.5.135), the resulting method (3.5.138) is the method of §2.4 of Chapter 2, as in (2.4.73). Thus, there is a close connection between projection methods and degenerate kernel methods. In particular, the coefficient matrices for the linear systems associated with $(\lambda - \mathcal{P}_n\mathcal{K})x_n = \mathcal{P}_n y$ and $(\lambda - \mathcal{P}_n\mathcal{K})\tilde{x}_n = y$ are the same if a Lagrange basis is used in implementing the collocation method or an orthonormal basis is used in implementing the Galerkin method. In computing the condition numbers of such matrices, we can regard the linear system as arising from either a degenerate kernel method or a projection method, whichever is the more useful.

3.6. Condition numbers

Almost all numerical methods for solving integral equations lead to the solution of linear systems. In this section we look at the conditioning of these linear systems, and we relate it to the conditioning of the original integral equation being solved. We begin by noting a few results on the behavior of the original integral equation.

Let $\mathcal{X}$ be a Banach space, and let $\mathcal{K}$ be a bounded linear operator from $\mathcal{X}$ to $\mathcal{X}$. Let $\rho(\mathcal{K})$ denote the set of all $\lambda \in \mathbf{C}$ for which $(\lambda - \mathcal{K})^{-1}$ is a bounded operator from $\mathcal{X}$ to $\mathcal{X}$, and let $\sigma(\mathcal{K})$ denote the complement in $\mathbf{C}$ of $\rho(\mathcal{K})$. The sets $\rho(\mathcal{K})$ and $\sigma(\mathcal{K})$ are called the *resolvent set* and the *spectrum* of $\mathcal{K}$, respectively. Using the geometric series theorem (cf. Theorem A.1 in the Appendix), it can be shown that $\rho(\mathcal{K})$ is an open set and that $\sigma(\mathcal{K})$ is a closed and bounded set.

Lemma 3.6.1. Let $\mathcal{K}$ be a bounded linear operator from $\mathcal{X}$ to $\mathcal{X}$, with $\mathcal{X}$ a Banach space. Let $\lambda \in \rho(\mathcal{K})$, and let $d(\lambda, \sigma(\mathcal{K}))$ denote the distance from λ to $\sigma(\mathcal{K})$. Then

$$\|(\lambda - \mathcal{K})^{-1}\| \geq \frac{1}{d(\lambda, \sigma(\mathcal{K}))} \tag{3.6.139}$$

Proof. Assume (3.6.139) is not true, but rather

$$\|(\lambda - \mathcal{K})^{-1}\| < \frac{1}{d(\lambda, \sigma(\mathcal{K}))} \tag{3.6.140}$$

Pick $\lambda_0 \in \sigma(\mathcal{K})$ with $|\lambda - \lambda_0| = d(\lambda, \sigma(\mathcal{K}))$, which is possible since $\sigma(\mathcal{K})$ is closed and bounded. Thus,

$$|\lambda - \lambda_0| < \frac{1}{\|(\lambda - \mathcal{K})^{-1}\|} \tag{3.6.141}$$

$$\|(\lambda - \mathcal{K}) - (\lambda_0 - \mathcal{K})\| < \frac{1}{\|(\lambda - \mathcal{K})^{-1}\|}$$

$$\|(\lambda - \mathcal{K})^{-1}[(\lambda - \mathcal{K}) - (\lambda_0 - \mathcal{K})]\|$$

$$\leq \|(\lambda - \mathcal{K})^{-1}\| \, \|(\lambda - \mathcal{K}) - (\lambda_0 - \mathcal{K})\| < 1 \tag{3.6.142}$$

Write

$$\lambda_0 - \mathcal{K} = (\lambda - \mathcal{K}) - [(\lambda - \mathcal{K}) - (\lambda_0 - \mathcal{K})]$$
$$= (\lambda - \mathcal{K})\{I - (\lambda - \mathcal{K})^{-1}[(\lambda - \mathcal{K}) - (\lambda_0 - \mathcal{K})]\}$$

The expression $I - (\lambda - \mathcal{K})^{-1}[(\lambda - \mathcal{K}) - (\lambda_0 - \mathcal{K})]$ has a bounded inverse by the geometric series theorem and (3.6.142). Therefore, $\lambda_0 - \mathcal{K}$ has a bounded inverse. This contradicts our assumption that there exists $\lambda_0 \in \sigma(\mathcal{K})$ for which (3.6.141) is true. Therefore, (3.6.139) must be valid. $\square$

The condition number of the equation $(\lambda - \mathcal{K})x = y$ is usually defined as

$$\mathrm{cond}(\lambda - \mathcal{K}) = \|\lambda - \mathcal{K}\| \, \|(\lambda - \mathcal{K})^{-1}\| \tag{3.6.143}$$

Thus as $\lambda \to \sigma(\mathcal{K})$, the equation $(\lambda - \mathcal{K})x = y$ is increasingly ill-conditioned. To see the sense of this definition, consider the solutions to the equations

$$(\lambda - \mathcal{K})x = y \quad \text{and} \quad (\lambda - \mathcal{K})x_\delta = y + \delta.$$

It is a standard argument to show

$$\frac{\|x - x_\delta\|}{\|x\|} \leq \mathrm{cond}(\lambda - \mathcal{K}) \frac{\|\delta\|}{\|y\|} \tag{3.6.144}$$

Moreover, there are nonzero values of x and δ for which this becomes an equality. As the condition number increases, the solution x is increasingly sensitive to some small changes δ in the data y. For a discussion of condition numbers for matrices, see Ref. [48, Section 8.4].

For the projection methods considered earlier in the chapter, we have both $\mathcal{P}_n\mathcal{K} \to \mathcal{K}$ and $(\lambda - \mathcal{P}_n\mathcal{K})^{-1} \to (\lambda - \mathcal{K})^{-1}$, and consequently

$$\|\lambda - \mathcal{P}_n\mathcal{K}\| \to \|\lambda - \mathcal{K}\|$$
$$\|(\lambda - \mathcal{P}_n\mathcal{K})^{-1}\| \to \|(\lambda - \mathcal{K})^{-1}\|$$

For the condition number of $(\lambda - \mathcal{P}_n \mathcal{K})x_n = \mathcal{P}_n y$, we then have

$$\text{cond}(\lambda - \mathcal{P}_n \mathcal{K}) \to \text{cond}(\lambda - \mathcal{K}) \quad \text{as } n \to \infty \qquad (3.6.145)$$

This means that $(\lambda - \mathcal{K})x = y$ and its approximating equation $(\lambda - \mathcal{P}_n \mathcal{K})x_n = \mathcal{P}_n y$ will be approximately the same in their conditioning. We will see, however, that this need not be true of the linear system associated with $(\lambda - \mathcal{P}_n \mathcal{K})x_n = \mathcal{P}_n y$. The same results are true for degenerate kernel methods, with an analogous argument.

3.6.1. Condition numbers for the collocation method

For the collocation method $(\lambda - \mathcal{P}_n \mathcal{K})x_n = \mathcal{P}_n y$, let $A_n c = \gamma$ denote the associated linear system, as given in (3.1.5):

$$\sum_{j=1}^{d} c_j \left\{ \lambda \phi_j(t_i) - \int_D K(t_i, s)\phi_j(s)\, ds \right\} = y(t_i), \quad i = 1, \ldots, d_n$$

$$(3.6.146)$$

This is associated with

$$x_n(t) = \sum_{j=1}^{d} c_j \phi_j(t), \quad t \in D \qquad (3.6.147)$$

with $\mathcal{X}_n = \text{span}\{\phi_1, \ldots, \phi_d\}$. We need to bound $\text{cond}(A_n) = \|A_n\| \|A_n^{-1}\|$. In the following, we concentrate on measuring the size of $\|A_n^{-1}\|$. It is usually straightforward to bound $\|A_n\|$, and it seldom becomes large (although it can).

We assume

$$\det(\Gamma_n) \neq 0, \qquad \Gamma_n = [\phi_j(t_i)]_{i,j=1}^{d_n}$$

To bound $\|A_n^{-1}\|$, consider solving $A_n \alpha = \gamma$ for arbitrary $\gamma \in \mathbf{R}^d$ (or $\mathbf{C}^d$). We use $\mathcal{X} = C(D)$ with $\|\cdot\|_\infty$, and for $\mathbf{R}^d$, we use the vector norm $\|\cdot\|_\infty$. We bound $\|\alpha\|_\infty$ in terms of $\|\gamma\|_\infty$ and thereby obtain a bound for $\|A_n^{-1}\|$.

Given γ, pick $y \in C(D)$ with $\|y\|_\infty = \|\gamma\|_\infty$ and $y(t_i) = \gamma_i$, $i = 1, \ldots, d_n$. The existence of such a function $y(t)$ is usually straightforward, generally being accomplished with a suitable piecewise linear function. We leave this as an exercise for the reader. Using this y, let $x_n = (\lambda - \mathcal{P}_n \mathcal{K})^{-1} y$, and write

$$x_n(t) = \sum_{j=1}^{d} \alpha_j \phi_j(t)$$

Taking bounds,

$$\|x_n\|_\infty \leq \|(\lambda - \mathcal{P}_n \mathcal{K})^{-1}\| \, \|\mathcal{P}_n\| \, \|y\|_\infty$$

$$= \|(\lambda - \mathcal{P}_n \mathcal{K})^{-1}\| \, \|\mathcal{P}_n\| \, \|\gamma\|_\infty \qquad (3.6.148)$$

The solution α of the linear system is related to x_n at the node points by

$$x_n(t_i) = \sum_{j=1}^{d} \alpha_j \phi_j(t_i), \quad i = 1, \ldots, d_n$$

In matrix form,

$$\Gamma_n \alpha = \underline{x}_n, \quad \underline{x}_n = [x_n(t_1), \ldots, x_n(t_d)]^{\mathrm{T}}$$

Then

$$\alpha = \Gamma_n^{-1} \underline{x}_n$$

$$\|\alpha\|_\infty \leq \left\| \Gamma_n^{-1} \right\| \, \|\underline{x}_n\|_\infty \leq \left\| \Gamma_n^{-1} \right\| \, \|x_n\|_\infty \qquad (3.6.149)$$

The matrix norm used is the *row norm*: for a general $d \times d$ matrix B,

$$\|B\| = \max_{i=1,\ldots,d} \sum_{j=1}^{d} |B_{i,j}|$$

Combining (3.6.149) and (3.6.148),

$$\left\| A_n^{-1} \gamma \right\|_\infty = \|\alpha\|_\infty \leq \left\| \Gamma_n^{-1} \right\| \, \|(\lambda - \mathcal{P}_n \mathcal{K})^{-1}\| \, \|\mathcal{P}_n\| \, \|\gamma\|_\infty$$

This implies our main result:

$$\left\| A_n^{-1} \right\| \leq \|\mathcal{P}_n\| \, \left\| \Gamma_n^{-1} \right\| \, \|(\lambda - \mathcal{P}_n \mathcal{K})^{-1}\| \qquad (3.6.150)$$

The term $\|(\lambda - \mathcal{P}_n \mathcal{K})^{-1}\|$ is approximately $\|(\lambda - \mathcal{K})^{-1}\|$ for large n, and we do not consider it further. The terms $\|\mathcal{P}_n t\|$ and $\left\| \Gamma_n^{-1} \right\|$ are both important, and they must be examined for each collocation method.

We obtain an important special case by using the Lagrange basis:

$$\phi_i = \ell_i, \quad i = 1, \ldots, d_n$$

For this important special case, we will use $B_n c = \gamma$ to denote the linear system, rather than using $A_n c = \gamma$ as above. For this case, $\Gamma_n = I$ and thus

$$\left\| B_n^{-1} \right\| \leq \|\mathcal{P}_n\| \, \|(\lambda - \mathcal{P}_n \mathcal{K})^{-1}\| \qquad (3.6.151)$$

Also note that the results for the collocation method carry across to the linear system associated with the degenerate kernel integral equation $(\lambda - \mathcal{P}_n \mathcal{K})x_n = y$, which has a degenerate kernel defined as in (2.3.40) of Chapter 2.

Example. Recall the use of piecewise linear interpolation in §3.2, in (3.2.41) and following. With it, $\|\mathcal{P}_n\| = 1$ and $\Gamma_n = I$. Thus $\|B_n^{-1}\| \le \|(\lambda - \mathcal{P}_n \mathcal{K})^{-1}\|$, and the matrix B_n is well-conditioned if the original integral equation is well-conditioned.

With other forms of piecewise polynomial interpolation of higher degree, the results will be much the same. The value of $\|\mathcal{P}_n\|$ will increase, but not by much. For example, with piecewise quadratic interpolation such as was used in Theorem 3.4.3, we have $\|\mathcal{P}_n\| = 1.25$.

Example. Recall the use of trigonometric interpolation to solve the 2π-periodic integral equation (3.2.50), and let $\mathcal{X} = C_p(2\pi)$ include all continuous complex-valued 2π-periodic functions. From (3.2.55), $\|\mathcal{P}_n\| = O(\log n)$. For the basis of $\mathcal{X}_n$, use the complex exponentials

$$\phi_j(t) = e^{ijt}, \quad -\infty < t < \infty, \quad j = -n, \dots, n$$

The interpolation nodes are $t_j = jh$, $j = 0, \dots, 2n$, $h = 2\pi/(2n+1)$. For the matrix Γ_n, we can show

$$\Gamma_n^* \Gamma_n = (2n+1)I \tag{3.6.152}$$

so that

$$\frac{1}{\sqrt{2n+1}} \Gamma_n$$

is a unitary matrix. This implies that in the matrix norm induced by the Euclidean vector norm,

$$\|\Gamma_n^{-1}\| = \frac{1}{2n+1} \|\Gamma_n^*\| = \frac{1}{\sqrt{2n+1}}$$

Using this and repeating the argument in (3.6.149),

$$\|\alpha\|_\infty \le \|\alpha\|_2 \le \frac{1}{\sqrt{2n+1}} \|\underline{x}_n\|_2 \le \|\underline{x}_n\|_\infty \le \|x_n\|_\infty$$

$$\|\alpha\|_\infty \le \|x_n\|_\infty$$

Combined with (3.6.148), we have

$$\|A_n^{-1}\| \leq \|\mathcal{P}_n\|\|(\lambda - \mathcal{P}_n\mathcal{K})^{-1}\| = O(\log n)\|(\lambda - \mathcal{P}_n\mathcal{K})^{-1}\| \qquad (3.6.153)$$

Thus $\|A_n^{-1}\|$ can grow, but not too rapidly, as n increases.

For A_n, it can be shown that $\|A_n\| = O(n)$. Therefore,

$$\text{cond}(A_n) = O(n \log n)\|(\lambda - \mathcal{P}_n\mathcal{K})^{-1}\| \qquad (3.6.154)$$

The collocation method with trigonometric interpolation is reasonably well-behaved as n increases. The size of n used with the trigonometric collocation method is generally small, certainly less than 100, and therefore the value of $\text{cond}(A_n)$ is usually not much worse than $\text{cond}(\lambda - \mathcal{P}_n\mathcal{K})$.

Example. Let $\mathcal{X} = C[-1, 1]$, and let $\mathcal{X}_n$ be the set of polynomials of degree less than n. For the interpolation nodes, use the zeros of the Chebyshev polynomial

$$T_n(t) = \cos(n \cos^{-1} t), \qquad -1 \leq t \leq 1$$

These are

$$t_j = \cos\left[\frac{2j-1}{2n}\pi\right], \qquad j = 1, \ldots, n$$

With this choice of nodes,

$$\|\mathcal{P}_n\| \leq 1 + \frac{1}{\pi}\log(n+1)$$

For example, see Rivlin [467, p. 13]. This choice of interpolation nodes is essentially the best possible choice when doing polynomial interpolation of increasing degree.

If we use the Lagrange basis functions of (3.1.9), then $\Gamma_n = I$, and we can apply (3.6.151) to obtain

$$\|B_n^{-1}\| \leq O(\log n)\|(\lambda - \mathcal{P}_n\mathcal{K})^{-1}\| \qquad (3.6.155)$$

This is not at all badly behaved, as it increases very slowly as n increases. However, if we choose as our basis the monomials

$$\phi_i(t) = t^{i-1}, \qquad i = 1, \ldots, n$$

then Γ_n is a Vandermonde matrix. In this case, $\|\Gamma_n^{-1}\|$ is known to grow rapidly with n, and the matrix A_n is likely to be ill-conditioned. For more information on Vandermonde matrices and their conditioning, see Gautschi [212].

3.6.2. Condition numbers based on the iterated collocation solution

From earlier at the end of §3.5, we noted that the linear system for the collocation method is exactly the same as that for the iterated collocation solution. In particular, the linear system takes the form (3.4.130),

$$\lambda \hat{x}_n(t_i) - \sum_{j=1}^{d} \hat{x}_n(t_j) \int_D K(t_i, s) \ell_j(s) \, ds = y(t_i), \quad i = 1, \ldots, d_n$$

This is associated with $(\lambda - \mathcal{K}\mathcal{P}_n)\hat{x}_n = y$, and the linear system is the one with matrix B_n, as referred to earlier in (3.6.151). As above, we seek to relate $\|B_n^{-1}\|$ to $(\lambda - \mathcal{K}\mathcal{P}_n)^{-1}$.

Consider the linear system by $B_n \alpha = \gamma$, with γ given. As before, choose $y \in C(D)$ with the properties

$$\|y\|_\infty = \|\gamma\|_\infty \quad \text{and} \quad y(t_i) = \gamma_i, \; i = 1, \ldots, d_n$$

Then introduce

$$\hat{x}_n = (\lambda - \mathcal{K}\mathcal{P}_n)^{-1} y$$

$$\|\hat{x}_n\|_\infty \le \|(\lambda - \mathcal{K}\mathcal{P}_n)^{-1}\| \, \|y\|_\infty$$

Let $\underline{\hat{x}}_n \equiv [\hat{x}_n(t_1), \ldots, \hat{x}_n(t_d)]^\mathrm{T}$. Then

$$\|\alpha\|_\infty = \|\underline{\hat{x}}_n\|_\infty \le \|\hat{x}_n\|_\infty \le \|(\lambda - \mathcal{K}\mathcal{P}_n)^{-1}\| \, \|\gamma\|_\infty$$

This shows

$$\left\| B_n^{-1} \right\| \le \|(\lambda - \mathcal{K}\mathcal{P}_n)^{-1}\| \qquad (3.6.156)$$

The bound can be converted to one using $\|(\lambda - \mathcal{P}_n\mathcal{K})^{-1}\|$ using (3.4.84). However, it is often possible to directly obtain better bounds for $\|(\lambda - \mathcal{K}\mathcal{P}_n)^{-1}\|$. This is especially useful for those cases in which $\|\mathcal{P}_n\|$ increases with n, but for which $\mathcal{K}\mathcal{P}_n$ is much better behaved. An example is the preceding case of polynomial interpolation at the Chebyshev zeros; see Sloan [504] and Sloan and Smith [514].

3.6.3. Condition numbers for the Galerkin method

We begin with the special case of the Galerkin method in which the linear system is obtained by using an orthonormal basis $\{\psi_i \mid i = 1, \ldots, d_n\}$ for the approximating subspace $\mathcal{X}_n$. The solution of $(\lambda - \mathcal{P}_n\mathcal{K})x_n = \mathcal{P}_n y$ takes the form

$$x_n(t) = \sum_{j=1}^{d} \beta_j \psi_j(t)$$

$$\lambda \beta_i - \sum_{j=1}^{d} \beta_j (\mathcal{K}\psi_j, \psi_i) = (y, \psi_i), \quad i = 1, \ldots, d_n \qquad (3.6.157)$$

Denote this linear system by $B_n\beta = \gamma$, with $\beta, \gamma \in \mathbf{R}^d$ (or $\mathbf{C}^d$), $\gamma = [(y, \psi_1), \ldots, (y, \psi_d)]^T$.

For the finite dimensional space $\mathbf{R}^d$, use the standard Euclidean norm

$$\|\gamma\|_2 = \left[\sum_{j=1}^d |\gamma_j|^2\right]^{\frac{1}{2}}$$

Define $\mathcal{Q}_n : \mathbf{R}^d \to \mathcal{X}_n$ by

$$\mathcal{Q}_n\gamma(t) = \sum_{j=1}^d \gamma_j \psi_j(t), \quad t \in D, \quad \gamma \in \mathbf{R}^d$$

Then

$$\|\mathcal{Q}_n\gamma\| = \left[\sum_{j=1}^d |\gamma_j|^2\right]^{\frac{1}{2}} = \|\gamma\|_2$$

$$\|\mathcal{Q}_n\| = \left\|\mathcal{Q}_n^{-1}\right\| = 1 \tag{3.6.158}$$

To bound $\|B_n^{-1}\|$ we proceed in the same manner as with the collocation method. Consider the system $B_n\beta = \gamma$ for an arbitrary $\gamma \in \mathbf{R}^d$, and bound β in terms of γ. For the given γ, define $y = \mathcal{Q}_n\gamma$. Introduce

$$x_n = (\lambda - \mathcal{P}_n\mathcal{K})^{-1}\mathcal{P}_n y = (\lambda - \mathcal{P}_n\mathcal{K})^{-1}y$$

with the last part following from $\mathcal{P}_n y = y$. Then

$$\|x_n\| \le \|(\lambda - \mathcal{P}_n\mathcal{K})^{-1}\| \, \|y\| = \|(\lambda - \mathcal{P}_n\mathcal{K})^{-1}\| \, \|\gamma\|_2$$

In addition, $\beta = \mathcal{Q}_n^{-1}x_n$, and thus $\|\beta\|_2 = \|x_n\|$ from (3.6.158). Combined with the last formula,

$$\left\|B_n^{-1}\gamma\right\|_2 = \|\beta\|_2 \le \|(\lambda - \mathcal{P}_n\mathcal{K})^{-1}\| \, \|\gamma\|_2$$

This proves

$$\left\|B_n^{-1}\right\| \le \|(\lambda - \mathcal{P}_n\mathcal{K})^{-1}\| \tag{3.6.159}$$

The matrix norm being used is that induced by the Euclidean vector norm: for a general matrix B,

$$\|B\| = \sqrt{r_\sigma(B^*B)}$$

For a general square matrix A, the quantity $r_\sigma(A)$ is called the *spectral radius* of A, and it is defined by

$$r_\sigma(A) = \max_{\lambda \in \sigma(A)} |\lambda|$$

The result (3.6.159) shows that the linear system (3.6.157) is well-behaved when an orthonormal basis is used in the implementation of Galerkin's method.

Consider now the more general case of Galerkin's method, with the basis $\{\phi_1, \ldots, \phi_d\}$ not necessarily orthonormal. The associated linear system $A_n c = \gamma$ was given in (3.1.13) of §3.1. Using $\{\phi_i\}$ and the earlier orthonormal basis $\{\psi_i\}$, introduce the following matrices:

$$\Gamma_n = [(\phi_i, \phi_j)], \qquad D_n = [(\phi_i, \psi_j)], \qquad G_n = \Gamma_n^{-1} D_n \qquad (3.6.160)$$

D_n and G_n are "change of basis" matrices, for changing between representations of elements of $\mathcal{X}_n$ using the two bases $\{\phi_1, \ldots, \phi_d\}$ and $\{\psi_1, \ldots, \psi_d\}$. In particular,

$$\phi_i = \sum_{j=1}^{d} (\phi_i, \psi_j) \psi_j, \qquad \psi_i = \sum_{j=1}^{d} G_{j,i} \phi_i$$

The matrix Γ_n is a *Gram matrix*, and as such it is symmetric and positive definite. These matrices satisfy the following relations (where we have assumed all matrices are real to simplify some of the arguments):

$$\Gamma_n G_n = D_n, \qquad D_n^{\mathrm{T}} G_n = I, \qquad D_n D_n^{\mathrm{T}} = \Gamma_n \qquad (3.6.161)$$

The proofs of these results are left as an exercise for the reader.

The linear system $A_n c = \delta$ associated with $(\lambda - \mathcal{P}_n \mathcal{K}) x_n = \mathcal{P}_n y$ and the basis $\{\phi_1, \ldots, \phi_d\}$ is

$$\sum_{j=1}^{d} c_j \{\lambda(\phi_j, \phi_i) - (\mathcal{K}\phi_j, \phi_i)\} = (y, \phi_i), \quad i = 1, \ldots, d_n \qquad (3.6.162)$$

and the solution is

$$x_n(t) = \sum_{j=1}^{d} c_j \phi_j(t)$$

Using the matrices introduced above, the linear system can be rewritten in an equivalent form as $B_n \beta = \gamma$ with

$$B_n = G_n^{\mathrm{T}} A_n D_n^{-\mathrm{T}}, \qquad \gamma = G_n^{\mathrm{T}} \delta, \qquad \beta = D_n^{\mathrm{T}} c$$

with $D_n^{-\mathrm{T}} \equiv (D_n^{\mathrm{T}})^{-1}$. When this is combined with the earlier material for $B_n \beta = \gamma$,

$$
\begin{aligned}
A_n^{-1} \gamma &= c \\
&= D_n^{-\mathrm{T}} \beta \\
&= D_n^{-\mathrm{T}} \mathcal{Q}_n x_n \\
&= D_n^{-\mathrm{T}} \mathcal{Q}_n (\lambda - \mathcal{P}_n \mathcal{K})^{-1} y \\
&= D_n^{-\mathrm{T}} \mathcal{Q}_n (\lambda - \mathcal{P}_n \mathcal{K})^{-1} \mathcal{Q}_n y \\
&= D_n^{-\mathrm{T}} \mathcal{Q}_n (\lambda - \mathcal{P}_n \mathcal{K})^{-1} \mathcal{Q}_n G_n^{\mathrm{T}} \delta
\end{aligned}
$$

Using $G_n^T = D_n^{-1}$, we have

$$A_n^{-1} = D_n^{-T} Q_n (\lambda - \mathcal{P}_n \mathcal{K})^{-1} Q_n D_n^{-1} \qquad (3.6.163)$$

Now

$$\left\| D_n^{-T} \right\| = \left\| D_n^{-1} \right\| = \sqrt{r_\sigma \left(D_n^{-T} D_n^{-1} \right)} = \sqrt{r_\sigma \left(\Gamma_n^{-1} \right)} = \left\| \Gamma_n^{-1} \right\|^{\frac{1}{2}}$$

Using this, taking bounds in (3.6.163), and using (3.6.158),

$$\left\| A_n^{-1} \right\| \leq \left\| \Gamma_n^{-1} \right\| \, \left\| (\lambda - \mathcal{P}_n \mathcal{K})^{-1} \right\| \qquad (3.6.164)$$

Note also that this result is the same as (3.6.150) for the collocation method, since $\|\mathcal{P}_n\| = 1$ for orthogonal projections.

We can also bound $\|A_n\|$ in a manner similar to the above. It can be shown that

$$A_n = G_n^{-T} B_n D_n^T = D_n B_n D_n^T$$

$$\|A_n\| \leq \|\Gamma_n\| \, \|\lambda - \mathcal{P}_n \mathcal{K}\| \qquad (3.6.165)$$

Combining (3.6.164) and (3.6.165),

$$\mathrm{cond}(A_n) \leq \mathrm{cond}(\Gamma_n)\mathrm{cond}(\lambda - \mathcal{P}_n \mathcal{K}) \qquad (3.6.166)$$

Again the matrix Γ_n plays a major role in the conditioning of A_n.

Also note that the results for the Galerkin method carry across to the linear system associated with the degenerate kernel integral equation $(\lambda - \mathcal{P}_n \mathcal{K})x_n = y$, which has a degenerate kernel defined as in (2.4.73) of Chapter 2.

Example. Recall the example (3.3.67) in which piecewise continuous linear functions are being used for the approximating functions. The Lagrange basis functions $\{\ell_1(t), \ldots, \ell_n(t)\}$ of (3.2.37) were used in that example, and here we modify them by rescaling. Define

$$\phi_i(t) = c_i \ell_i(t), \qquad c_i = \begin{cases} \sqrt{\dfrac{3}{h}}, & i = 0 \text{ or } n \\[2mm] \sqrt{\dfrac{3}{2h}}, & i = 1, \ldots, n-1 \end{cases}$$

These are not orthogonal; but as noted earlier, they are a simple and easy-to-use basis.

For the analysis of conditioning, we need to calculate Γ_n and find its condition number. Note that

$$\|\phi_i\| = 1, \qquad (\phi_i, \phi_j) = 0 \quad \text{for } |i - j| \geq 2$$

$$(\phi_i, \phi_{i-1}) = (\phi_{i-1}, \phi_i) = \begin{cases} \dfrac{\sqrt{2}}{4}, & i = 1, n \\[2mm] \dfrac{1}{4}, & i = 2, \ldots, n - 1 \end{cases}$$

The matrix Γ_n is tridiagonal and diagonally dominant. Using the Gerschgorin Circle Theorem (cf. [48, Section 9.1]) to estimate the eigenvalues λ of Γ_n, we have

$$\frac{3 - \sqrt{2}}{4} \leq \lambda \leq \frac{5 + \sqrt{2}}{4}$$

Thus,

$$\|\Gamma_n\| \leq \frac{5 + \sqrt{2}}{4}, \qquad \|\Gamma_n^{-1}\| \leq \frac{4}{3 - \sqrt{2}}$$

$$\text{cond}(A_n) \leq \frac{5 + \sqrt{2}}{3 - \sqrt{2}} \text{cond}(\lambda - \mathcal{P}_n\mathcal{K}) \doteq 4.04 \, \text{cond}(\lambda - \mathcal{P}_n\mathcal{K}) \quad (3.6.167)$$

This shows that the conditioning of the Galerkin system using piecewise linear approximants is not much worse than that of the original operator equation $(\lambda - \mathcal{P}_n\mathcal{K})x_n = \mathcal{P}_n y$.

Example. Let $\mathcal{X} = L^2(0, 1)$, and let $\mathcal{X}_n$ be the set of polynomials of degree less than n. For a basis, use

$$\phi_i(t) = t^{i-1}, \quad i = 0, 1, \ldots, n - 1$$

Then the matrix Γ_n becomes the well-known *Hilbert matrix* of order n. It is well known that $\text{cond}(\Gamma_n)$ increases very rapidly with n. For more detailed information see [48, p. 533]. It is clear from this that the above monomial basis is not a good choice. A much better one is to use the normalized Legendre polynomials, chosen with $[0, 1]$ as the interval of orthogonality. For this choice, $\Gamma_n = I$, and we have the excellent stability result of (3.6.159).

Discussion of the literature

Projection methods have been around for a long time, and their general abstract treatment goes back at least to the fundamental 1948 paper of Kantorovich

[303]. In this paper Kantorovich gave a general schema for defining and analyzing approximate methods for solving linear operator equations, and he analyzed projection methods within this framework. An updated form of this general schema is given in [304], and it is more general than that presented here. The treatment given in this chapter is much simpler than that of [303] because we analyze only projection methods. Analyses similar to ours can also be found in the books of Ref. [39, pp. 50–87], Baker [73, §4.11, 4.12], Fenyö and Stolle [200, Chap. 20], Hackbusch [251, §4.3–4.6], Kress [325, Chap. 13], and Reinhardt [458]. Other analyses that have influenced our treatment are given in the papers of Noble [405], Phillips [424], and Prenter [433]. The paper of Elliott [176] contains a collocation method based on the properties of Chebyshev polynomials and Chebyshev expansions, and this often leads to excellent convergence with a linear system of relatively small order.

The projection solution has been modified in several ways. The first of these is the Kantorovich regularization of $(\lambda - \mathcal{K})x = y$, as presented in §3.5, which leads to numerical methods that are equivalent to degenerate kernel methods. A more important modification is the *Sloan iterate*, which is presented in §3.4. This iteration always leads to a faster rate of convergence when used with Galerkin methods for solving $(\lambda - \mathcal{K})x = y$ with $\mathcal{K}$ compact; the fundamental paper here is Sloan [503]. With collocation methods, the Sloan iterate does not always converge more rapidly, but when it does, it leads to superconvergence results for the original collocation solution at the collocation node points. This is discussed in detail in §3.4.3 and §3.4.4. The first results of this kind seem to be those of Chandler [105], with related results given by Chatelin and Lebbar [114] and Richter [464]. In §4.3 and §4.4 of the next chapter, we discuss a further discretization of projection methods, in which the integrals occurring in such methods are evaluated numerically.

Projection methods have also been used to solve nonlinear integral equations $x = \mathcal{K}(x)$, with $\mathcal{K}$ a completely continuous nonlinear operator from an open domain $D \subset \mathcal{X}$ to $\mathcal{X}$, with $\mathcal{X}$ a Banach space. The fundamental work in this area is that of Krasnoselskii [319], and the Sloan iterate in the nonlinear case is analyzed in Atkinson and Potra [66]. A general discussion of the numerical solution of such nonlinear problems is given in Rall [444], and a survey of numerical methods for nonlinear integral equations is given in Ref. [51].

Not many results have been given on the stability of the linear system associated with projection methods. Some early results were given in Kantorovich [303]. The first important results appear to be those of Phillips [424], and this was extended in Ref. [39, p. 73].

4

The Nyström method

The Nyström method was originally introduced to handle approximations based on numerical integration of the integral operator in the equation

$$\lambda x(t) - \int_D K(t,s)x(s)\,ds = y(t), \quad t \in D \tag{4.0.1}$$

The resulting solution is found first at the set of quadrature node points, and then it is extended to all points in D by means of a special, and generally quite accurate, interpolation formula. The numerical method is much simpler to implement on a computer, but the error analysis is more sophisticated than for the methods of the preceding two chapters. The resulting theory has taken an abstract form that includes error analyses of both projection methods and degenerate kernel methods, although such methods are probably still best understood as distinct methods of interest in their own right.

In this chapter we begin by giving the original Nyström method, which assumes the kernel function is continuous. We give a preliminary error analysis, and then this is generalized to the theory of *collectively compact operator approximations*. In the following section we extend the Nyström method to the case of kernel functions that are not continuous, based on using *product integration*. Such methods are used in dealing with applications to boundary integral equations, which is considered in a later chapter. The final two sections introduce and analyze *discrete collocation* and *discrete Galerkin* methods, respectively.

4.1. The Nyström method for continuous kernel functions

Let a numerical integration scheme be given:

$$\int_D g(s)\,ds \approx \sum_{j=1}^{q_n} w_{n,j}\,g(t_{n,j}), \quad g \in C(D) \tag{4.1.2}$$

with an increasing sequence of values of n. We assume that for every $g \in C(D)$, the numerical integrals converge to the true integral as $n \to \infty$. This implies, by the principle of uniform boundedness (cf. A.3 in the Appendix) that

$$c_I \equiv \sup_{n \geq 1} \sum_{j=1}^{q_n} |w_{n,j}| < \infty \qquad (4.1.3)$$

To simplify the notation, we omit the subscript n, so that $w_{n,j} \equiv w_j$, $t_{n,j} \equiv t_j$; but the presence of n is to be understood implicitly. On occasion, we also use $q \equiv q_n$.

Let $K(t, s)$ be continuous for all $t, s \in D$, where D is a closed and bounded set in $\mathbf{R}^m$ for some $m \geq 1$, as in earlier chapters. Usually, in fact, we will want $K(t, s)$ to be several times continuously differentiable. Using the above quadrature scheme, approximate the integral in (4.0.1), obtaining a new equation:

$$\lambda x_n(t) - \sum_{j=1}^{q_n} w_j K(t, t_j) x_n(t_j) = y(t), \quad t \in D \qquad (4.1.4)$$

We write this as an exact equation with a new unknown function $x_n(t)$. To find the solution at the node points, let t run through the quadrature node points t_i. This yields

$$\lambda x_n(t_i) - \sum_{j=1}^{q_n} w_j K(t_i, t_j) x_n(t_j) = y(t_i), \quad i = 1, \ldots, q_n \qquad (4.1.5)$$

which is a linear system of order q_n. The unknown is a vector

$$\underline{x}_n \equiv [x_n(t_1), \ldots, x_n(t_q)]^{\mathrm{T}}$$

Each solution $x_n(t)$ of (4.1.4) furnishes a solution to (4.1.5): merely evaluate $x_n(t)$ at the node points. The converse is also true. To each solution $\underline{z} \equiv [z_1, \ldots, z_q]^{\mathrm{T}}$ of (4.1.5), there is a unique solution of (4.1.4) that agrees with z at the node points. If one solves for $x_n(t)$ in (4.1.4), then $x_n(t)$ is determined by its values at the node points $\{t_j\}$. Therefore, when given a solution $\underline{z}$ to (4.1.5), define

$$z(t) = \frac{1}{\lambda}\left[y(t) + \sum_{j=1}^{q_n} w_j K(t, t_j) z_j\right], \quad t \in D \qquad (4.1.6)$$

This is an interpolation formula. In fact,

$$z(t_i) = \frac{1}{\lambda}\left[y(t_i) + \sum_{j=1}^{q_n} w_j K(t_i, t_j) z_j\right]$$
$$= z_i$$

for $i = 1, \ldots, q_n$. The last step follows from $\underline{z}$ being a solution to (4.1.5). Using this interpolation result in (4.1.6), we have that $z(t)$ solves (4.1.4). The uniqueness of the relationship between $\underline{z}$ and $z(t)$ follows from the solutions $x_n(t)$ of (4.1.4) being completely determined by their values at the nodes $\{t_i\}$. The formula (4.1.6) is called the *Nyström interpolation formula*. In the original paper of Nyström [409], the author uses a highly accurate Gaussian quadrature formula with a very small number of quadrature nodes (for example, $q = 3$). He then uses (4.1.6) to extend the solution to all other $t \in D$ while retaining the accuracy found in the solution at the node points. The formula (4.1.6) is usually a very good interpolation formula.

Example. Consider the integral equation

$$\lambda x(t) - \int_0^1 e^{st} x(s)\, ds = y(t), \quad 0 \le t \le 1 \tag{4.1.7}$$

with $\lambda = 2$ and $x(t) = e^t$. Since $\|\mathcal{K}\| = e - 1 \doteq 1.73$, the geometric series theorem (cf. Theorem A.1 in the Appendix) implies the integral equation is uniquely solvable for any given $y \in C[0, 1]$.

Consider first using the three-point Simpson rule to approximate (4.1.7), with nodes $\{0, 0.5, 1\}$. Then the errors are

$$\begin{bmatrix} x(0) \\ x(.5) \\ x(1) \end{bmatrix} - \begin{bmatrix} x_3(0) \\ x_3(.5) \\ x_3(1) \end{bmatrix} = \begin{bmatrix} -0.0047 \\ -0.0080 \\ -0.0164 \end{bmatrix} \tag{4.1.8}$$

which are reasonably small errors. For comparison, use Gauss-Legendre quadrature with three nodes,

$$\int_0^1 g(t)\, dt \approx \frac{1}{18}[5g(t_1) + 8g(t_2) + 5g(t_3)]$$

$$t_1 = \frac{1 - \sqrt{0.6}}{2} \doteq 0.11270167, \quad t_2 = 0.5,$$

$$t_3 = \frac{1 + \sqrt{0.6}}{2} \doteq 0.88729833$$

The symbol $\doteq$ usually means "approximately equals;" and in this case the approximation is due to rounding error. The error in solving (4.1.7) with the Nyström method is now

$$\begin{bmatrix} x(t_1) \\ x(t_2) \\ x(t_3) \end{bmatrix} - \begin{bmatrix} x_3(t_1) \\ x_3(t_2) \\ x_3(t_3) \end{bmatrix} = \begin{bmatrix} 2.10 \times 10^{-5} \\ 3.20 \times 10^{-5} \\ 6.32 \times 10^{-5} \end{bmatrix} \tag{4.1.9}$$

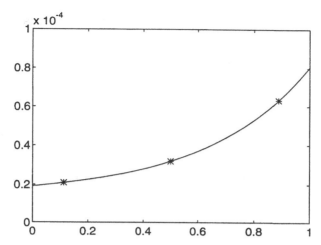

Figure 4.1. Error in Nyström interpolation for (4.1.7).

which is much smaller than with Simpson's rule when using an equal number of node points. Generally, Gaussian quadrature is much superior to Simpson's rule; but it results in the answers being given at the Gauss-Legendre nodes, which is usually not a convenient choice for subsequent uses of the answers.

Quadratic interpolation can be used to extend the numerical solution to all other $t \in [0, 1]$, but it generally results in much larger errors. For example,

$$x(1.0) - \mathcal{P}_2 x_3(1.0) = 0.0158$$

where $\mathcal{P}_2 x_3(t)$ denotes the quadratic polynomial interpolating the Nyström solution at the Gaussian quadrature node points given above. In contrast, the Nyström formula (4.1.6) gives errors that are consistent in size with those in (4.1.9). For example,

$$x(1.0) - x_3(1.0) = 8.08 \times 10^{-5}$$

A graph of the error in $x_3(t)$ over $[0, 1]$ is given in Figure 4.1. For a detailed discussion of the accuracy of Nyström interpolation, see Wright [574].

4.1.1. Properties and error analysis of the Nyström method

The Nyström method is implemented with the finite linear system (4.1.5), but the formal error analysis is done using the functional equation (4.1.4). We write the integral equation (4.0.1) in abstract form as $(\lambda - \mathcal{K})x = y$ as in earlier chapters, and we write the numerical integral (4.1.4) as $(\lambda - \mathcal{K}_n)x_n = y$.

The Banach space for our initial error analysis is $\mathcal{X} = C(D)$. The numerical integral operator

$$\mathcal{K}_n x(t) \equiv \sum_{j=1}^{q_n} w_j K(t, t_j) x(t_j), \quad t \in D, \; x \in C(D) \qquad (4.1.10)$$

is a bounded, finite rank linear operator on $C(D)$ to $C(D)$, with

$$\|\mathcal{K}_n\| = \max_{t \in D} \sum_{j=1}^{q_n} |w_j K(t, t_j)| \qquad (4.1.11)$$

The error analyses of Chapters 2 and 3 depended on showing $\|\mathcal{K} - \mathcal{K}_n\|$ converges to zero as n increases, with $\mathcal{K}_n$ the approximation to the integral operator $\mathcal{K}$. This cannot be done here; and in fact,

$$\|\mathcal{K} - \mathcal{K}_n\| \geq \|\mathcal{K}\| \qquad (4.1.12)$$

We leave the proof of this as a problem for the reader. Because of this result, the standard type of perturbation analysis used earlier will need to be modified. First, we look at quantities that do converge to zero as $n \to \infty$.

Lemma 4.1.1. Let D be a closed, bounded set in $\mathbf{R}^m$, some $m \geq 1$, and let $K(t, s)$ be continuous for $t, s \in D$. Let the quadrature scheme (4.1.2) be convergent for all continuous functions on D. Define

$$E_n(t, s) = \int_D K(t, v)K(v, s)\, dv - \sum_{j=1}^{q_n} w_j K(t, t_j) K(t_j, s),$$

$$t, s \in D, \; n \geq 1 \qquad (4.1.13)$$

It is the numerical integration error for the integrand $K(t, \cdot)K(\cdot, s)$. Then for $z \in C(D)$,

$$(\mathcal{K} - \mathcal{K}_n)\mathcal{K}z(t) = \int_D E_n(t, s)z(s)\, ds \qquad (4.1.14)$$

$$(\mathcal{K} - \mathcal{K}_n)\mathcal{K}_n z(t) = \sum_{j=1}^{q_n} w_j E_n(t, t_j)z(t_j) \qquad (4.1.15)$$

In addition,

$$\|(\mathcal{K} - \mathcal{K}_n)\mathcal{K}\| = \max_{t \in D} \int_D |E_n(t, s)|\, ds \qquad (4.1.16)$$

$$\|(\mathcal{K} - \mathcal{K}_n)\mathcal{K}_n\| = \max_{t \in D} \sum_{j=1}^{q_n} |w_j E_n(t, t_j)| \qquad (4.1.17)$$

Finally, the numerical integration error E_n converges to zero uniformly on D,

$$c_E \equiv \lim_{n \to \infty} \max_{t,s \in D} |E_n(t, s)| = 0 \qquad (4.1.18)$$

and thus

$$\|(\mathcal{K} - \mathcal{K}_n)\mathcal{K}\|, \quad \|(\mathcal{K} - \mathcal{K}_n)\mathcal{K}_n\| \to 0 \quad \text{as } n \to \infty \qquad (4.1.19)$$

Proof. The proofs of (4.1.14) and (4.1.15) are straightforward manipulations, and we omit them. The quantity $(\mathcal{K} - \mathcal{K}_n)\mathcal{K}$ is an integral operator on $C(D)$, by (4.1.14); therefore, we have (4.1.16) for its bound. The proof of (4.1.17) is also straightforward, and we omit it.

To prove (4.1.18), we begin by showing that $\{E_n(t, s) \mid n \geq 1\}$ is a uniformly bounded and equicontinuous family that is pointwise convergent to 0 on the closed bounded set D; and then $E_n(t, s) \to 0$ uniformly on D by the Arzela-Ascoli theorem. By the assumption that the quadrature rule of (4.1.2) converges for all continuous functions g on D, we have that for each $t, s \in D$, $E_n(t, s) \to 0$ as $n \to \infty$.

To prove boundedness,

$$|E_n(t, s)| \leq (c_D + c_I)c_K^2$$

with

$$c_D = \int_D ds, \qquad c_K = \max_{t,s \in D} |K(t, s)|$$

and c_I the constant from (4.1.3). For equicontinuity,

$$|E_n(t, s) - E_n(\tau, \sigma)| \leq |E_n(t, s) - E_n(\tau, s)| + |E_n(\tau, s) - E_n(\tau, \sigma)|$$

$$|E_n(t, s) - E_n(\tau, s)| \leq c_K(c_D + c_I) \max_{s \in D} |K(t, s) - K(\tau, s)|$$

$$|E_n(\tau, s) - E_n(\tau, \sigma)| \leq c_K(c_D + c_I) \max_{t \in D} |K(t, s) - K(t, \sigma)|$$

By the uniform continuity of $K(t, s)$ on the closed bounded set D, this shows the equicontinuity of $\{E_n(t, s)\}$. This also completes the proof of (4.1.18).

For (4.1.19) and (4.1.20),

$$\|(\mathcal{K} - \mathcal{K}_n)\mathcal{K}\| \leq c_D \max_{t,s \in D} |E_n(t, s)| \qquad (4.1.20)$$

$$\|(\mathcal{K} - \mathcal{K}_n)\mathcal{K}_n\| \leq c_I \max_{t,s \in D} |E_n(t, s)| \qquad (4.1.21)$$

This completes the proof.   ☐

To carry out an error analysis for the Nyström method (4.1.4)–(4.1.6), we
need the following perturbation theorem. It furnishes an alternative to the
arguments of earlier chapters that were based on a straightforward use of the
geometric series theorem.

Theorem 4.1.1. *Let $\mathcal{X}$ be a Banach space, let $\mathcal{S}, \mathcal{T}$ be bounded operators on
$\mathcal{X}$ to $\mathcal{X}$, and let $\mathcal{S}$ be compact. For given $\lambda \neq 0$, assume $\lambda I - \mathcal{T} : \mathcal{X} \overset{1-1}{\to} \mathcal{X}$,
which implies $(\lambda I - \mathcal{T})^{-1}$ exists as a bounded operator on $\mathcal{X}$ to $\mathcal{X}$. Finally,
assume*

$$\|(\mathcal{T} - \mathcal{S})\mathcal{S}\| < \frac{|\lambda|}{\|(\lambda I - \mathcal{T})^{-1}\|} \tag{4.1.22}$$

Then $(\lambda I - \mathcal{S})^{-1}$ exists and is bounded on $\mathcal{X}$ to $\mathcal{X}$, with

$$\|(\lambda I - \mathcal{S})^{-1}\| \leq \frac{1 + \|(\lambda I - \mathcal{T})^{-1}\|\,\|\mathcal{S}\|}{|\lambda| - \|(\lambda I - \mathcal{T})^{-1}\|\,\|(\mathcal{T} - \mathcal{S})\mathcal{S}\|} \tag{4.1.23}$$

If $(\lambda - \mathcal{T})x = y$ and $(\lambda - \mathcal{S})z = y$, then

$$\|x - z\| \leq \|(\lambda I - \mathcal{S})^{-1}\|\,\|\mathcal{T}x - \mathcal{S}x\| \tag{4.1.24}$$

Proof. Consider that if $(\lambda I - \mathcal{S})^{-1}$ were to exist, then it would satisfy the
identity

$$(\lambda I - \mathcal{S})^{-1} = \frac{1}{\lambda}\{I + (\lambda I - \mathcal{S})^{-1}\mathcal{S}\} \tag{4.1.25}$$

Without any motivation at this point, consider the approximation

$$(\lambda I - \mathcal{S})^{-1} \approx \frac{1}{\lambda}\{I + (\lambda I - \mathcal{T})^{-1}\mathcal{S}\} \tag{4.1.26}$$

To check this approximation, compute

$$\frac{1}{\lambda}\{I + (\lambda I - \mathcal{T})^{-1}\mathcal{S}\}(\lambda I - \mathcal{S}) = \left\{I + \frac{1}{\lambda}(\lambda I - \mathcal{T})^{-1}(\mathcal{T} - \mathcal{S})\mathcal{S}\right\} \tag{4.1.27}$$

The right side is invertible by the geometric series theorem, because (4.1.22)
implies

$$\frac{1}{|\lambda|}\|(\lambda I - \mathcal{T})^{-1}\|\,\|(\mathcal{T} - \mathcal{S})\mathcal{S}\| < 1$$

In addition, the geometric series theorem implies, after simplification, that

$$\|[\lambda I + (\lambda I - \mathcal{T})^{-1}(\mathcal{T} - \mathcal{S})\mathcal{S}]^{-1}\| \leq \frac{1}{|\lambda| - \|(\lambda I - \mathcal{T})^{-1}\|\|(\mathcal{T} - \mathcal{S})\mathcal{S}\|}$$

(4.1.28)

Since the right side of (4.1.27) is invertible, the left side is also invertible. This implies that $\lambda I - \mathcal{S}$ is one-to-one, as otherwise the left side would not be invertible. Since $\mathcal{S}$ is compact, the Fredholm alternative theorem (cf. Theorem 1.3.1 of §1.3, Chapter 1) implies $(\lambda I - \mathcal{S})^{-1}$ exists and is bounded on $\mathcal{X}$ to $\mathcal{X}$. In particular,

$$(\lambda I - \mathcal{S})^{-1} = [\lambda I + (\lambda I - \mathcal{T})^{-1}(\mathcal{T} - \mathcal{S})\mathcal{S}]^{-1}\{I + (\lambda I - \mathcal{T})^{-1}\mathcal{S}\}$$

(4.1.29)

The bound (4.1.23) follows directly from this and (4.1.28).

For the error $x - z$, rewrite $(\lambda - \mathcal{T})x = y$ as

$$(\lambda - \mathcal{S})x = y + (\mathcal{T} - \mathcal{S})x$$

Subtract $(\lambda - \mathcal{S})z = y$ to get

$$(\lambda - \mathcal{S})(x - z) = (\mathcal{T} - \mathcal{S})x \qquad (4.1.30)$$

$$x - z = (\lambda - \mathcal{S})^{-1}(\mathcal{T} - \mathcal{S})x \qquad (4.1.31)$$

$$\|x - z\| \leq \|(\lambda - \mathcal{S})^{-1}\|\|(\mathcal{T} - \mathcal{S})x\|$$

which proves (4.1.24).   □

Using this theorem, we can give a complete convergence analysis for the Nyström method (4.1.4)–(4.1.6).

Theorem 4.1.2. *Let D be a closed, bounded set in $\mathbf{R}^m$, some $m \geq 1$, and let $K(t, s)$ be continuous for $t, s \in D$. Assume the quadrature scheme (4.1.2) is convergent for all continuous functions on D. Further, assume that the integral equation (4.0.1) is uniquely solvable for given $y \in C(D)$, with $\lambda \neq 0$. Then for all sufficiently large n, say $n \geq N$, the approximate inverses $(\lambda - \mathcal{K}_n)^{-1}$ exist and are uniformly bounded,*

$$\|(\lambda - \mathcal{K}_n)^{-1}\| \leq \frac{1 + \|(\lambda - \mathcal{K})^{-1}\|\|\mathcal{K}_n\|}{|\lambda| - \|(\lambda - \mathcal{K})^{-1}\|\|(\mathcal{K} - \mathcal{K}_n)\mathcal{K}_n\|} \leq c_s, \quad n \geq N$$

(4.1.32)

with a suitable constant $c_s < \infty$. For the equations $(\lambda - \mathcal{K})x = y$ and $(\lambda - \mathcal{K}_n)x_n = y$,

$$\|x - x_n\|_\infty \leq \|(\lambda - \mathcal{K}_n)^{-1}\| \|(\mathcal{K} - \mathcal{K}_n)x\|_\infty$$

$$\leq c_s \|(\mathcal{K} - \mathcal{K}_n)x\|_\infty, \quad n \geq N \qquad (4.1.33)$$

Proof. This is a simple application of the preceding theorem, with $\mathcal{S} = \mathcal{K}_n$ and $\mathcal{T} = \mathcal{K}$. From Lemma 4.1.1, we have $\|(\mathcal{K} - \mathcal{K}_n)\mathcal{K}_n\| \to 0$, and therefore (4.1.22) is satisfied for all sufficiently large n, say $n \geq N$. From (4.1.11), the boundedness of $K(t, s)$ over D, and (4.1.3),

$$\|\mathcal{K}_n\| \leq c_I c_K, \quad n \geq 1$$

Then

$$c_s \equiv \sup_{n \geq N} \frac{1 + \|(\lambda - \mathcal{K})^{-1}\| \|\mathcal{K}_n\|}{|\lambda| - \|(\lambda - \mathcal{K})^{-1}\| \|(\mathcal{K} - \mathcal{K}_n)\mathcal{K}_n\|} < \infty \qquad (4.1.34)$$

This completes the proof.   □

This last theorem gives complete information for analyzing the convergence of the Nyström method (4.1.4)–(4.1.6). The term $\|(\mathcal{K} - \mathcal{K}_n)\mathcal{K}_n\|$ can be analyzed from (4.1.17) by analyzing the numerical integration error $E_n(t, s)$ of (4.1.13). From the error bound (4.1.33), the speed with which $\|x - x_n\|_\infty$ converges to zero is bounded by that of the numerical integration error

$$\|(\mathcal{K} - \mathcal{K}_n)x\|_\infty = \max_{t \in D} \left| \int_D K(t, s)x(s)\, ds - \sum_{j=1}^{q_n} w_j K(t, t_j)x(t_j) \right|$$

$$(4.1.35)$$

In fact, the error $\|x - x_n\|_\infty$ converges to zero with exactly this speed. Recall from applying (4.1.30) that

$$(\lambda - \mathcal{K}_n)(x - x_n) = (\mathcal{K} - \mathcal{K}_n)x \qquad (4.1.36)$$

From bounding this,

$$\|(\mathcal{K} - \mathcal{K}_n)x\|_\infty \leq \|\lambda - \mathcal{K}_n\| \|x - x_n\|_\infty$$

When combined with (4.1.33), this shows the assertion that $\|x - x_n\|_\infty$ and $\|(\mathcal{K} - \mathcal{K}_n)x\|_\infty$ converge to zero with the same speed.

There is a very large literature on bounding and estimating the errors for the common numerical integration rules. Thus the speed of convergence with which $\|x - x_n\|_\infty$ converges to zero can be determined by using results on the speed of convergence of the integration rule (4.1.2) when it is applied to the integral

$$\int_D K(t, s)x(s)\, ds$$

Example. Consider the trapezoidal numerical integration rule

$$\int_a^b g(s)\, ds \approx h \sum_{j=0}^n {}'' g(t_j) \tag{4.1.37}$$

with $h = (b - a)/n$ and $t_j = a + jh$ for $j = 0, \ldots, n$. The notation Σ'' means the first and last terms are to be halved before summing. For the error,

$$\int_a^b g(s)\, ds - h \sum_{j=0}^n {}'' g(t_j) = -\frac{h^2(b-a)}{12} g''(\xi_n), \quad g \in C^2[a, b], \quad n \geq 1 \tag{4.1.38}$$

with ξ_n some point in $[a, b]$. There is also the asymptotic error formula

$$\int_a^b g(s)\, ds - h \sum_{j=0}^n {}'' g(t_j) = -\frac{h^2}{12}[g'(b) - g'(a)] + O(h^4), \quad g \in C^4[a, b] \tag{4.1.39}$$

and we make use of it in a later example.

When this is applied to the integral equation

$$\lambda x(t) - \int_a^b K(t, s)x(s)\, ds = y(t), \quad a \leq t \leq b \tag{4.1.40}$$

we obtain the approximating linear system

$$\lambda x_n(t_i) - h \sum_{j=0}^n {}'' K(t_i, t_j)x_n(t_j) = y(t_i), \quad i = 0, 1, \ldots, n \tag{4.1.41}$$

which is of order $q_n = n + 1$. The Nyström interpolation formula is given by

$$x_n(t) = \frac{1}{\lambda}\left[y(t) + h \sum_{j=0}^n {}'' K(t, t_j)x_n(t_j) \right], \quad a \leq t \leq b \tag{4.1.42}$$

The speed of convergence is based on the numerical integration error

$$(\mathcal{K} - \mathcal{K}_n)x(s) = -\frac{h^2(b-a)}{12}\left[\frac{\partial^2 K(t, s)x(s)}{\partial s^2} \right]_{s=\xi_n(t)} \tag{4.1.43}$$

with $\xi_n(t) \in [a, b]$. From (4.1.39), the asymptotic integration error is

$$(\mathcal{K} - \mathcal{K}_n)x(s) = -\frac{h^2}{12}\left[\frac{\partial K(t, s)x(s)}{\partial s}\right]_{s=a}^{s=b} + O(h^4) \qquad (4.1.44)$$

From (4.1.43), we see the Nyström method converges with an order of $O(h^2)$, provided $K(t, s)x(s)$ is twice continuously differentiable with respect to s, uniformly in t.

Example. Consider the use of the Gauss-Legendre quadrature formula

$$\int_a^b g(s)\,ds \approx \sum_{j=1}^n w_j g(t_j) \qquad (4.1.45)$$

with respect to the interval $[a, b]$. For an introduction to such quadrature, see [48, p. 276]. The application of Gaussian quadrature to solving (4.1.40) leads to a very rapidly convergent Nyström method, which we illustrate in a following example. The associated Nyström linear system (4.1.5) has order $q_n = n$.

Unlike (4.1.38) for the trapezoidal rule (and other composite quadrature rules), there is not a convenient and intuitive error formula for Gaussian quadrature. We can show, however, that

$$\left|\int_a^b g(s)\,ds - \sum_{j=1}^n w_j g(t_j)\right| \leq e_n M_{2n}$$

with

$$M_\ell = \frac{1}{\ell!}\left(\frac{b - a}{2}\right)^{\ell+1} \left\|g^{(\ell)}\right\|_\infty$$

and

$$e_n \approx \frac{\pi}{4^n} \quad \text{as } n \to \infty$$

in an asymptotic sense. For many infinitely differentiable integrands, M_ℓ is bounded or tends to zero as $\ell \to \infty$. This gives some intuition for why Gauss-Legendre quadrature is so very accurate, as is illustrated in the examples given below. For a further discussion, see [48, Section 5.3] and the references given therein.

Example. Consider again the integral equation

$$\lambda x(t) - \int_0^1 e^{st} x(s)\,ds = y(t), \quad 0 \leq t \leq 1 \qquad (4.1.46)$$

Table 4.1. *Nyström-Trapezoidal method*
for (4.1.46)

n	E_1	Ratio	E_2	Ratio
2	5.35E−3		5.44E−3	
4	1.35E−3	3.9	1.37E−3	4.0
8	3.39E−4	4.0	3.44E−4	4.0
16	8.47E−5	4.0	8.61E−5	4.0

Table 4.2. *Nyström-Gaussian method for (4.1.46)*

n	E_1	Ratio	E_2	Ratio
1	4.19E−3		9.81E−3	
2	1.22E−4	34	2.18E−4	45
3	1.20E−6	100	1.86E−6	117
4	5.90E−9	200	8.47E−9	220
5	1.74E−11	340	2.39E−11	354

with $x(t) = e^t$. But now let $\lambda = 50$, to better compare it with the earlier example in Table 2.1 of Chapter 2. In Table 4.1 we give numerical results when using the trapezoidal rule, and in Table 4.2 we give results when using Gaussian quadrature. In the latter two tables we use the notation

$$E_1 = \max_i |x(t_i) - x_n(t_i)|$$

$$E_2 = \|x - x_n\|_\infty$$

For E_2, $x_n(s)$ is obtained using Nyström interpolation. The results for the trapezoidal rule show clearly the $O(h^2)$ behavior of the error. Also, as asserted earlier, the use of Gaussian quadrature leads to very rapid convergence of x_n to x.

An asymptotic error estimate

In those cases for which the quadrature formula has an asymptotic error formula, as in (4.1.39), we can give an asymptotic estimate of the error in solving the integral equation using the Nyström method. Returning to (4.1.36), we can write

$$x - x_n = (\lambda - \mathcal{K}_n)^{-1}(\mathcal{K} - \mathcal{K}_n)x$$

$$= e_n + R_n \tag{4.1.47}$$

with

$$e_n = (\lambda - \mathcal{K})^{-1}(\mathcal{K} - \mathcal{K}_n)x$$

$$R_n = [(\lambda - \mathcal{K}_n)^{-1} - (\lambda - \mathcal{K})^{-1}](\mathcal{K} - \mathcal{K}_n)x$$

$$= (\lambda - \mathcal{K}_n)^{-1}(\mathcal{K}_n - \mathcal{K})(\lambda - \mathcal{K})^{-1}(\mathcal{K} - \mathcal{K}_n)x \qquad (4.1.48)$$

The term R_n will generally converge to zero more rapidly than will the term e_n, although showing this is dependent on the quadrature rule being used. Assuming the latter to be true, we have

$$x - x_n \approx e_n \qquad (4.1.49)$$

with e_n satisfying the original integral equation with the integration error $(\mathcal{K} - \mathcal{K}_n)x$ as the right hand side,

$$(\lambda - \mathcal{K})e_n = (\mathcal{K} - \mathcal{K}_n)x \qquad (4.1.50)$$

At this point, one needs to consider the quadrature rule in more detail.

Example. Consider the earlier example (4.1.37)–(4.1.44) of the Nyström method with the trapezoidal rule. Assume further that $K(t, s)$ is four times continuously differentiable with respect to both s and t, and assume $x \in C^4[a, b]$. Then from the asymptotic error formula (4.1.44), we can decompose the right side $(\mathcal{K} - \mathcal{K}_n)x$ of (4.1.50) into two terms, of sizes $O(h^2)$ and $O(h^4)$. Introduce the function $\gamma(s)$ satisfying the integral equation

$$\lambda\gamma(t) - \int_a^b K(t, s)\gamma(s)\,ds = -\frac{1}{12}\left[\frac{\partial K(t, s)x(s)}{\partial s}\right]_{s=a}^{s=b}, \qquad a \le t \le b$$

Then the error term e_n in (4.1.49)–(4.1.50) is dominated by $\gamma(t)h^2$. By a similar argument, it can also be shown that the term $R_n = O(h^4)$. Thus, we have the asymptotic error estimate

$$x - x_n \approx \gamma(t)h^2 \qquad (4.1.51)$$

for the Nyström method with the trapezoidal rule.

Conditioning of the linear system

Let A_n denote the matrix of coefficients for the linear system (4.1.5):

$$(A_n)_{i,j} = \lambda\delta_{i,j} - w_j K(t_i, t_j)$$

We want to bound $\text{cond}(A_n) = \|A_n\|\|A_n^{-1}\|$.

For general $z \in C(D)$,

$$\max_{i=1,\dots,q_n} \left| \lambda z(t_i) - \sum_{j=1}^{q_n} w_j K(t_i, t_j) z(t_j) \right| \le \sup_{t \in D} \left| \lambda z(t) - \sum_{j=1}^{q_n} w_j K(t, t_j) z(t_j) \right|$$

This shows

$$\|A_n\| \le \|\lambda - \mathcal{K}_n\| \tag{4.1.52}$$

For A_n^{-1},

$$\left\| A_n^{-1} \right\| = \sup_{\substack{\gamma \in \mathbf{R}^{q_n} \\ \|\gamma\|_\infty = 1}} \left\| A_n^{-1} \gamma \right\|_\infty$$

For such γ, let $z = A_n^{-1}\gamma$ or $\gamma = A_n z$. Pick $y \in C(D)$ such that

$$y(t_i) = \gamma_i, \quad i = 1, \dots, q_n$$

and $\|y\|_\infty = \|\gamma\|_\infty$. Let $x_n = (\lambda - \mathcal{K}_n)^{-1} y$, or equivalently, $(\lambda - \mathcal{K}_n)x_n = y$. Then from the earlier discussion of the Nyström method,

$$x_n(t_i) = z_i, \quad i = 1, \dots, q_n$$

Then

$$\left\| A_n^{-1} \gamma \right\|_\infty = \|z\|_\infty \le \|x_n\|_\infty \le \|(\lambda - \mathcal{K}_n)^{-1}\| \|y\|_\infty = \|(\lambda - \mathcal{K}_n)^{-1}\| \|\gamma\|_\infty$$

This proves

$$\left\| A_n^{-1} \right\|_\infty \le \|(\lambda - \mathcal{K}_n)^{-1}\| \tag{4.1.53}$$

Combining these results,

$$\mathrm{cond}(A_n) \le \|\lambda - \mathcal{K}_n\| \|(\lambda - \mathcal{K}_n)^{-1}\| \equiv \mathrm{cond}(\lambda - \mathcal{K}_n) \tag{4.1.54}$$

Thus if the operator equation $(\lambda - \mathcal{K}_n)x_n = y$ is well-conditioned, then so is the linear system associated with it. To examine the conditioning of $(\lambda - \mathcal{K}_n)x_n = y$, refer back to the introductory discussion of the conditioning of the systems associated with projection methods, given in §3.6. Also note that (4.1.32) gives bounds on $\|(\lambda - \mathcal{K}_n)^{-1}\|$ in terms of $\|(\lambda - \mathcal{K})^{-1}\|$.

4.1.2. Collectively compact operator approximations

The error analysis of the Nyström method was developed mainly during the period 1940 to 1970, and a number of researchers were involved. Initially, the only goal was to show that the method was stable and convergent, and perhaps, to obtain computable error bounds. As this was accomplished, a second goal emerged of creating an abstract framework for the method and its error analysis, a framework in the language of functional analysis that referred only to mapping properties of the approximate operators and not to properties of the particular integral operator, function space, or quadrature scheme being used. The final framework developed is due primarily to Phil Anselone, who gave it the name "theory of collectively compact operator approximations." A complete presentation of it is given in his book [16], and we present only a portion of it here.

Within a functional analysis framework, how does one characterize the numerical integral operators $\{\mathcal{K}_n \mid n \geq 1\}$? We want to know the characteristic properties of these operators, which imply that $\|(\mathcal{K} - \mathcal{K}_n)\mathcal{K}_n\| \to 0$ as $n \to \infty$. Then the earlier Theorem 4.1.1 will remain valid, and the Nyström method and its error analysis can be extended to other situations, some of which are discussed in later sections.

We assume that $\{\mathcal{K}_n \mid n \geq 1\}$ satisfies the following properties.

A1. $\mathcal{X}$ is a Banach space, and $\mathcal{K}$ and $\mathcal{K}_n$, $n \geq 1$ are linear operators on $\mathcal{X}$ into $\mathcal{X}$.
A2. $\mathcal{K}_n x \to \mathcal{K}x$ as $n \to \infty$, for all $x \in \mathcal{X}$.
A3. The set $\{\mathcal{K}_n \mid n \geq 1\}$ is *collectively compact*, which means that the set

$$\mathcal{S} = \{\mathcal{K}_n x \mid n \geq 1 \text{ and } \|x\| \leq 1\} \qquad (4.1.55)$$

has compact closure in $\mathcal{X}$.

These assumptions are an excellent abstract characterization of the numerical integral operators introduced earlier in (4.1.10) of this chapter. We refer to a family $\{\mathcal{K}_n\}$ that satisfies **A1–A3** as a *collectively compact family of pointwise convergent operators*.

Lemma 4.1.2. Assume the above properties **A1–A3**. Then

(1) $\mathcal{K}$ is compact,
(2) $\{\mathcal{K}_n \mid n \geq 1\}$ is uniformly bounded,
(3) for any compact operator $\mathcal{M} : \mathcal{X} \to \mathcal{X}$,

$$\|(\mathcal{K} - \mathcal{K}_n)\mathcal{M}\| \to 0 \quad \text{as } n \to \infty$$

(4) $\|(\mathcal{K} - \mathcal{K}_n)\mathcal{K}_n\| \to 0$ as $n \to \infty$.

Proof.

(1) To show $\mathcal{K}$ is compact, it is sufficient to show that the set

$$\{\mathcal{K}x \mid \|x\| \leq 1\}$$

has compact closure in $\mathcal{X}$. By **A2**, this last set is contained in $\bar{\mathcal{S}}$, and it is compact by **A3**.

(2) This follows from the definition of operator norm and the boundedness of the set $\bar{\mathcal{S}}$.

(3) Using the definition of operator norm,

$$\|(\mathcal{K} - \mathcal{K}_n)\mathcal{M}\| = \sup_{\|x\| \leq 1} \|(\mathcal{K} - \mathcal{K}_n)\mathcal{M}x\| = \sup_{z \in \mathcal{M}(B)} \|(\mathcal{K} - \mathcal{K}_n)z\|$$

$$(4.1.56)$$

with $B = \{x \mid \|x\| \leq 1\}$. From the compactness of $\mathcal{M}$, the set $\mathcal{M}(B)$ has compact closure. Using Lemma 3.1.1, we find that the last quantity in (4.1.56) goes to zero as $n \rightarrow \infty$.

(4) Again, using the definition of operator norm,

$$\|(\mathcal{K} - \mathcal{K}_n)\mathcal{K}_n\| = \sup_{\|x\| \leq 1} \|(\mathcal{K} - \mathcal{K}_n)\mathcal{K}_n\| = \sup_{z \in S} \|(\mathcal{K} - \mathcal{K}_n)z\| \quad (4.1.57)$$

Using **A3**, S has compact closure, and then using Lemma 3.1.1, we find the last quantity in (4.1.57) goes to zero as $n \rightarrow \infty$. □

As a consequence of this lemma, we can apply Theorem 4.1.3 to any set of approximating equations $(\lambda - \mathcal{K}_n)x_n = y$ where the set $\{\mathcal{K}_n\}$ satisfies **A1–A3**. This extends the idea of the Nyström method, and examples of this are given in the remaining sections of this chapter.

Returning to the proof of Theorem 4.1.1, we can better motivate an argument used there. With $S = \mathcal{K}_n$ and $T = \mathcal{K}$, the statements (4.1.25) and (4.1.26) become

$$(\lambda - \mathcal{K}_n)^{-1} = \frac{1}{\lambda}[I + (\lambda - \mathcal{K}_n)^{-1}\mathcal{K}_n] \quad (4.1.58)$$

$$(\lambda - \mathcal{K}_n)^{-1} \approx \frac{1}{\lambda}[I + (\lambda - \mathcal{K})^{-1}\mathcal{K}_n] \quad (4.1.59)$$

Since $\mathcal{K}_n$ is not norm convergent to $\mathcal{K}$, we cannot expect $(\lambda - \mathcal{K})^{-1} \approx (\lambda - \mathcal{K}_n)^{-1}$ to be a good approximation. However, it becomes a much better approximation when the operators are restricted to act on a compact subset of $\mathcal{X}$. Since the family $\{\mathcal{K}_n\}$ is collectively compact, (4.1.59) is a good approximation of (4.1.58).

4.2. Product integration methods

We now consider the numerical solution of integral equations of the second kind in which the kernel function $K(t, s)$ is not continuous, but for which the associated integral operator $\mathcal{K}$ is still compact on $C(D)$ into $C(D)$. The main ideas we present will extend to functions in any finite number of variables, but it is more intuitive to first present these ideas for integral equations for functions of a single variable,

$$\lambda x(t) - \int_a^b K(t, s)x(s)\, ds = y(t), \quad a \leq t \leq b \tag{4.2.60}$$

In this setting most such discontinuous kernel functions $K(t, s)$ have an infinite singularity, and the most important examples are $\log|t - s|$, $|t - s|^{\gamma - 1}$ for some $\gamma > 0$ and variants of them.

We introduce the idea of product integration by considering the special case of

$$\lambda x(t) - \int_a^b L(t, s) \log|s - t| x(s)\, ds = y(t), \quad a \leq t \leq b \tag{4.2.61}$$

with the kernel

$$K(t, s) = L(t, s) \log|s - t| \tag{4.2.62}$$

We assume that $L(t, s)$ is a well-behaved function (that is, it is several times continuously differentiable), and initially we assume the unknown solution $x(t)$ is also well-behaved. To solve (4.2.61), we define a method called the *product trapezoidal rule*.

Let $n \geq 1$ be an integer, $h = (b - a)/n$, and $t_j = a + jh$, $j = 0, 1, \ldots, n$. For general $x \in C[a, b]$, define

$$[L(t, s)x(s)]_n = \frac{1}{h}[(t_j - s)L(t, t_{j-1})x(t_{j-1}) + (s - t_{j-1})L(t, t_j)x(t_j)], \tag{4.2.63}$$

for $t_{j-1} \leq s \leq t_j$, $j = 1, \ldots, n$ and $a \leq t \leq b$. This is piecewise linear in s, and it interpolates $L(t, s)x(s)$ at $s = t_0, \ldots, t_n$, for all $t \in [a, b]$. Define a numerical approximation to the integral operator in (4.2.61) by

$$\mathcal{K}_n x(t) = \int_a^b [L(t, s)x(s)]_n \log|s - t|\, ds, \quad a \leq t \leq b \tag{4.2.64}$$

This can also be written as

$$\mathcal{K}_n x(t) = \sum_{j=0}^n w_j(t)L(t, t_j)x(t_j), \quad x \in C[a, b] \tag{4.2.65}$$

with weights

$$w_0(t) = \frac{1}{h} \int_{t_0}^{t_1} (t_1 - s) \log |t - s| \, ds,$$

$$(4.2.66)$$

$$w_n(t) = \frac{1}{h} \int_{t_{n-1}}^{t_n} (s - t_{n-1}) \log |t - s| \, ds$$

$$w_j(t) = \frac{1}{h} \int_{t_{j-1}}^{t_j} (s - t_{j-1}) \log |t - s| \, ds + \frac{1}{h} \int_{t_j}^{t_{j+1}} (t_{j+1} - s) \log |t - s| \, ds$$

$$(4.2.67)$$

To approximate the integral equation (4.2.61), we use

$$\lambda x_n(t) - \sum_{j=0}^{n} w_j(t) L(t, t_j) x_n(t_j) = y(t), \quad a \le t \le b \qquad (4.2.68)$$

As with the Nyström method (4.1.4)–(4.1.6), this is equivalent to first solving the linear system

$$\lambda x_n(t_i) - \sum_{j=0}^{n} w_j(t_i) L(t_i, t_j) x_n(t_j) = y(t_i), \quad i = 0, \ldots, n \qquad (4.2.69)$$

and then using the Nyström interpolation formula

$$x_n(t) = \frac{1}{\lambda} \left[y(t) + \sum_{j=0}^{n} w_j(t) L(t, t_j) x_n(t_j) \right], \quad a \le t \le b \qquad (4.2.70)$$

We leave it to the reader to check these assertions, since it is quite similar to what was done for the original Nyström method. With this method, we approximate those parts of the integrand in (4.2.61) that can be well-approximated by piecewise linear interpolation, and we integrate exactly the remaining more singular parts of the integrand.

Example. Solve

$$\lambda x(t) - \int_0^1 \log |s - t| x(s) \, ds = y(t), \quad 0 \le t \le 1 \qquad (4.2.71)$$

where the function $y(t)$ has been so chosen that $x(t) = e^t$. Numerical results are given in Table 4.3 for $\lambda = 1$ and -1. We discuss the rate of convergence of x_n later in the section. But it is clear from the table that the rate of convergence in this case is $O(h^2)$, which is reasonable since that is the rate of convergence for piecewise linear interpolation.

Table 4.3. *The product trapezoidal rule*
for (4.2.71)

n	λ = 1		λ = −1	
	$\|x - x_n\|_\infty$	*Ratio*	$\|x - x_n\|_\infty$	*Ratio*
10	1.16E−3		5.56E−3	
20	2.78E−4	4.17	1.42E−3	3.92
40	6.80E−5	4.09	3.57E−4	3.98

Rather than using piecewise linear interpolation, other more accurate interpolation schemes could have been used to obtain a more rapidly convergent numerical method. Later in the section we consider the use of piecewise quadratic interpolation. We have also used evenly spaced node points $\{t_i\}$, but this is not necessary. The use of such evenly spaced nodes is an important case, but we will see later in the section that special choices of nonuniformly spaced node points are often needed for solving an integral equation such as (4.2.61).

Other singular kernel functions can be handled in a manner analogous to what has been done for (4.2.61). Consider the equation

$$\lambda x(t) - \int_a^b L(t, s) H(t, s) x(s) \, ds = y(t), \quad a \le t \le b \qquad (4.2.72)$$

in which $H(t, s)$ is singular, with $L(t, s)$ and $x(t)$ as before. Another important case is to take

$$H(t, s) = \frac{1}{|t - s|^{1-\gamma}}$$

for some $\gamma > 0$. To approximate (4.2.72), use the earlier approximation (4.2.63). Then

$$\mathcal{K}_n x(t) = \int_a^b [L(t, s) x(s)]_n H(t, s) \, ds, \quad a \le t \le b \qquad (4.2.73)$$

All arguments proceed exactly as before. To evaluate $\mathcal{K}_n x(t)$, we need to evaluate the analogs of the weights in (4.2.66)–(4.2.67), where $\log |t - s|$ is replaced by $H(t, s)$. We assume these weights can be calculated in some practical manner, probably analytically. We consider further generalizations later in the section.

4.2.1. Computation of the quadrature weights

The weights $w_j(t_i)$ can often be calculated analytically, and when evenly spaced node points are used, simplifications are possible that lead to faster evaluation of the weights. Here we consider the case of $H(t, s) = \log |t - s|$.

Introduce

$$\alpha_j(t_i) = \frac{1}{h} \int_{t_{j-1}}^{t_j} (t_j - s) \log |t_i - s| \, ds,$$

$$\beta_j(t_i) = \frac{1}{h} \int_{t_{j-1}}^{t_j} (s - t_{j-1}) \log |t_i - s| \, ds$$

for $j = 1, \ldots, n$ and $i = 0, \ldots, n$. Then the weights (4.2.66)–(4.2.67) at the node points $\{t_i\}$ are given by

$$w_0(t_i) = \alpha_1(t_i), \qquad w_n(t_i) = \beta_n(t_i)$$

$$w_j(t_i) = \beta_j(t_i) + \alpha_{j+1}(t_i), \qquad j = 1, \ldots, n - 1$$

Introduce

$$\psi_0(k) = \int_0^1 \log |k - \mu| \, d\mu, \qquad \psi_1(k) = \int_0^1 \mu \log |k - \mu| \, d\mu$$

$$(4.2.74)$$

for $k = 0, \pm 1, \ldots$. Using these integrals,

$$\alpha_j(t_i) = \frac{h}{2} \log h + h[\psi_0(i - j + 1) - \psi_1(i - j + 1)]$$

$$\beta_j(t_i) = \frac{h}{2} \log h + h\psi_1(i - j + 1)$$

These must be calculated for $i = 0, 1, \ldots, n$ and $j = 1, \ldots, n$, and this implies

$$-n + 1 \leq i - j + 1 \leq n$$

If we calculate a table of values of $\psi_0(k)$ and $\psi_1(k)$ for $-(N - 1) \leq k \leq N$, then we can calculate the weights $\{w_j(t_i)\}$ for all $n = 1, 2, \ldots, N$. The cost of calculating such a table is relatively cheap. The number of arithmetic operations is of the order $O(n)$, and this is much less than might be expected based on there being $O(n^2)$ weights $w_j(t_i)$ needed in setting up the linear system (4.2.69). The only potential difficulty with this approach is that the standard formulas for (4.2.74) often lead to loss of significance errors in their calculation when k becomes large, and care must be taken to deal with this.

It should be clear that a similar discussion can be given for any case in which $H(t, s)$ depends on only $|t - s|$, for example, $H(t, s) = |t - s|^{\gamma - 1}$, $\gamma > 0$.

4.2.2. Error analysis

We consider the equation (4.2.72), with $L(t, s)$ assumed to be continuous. Further, we assume the following for $H(t, s)$:

$$c_H \equiv \sup_{a \leq t \leq b} \int_a^b |H(t, s)| \, ds < \infty \qquad (4.2.75)$$

$$\lim_{h \to 0^+} \omega_H(h) = 0 \qquad (4.2.76)$$

where

$$\omega_H(h) \equiv \sup_{\substack{|t - \tau| \leq h \\ a \leq t, \tau \leq b}} \int_a^b |H(t, s) - H(\tau, s)| \, ds$$

These can be shown to be true for both $\log|t - s|$ and $|t - s|^{\gamma - 1}$, $\gamma > 0$. Such assumptions were used earlier in §1.2 in showing compactness of integral operators on $C[a, b]$, and we refer to that earlier material [cf. (1.2.18)–(1.2.19) and following].

Theorem 4.2.1. *Assume the function* $H(t, s)$ *satisfies (4.2.75)–(4.2.76), and assume* $L(t, s)$ *is continuous for* $a \leq t, s \leq b$. *For a given* $y \in C[a, b]$, *assume the integral equation*

$$\lambda x(t) - \int_a^b L(t, s) H(t, s) x(s) \, ds = y(t), \qquad a \leq t \leq b$$

is uniquely solvable. Consider the numerical approximation (4.2.73), with $[L(t, s)x(s)]_n$ *defined with piecewise linear interpolation, as in (4.2.63). Then for all sufficiently large* n, *say* $n \geq N$, *the equation (4.2.73) is uniquely solvable, and the inverse operators are uniformly bounded for such* n. *Moreover,*

$$\|x - x_n\|_\infty \leq c\|\mathcal{K}x - \mathcal{K}_n x\|_\infty, \qquad n \geq N \qquad (4.2.77)$$

for suitable $c > 0$.

Proof. We show that the operators $\{\mathcal{K}_n\}$ of (4.2.73) are a collectively compact and pointwise convergent family on $C[a, b]$ to $C[a, b]$. This will prove the abstract assumptions **A1–A3** in the subsection 4.1.2. By using Lemma 4.1.2, we can then apply Theorem 4.1.2. We note that **A1** is obvious from the definitions of $\mathcal{K}$ and $\mathcal{K}_n$.

Let $\mathcal{S} = \{\mathcal{K}_n \mid n \geq 1 \text{ and } \|x\|_\infty \leq 1\}$. For bounds on $\|\mathcal{K}_n x\|_\infty$, first note that the piecewise linear interpolant z_n of a function $z \in C[a, b]$ satisfies

$$\|z_n\|_\infty \leq \|z\|_\infty$$

With this, it is straightforward to show

$$\|\mathcal{K}_n x\|_\infty \le c_L c_H, \quad x \in C[a,b], \quad \|x\|_\infty \le 1$$

with

$$c_L \equiv \max_{a \le t, s \le b} |L(t,s)|$$

This also shows the uniform boundedness of $\{\mathcal{K}_n\}$, with

$$\|\mathcal{K}_n\| \le c_L c_H, \quad n \ge 1$$

For equicontinuity of $\mathcal{S}$, write

$$\mathcal{K}_n x(t) - \mathcal{K}_n x(\tau)$$

$$= \int_a^b [L(t,s)x(s)]_n H(t,s)\,ds - \int_a^b [L(\tau,s)x(s)]_n H(\tau,s)\,ds$$

$$= \int_a^b [\{L(t,s) - L(\tau,s)\}x(s)]_n H(t,s)\,ds$$

$$+ \int_a^b [L(\tau,s)x(s)]_n \{H(t,s) - H(\tau,s)\}\,ds$$

This uses the linearity in z of the piecewise linear interpolation being used in defining $[z(s)]_n$. The assumptions on $H(t,s)$ and $L(t,s)$, together with $\|x\|_\infty \le 1$, now imply

$$\left| \int_a^b [\{L(t,s) - L(\tau,s)\}x(s)]_n H(t,s)\,ds \right|$$

$$\le c_H \|x\|_\infty \max_{a \le s \le b} |L(t,s) - L(\tau,s)|$$

Also,

$$\left| \int_a^b [L(\tau,s)x(s)]_n \{H(t,s) - H(\tau,s)\}\,ds \right| \le c_L \|x\|_\infty \omega_H(|t - \tau|)$$

Combining these results shows the desired equicontinuity of $\mathcal{S}$, and it completes the proof of the abstract property **A3** needed in applying the collectively compact operator framework.

For showing **A2**, recall that for piecewise linear interpolation,

$$|z(s) - z_n(s)| \le \omega(z,h), \quad z \in C[a,b]$$

with $\omega(z,h)$ the modulus of continuity of z on $[a,b]$. With this,

$$|\mathcal{K}x(t) - \mathcal{K}_n x(t)| \le c_H \omega(L(t,\cdot)x(\cdot), h)$$

$$\|\mathcal{K}x - \mathcal{K}_n x\|_\infty \le c_H \max_{a \le t \le b} \omega(L(t,\cdot)x(\cdot), h)$$

and the latter converges to zero by the uniform continuity of $L(t, s)x(s)$ over the square $a \leq t, s \leq b$. This proves $\mathcal{K}_n x \rightarrow \mathcal{K}x$ as $n \rightarrow \infty$, for all $x \in C[a, b]$; and **A2** is proven.

We apply Lemma 4.1.2 and Theorem 4.1.2 to complete the proof of the theorem. The constant c is the uniform bound on $\|(\lambda - \mathcal{K}_n)^{-1}\|$ for $n \geq N$. □

Let $z_n(s)$ denote the piecewise linear interpolant of $z(s)$, as used above in defining the product trapezoidal rule. It is a well-known standard result that

$$|z(s) - z_n(s)| \leq \frac{h^2}{8} \|z''\|_\infty, \quad z \in C^2[a, b]$$

Thus if $L(t, \cdot) \in C^2[a, b]$, $a \leq t \leq b$, and if $x \in C^2[a, b]$, then (4.2.77) implies

$$\|x - x_n\|_\infty \leq \frac{ch^2}{8} \max_{a \leq t, s \leq b} \left| \frac{\partial^2 L(t, s)x(s)}{\partial s^2} \right|, \quad n \geq N \qquad (4.2.78)$$

This rate of convergence is illustrated with equation (4.2.71) and Table 4.3.

The above ideas for solving (4.2.72) will generalize easily to higher degrees of piecewise polynomial interpolation. All elements of the above proof also generalize, and we obtain a theorem analogous to Theorem 4.2.1. In particular, suppose $[L(\tau, s)x(s)]_n$ is defined using piecewise polynomial interpolation of degree $m \geq 0$. Assume $L(t, \cdot) \in C^{m+1}[a, b]$, $a \leq t \leq b$, and $x \in C^{m+1}[a, b]$. Then

$$\|x - x_n\|_\infty \leq ch^{m+1} \max_{a \leq t, s \leq b} \left| \frac{\partial^{m+1} L(t, s)x(s)}{\partial s^{m+1}} \right|, \quad n \geq N \qquad (4.2.79)$$

for a suitable constant $c > 0$. When using piecewise quadratic interpolation, the method (4.2.73) is called the *product Simpson rule*, and according to (4.2.79), its rate of convergence is at least $O(h^3)$.

4.2.3. *Generalizations to other kernel functions*

Many singular integral equations are not easily written in the form (4.2.72) with a function $L(t, s)$ that is smooth and a function $H(t, s)$ for which weights such as those in (4.2.66)–(4.2.67) can be easily calculated. For such equations we assume instead that the singular kernel function $K(t, s)$ can be written in the form

$$K(t, s) = \sum_{j=1}^{r} L_j(t, s)H_j(t, s) \qquad (4.2.80)$$

with each $L_j(t, s)$ and $H_j(t, s)$ satisfying the properties listed above for $L(t, s)$ and $H(t, s)$. We now have an integral operator written as a sum of integral operators of the form used in (4.2.72):

$$\mathcal{K}x(t) = \sum_{j=1}^{r} \mathcal{K}_j x(t) = \sum_{j=1}^{r} \int_a^b L_j(t, s) H_j(t, s) x(s) \, ds, \quad x \in C[a, b]$$

Example. Consider the integral equation

$$x(t) - \int_0^\pi x(s) \log |\cos t - \cos s| \, ds = 1, \quad 0 \le t \le \pi \qquad (4.2.81)$$

One possibility for the kernel function $K(t, s) = \log |\cos t - \cos s|$ is to write

$$K(t, s) = \underbrace{|t - s|^{\frac{1}{2}} \log |\cos t - \cos s|}_{=L(t,s)} \underbrace{|t - s|^{-\frac{1}{2}}}_{=H(t,s)}$$

Unfortunately, this choice of $L(t, s)$ is continuous without being differentiable, and the function $L(t, s)$ needs to be differentiable to have the numerical method converge with sufficient speed. A better choice is to use

$$K(t, s) = \log \left| 2 \sin \frac{1}{2}(t - s) \sin \frac{1}{2}(t + s) \right|$$

$$= \log \left\{ \frac{2 \sin \frac{1}{2}(t - s) \sin \frac{1}{2}(t + s)}{(t - s)(t + s)(2\pi - t - s)} \right\}$$

$$+ \log |t - s| + \log(t + s) + \log(2\pi - t - s) \qquad (4.2.82)$$

This is of the form (4.2.80) with $H_1 = L_2 = L_3 = L_4 \equiv 1$ and

$$L_1(t, s) = \log \left\{ \frac{2 \sin \frac{1}{2}(t - s) \sin \frac{1}{2}(t + s)}{(t - s)(t + s)(2\pi - t - s)} \right\}$$

$$H_2(t, s) = \log |t - s|, \qquad H_3(t, s) = \log(t + s),$$

$$H_4(t, s) = \log(2\pi - t - s)$$

This is of the form (4.2.80). The function $L_1(t, s)$ is infinitely differentiable on $[0, 2\pi]$, and the functions H_2, H_3, and H_4 are singular functions for which the needed integration weights are easily calculated.

We solve (4.2.81) with both the product trapezoidal rule and the product Simpson rule, and error results are given in Tables 4.4 and 4.5. The decomposition (4.2.82) is used to define the approximating operators. With the operator

Table 4.4. *Product trapezoidal
rule for (4.2.81)*

n	$\|x - x_n\|_\infty$	Ratio
2	9.50E−3	
4	2.49E−3	3.8
8	6.32E−4	3.9
16	1.59E−4	4.0
32	3.98E−5	4.0

Table 4.5. *Product Simpson rule
for (4.2.81)*

n	$\|x - x_n\|_\infty$	Ratio
2	2.14E−4	
4	1.65E−5	13.0
8	1.13E−6	14.6
16	7.25E−8	15.6
32	4.56E−9	15.9

with kernel $L_1(t, s)H_1(t, s)$, we use the regular Simpson rule. The true solution of the equation is

$$x(t) \equiv \frac{1}{1 + \pi \log 2} \doteq 0.31470429802$$

Note that the error for the product trapezoidal rule is consistent with (4.2.78). But for the product Simpson rule we appear to have an error behavior of $O(h^4)$, whereas that predicted by (4.2.79) is only $O(h^3)$. This is discussed further below.

4.2.4. Improved error results for special kernels

If we consider again the error formula (4.2.77), the error result (4.2.79) was based on applying standard error bounds for polynomial interpolation to bounding the numerical integration error $\|\mathcal{K}x - \mathcal{K}_n x\|_\infty$. We know that for many ordinary integration rules (for example, Simpson's rule), there is an improvement in the speed of convergence over that predicted by the polynomial interpolation error, and this improvement is made possible by fortuitous cancellation of errors when integrating. Thus it is not surprising that the same type of cancellation occurs with the error $\|\mathcal{K}x - \mathcal{K}_n x\|_\infty$ in product integration, as is illustrated in Table 4.5.

For the special cases of $H(t, s)$ equal to $\log |t - s|$ and $|t - s|^{\gamma - 1}$, deHoog and Weiss [164] improved on the bound (4.2.79). In [164] they first extended known asymptotic error formulas for ordinary composite integration rules to product integration formulas, and these results were further extended to estimate $\mathcal{K}x - \mathcal{K}_n x$ for product integration methods of solving singular integral equations. For the case of the product Simpson's rule, their results state that if $x \in C^4[a, b]$, then

$$\|\mathcal{K}x - \mathcal{K}_n x\|_\infty \leq \begin{cases} ch^4 \log h, & H(t, s) = \log |t - s| \\ ch^{3+\gamma}, & H(t, s) = |t - s|^{\gamma - 1} \end{cases} \qquad (4.2.83)$$

This is in agreement with the results in Table 4.5.

4.2.5. Product integration with graded meshes

The rate of convergence results (4.2.79) and (4.2.83) both assume that the unknown solution $x(t)$ possesses several continuous derivatives. In fact, $x(t)$ seldom is smoothly differentiable, but rather has somewhat singular behavior in the neighborhood of the endpoints of the interval $[a, b]$ on which the integral equation is being solved. In the following, this is made more precise, and we also give a numerical method that restores the speed of convergence seen above with smoothly differentiable unknown solution functions.

To examine the differentiability of the solution $x(t)$ of a general integral equation $(\lambda - \mathcal{K})x = y$, the differentiability of the kernel function $K(t, s)$ allows the smoothness of $y(t)$ to be carried over to that of $x(t)$: use

$$\frac{d^j x(t)}{dt^j} = \frac{1}{\lambda}\left[\frac{d^j y(t)}{dt^j} + \int_a^b \frac{\partial^j K(t, s)}{\partial t^j} x(s)\, ds \right]$$

But if the kernel function is not differentiable, then the integral operator need not be smoothing. To see that the integral operator $\mathcal{K}$ with kernel $\log |t - s|$ is not smoothing in the manner that is true with differentiable kernel functions, let $x_0(t) \equiv 1$ on the interval $[0, 1]$, and calculate $\mathcal{K}x_0(t)$:

$$\mathcal{K}x_0(t) = \int_0^1 \log |t - s|\, ds = t \log t + (1 - t)\log(1 - t) - 1, \quad 0 \leq t \leq 1 \qquad (4.2.84)$$

The function $\mathcal{K}x_0(t)$ is not continuously differentiable on $[0, 1]$, whereas the function $x_0(t)$ is a C^∞ function. This formula also contains the typical type of singular behavior that appears in the solution when solving a second kind integral equation with a kernel function $K(t, s) = L(t, s)\log |t - s|$.

A number of people have examined the behavior of solutions of (4.2.72) for the case that $H(t, s)$ equals either $\log |t - s|$ or $|t - s|^{\gamma-1}$, some $0 < \gamma < 1$. These include Chandler [105], Graham [227], Pitkäranta [425], Richter [462], and Schneider [493]. We give the main result of Schneider [493]. As notation, introduce the following spaces:

$$C^{(0,\beta)}[a, b] = \left\{ g \in C[a, b] \mid d_\beta(g) \equiv \sup_{a \le t, \tau \le b} \frac{|g(t) - g(\tau)|}{|t - \tau|^\beta} < \infty \right\},$$

$$0 < \beta < 1 \qquad (4.2.85)$$

$$C^{(0,1)}[a, b] = \left\{ g \in C[a, b] \Bigg| \sup_{a \le t, \tau \le b} \frac{|g(t) - g(\tau)|}{|t - \tau| \log |B/(t - \tau)|} < \infty \right\}$$

for some $B > b - a$. For $0 < \beta < 1$, these are the standard Hölder spaces.

Theorem 4.2.2. *Let $k \ge 0$ be an integer, and let $0 < \gamma \le 1$. Assume $y \in C^{(0,\gamma)}[a, b]$, $y \in C^k(a, b)$, and*

$$(t - a)^i (b - t)^i y^{(i)}(t) \in C^{(0,\gamma)}[a, b], \quad i = 1, \ldots, k$$

Also assume $L \in C^{k+1}(D)$ with $D = [a, b] \times [a, b]$. Finally, assume the integral equation

$$\lambda x(t) - \int_a^b L(t, s) g_\gamma(t - s) x(s) \, ds = y(t), \quad a \le t \le b, \qquad (4.2.86)$$

$$g_\gamma(u) \equiv \begin{cases} u^{\gamma-1}, & 0 < \gamma < 1 \\ \log |u|, & \gamma = 1 \end{cases}$$

is uniquely solvable. Then

(a) *The solution $x(t)$ satisfies $x \in C^{(0,\gamma)}[a, b]$, $x \in C^k(a, b)$, and*

$$x_i(t) \equiv (t - a)^i (b - t)^i x^{(i)}(t) \in C^{(0,\gamma)}[a, b], \quad i = 1, \ldots, k \quad (4.2.87)$$

Further, $x_i(a) = x_i(b) = 0, i = 1, \ldots, k$

(b) *For $0 < \gamma < 1$,*

$$|x^{(i)}(t)| \le c_i (t - a)^{\gamma-i}, \quad a < t \le \frac{1}{2}(a + b), \quad i = 1, \ldots, k \quad (4.2.88)$$

With $\gamma = 1$, for any $\epsilon \in (0, 1)$,

$$|x^{(i)}(t)| \le c_i (t - a)^{1-\epsilon-i}, \quad a < t \le \frac{1}{2}(a + b), \quad i = 1, \ldots, k$$

$$(4.2.89)$$

with c_i dependent on ϵ. Analogous results are true for t in a neighborhood of b, with $t - a$ replaced by $b - t$.

Proof. See Ref. [493, p. 63]. In addition, more detail on the asymptotic behavior of $x(t)$ for t near to either a or b is given in the same reference and in Graham [227], bringing in functions of the type seen on the right side of (4.2.84) for the case of logarithmic kernel functions. □

This theorem says we should expect endpoint singularities in $x(t)$ of the form $(t - a)^\gamma$ and $(b - t)^\gamma$ for the case $H(t, s) = |t - s|^{\gamma - 1}, 0 < \gamma < 1$, with corresponding results for the logarithmic kernel. Thus the approximation of the unknown $x(t)$ should be based on such behavior. We do so by introducing the concept of a *graded mesh*, an idea developed in Rice [461] for the types of singular functions considered here.

We first develop the idea of a graded mesh for functions on [0, 1] with the singular behavior in the function occurring at 0, and then extend the construction to other situations by a simple change of variables. The singular behavior in which we are interested is $x(t) = t^\gamma, \gamma > 0$. For a given integer $n \geq 1$, define

$$t_j = \left(\frac{j}{n}\right)^q, \quad j = 0, 1, \ldots, n \qquad (4.2.90)$$

with the real number $q \geq 1$ to be specified later. For $q > 1$, this is an example of a *graded mesh*, and it is the one introduced and studied in Rice [461]. For a given integer $m \geq 0$, let a partition of [0, 1] be given:

$$0 \leq \mu_0 < \cdots < \mu_m \leq 1 \qquad (4.2.91)$$

Define interpolation nodes on each subinterval $[t_{j-1}, t_j]$ by

$$t_{ji} = t_{j-1} + \mu_i h_j, \quad i = 0, 1, \ldots, m, \quad h_j \equiv t_j - t_{j-1}$$

Let $\mathcal{P}_n x(t)$ be the piecewise polynomial function that is of degree $\leq m$ on each subinterval $[t_{j-1}, t_j]$ and that interpolates $x(t)$ at the nodes $\{t_{j0}, \ldots, t_{jm}\}$ on that subinterval. To be more explicit, let

$$L_i(\mu) = \prod_{\substack{k=0 \\ k \neq i}}^{m} \frac{\mu - \mu_k}{\mu_i - \mu_k}, \quad i = 0, 1, \ldots, m$$

which are the basis functions associated with intepolation at the nodes of

(4.2.91). Then

$$\mathcal{P}_n x(t) = \sum_{i=0}^{m} L_i \left(\frac{t - t_{j-1}}{h_j} \right) x(t_{ji}), \quad t_{j-1} \le t \le t_j, \quad j = 1, \ldots, n$$

(4.2.92)

If $\mu_0 > 0$ or $\mu_m < 1$, then $\mathcal{P}_n x(t)$ is likely to be discontinuous at the interior breakpoints $t_1, \ldots, t_{n-1}$. We now present the main result from Rice [461].

Lemma 4.2.3. Let n, m, $\{t_j\}$, $\{t_{ji}\}$, and $\mathcal{P}_n$ be as given in the preceding paragraph. For $0 < \gamma < 1$, assume $x \in C^{(0,\gamma)}[0, 1] \cap C^{m+1}(0, 1]$, with

$$|x^{(m+1)}(t)| \le c_{\gamma,m}(x) t^{\gamma-(m+1)}, \quad 0 < t \le 1$$

(4.2.93)

Then for

$$q \ge \frac{m+1}{\gamma}$$

(4.2.94)

we have,

$$\|x - \mathcal{P}_n x\|_\infty \le \frac{c}{n^{m+1}}$$

(4.2.95)

with c a constant independent of n. (We use c as a generic constant, varying from place to place.) For $1 \le p < \infty$, let

$$q > \frac{p(m+1)}{1 + p\gamma}$$

(4.2.96)

Then

$$\|x - \mathcal{P}_n x\|_p \le \frac{c}{n^{m+1}}$$

(4.2.97)

with $\|\cdot\|_p$ denoting the standard p-norm for $L^p(0, 1)$. [In the language of Rice [461], the function $x(t)$ is said to be of *Type*$(\gamma, m + 1)$].

Proof. (a) To introduce some intuition into this important topic, we first consider the proof for the special case of $x(t) = t^\gamma$. We begin by showing (4.2.95). Let $0 \le t \le t_1$. Then

$$x(t) - \mathcal{P}_n x(t) = t^\gamma - \sum_{i=0}^{m} L \left(\frac{t}{t_1} \right) (\mu_i t_1)^\gamma$$

$$= t_1^\gamma \left[\left(\frac{t}{t_1} \right)^\gamma - \sum_{i=0}^{m} L \left(\frac{t}{t_1} \right) \mu_1^\gamma \right]$$

$$\le c_{m,\gamma} t_1^\gamma$$

(4.2.98)

with

$$c_{m,\gamma} = \max_{0 \le u \le 1} \left| u^\gamma - \sum_{i=0}^m L(u)\mu_1^\gamma \right|$$

We want to have

$$|x(t) - \mathcal{P}_n x(t)| \le \frac{c}{n^{m+1}}, \quad 0 \le t \le t_1 \qquad (4.2.99)$$

for some $c > 0$. From (4.2.5), this will be true if

$$t_1^\gamma = \frac{1}{n^{q\gamma}} \le \frac{c}{n^{m+1}}$$

and the latter is true if the condition (4.2.94) on q is satisfied.

Now assume $t_{j-1} \le t \le t_j$ for some $2 \le j \le n$. Using the standard error formula for polynomial interpolation of degree m, we have

$$x(t) - \mathcal{P}_n x(t) = \frac{(t - t_{j0}) \cdot (t - t_{jm})}{(m+1)!} x^{(m+1)}(\xi_j(t)) \qquad (4.2.100)$$

with $t_{j-1} \le \xi_j(t) \le t_j$. Introduce $t - t_{j-1} = uh_j$ and apply the above to $x(t) = t^\gamma$ to obtain

$$x(t) - \mathcal{P}_n x(t) = h_j^{m+1} \binom{\gamma}{m+1} \omega_m(u)\xi_j^{\gamma-m-1}$$

with

$$\omega_m(u) = u\left(u - \frac{1}{m}\right) \cdot \left(u - \frac{m-1}{m}\right)(u-1)$$

Denote its maximum over $[0, 1]$ by $\|\omega_m\|_\infty$. Then

$$|x(t) - \mathcal{P}_n x(t)| \le h_j^{m+1} \left| \binom{\gamma}{m+1} \right| \|\omega_m\|_\infty t_{j-1}^{\gamma-m-1} \qquad (4.2.101)$$

We now need the bound

$$h_j \le \frac{q}{n} t_j^{1-\frac{1}{q}}, \quad j = 2, \dots, n \qquad (4.2.102)$$

the proof of which we leave to the reader. Note also that

$$t_{j-1}^{-1} = t_j^{-1}\left(\frac{t_j}{t_{j-1}}\right) = t_j^{-1}\left(\frac{j}{j-1}\right)^q \le 2^q t_j^{-1}, \quad j = 2, \dots, n$$

$$(4.2.103)$$

Combining these results with (4.2.101), we have

$$|x(t) - \mathcal{P}_n x(t)| \le \frac{c}{n^{m+1}} t_j^{\gamma-(m+1)/q}$$

with

$$c = \left| \binom{\gamma}{m+1} \right| \|\omega_m\|_\infty q^{m+1} 2^{(m+1-\gamma)q}$$

The desired result (4.2.95) follows if the exponent $\gamma - (m+1)/q \ge 0$, and this is exactly the condition (4.2.94). This completes the proof of (4.2.95).

To prove (4.2.97) is quite similar. On the interval $[0, t_1]$,

$$E_1 = \int_0^{t_1} |x(t) - \mathcal{P}_n x(t)|^p \, dt = t_1^{\gamma p+1} \int_0^1 \left| u^\gamma - \sum_{i=0}^m L(u)\mu_1^\gamma \right| du$$

where we have used the change of variables $t = ut_1$, $0 \le u \le 1$. Then for the p-norm on $[0, t_1]$,

$$E_1^{1/p} = ct_1^{\gamma+\frac{1}{p}} = \frac{c}{n^{q(\gamma+\frac{1}{p})}}$$

and we want this to be $O(1/n^{m+1})$. This will follow if

$$q\left(\gamma + \frac{1}{p}\right) \ge m+1$$

$$q \ge \frac{p(m+1)}{1+p\gamma}$$

On the interval $[t_{j-1}, t_j]$, $2 \le j \le n$, by using the standard interpolation error formula (4.2.100),

$$E_j = \int_{t_{j-1}}^{t_j} |x(t) - \mathcal{P}_n x(t)|^p \, dt = \int_{t_{j-1}}^{t_j} \left| h_j^{m+1} \binom{\gamma}{m+1} \omega_m(u)\xi_j^{\gamma-m-1} \right|^p dt$$

Use the change of variables $t - t_{j-1} = uh_j$ and bound ξ_j, to get

$$E_j \le h_j^{p(m+1)+1} \left| \binom{\gamma}{m+1} \right|^p \|\omega_m\|_p^p t_{j-1}^{-p(m+1-\gamma)}$$

Use the earlier bounds (4.2.102) and (4.2.103) to get

$$E_j \le \frac{c^p}{n^{(m+1)p}} h_j t_j^{p(\gamma-\frac{m+1}{q})}$$

$$c = q^{m+1} \left| \binom{\gamma}{m+1} \right| 2^{m+1-\gamma} \|\omega_m\|_p$$

The p-norm of $x - \mathcal{P}_n x$ on $[t_2, 1]$ is given by

$$\left[\sum_{j=2}^{n} E_j \right]^{\frac{1}{p}} \leq \frac{c}{n^{m+1}} \left[\sum_{j=2}^{n} h_j t_j^{p(\gamma - \frac{m+1}{q})} \right]^{\frac{1}{p}} \tag{4.2.104}$$

If the exponent of t_j satisfies

$$\delta \equiv p \left(\gamma - \frac{m+1}{q} \right) > -1 \tag{4.2.105}$$

then

$$\left[\sum_{j=2}^{n} h_j t_j^{p(\gamma - \frac{m+1}{q})} \right]^{\frac{1}{p}}$$

is bounded by using an integral comparison. The most important case is when $-1 < \delta < 0$, and in that case we have

$$\left[\sum_{j=2}^{n} h_j t_j^{p(\gamma - \frac{m+1}{q})} \right]^{\frac{1}{p}} \leq \left[\int_{t_1}^{1} t^\delta dt \right]^{\frac{1}{p}} < \left(\frac{1}{1 + \delta} \right)^{\frac{1}{p}}$$

The condition (4.2.105) is equivalent to (4.2.96). When these results are combined with (4.2.104), we have the result (4.2.97).

(b) Consider now the more general case of a function $x(t)$ satisfying the given hypotheses. We will prove only the bound (4.2.95), as the proof of (4.2.97) is similar.

As before, consider first $0 \leq t \leq t_1$. Using the exactness of interpolation for constant functions,

$$x(t) - \mathcal{P}_n x(t) = [x(t) - x(0)] - \sum_{i=0}^{m} L_j \left(\frac{t}{t_1} \right) (x(t_i) - x(0))$$

$$|x(t) - \mathcal{P}_n x(t)| \leq c \max_{0 \leq s \leq t_1} |x(s) - x(0)|$$

$$c = 1 + \max_{0 \leq u \leq 1} \sum_{i=0}^{m} |L_j(u)|$$

Using the assumption $x \in C^{(0,\gamma)}[0, 1]$, we have

$$|x(t) - \mathcal{P}_n x(t)| \leq c \, d_\gamma(x) t_1^\gamma$$

with $d_\gamma(x)$ the Hölder constant for x [cf. (4.2.85)]. Proceed as following (4.2.99) to show this satisfies (4.2.95) when q satisfies (4.2.94).

Now consider $t_{j-1} \leq t \leq t_j$, $j = 2, \ldots, n$. Using the interpolation error formula (4.2.100) and applying the assumption (4.2.93), we have

$$|x(t) - \mathcal{P}_n x(t)| \leq c_{\gamma,m}(x) h_j^{m+1} \left| \binom{\gamma}{m+1} \right| \|\omega_m\|_\infty t_{j-1}^{\gamma-m-1}$$

Proceed exactly as following (4.2.101) to complete the proof of (4.2.95). □

Application to integral equations

The earlier product integration methods were based on using interpolation on a uniform subdivision of the interval $[a, b]$. Now we use the same form of interpolation, but base it on a graded mesh for $[a, b]$. Given an even $n \geq 2$, define

$$t_j = a + \left(\frac{2j}{n}\right)^q \left(\frac{b-a}{2}\right), \qquad t_{n-j} = b + a - t_j, \quad j = 0, 1, \ldots, \frac{n}{2}$$

Use the partition (4.2.91) as the basis for polynomial interpolation of degree m on each of the intervals $[t_{j-1}, t_j]$, for $j = 1, \ldots, \frac{1}{2}n$, just as was done in (4.2.92); use the partition

$$0 \leq 1 - \mu_m < \cdots < 1 - \mu_0 \leq 1$$

when defining the interpolation on the subintervals $[t_{j-1}, t_j]$ of the remaining half $[\frac{1}{2}(a+b), b]$. In the integral equation (4.2.86), replace $[L(t, s)x(s)]$ with $[L(t, s)x(s)]_n$ using the interpolation just described. For the resulting approximation

$$\lambda x_n(t) - \int_a^b [L(t, s)x_n(s)]_n g_\gamma(t - s)\, ds = y(t), \quad a \leq t \leq b, \quad (4.2.106)$$

we have the following convergence result.

Theorem 4.2.3. *Consider again the integral equation (4.2.86) with the same assumptions as for Theorem 4.2.2, but with the integer k replaced by $m + 1$, where m is the integer used in defining the interpolation of the preceding paragraph. Then the approximating equation (4.2.106) is uniquely solvable for all sufficiently large n, say $n \geq N$, and the inverse operator for the equation is uniformly bounded for $n \geq N$. If $0 < \gamma < 1$, then choose the grading exponent q to satisfy*

$$q \geq \frac{m+1}{\gamma} \qquad (4.2.107)$$

Table 4.6. *Graded product trapezoidal solution of (4.2.110)*

	q = 1		q = 2		q = 3		q = 4	
n	Error	Order	Error	Order	Error	Order	Error	Order
8	7.98E−2		3.95E−2		2.20E−2		3.35E−2	
16	5.46E−2	0.55	1.91E−2	1.05	7.34E−3	1.58	9.97E−3	1.75
32	3.79E−2	0.53	9.38E−3	1.03	2.51E−3	1.55	2.80E−3	1.83
64	2.65E−2	0.52	4.64E−3	1.02	8.64E−4	1.54	7.50E−4	1.90
128	1.86E−2	0.51	2.31E−3	1.01	3.00E−4	1.53	1.95E−4	1.94
256	1.31E−2	0.51	1.15E−3	1.01	1.05E−4	1.51	4.99E−5	1.97

If $\gamma = 1$, then choose

$$q > m + 1 \qquad (4.2.108)$$

With such choices, the approximate solution x_n satisfies

$$\|x - x_n\|_\infty \leq \frac{c}{n^{m+1}} \qquad (4.2.109)$$

Proof. The proof is a straightforward generalization of the method of proof used in Theorem 4.2.1, resulting in the error bound

$$\|x - x_n\|_\infty \leq c\|\mathcal{K}x - \mathcal{K}_n x\|_\infty, \qquad n \geq N$$

Combine Theorem 4.2.2 and Lemma 4.2.3 to complete the proof.

This theorem is from Schneider [494, Theorem 2], and he also allows for greater generality in the singularity in $x(t)$ than has been assumed here. In addition, he extends results of deHoog and Weiss [164], such as (4.2.83), to the use of graded meshes. □

Example. Consider the integral equation

$$\lambda x(t) - \int_0^1 \frac{x(s)}{|t - s|^{\frac{1}{2}}} \, ds = y(t), \qquad 0 \leq t \leq 1 \qquad (4.2.110)$$

with $y(t)$ so chosen that

$$x(t) = \sqrt{t} + \sqrt{1 - t}$$

Here $\gamma = \frac{1}{2}$, and we use $\lambda = 10$. In Table 4.6 we give errors and empirical orders of convergence for the product trapezoidal rule. The columns labeled

$Error \equiv Error(n)$ denote the maximum of the errors in $x_n(t)$ for t a node point or a point midway between successive node points. The columns labeled *Order* give the estimated order of convergence, using the most recent computed values of x_n, based on

$$Order = \frac{1}{\log 2} \log \frac{Error(\frac{1}{2}n)}{Error(n)}$$

Empirically,

$$Error(n) = O(n^{-Order})$$

For an optimal rate of convergence of $O(n^{-2})$, the needed grading parameter for the mesh is $q = 4$. We also give results for the smaller values of $q = 1, 2, 3$; and the empirical orders are consistent with results obtained in the proof of Lemma 4.2.3. The results show that smaller values of q are not adequate in this case.

Graded meshes are used with other problems in which there is some kind of singular behavior in the functions being considered. Later we will consider their use in solving boundary integral equations for the planar Laplace's equation for regions in which the boundary has corners. In Chapter 9 we look at using multivariable generalizations of graded meshes for solving boundary integral equations on polyhedral surfaces.

The relationship of product integration and collocation methods

Recall the discussion of the collocation method from Chapter 3. It turns out that collocation methods can be regarded as product integration methods, and occasionally there is an advantage to doing so.

Recalling the discussion of Chapter 3, let $\mathcal{P}_n$ be the interpolatory projection operator from $C(D)$ onto the interpolatory approximating space $\mathcal{X}_n$. Then the collocation solution of $(\lambda - \mathcal{K})x = y$ can be regarded abstractly as $(\lambda - \mathcal{P}_n\mathcal{K})x_n = \mathcal{P}_n y$, and the iterated collocation solution

$$\hat{x}_n = \frac{1}{\lambda}[y + \mathcal{K}x_n]$$

is the solution of the equation

$$(\lambda - \mathcal{K}\mathcal{P}_n)\hat{x}_n = y \tag{4.2.111}$$

Define a numerical integral operator by

$$\mathcal{K}_n x(t) = \mathcal{K}\mathcal{P}_n x(t) = \int_D K(t, s)(\mathcal{P}_n x)(s)\, ds \tag{4.2.112}$$

which is product integration with $L(t, s) \equiv 1$ and $H(t, s) = K(t, s)$. Thus the iterated collocation solution $\hat{x}_n$ of (4.2.111) is simply the Nyström solution when defining $\mathcal{K}_n$ using the simple product integration formula of (4.2.112). Since the collocation solution $x_n = \mathcal{P}_n \hat{x}_n$, we can use results from the error analysis of product integration methods to analyze collocation methods.

4.3. Discrete collocation methods

The actual solution processes for the projection methods of Chapter 3 and the degenerate kernel methods of Chapter 2 lead to algebraic linear systems in which each coefficient of the system is an integral. For example, see (2.3.41) for interpolatory degenerate kernels, (3.1.5) for collocation methods, and (3.1.13) for Galerkin methods. The integrals in these systems are almost always evaluated numerically. Thus we solve a linear system that has been modified by using numerical integration, and we obtain a numerical solution z_n, which is a modification of the solution x_n that would have been obtained if exact integration had been used. When this is done with the collocation or Galerkin methods, we refer to the modified numerical solution z_n as a *discrete collocation solution* or a *discrete Galerkin solution*, respectively; and generally, we refer to it as a *discrete projection solution*. In this and the following section, we give an error analysis for such discrete projection solutions.

The traditional way of analyzing the effect of the error in such numerical integrations has been to regard the resulting linear system as a perturbation of the original linear system and then apply perturbation theorems to compare the solutions of the two systems. An example of this was carried out in the final subsection of §2.3 in Chapter 2, for degenerate kernels defined by interpolation of the kernel function $K(t, s)$. For an extensive discussion of such error analyses, see Golberg [219, Chap. 3]. In contrast, in this and the following section we regard the projection methods with numerical integration as new numerical methods, and we give a corresponding direct error analysis.

Recall the introductory material on projection methods, following (3.1.4). To solve $(\lambda - \mathcal{K})x = y$ approximately, we write

$$x_n(t) = \sum_{j=1}^{d_n} c_j \phi_j(t), \quad t \in D \tag{4.3.113}$$

and we find $\{c_1 \ldots, c_d\}$, $d \equiv d_n$, by solving the linear system

$$\sum_{j=1}^{d_n} c_j \left\{ \lambda \phi_j(t_i) - \int_D K(t_i, s)\phi_j(s)\, ds \right\} = y(t_i), \quad i = 1, \ldots, d_n$$

$$\tag{4.3.114}$$

Moreover, these two formulas are equivalent to solving the abstract problem

$$(\lambda - \mathcal{P}_n \mathcal{K}) x_n = \mathcal{P}_n y, \quad x_n \in \mathcal{X} = C(D) \qquad (4.3.115)$$

with $\mathcal{P}_n$ the interpolatory projection onto $\mathcal{X}_n = \mathrm{span}\{\phi_1, \ldots, \phi_d\}$ with respect to the interpolation nodes $\{t_i\}$.

To approximate the integrals in (4.3.114), we introduce a sequence of numerical integral operators $\mathcal{K}_n$. For $n \geq 1$, define

$$\mathcal{K}_n x(t) = \sum_{k=1}^{q_n} w_k(t) x(\tau_k), \quad t \in D, \quad x \in C(D) \qquad (4.3.116)$$

If the kernel function $K(t, s)$ is continuous, then generally we would have a formula of the form

$$\mathcal{K}_n x(t) = \sum_{k=1}^{q_n} w_k K(t, \tau_k) x(\tau_k), \quad t \in D, \quad x \in C(D)$$

based on some numerical integration formula

$$\int_D g(t) \, dt \approx \sum_{k=1}^{q_n} w_k g(\tau_k), \quad g \in C(D) \qquad (4.3.117)$$

The more general formula (4.3.116) includes numerical integral operators based on product integration. Regardless of how $\mathcal{K}_n$ is defined, we assume that $\{\mathcal{K}_n\}$ is a collectively compact family that is pointwise convergent to $\mathcal{K}$ on $C(D)$, and we assume $\mathcal{K}_n x(t)$ depends on $\{x(\tau_1), \ldots, x(\tau_q)\}$, $q \equiv q_n$.

The formula (4.3.116) is applied to each of the integrals in (4.3.114). This leads to the following numerical method. Let

$$z_n(t) = \sum_{j=1}^{d_n} \gamma_j \phi_j(t), \quad t \in D \qquad (4.3.118)$$

with $\{\gamma_1, \ldots, \gamma_d\}$ determined as the solution of the linear system

$$\sum_{j=1}^{d_n} \gamma_j \left\{ \lambda \phi_j(t_i) - \sum_{k=1}^{q_n} w_k(t_i) \phi_j(\tau_k) \right\} = y(t_i), \quad i = 1, \ldots, d_n \qquad (4.3.119)$$

We call this the *discrete collocation method*, and the function $z_n(t)$ is called the discrete collocation solution.

The discrete collocation method can be put into a particularly simple abstract form, one closely related to (4.3.115) for the original collocation method. The function z_n satisfies the equation

$$(\lambda - \mathcal{P}_n \mathcal{K}_n) z_n = \mathcal{P}_n y, \quad z_n \in \mathcal{X} = C(D) \tag{4.3.120}$$

To prove this, first note that

$$z_n = \frac{1}{\lambda} \mathcal{P}_n [y + \mathcal{K}_n z_n] \in \mathcal{X}_n$$

Thus $\mathcal{P}_n z_n = z_n$, and (4.3.120) can be written as

$$\mathcal{P}_n [(\lambda - \mathcal{K}_n) z_n - y] = 0 \tag{4.3.121}$$

with z_n as in (4.3.118). The equation (4.3.121) says that $(\lambda - \mathcal{K}_n) z_n - y$ must be zero at all the collocation node points $t_1, \ldots, t_d$, and that is exactly the system (4.3.119). The argument is reversible, thus showing the equivalence of the formulations.

We can analyze the discrete collocation method by analyzing (4.3.120). But before doing so, we introduce the *iterated discrete collocation solution*. Define

$$\hat{z}_n = \frac{1}{\lambda} [y + \mathcal{K}_n z_n] \tag{4.3.122}$$

This is in analogy with the iterated collocation solution $\hat{x}_n$ of (3.4.80). As in §3.4, we have

$$\mathcal{P}_n \hat{z}_n = z_n$$

and

$$(\lambda - \mathcal{K}_n \mathcal{P}_n) \hat{z}_n = y \tag{4.3.123}$$

We begin our analysis of discrete collocation methods with an important special case.

Theorem 4.3.1. *Assume that $\{\tau_k\} \subset \{t_i\}$ for the quadrature nodes and collocation nodes. Then the iterated discrete collocation solution $\hat{z}_n$ is the Nyström solution associated with the numerical integral operator (4.3.116), and it satisfies the equation*

$$(\lambda - \mathcal{K}_n) \hat{z}_n = y$$

Proof. The proof is almost immediate. From the assumptions, we have

$$\mathcal{P}_n x(\tau_k) = x(\tau_k), \quad k = 1, \ldots, q_n$$

from the definition of $\mathcal{P}_n x$. Then

$$\mathcal{K}_n \mathcal{P}_n x(t) = \sum_{k=1}^{q_n} w_k(t)(\mathcal{P}_n x)(\tau_k)$$

$$= \sum_{k=1}^{q_n} w_k(t) x(\tau_k)$$

$$= \mathcal{K}_n x(t)$$

Substitute this into (4.3.123) to complete the proof. $\square$

This theorem is important in that many suggested discrete collocation methods have had $\{\tau_k\} \subset \{t_i\}$. Such methods can be completely analyzed by using the framework of §4.1 for Nyström methods. Thus if the Nyström method is convergent, we have

$$\|x - \hat{z}_n\|_\infty \leq c \|\mathcal{K}x - \mathcal{K}_n x\|_\infty \qquad (4.3.124)$$

from (4.1.33) of Theorem 4.1.2. It is noteworthy that the error does not involve the interpolation error $\|x - \mathcal{P}_n x\|_\infty$, even though $\hat{z}_n$ is a discrete *collocation* solution. The absence of the interpolation error is also in contrast to the formula (3.4.88) for the iterated collocation solution $\hat{x}_n$.

Example. As a case that has been widely studied, recall the collocation method for piecewise polynomial collocation at Gauss-Legendre nodes; see the discussion following (3.4.115) in §3.4 of Chapter 3. For it, we have $\mathcal{X} = C[a, b]$, $h = (b - a)/n$, $\xi_k = a + kh$ for $k = 0, \ldots, n$, and $\mathcal{X}_n = \{g \in L^\infty \mid g(t)$ a polynomial of degree $\leq r$ on each subinterval $(\xi_{k-1}, \xi_k)\}$. The collocation points are the zeros of the Gauss-Legendre polynomial of degree $r + 1$, transformed to each of the subintervals $[\xi_{k-1}, \xi_k]$. Thus $d_n = n(r + 1)$. For the rate of convergence, we proved

$$\|x - \hat{x}_n\|_\infty = O(h^{2r+2}), \quad x \in C^{2r+2}[a, b] \qquad (4.3.125)$$

for the iterated collocation solution $\hat{x}_n$; see (3.4.126).

For a quadrature rule for the integrals of this method, we apply the $(r + 1)$-point Gauss-Legendre quadrature rule to the integrals over each of the subintervals $[\xi_{k-1}, \xi_k]$. Let

$$\int_0^1 G(\sigma) \, d\sigma \approx \sum_{j=0}^r \omega_j G(\sigma_j)$$

denote the $(r+1)$-point Gauss-Legendre quadrature rule relative to the interval $[0, 1]$. It has a degree of precision of $2r+1$. Then our integration rule (4.3.117) for $[a, b]$ is

$$\int_a^b g(t)\,dt \approx h \sum_{k=1}^n \sum_{j=0}^r \omega_j g(\xi_{k-1} + \sigma_j h) \tag{4.3.126}$$

Thus $q_n = n(r+1) = d_n$, and the quadrature nodes coincide with the collocation nodes. For the error in (4.3.126) it is straightforward to show it is $O(h^{2r+2})$, provided $g \in C^{2r+2}[a, b]$. Combining this with (4.3.124), we have

$$\|x - \hat{z}_n\|_\infty = O(h^{2r+2})$$

provided $x \in C^{2r+2}[a, b]$ and $K(t, \cdot) \in C^{2r+2}[a, b]$, uniformly for $a \le t \le b$. Comparing with (4.3.125), we have that $\hat{z}_n$ is as rapidly convergent as $\hat{x}_n$.

4.3.1. Convergence analysis for $\{\tau_k\} \not\subset \{t_i\}$

In analogy with what was shown previously in §3.4 for the collocation and iterated collocation solutions, the existence of $(\lambda - \mathcal{P}_n \mathcal{K}_n)^{-1}$ and $(\lambda - \mathcal{K}_n \mathcal{P}_n)^{-1}$ are closely related. Simply apply Lemma 3.4.1 in §3.4. We see that if $(\lambda - \mathcal{P}_n \mathcal{K}_n)^{-1}$ exists, then so does $(\lambda - \mathcal{K}_n \mathcal{P}_n)^{-1}$, and

$$(\lambda - \mathcal{K}_n \mathcal{P}_n)^{-1} = \frac{1}{\lambda}[I + \mathcal{K}_n(\lambda - \mathcal{P}_n \mathcal{K}_n)^{-1}\mathcal{P}_n] \tag{4.3.127}$$

Conversely, if $(\lambda - \mathcal{K}_n \mathcal{P}_n)^{-1}$ exists, then so does $(\lambda - \mathcal{P}_n \mathcal{K}_n)^{-1}$, and

$$(\lambda - \mathcal{P}_n \mathcal{K}_n)^{-1} = \frac{1}{\lambda}[I + \mathcal{P}_n(\lambda - \mathcal{K}_n \mathcal{P}_n)^{-1}\mathcal{K}_n] \tag{4.3.128}$$

By combining these, we also have

$$(\lambda - \mathcal{P}_n \mathcal{K}_n)^{-1}\mathcal{P}_n = \mathcal{P}_n(\lambda - \mathcal{K}_n \mathcal{P}_n)^{-1}$$

We can choose to show the existence of either $(\lambda - \mathcal{P}_n \mathcal{K}_n)^{-1}$ or $(\lambda - \mathcal{K}_n \mathcal{P}_n)^{-1}$, whichever is the more convenient, and the existence of the other inverse will follow immediately. Bounds on one inverse in terms of the other can also be given by using the above formulas. We choose to analyze (4.3.123) for the iterated discrete collocation method.

Theorem 4.3.2. *Assume the family $\{\mathcal{K}_n\}$ of (4.3.116) is collectively compact and pointwise convergent on $\mathcal{X} = C(D)$. Let $\{\mathcal{P}_n\}$ be a family of interpolatory* *projection operators on $C(D)$ to $C(D)$, and assume*

$$\mathcal{P}_n x \to x \quad as \ n \to \infty \tag{4.3.129}$$

for all $x \in C(D)$. Finally, assume the integral equation

$$\lambda x(t) - \int_D K(t,s)x(s)\,ds = y(t), \quad t \in D$$

is uniquely solvable for all $y \in C(D)$. Then $(\lambda - \mathcal{K}_n \mathcal{P}_n)^{-1}$ exists for all sufficiently large n, say $n \geq N$, and is uniformly bounded in n. For the solution $\hat{z}_n$ of (4.3.123),

$$\|x - \hat{z}_n\|_\infty \leq \|(\lambda - \mathcal{K}_n \mathcal{P}_n)^{-1}\| \|\mathcal{K}x - \mathcal{K}_n \mathcal{P}_n x\|_\infty, \quad n \geq N \tag{4.3.130}$$

Proof. We show $\{\mathcal{K}_n \mathcal{P}_n\}$ is collectively compact and pointwise convergent on $C(D)$. The remainder of the proof then follows by Theorem 4.1.2.

From (4.3.129) and the principle of uniform boundedness (cf. Theorem A.3 in the Appendix),

$$c_P \equiv \sup \|\mathcal{P}_n\| < \infty \tag{4.3.131}$$

Similarly, from the pointwise convergence of $\{\mathcal{K}_n\}$,

$$c_K \equiv \sup \|\mathcal{K}_n\| < \infty$$

Together, these imply the uniform boundedness of $\{\mathcal{K}_n \mathcal{P}_n\}$, with a bound of $c_K c_P$. For the pointwise convergence on $C(D)$,

$$\|\mathcal{K}x - \mathcal{K}_n \mathcal{P}_n x\|_\infty \leq \|\mathcal{K}x - \mathcal{K}_n x\|_\infty + \|\mathcal{K}_n(x - \mathcal{P}_n x)\|_\infty$$
$$\leq \|\mathcal{K}x - \mathcal{K}_n x\|_\infty + c_K \|x - \mathcal{P}_n x\|_\infty$$

The desired convergence now follows from that of $\{\mathcal{K}_n x\}$ and $\{\mathcal{P}_n x\}$.

To show the collective compactness of $\{\mathcal{K}_n \mathcal{P}_n\}$, we must show that

$$\mathcal{F} \equiv \{\mathcal{K}_n \mathcal{P}_n x \mid n \geq 1 \ \text{and} \ \|x\|_\infty \leq 1\}$$

has compact closure in $C(D)$. Using (4.3.131), we have

$$\mathcal{F} \subset \{\mathcal{K}_n x \mid n \geq 1 \ \text{and} \ \|x\|_\infty \leq c_P\}$$

and the latter has compact closure from the collective compactness of $\{\mathcal{K}_n\}$.

From Theorem 4.1.2, $(\lambda - \mathcal{K}_n \mathcal{P}_n)^{-1}$ exists for all sufficiently large n, say $n \geq N$; and there is $c_I < \infty$ for which

$$\|(\lambda - \mathcal{K}_n \mathcal{P}_n)^{-1}\| \leq c_I, \quad n \geq N$$

In addition, (4.3.130) follows from the identity

$$x - \hat{z}_n = (\lambda - \mathcal{K}_n \mathcal{P}_n)^{-1}(\mathcal{K}x - \mathcal{K}_n \mathcal{P}_n x)$$

In some cases, it is possible to show

$$\|\mathcal{K} - \mathcal{P}_n \mathcal{K}_n\| \to 0 \quad \text{as } n \to \infty \tag{4.3.132}$$

and this permits a direct analysis of the discrete collocation method. Then the iterated discrete collocation method can be analyzed as a consequence of that of the discrete collocation method. □

Corollary 4.3.1. *For the iterated discrete collocation solution $\hat{z}_n$, with $n \geq N$,*

$$\|x - \hat{z}_n\|_\infty \leq c_I [\|\mathcal{K}x - \mathcal{K}_n x\|_\infty + \|\mathcal{K}_n(x - \mathcal{P}_n x)\|_\infty] \tag{4.3.133}$$

At worst,

$$\|x - \hat{z}_n\|_\infty \leq c_I [\|\mathcal{K}x - \mathcal{K}_n x\|_\infty + c_K \|x - \mathcal{P}_n x\|_\infty] \tag{4.3.134}$$

For the discrete collocation solution z_n,

$$\|x - z_n\|_\infty \leq c_I [c_P \|\mathcal{K}x - \mathcal{K}_n x\|_\infty + (1 + c_P c_K) \|x - \mathcal{P}_n x\|_\infty] \tag{4.3.135}$$

Proof. The bound (4.3.133) was given in the proof of the theorem, and (4.3.134) is a straightforward consequence. For the error in z_n,

$$x - z_n = x - \mathcal{P}_n \hat{z}_n = [x - \mathcal{P}_n x] + \mathcal{P}_n [x - \hat{z}_n]$$

$$\|x - z_n\|_\infty \leq \|x - \mathcal{P}_n x\|_\infty + c_P \|x - \hat{z}_n\|_\infty$$

Then apply (4.3.134) to obtain (4.3.135). □

Where possible and practical, we usually choose the numerical integration operators $\{\mathcal{K}_n\}$ so as to preserve the original rates of convergence of x_n and $\hat{x}_n$. The bounds (4.3.133) and (4.3.135) give the accuracy needed in the numerical integration. The most interesting applications of this arise in the study of multivariable integral equations, and we defer until Chapter 5 the application of the above results. A fairly extensive discussion for both one variable and multivariable problems is given in Atkinson and Flores [60, §4].

4.4. Discrete Galerkin methods

Discrete Galerkin methods result from the numerical integration of all integrals in the linear system associated with the Galerkin method. We begin by reviewing some results from Chapter 3 on Galerkin methods, and then we look at the effect of replacing the integrals with numerical integrals.

Let $\mathcal{X}_n = \text{span}\{\phi_1, \ldots, \phi_d\}$, $d \equiv d_n$, be a subspace of $\mathcal{X} = C(D)$, and let $\mathcal{P}_n$ be the orthogonal projection of $\mathcal{X}$ onto $\mathcal{X}_n$ with respect to the inner product

$$(f, g) = \int_D f(t)g(t)\,dt \tag{4.4.136}$$

In abstract form, the Galerkin method is

$$(\lambda - \mathcal{P}_n\mathcal{K})x_n = \mathcal{P}_n y, \quad y \in C(D) \tag{4.4.137}$$

To obtain x_n, let

$$x_n(t) = \sum_{i=1}^{d_n} c_i \phi_i(t), \quad t \in D \tag{4.4.138}$$

The coefficients $\{c_1, \ldots, c_d\}$ are obtained by solving the linear system

$$\sum_{j=1}^{d_n} c_j \{\lambda(\phi_j, \phi_i) - (\mathcal{K}\phi_j, \phi_i)\} = (y, \phi_i), \quad i = 1, \ldots, d_n \tag{4.4.139}$$

There are two types of integrals to be evaluated in (4.4.139):

1. the inner product $(\cdot, \cdot)$, and
2. the integral operator $\mathcal{K}$.

To approximate the integral operator, we proceed exactly as with the discrete collocation method. Assume $\mathcal{K}x \approx \mathcal{K}_n x$, with the latter as in (4.3.116):

$$\mathcal{K}_n x(t) = \sum_{k=1}^{q_n} w_k(t)x(\tau_k), \quad t \in D, \ x \in C(D) \tag{4.4.140}$$

We assume $\{\mathcal{K}_n\}$ is a collectively compact and pointwise convergent family on $C(D)$ to $C(D)$.

For the inner product, we use a second numerical integration formula. Consider a formula that uses the same node points as in (4.4.140):

$$\int_D g(t)\,dt \approx \sum_{k=1}^{q_n} v_k g(\tau_k), \quad g \in C(D)$$

We assume it is a convergent quadrature method for all $g \in C(D)$. Using it, define a *discrete semidefinite inner product*:

$$(f, g)_n = \sum_{k=1}^{q_n} v_k f(\tau_k) g(\tau_k), \quad f, g \in C(D) \tag{4.4.141}$$

and a *discrete seminorm*

$$\|g\|_n = \sqrt{(g, g)_n}, \quad g \in C(D) \tag{4.4.142}$$

We make the following assumptions:

H1. $q_n \geq d_n$
H2. Rank $\Phi_n = d_n$, with

$$(\Phi_n)_{ik} = \phi_i(\tau_k), \quad i = 1, \ldots, d_n, \quad k = 1, \ldots, q_n \tag{4.4.143}$$

H3. $v_k > 0$ for $k = 1, \ldots, q_n$
H4. If $\|g\|_n = 0$ and $g \in \mathcal{X}_n$, then $g = 0$.

These assumptions are not independent. For example, **H1–H3** will imply **H4**. However, we introduce all of these hypotheses to make the discussion somewhat easier to follow. With these assumptions, the quantity $\|\cdot\|_n$ is a seminorm on $C(D)$, and it is a norm on $\mathcal{X}_n$.

For most quadrature schemes (4.4.140)–(4.4.141) and approximating subspaces $\mathcal{X}_n$ that are used, showing the hypotheses is usually very straightforward, with **H2** being possibly the only difficult one. It can be proven by showing that some subset of d_n of the quadrature nodes, say $\{\tau_1, \ldots, \tau_d\}$, has the *interpolation property*: for every set of data $\{\zeta_1, \ldots, \zeta_d\}$, there is a unique element $\phi \in \mathcal{X}_n$ with

$$\phi(\tau_i) = \zeta_i, \quad i = 1, \ldots, d_n$$

If this is true, then **H2** is true. We leave the proof to the reader. In general, these hypotheses appear to be true for all cases of practical interest.

Our discrete Galerkin method for (4.4.138)–(4.4.139) is to take

$$z_n(t) = \sum_{j=1}^{d_n} \gamma_j \phi_j(t), \quad t \in C(D) \tag{4.4.144}$$

with $\{\gamma_1, \ldots, \gamma_d\}$ determined from solving the linear system

$$\sum_{j=1}^{d_n} \gamma_j \{\lambda(\phi_j, \phi_i)_n - (\mathcal{K}_n \phi_j, \phi_i)_n\} = (y, \phi_i)_n, \quad i = 1, \ldots, d_n \tag{4.4.145}$$

We also define the *iterated discrete Galerkin solution*:

$$\hat{z}_n = \frac{1}{\lambda}[y + \mathcal{K}_n z_n] \tag{4.4.146}$$

4.4.1. The discrete orthogonal projection operator

To aid in the analysis of the discrete Galerkin method, we introduce a "discrete analog" of the orthogonal projection operator $\mathcal{P}_n$. One of the ways of defining the orthogonal projection $\mathcal{P}_n$ is as follows. For each $x \in C(D)$, let $w = \mathcal{P}_n x$ be the unique element in $\mathcal{X}_n$ that satisfies

$$(w, \phi) = (x, \phi), \quad \text{all } \phi \in \mathcal{X}_n \tag{4.4.147}$$

This can be shown to be equivalent to the earlier definition of $\mathcal{P}_n$ in (3.1.14) of Chapter 3, and we leave showing this to the reader.

For $x \in C(D)$, define $w = \mathcal{Q}_n x$ to be the unique element in $\mathcal{X}_n$ that satisfies

$$(w, \phi)_n = (x, \phi)_n, \quad \text{all } \phi \in \mathcal{X}_n \tag{4.4.148}$$

We must show there exists a unique such $w \in \mathcal{X}_n$.

Given $x \in C(D)$, let

$$w = \sum_{j=1}^{d_n} \beta_j \phi_j \tag{4.4.149}$$

In the condition (4.4.148), let $\phi = \phi_i$, $i = 1, \ldots, d_n$. Then we obtain the linear system

$$\sum_{j=1}^{d_n} \beta_j (\phi_j, \phi_i)_n = (x, \phi_i)_n, \quad i = 1, \ldots, d_n \tag{4.4.150}$$

Denote this linear system by $G_n \beta = [(x, \phi_1)_n, \ldots, (x, \phi_d)_n]^T$. The matrix G_n is an example of a Gram matrix, and it is both symmetric and positive definite. To show it is positive definite, write

$$\alpha^T G_n \alpha = \sum_{i=1}^{d_n} \sum_{j=1}^{d_n} \alpha_i \alpha_j (\phi_j, \phi_i)_n = \|y\|_n^2, \quad \alpha \in \mathbf{R}^{d_n}$$

with

$$y = \sum_{j=1}^{d_n} \alpha_j \phi_j$$

Thus $\alpha^{\mathrm{T}} G_n \alpha \geq 0$ for all α, and $\alpha^{\mathrm{T}} G_n \alpha = 0$ if and only if $\alpha = 0$, by **H4** and the independence of $\{\phi_1, \ldots, \phi_d\}$. This proves the unique solvability of (4.4.150), and thus it shows the existence and uniqueness of $Q_n x$ for all $x \in C(D)$. Restating the definition: $Q_n x \in X_n$ and

$$(Q_n x, \phi)_n = (x, \phi)_n, \quad \text{all } \phi \in X_n \tag{4.4.151}$$

Introduce the diagonal matrix W_n, with

$$(W_n)_{ii} = v_i, \quad i = 1, \ldots, q_n$$

using the weights from the quadrature formula (4.4.141) defining the discrete inner product. Then it is relatively straightforward to show

$$G_n = \Phi_n W_n \Phi_n^{\mathrm{T}} \tag{4.4.152}$$

In addition, the linear system (4.4.150) can be written

$$G_n \beta = \Phi_n W_n \mathcal{R}_n x \tag{4.4.153}$$

with $\mathcal{R}_n : C(D) \overset{onto}{\to} \mathbf{R}^{q_n}$ the *restriction operator*, defined by

$$\mathcal{R}_n x = [x(\tau_1), \ldots, x(\tau_q)]^{\mathrm{T}}$$

Since G_n is nonsingular,

$$\beta = G_n^{-1} \Phi_n W_n \mathcal{R}_n x$$

When combined with (4.4.149), we have that Q_n is linear. Moreover, by its definition $Q_n x = x$ for all $x \in X_n$, and therefore, Q_n is a projection.

Lemma 4.4.1.

(a) Q_n is self-adjoint on $C(D)$ with respect to the discrete inner product $(\cdot, \cdot)_n$:

$$(Q_n f, g)_n = (f, Q_n g)_n, \quad f, g \in C(D) \tag{4.4.154}$$

(b) For all $g \in C(D)$,

$$\|Q_n g\|_n \leq \|g\|_n \tag{4.4.155}$$

(c) Assume the family $\{Q_n \mid n \geq 1\}$ is uniformly bounded on $C(D)$, using the usual operator norm induced by the uniform norm $\|\cdot\|_\infty$. Then

$$\|g - Q_n g\|_\infty \leq c \inf_{\phi \in X_n} \|g - \phi\|_\infty \tag{4.4.156}$$

with c independent of n and g.

(d) When $q_n = d_n$, the discrete projection $\mathcal{Q}_n$ is an interpolating projection operator:

$$(\mathcal{Q}_n g)(\tau_i) = g(\tau_i), \quad i = 1, \ldots, q_n, \ g \in C(D) \qquad (4.4.157)$$

Proof.

(a) Use (4.4.151) with $x = f$ and $\phi = \mathcal{Q}_n g$ to get

$$(\mathcal{Q}_n f, \mathcal{Q}_n g)_n = (f, \mathcal{Q}_n g)_n$$

Then use (4.4.151) with $x = \mathcal{Q}_n f$ and $\phi = g$ to get

$$(\mathcal{Q}_n g, \mathcal{Q}_n f)_n = (g, \mathcal{Q}_n f)_n$$

Combine these and use the symmetry of the discrete inner product to show (4.4.154).

(b)

$$\begin{aligned}
\|\mathcal{Q}_n g\|_n^2 &= (\mathcal{Q}_n g, \mathcal{Q}_n g)_n \\
&= (g, \mathcal{Q}_n g)_n, && \text{using } \phi = \mathcal{Q}_n g \text{ in (4.4.151)} \\
&\leq \|g\|_n \|\mathcal{Q}_n g\|_n, && \text{using the Cauchy-Schwarz inequality}
\end{aligned}$$

Cancel $\|\mathcal{Q}_n g\|_n$ to complete the proof of (4.4.155).

(c) Let

$$m = \sup_{n \geq 1} \|\mathcal{Q}_n\|_\infty < \infty \qquad (4.4.158)$$

which is finite by assumption. Then for any $\phi \in \mathcal{X}_n$,

$$g - \mathcal{Q}_n g = g - \phi + \mathcal{Q}_n \phi - \mathcal{Q}_n g = (I - \mathcal{Q}_n)(g - \phi)$$
$$\|g - \mathcal{Q}_n g\|_\infty \leq (1 + m)\|g - \phi\|_\infty$$

Since ϕ is arbitrary, this proves (4.4.156).

(d) Note that in this instance Φ_n is square and nonsingular (by **H2**). Recall from (4.4.149) that $\mathcal{Q}_n g$ satisfies

$$\mathcal{Q}_n g = \sum_{j=1}^{d_n} \beta_j \phi_j$$

Evaluating $\mathcal{Q}_n g$ at the nodes $\{\tau_i\}$, we get

$$
\begin{aligned}
\mathcal{R}_n \mathcal{Q}_n g &= \Phi_n^T \beta, && \text{from the definition of } \Phi_n \\
&= \Phi_n^T G_n^{-1} \Phi_n W_n \mathcal{R}_n x, && \text{from (4.4.153)} \\
&= \Phi_n^T \left[\Phi_n W_n \Phi_n^T \right]_n^{-1} \Phi_n W_n \mathcal{R}_n x, && \text{from (4.4.152)} \\
&= \mathcal{R}_n x
\end{aligned}
$$

This completes the proof of (4.4.157). □

To prove that the operators $\{\mathcal{Q}_n\}$ are uniformly bounded, as in (4.4.158), can be difficult. This is discussed at some length in Atkinson and Bogomolny [56, p. 607]. In that paper it is shown that (4.4.158) is true in general when $\mathcal{X}_n$ is defined as a finite element type of approximating subspace, for example, piecewise polynomial functions. For subspaces $\mathcal{X}_n$ of globally defined smooth functions, we understood far less about $\mathcal{Q}_n$. For example, consider $\mathcal{X}_n$ to be a set of polynomials or trigonometric polynomials. For such choices of $\mathcal{X}_n$, the assumption (4.4.158) is most likely not true, and we conjecture for trigonometric polynomials that

$$
\|\mathcal{Q}_n\| = O(\log n)
$$

This is based on the same known rate of growth for $\|\mathcal{P}_n\|$ with $\mathcal{P}_n$ the interpolation projection operator for $\mathcal{X}_n$ the trigonometric polynomials of degree $\leq n$.

4.4.2. An abstract formulation

Using $\mathcal{Q}_n$, we can rewrite (4.4.144)–(4.4.145) in a simpler form. In fact, z_n is the solution to

$$
(\lambda - \mathcal{Q}_n \mathcal{K}_n) z_n = \mathcal{Q}_n y, \quad z_n \in C(D) \tag{4.4.159}
$$

To show the equivalence, first note that this equation implies

$$
z_n = \frac{1}{\lambda} [\mathcal{Q}_n y + \mathcal{Q}_n \mathcal{K}_n z_n] \in C(D)
$$

This implies $z_n = \mathcal{Q}_n z_n$, and thus (4.4.159) can be rewritten as

$$
\mathcal{Q}_n [y - (\lambda - \mathcal{K}_n) z_n] = 0, \quad z_n \in \mathcal{X}_n
$$

From the definition (4.4.151),

$$
\mathcal{Q}_n r = 0 \quad \text{if and only if} \quad (r, \phi_i) = 0, \quad i = 1, \ldots, d_n
$$

because $\{\phi_1, \ldots, \phi_d\}$ is a basis for $\mathcal{X}_n$. Thus (4.4.159) is equivalent to (4.4.144)–(4.4.145). The argument is reversible, thus showing the equivalence of the two formulations.

We can now use the same kind of analysis as was used for the discrete collocation method. For the iterated discrete Galerkin solution of (4.4.146), we have

$$\mathcal{Q}_n \hat{z}_n = \frac{1}{\lambda} [\mathcal{Q}_n y + \mathcal{Q}_n \mathcal{K}_n z_n] = z_n \qquad (4.4.160)$$

Substituting back into (4.4.146), we have $\hat{z}_n$ satisfies

$$(\lambda - \mathcal{K}_n \mathcal{Q}_n) \hat{z}_n = y \qquad (4.4.161)$$

In analogy with Theorem 4.3.1 for discrete collocation methods, we have the following.

Theorem 4.4.1. *Assume that $d_n = q_n$ for the number d_n of basis functions in $\mathcal{X}_n$ and the number q_n of quadrature nodes $\{\tau_k\}$. Moreover, assume **H2**. Then the iterated discrete Galerkin solution $\hat{z}_n$ is the Nyström solution associated with the numerical integral operator $\mathcal{K}_n$ of (4.4.140), and it satisfies the equation*

$$(\lambda - \mathcal{K}_n) \hat{z}_n = y \qquad (4.4.162)$$

In this case, $z_n(t)$ and $\hat{z}_n(t)$ agree at the quadrature node points $\{\tau_k\}$.

Proof. The proof uses Lemma 4.4.1(d). For general $z \in C(D)$,

$$\mathcal{K}_n \mathcal{Q}_n z(t) = \sum_{j=1}^{q_n} w_j(t)(\mathcal{Q}_n z)(\tau_j) = \sum_{j=1}^{q_n} w_j(t) z(\tau_j) = \mathcal{K}_n z(t)$$

Apply this to (4.4.161) to complete the proof of (4.4.162). The remainder of the proof follows from Lemma 4.4.1(d) and (4.4.160).   □

A number of the important examples of the discrete Galerkin method have $d_n = q_n$, and thus their error analysis is simply that of a Nyström method.

Example. Consider the integral equation

$$\lambda x(t) - \int_0^{2\pi} K(t,s) x(s) \, ds = y(t), \quad 0 \le t \le 2\pi \qquad (4.4.163)$$

in which we assume $y \in C_p(2\pi)$, the space of 2π-periodic functions. Also assume $K(t,s)$ is 2π-periodic with respect to both t and s. This equation was considered previously in Chapter 3; cf. (3.2.50)–(3.2.59) for collocation methods and (3.4.83)–(3.3.76) for Galerkin methods. We consider two different approximating subspaces $\mathcal{X}_n$.

(a) For an odd integer $n \geq 1$, let $h = 2\pi/n$, $\tau_j = jh$ for $j = 1, \ldots, n$. Let $\mathcal{X}_n$ be the periodic piecewise linear functions with breakpoints $\{0, h, 2h, \ldots, nh\}$. The dimension of $\mathcal{X}_n$ is $d_n = n$. For the quadrature methods, define both the inner product norm $(\cdot, \cdot)_n$ and the numerical integral operator $\mathcal{K}_n$ by using the rectangle rule:

$$\int_0^{2\pi} g(s)\, ds \approx h \sum_{j=1}^n g(\tau_j) \tag{4.4.164}$$

This is a simplification of the trapezoidal rule when applied to 2π-periodic functions $g(t)$. With this choice of approximation functions and numerical integration, the iterated discrete Galerkin method for (4.4.163) is simply the Nyström method with the quadrature rule (4.4.164).

The rate of convergence of the Galerkin and iterated Galerkin methods is

$$\|x - x_n\|_\infty = O(\|x - \mathcal{P}_n x\|_\infty) = O(h^2) \tag{4.4.165}$$

$$\|x - \hat{x}_n\|_\infty = O(\|\mathcal{K}x - \mathcal{K}\mathcal{P}_n x\|_\infty) = O(h^4) \tag{4.4.166}$$

provided $x(t)$ and $K(t, s)$ are sufficiently differentiable. From the error analysis for the Nyström method, we know the speed of convergence of $\hat{z}_n$ is determined by the numerical integration error $\|\mathcal{K}x - \mathcal{K}_n x\|_\infty$; cf. Theorem 4.1.2. For periodic functions, the rectangle rule (which in this case is the trapezoidal rule) is extremely rapid in its convergence, provided that $K(t, s)x(s)$ is sufficiently differentiable with respect to s, uniformly in t; cf. [48, p. 288]. If $K(t, s)x(s)$ is C^∞ with respect to s, uniformly in t, then

$$\|x - \hat{z}_n\|_\infty = O(\|\mathcal{K}x - \mathcal{K}_n x\|_\infty) = O(h^p) \tag{4.4.167}$$

for every $p > 0$. Since z_n and $\hat{z}_n$ agree at the node points, we have the same rate of convergence for the discrete Galerkin solution z_n at the node points. Contrast these results with (4.4.165) and (4.4.166).

(b) For the integer n of (a), let $m = \frac{1}{2}(n-1)$. Let $\mathcal{X}_n$ be the set of trigonometric polynomials of degree $\leq m$, so $d_n = n$. The Galerkin system in this case is given in (3.3.76) in Chapter 3 (with m replaced by n). As in (a), use the rectangular rule (4.4.164) to define the discrete Galerkin and iterated discrete Galerkin solutions.

The convergence results for the Galerkin method are much better in this case. If $K(t, s)x(s)$ belongs to $C_p^{p,\alpha}(2\pi)$ with respect to s, uniformly in t, then

$$\|x - x_n\|_\infty \leq \frac{c \log n}{n^{p+\alpha}}, \quad x \in C_p^{p,\alpha}(2\pi) \tag{4.4.168}$$

by applying (3.3.78) and (3.2.57) to the standard error formula $\|x - x_n\|_\infty = O(\|x - \mathcal{P}_n x\|_\infty)$. For the discrete Galerkin method, we know that $\hat{z}_n$ satisfies exactly the same Nyström method as before in (a). To obtain z_n from $\hat{z}_n$, use (4.4.160) and results on trigonmometric interpolation from (3.2.57)–(3.2.58) in Chapter 3. Comparing (4.4.166) and (4.4.168), the Galerkin and discrete Galerkin solutions are now comparable in their speeds of convergence.

The results in (a) are surprising, as they illustrate that the speed of convergence of the underlying projections need not be important in determining the speed of convergence of the iterated discrete Galerkin method or of the discrete Galerkin solution at the quadrature node points. Also, the same Nyström method can be regarded as arising from quite different Galerkin methods, as in (a) and (b).

For further discussions of the discrete Galerkin method for the case $d_n = q_n$, see Refs. [56] and [68]. The latter also contains a definition and analysis of higher-order methods for integral equations in which the kernel function is a Green's function for a boundary value problem for an ordinary differential equation.

Theorem 4.4.2. *Assume the family $\{\mathcal{K}_n\}$ of (4.4.140) is collectively compact and pointwise convergent to $\mathcal{K}$ on $C(D)$ to $C(D)$. Assume for the discrete orthogonal projection operators $\mathcal{Q}_n$ that*

$$m = \sup \|\mathcal{Q}_n\| < \infty \tag{4.4.169}$$

when regarded as operators on $C(D)$ to $C(D)$ with the uniform norm. Further assume that for every $g \in C(D)$,

$$\sup_{\phi \in \mathcal{X}_n} \|g - \phi\|_\infty \to 0 \quad as \ n \to \infty \tag{4.4.170}$$

Finally, assume the integral equation

$$\lambda x(t) - \int_D K(t, s)x(s)\, ds = y(t), \quad t \in D$$

is uniquely solvable for all $y \in C(D)$. Then

(a) *The families* $\{\mathcal{K}_n \mathcal{Q}_n\}$ *and* $\{\mathcal{Q}_n \mathcal{K}_n\}$ *are collectively compact and pointwise convergent families on* $C(D)$ *to* $C(D)$, *converging to* $\mathcal{K}$.

(b) *The inverse operators* $(\lambda - \mathcal{K}_n \mathcal{Q}_n)^{-1}$ *and* $(\lambda - \mathcal{Q}_n \mathcal{K}_n)^{-1}$ *exist for all sufficiently large* n, *say* $n \geq N$, *and they are uniformly bounded in* n.

(c) *For the errors in* z_n *and* $\hat{z}_n$, *use*

$$(\lambda - \mathcal{Q}_n \mathcal{K}_n)(x - z_n) = (I - \mathcal{Q}_n)x + \mathcal{Q}_n(\mathcal{K} - \mathcal{K}_n)x \qquad (4.4.171)$$

$$(\lambda - \mathcal{K}_n \mathcal{Q}_n)(x - \hat{z}_n) = \mathcal{K}x - \mathcal{K}_n \mathcal{Q}_n x$$

$$= (\mathcal{K}x - \mathcal{K}_n x) + \mathcal{K}_n(I - \mathcal{Q}_n)x \qquad (4.4.172)$$

and both z_n *and* $\hat{z}_n$ *converge to* x *as* $n \to \infty$.

Proof. For the pointwise convergence of $\{\mathcal{K}_n \mathcal{Q}_n\}$ to $\mathcal{K}$, let $z \in C(D)$ be arbitrary. Then

$$\mathcal{K}z - \mathcal{K}_n \mathcal{Q}_n z = [\mathcal{K}z - \mathcal{K}_n z] + \mathcal{K}_n[z - \mathcal{Q}_n z] \qquad (4.4.173)$$

$$\|\mathcal{K}z - \mathcal{K}_n \mathcal{Q}_n z\|_\infty \leq \|\mathcal{K}z - \mathcal{K}_n z\|_\infty + \|\mathcal{K}_n\| \|z - \mathcal{Q}_n z\|_\infty \qquad (4.4.174)$$

The terms $\|\mathcal{K}_n\|$ are uniformly bounded by the collective compactness of $\{\mathcal{K}_n\}$. The term $\|\mathcal{K}z - \mathcal{K}_n z\|_\infty$ converges to zero by the assumptions on $\{\mathcal{K}_n\}$, and $\|z - \mathcal{Q}_n z\|_\infty$ converges to zero by Lemma 4.4.1(c). The proof of the pointwise convergence of $\{\mathcal{Q}_n \mathcal{K}_n\}$ is similar.

The collective compactness of $\{\mathcal{K}_n \mathcal{Q}_n\}$ and $\{\mathcal{Q}_n \mathcal{K}_n\}$ are shown by standard arguments of the type used in proving Theorem 4.3.2. We leave these to the reader. The existence and uniform boundedness of $(\lambda - \mathcal{K}_n \mathcal{Q}_n)^{-1}$ and $(\lambda - \mathcal{Q}_n \mathcal{K}_n)^{-1}$ are straightforward consequences of Theorem 4.1.2. The formulas (4.4.171) and (4.4.172) are standard manipulations of the error formulas, and we omit them.

We note that $(\lambda - \mathcal{K}_n \mathcal{Q}_n)^{-1}$ and $(\lambda - \mathcal{Q}_n \mathcal{K}_n)^{-1}$ are linked, via Lemma 3.4.1, in the same manner as was true for discrete collocation methods in (4.3.127) and (4.3.128). More precisely,

$$(\lambda - \mathcal{K}_n \mathcal{Q}_n)^{-1} = \frac{1}{\lambda}[I + \mathcal{K}_n(\lambda - \mathcal{Q}_n \mathcal{K}_n)^{-1}\mathcal{Q}_n]$$

$$(\lambda - \mathcal{Q}_n \mathcal{K}_n)^{-1} = \frac{1}{\lambda}[I + \mathcal{Q}_n(\lambda - \mathcal{K}_n \mathcal{Q}_n)^{-1}\mathcal{K}_n]$$

Thus one of the inverses can be bounded in terms of the other. □

The formulas (4.4.173) and (4.4.174) can be used to obtain bounds on rates of convergence for particular choices of approximating subspaces $\mathcal{X}_n$ and numerical integration schemes $\mathcal{K}_n$ and $(\cdot, \cdot)_n$.

Lemma 4.4.2. Assume that the numerical integration operator $\mathcal{K}_n$ is defined using the same quadrature scheme as is used in defining the discrete inner product $(\cdot, \cdot)_n$:

$$\mathcal{K}_n x(t) = \sum_{j=1}^{q_n} v_j K(t, \tau_j) x(\tau_j), \quad x \in C(D)$$

Let $k_t(s) \equiv K(t, s), s, t \in D$. Then

$$\mathcal{K}x(t) - \mathcal{K}_n \mathcal{Q}_n x(t) = [\mathcal{K}x(t) - \mathcal{K}_n x(t)] + ((I - \mathcal{Q}_n)k_t, (I - \mathcal{Q}_n)x)_n$$

$$(4.4.175)$$

and

$$|\mathcal{K}x(t) - \mathcal{K}_n \mathcal{Q}_n x(t)| \leq |\mathcal{K}x(t) - \mathcal{K}_n x(t)|$$
$$+ \|(I - \mathcal{Q}_n)k_t\|_{n,1} \|(I - \mathcal{Q}_n)x\|_{n,\infty}$$

$$(4.4.176)$$

The seminorms are

$$\|g\|_{n,1} = \sum_{j=1}^{q_n} v_j |g(\tau_j)|, \quad \|gt\|_{n,\infty} = \max_{1 \leq j \leq q_n} |g(\tau_j)|$$

Proof. Recall the formula (4.4.172). From Lemma 4.4.1(a) and the fact that $\mathcal{Q}_n$ is a projection,

$$\mathcal{K}_n(I - \mathcal{Q}_n)x(t) = (k_t, (I - \mathcal{Q}_n)x)_n$$
$$= (k_t, (I - \mathcal{Q}_n)^2 x)_n$$
$$= ((I - \mathcal{Q}_n)k_t, (I - \mathcal{Q}_n)x)_n$$

This shows (4.4.175); and (4.4.176) is a straightforward calculation. $\square$

Theorem 4.4.3. Assume that the numerical integration operator $\mathcal{K}_n$ is defined using the same quadrature scheme as is used in defining the discrete inner product $(\cdot, \cdot)_n$. Assume that for some $l > 0$, the differentiability of the kernel function $K(t, s)$ is such as to have

$$\mathcal{K}x(t) - \mathcal{K}_n x(t) = O(n^{-l}) \qquad (4.4.177)$$

uniformly in t, for all sufficiently smooth $x(t)$. Assume that for some $r > 0$,

$$\rho_n(g) \equiv \inf_{\phi \in \mathcal{X}_n} \|g - \phi\|_\infty = O(n^{-r}) \qquad (4.4.178)$$

for all sufficiently smooth $g \in C(D)$. For the discrete orthogonal projection *operators, assume*

$$m \equiv \sup_{n \geq 1} \|Q_n\| < \infty \qquad (4.4.179)$$

Finally, assume the integral equation

$$\lambda x(t) - \int_D K(t,s)x(s)\,ds = y(t), \quad t \in D$$

is uniquely solvable for all $y \in C(D)$. Then for the discrete and iterated discrete Galerkin solutions of this integral equation,

$$\|x - z_n\|_\infty = O(n^{-\min\{r,l\}}) \qquad (4.4.180)$$

$$\|x - \hat{z}_n\|_\infty = O(n^{-\min\{2r,l\}}) \qquad (4.4.181)$$

provided $K(t,s)$ is sufficiently differentiable with respect to s, uniformly in t.

Proof. From Theorem 4.4.2 and (4.4.171), and using Lemma 4.4.1(c),

$$\|x - z_n\|_\infty \leq c[\|(I - Q_n)x\|_\infty + \|Q_n\| \|(\mathcal{K} - \mathcal{K}_n)x\|_\infty]$$
$$\leq O(n^{-r}) + \|Q_n\|O(n^{-l}) \qquad (4.4.182)$$

The result (4.4.180) then follows from (4.4.179). The formula (4.4.181) follows from (4.4.176) in Lemma 4.4.2. $\qquad \square$

Compare these results to that of the original Galerkin method in this situation:

$$\|x - x_n\|_\infty = O(\|x - \mathcal{P}_n x\|_\infty) = O(h^r) \qquad (4.4.183)$$

$$\|x - \hat{x}_n\|_\infty = O(\|\mathcal{K}x - \mathcal{K}\mathcal{P}_n x\|_\infty) = O(h^{2r}) \qquad (4.4.184)$$

For $\hat{z}_n$ to retain the speed of convergence associated with the iterated Galerkin solution $\hat{x}_n$, formula (4.4.181) says the numerical integration scheme must satisfy $l \geq 2r$. But for the discrete Galerkin solution we need only the bound $l \geq r$. The result (4.4.181) generalizes a result of Chandler [105] for one-dimensional integral equations and piecewise polynomial subspaces.

The bound in (4.4.182) allows $\|Q_n\|$ to grow with n, and we will still have convergence of z_n provided $\|Q_n\|$ grows less rapidly than n^l. Of course, we would still need to show the existence and uniform boundedness of the operators $(\lambda - Q_n \mathcal{K}_n)^{-1}$ [or of $(\lambda - \mathcal{K}_n Q_n)^{-1}$] for sufficiently large n. For cases in which $\|Q_n\|$ grows with n, this remains an open problem. It is of particular interest for integral equations over surfaces that are smooth deformations of the unit sphere and for which spherical polynomials are used to define the approximating subspaces $\mathcal{X}_n$. We discuss this further in Chapters 5 and 9.

Aside from the case $q_n = d_n$, the most interesting applications of the above Theorems 4.4.2 and 4.4.3 are to integral equations for functions x of more than one variable. Consequently, we defer numerical examples of this material to the following chapter.

Discussion of the literature

The quadrature-based approximation (4.1.5) for solving $(\lambda - \mathcal{K})x = y$ has a long history, with David Hilbert having used it to develop some of his theoretical results on the solvability of Fredholm integral equations. The error analysis for (4.1.5) has been studied by many authors, with many of the analyses showing

$$\max_i |x(t_i) - x_n(t_i)| \to 0 \quad \text{as } n \to \infty$$

without making use of the Nyström interpolation formula (4.1.6). Such error analyses have related the linear system (4.1.5) to either a degenerate kernel approximation or a collocation solution of $(\lambda - \mathcal{K})x = y$. For such analyses, see Bückner [98, p. 459], Kantorovich [303], and Prenter [433].

It is interesting to note that Nyström [409] introduced his interpolation formula (4.1.6) to extend accurately the solution he obtained when using only a small number of Gaussian quadrature points, for example, $n = 3$. The Nyström interpolation formula is quite accurate, as it generally maintains at all points of the interval the accuracy obtained at the node points.

The problem of analyzing (4.1.5) has been a popular one in the research literature. Historically, the approximation of $(\lambda - \mathcal{K})x = y$ by means of a quadrature-based approximation $(\lambda - \mathcal{K}_n)x_n = y$ was more difficult to analyze since $\mathcal{K}_n$ does not converge to $\mathcal{K}$ in the operator norm for the space of continuous functions. A number of people made contributions to the development of the error analysis of $(\lambda - \mathcal{K}_n)x_n = y$, and we note especially Brakhage [85] and Mysovskih [392]. The final stage in developing a general framework for the error analysis was the creation by Phil Anselone of the theory of collectively compact operator approximations, and this is described in detail in his book Anselone [16]. This framework also led directly to a complete error analysis for the solution of the associated eigenvalue problem and to generalizations of the Nyström method to Fredholm integral equations with a weakly singular kernel function. Other related discussions of this theory can be found in Ref. [39, pp. 88–106]; Baker [73, pp. 356–374]; Fenyö and Stolle [200, Chap. 19]; Hackbusch [251, §4.7]; Kantorovich and Krylov [305]; Kress [325, Chap. 12]; Linz [344, §6.3]; Noble [405]; Stummel [538]; Thomas [541], [542]; and Vainikko [548], [549].

A program with automatic error control that implements the Nyström method for solving $(\lambda - \mathcal{K})x = y$ when the kernel function is nonsingular is given in Ref. [37], [38]. This uses the two-grid iteration method of §6.2.2 in Chapter 6. Another automatic program has been written by Hosae Lee [338], and it uses a multigrid iteration method that is similar to that discussed in §6.4.

The eigenvalue problem $\mathcal{K}x = \lambda x$ has been well-studied, and we refer the reader to Fenyö and Stolle [198, Chaps. 8, 9], Hille and Tamarkin [271], and König [315] for additional information on the behavior of the eigenvalues of $\mathcal{K}$. Various specialized results on the convergence of the eigenvalues of $\mathcal{K}_n$ to those of $\mathcal{K}$ go back to the early years of this century. For analyses that are both general and more modern, we refer the reader to Refs. [30], [36]; Anselone and Palmer [19]; Brakhage [86]; Cubillos [147], [148]; Osborn [411]; Spence [522]–[524]; and Vainikko [547]. A complete compilation and extension of these results is given in the comprehensive book of Chatelin [113].

The idea of product integration for approximating integral operators appears to have originated with Young [585]. In its present form using piecewise polynomial interpolation, it was given in Refs. [28], [31], [32]. These results were improved by deHoog and Weiss [164], again assuming the functions $x(t)$ being interpolated were sufficiently differentiable. Later it was recognized that with kernel functions containing algebraic and/or logarithmic singularities, the solution function would contain associated singular behavior around the endpoints of the interval of integration. See Graham [227], Richter [462], and Schneider [493] for various contributions on the behavior of the solutions of $(\lambda - \mathcal{K})x = y$. Due to the endpoint behavior of the solution, a graded mesh is needed in defining the piecewise polynomial interpolation (as was applied in §4.2.5), and this seems to have first been introduced by Schneider [494]. These ideas have since been extended to the solution of boundary integral equations on two-dimensional regions with boundaries that are only piecewise smooth, and we take this up in Chapter 8. As an alternative to the use of graded meshes, Cao and Xu [103] introduce additional basis functions that reflect the singular behavior of the solution function.

Discrete projection methods are further discretizations of projection methods, in which the needed integrations are performed numerically. Since collocation and Galerkin methods contain integrals, it is important that these integrations be performed with such accuracy as to preserve the speed of convergence of the original projection solution. But it is also necessary to make the numerical integrations as efficient as possible, to avoid unnecessary calculations. Most early discussions of projection methods ignored these considerations, leaving it to the user to decide in each individual case as to the form of numerical integration to be used. Early examinations of discrete Galerkin methods were given by Spence and Thomas [525] and Joe [293], and early discussions of discrete collocation methods were given by Brunner [94] and Joe [292]. Most of these papers regarded the fully discretized problem as a perturbation of the original projection method, which was the basis of the error analysis. In Atkinson and Bogomolny [56] and Atkinson and Potra [68] the discretized problem is regarded as a new approximation and is analyzed using the theory of collectively compact operator approximations. In Flores [201] and Golberg [219, Chap. 3] these same ideas are used to analyze discrete collocation methods. The latter is also an excellent survey of the entire area of discrete projection methods for solving $(\lambda - \mathcal{K})x = y$.

Nonlinear problems $\mathcal{K}(x) = \lambda x$, with $\mathcal{K}$ completely continuous, have also been solved with generalizations of the Nyström method. A general theory for such methods was given in Ref. [33], and it was subsequently simplified in the most important cases by Weiss [556]. Discrete collocation methods for such nonlinear problems are given in Atkinson and Flores [60], and discrete Galerkin methods are given in Atkinson and Potra [68]. A survey of numerical methods for solving $\mathcal{K}(x) = \lambda x$ is given in [51].

5
Solving multivariable integral equations

The material of earlier chapters was illustrated with integral equations for functions of a single variable. In this chapter we develop interpolation and numerical integration tools and use them with the projection and Nyström methods, developed in Chapters 3 and 4, to solve multivariable integral equations. Our principal interest will be the solution of integral equations defined on surfaces in $\mathbf{R}^3$, with an eye towards solving boundary integral equations on such surfaces. The solution of boundary integral equations on piecewise smooth surfaces is taken up in Chapter 9.

In §5.1 we develop interpolation and numerical integration formulas for multivariable problems, and these are applied to integral equations over planar regions in §5.2. The interpolation and integration results of §5.1 are extended to surface problems in §5.3. Methods for the numerical solution of integral equations on surfaces are given in §§5.4 and 5.5.

5.1. Multivariable interpolation and numerical integration

Interpolation for functions of more than one variable is a large topic with applications to many areas. In this section we consider only those aspects of the subject that we need for our work in the numerical solution of integral equations. To simplify the notation and to make more intuitive the development, we consider only functions of two variables. Generalizations to functions in more than two variables should be fairly straightforward for the reader.

Applications of multivariable interpolation are generally based on first breaking up a large planar region R into smaller ones of an especially simple form, and then polynomial interpolation is carried out over these smaller regions. The two most common shapes for these smaller regions are the rectangle and the triangle. The result on the original region R is a piecewise polynomial

interpolation function comparable to the piecewise polynomial functions of
one variable that we used in earlier chapters. Polynomial interpolation over
rectangular regions is generally based on using repeated one-variable interpolation formulas. We give a simple introduction to such interpolation, while
leaving some of its development and illustration to the reader. Most applications of multivariable interpolation are based on using triangular subregions,
and that will be our principal focus, beginning in the next subsection.

For a given function $g(x)$, the polynomial

$$p_n(x) = \sum_{i=0}^{n} g(x_i)\ell_i(x) \tag{5.1.1}$$

$$\ell_i(x) = \prod_{\substack{j=0 \\ j \neq i}}^{n} \frac{x - x_j}{x_i - x_j}, \quad i = 0, \ldots, n \tag{5.1.2}$$

interpolates $g(x)$ at $x = x_0, \ldots, x_n$. This formula for $p_n(x)$ is called Lagrange's
form of the interpolation polynomial. For its error, we have

$$g(x) - p_n(x) = \frac{\omega_n(x)}{(n+1)!} g^{(n+1)}(\xi_n) \tag{5.1.3}$$

with

$$\omega_n(x) = (x - x_0) \cdots (x - x_n)$$

for some $\xi_n \in [\min\{x_0, \ldots, x_n, x\}, \max\{x_0, \ldots, x_n, x\}]$. There is also the additional error formula

$$g(x) - p_n(x) = \omega_n(x)g[x_0, \ldots, x_n, x] \tag{5.1.4}$$

which uses the Newton divided difference of order $n + 1$. For a complete
development of this, see Ref. [48, Chap. 3, Sections 1, 2, 5].

Let $R = [a, b] \times [c, d]$, and assume $g \in C(R)$. Let $[a, b]$ and $[c, d]$ be
partitioned by

$$a \leq x_0 < \cdots < x_m \leq b$$
$$c \leq y_0 < \cdots < y_n \leq d \tag{5.1.5}$$

Define

$$p_{m,n}(x, y) = \sum_{i=0}^{m} \sum_{j=0}^{n} g(x_i, y_j)\ell_{i,m}(x)\ell_{j,n}(y) \tag{5.1.6}$$

with $\ell_{i,m}(x)$ and $\ell_{j,n}(y)$ defined suitably from (5.1.2), using the respective nodes $\{x_i\}$ and $\{y_j\}$. This polynomial $p_{m,n}(x, y)$ interpolates $g(x, y)$ at $(x, y) = (x_i, y_j)$, $i = 0, \dots, m$, $j = 0, \dots, n$; it is of degree $r = m + n$. The function $p_{m,n}(x, y)$ does not include all polynomials of degree $m + n$. All terms $x^k y^l$ in the expansion of (5.1.6) must satisfy $k \le m$ and $l \le n$ as well as $k + l \le m + n$. A popular form of (5.1.6) is to take $m = n = 1$, yielding the *bilinear interpolation polynomial*

$$
p_{1,1}(x, y) = \frac{(b - x)(d - y)}{(b - a)(d - c)} g(a, c) + \frac{(b - x)(y - c)}{(b - a)(d - c)} g(a, d)
$$
$$
+ \frac{(x - a)(d - y)}{(b - a)(d - c)} g(b, c) + \frac{(x - a)(y - c)}{(b - a)(d - c)} g(b, d) \quad (5.1.7)
$$

It interpolates $g(x, y)$ at the nodes $\{(a, c), (a, d), (b, c), (b, d)\}$. This is a quadratic polynomial, but no term of the form x^2 and y^2 is included in the expansion of $p_{1,1}(x, y)$.

For the error in $p_{m,n}(x, y)$ we can use (5.1.3) to obtain the following.

Theorem 5.1.1. *Let $R = [a, b] \times [c, d]$, and assume the interpolation nodes $\{(x_i, y_j)\}$ satisfy (5.1.5). Assume $\partial^{m+1} g(x, y)/\partial x^{m+1}$ and $\partial^{n+1} g(x, y)/\partial y^{n+1}$ exist and are continuous for all $(x, y) \in R$. Then for any $(x, y) \in R$,*

$$
|g(x, y) - p_{m,n}(x, y)| \le \frac{|\omega_m(x)|}{(m + 1)!} \max_{a \le \xi \le b} \left| \frac{\partial^{m+1} g(\xi, y)}{\partial \xi^{m+1}} \right|
$$
$$
+ \Lambda_m(x) \frac{|\omega_n(y)|}{(n + 1)!} \max_{(\xi, \eta) \in R} \left| \frac{\partial^{n+1} g(\xi, \eta)}{\partial \eta^{n+1}} \right| \quad (5.1.8)
$$

with

$$
\Lambda_m(x) = \sum_{i=0}^{m} |\ell_{i,m}(x)|
$$

Proof. Write the error as

$$
g(x, y) - p_{m,n}(x, y) = \left[g(x, y) - \sum_{i=0}^{m} g(x_i, y)\ell_{i,m}(x) \right]
$$
$$
+ \sum_{i=0}^{m} \ell_{i,m}(x) \left[g(x_i, y) - \sum_{j=0}^{n} g(x_i, y_j)\ell_{j,n}(y) \right]
$$

Take bounds and use (5.1.3) to complete the proof. An alternative form of the error can be obtained by decomposing the error first with respect to interpolation over y, rather than over x as above.   □

An application of the use of bilinear interpolation was given in §2.3. The integral equation (2.3.54) was solved numerically by using piecewise bilinear interpolation to define a degenerate kernel approximation of the integral operator in the equation; cf. (2.3.55). The error results (2.3.56)–(2.3.57) were obtained by applying the error analysis of the above lemma.

5.1.1. *Interpolation over triangles*

Let Δ denote a planar triangle in the xy-plane, and let $g(x, y)$ be a continuous function on Δ. Consider approximating g by a polynomial interpolant $p(x, y)$ of degree r, for some $r \geq 0$. How does one construct such a polynomial? We begin by noting that the polynomial

$$p(x, y) = \sum_{\substack{i,j \geq 0 \\ i+j \leq r}} c_{i,j} x^i y^j$$

has

$$f_r \equiv \frac{(r + 1)(r + 2)}{2} \tag{5.1.9}$$

degrees of freedom; thus, f_r interpolation conditions must be imposed on $p(x, y)$ to determine the coefficients $\{c_{i,j}\}$. We require

$$p(x_k, y_k) = g(x_k, y_k), \quad k = 1, \ldots, f_r$$

for some choice of f_r interpolation nodes $\{(x_k, y_k) \mid 1 \leq k \leq f_r\} \subset \Delta$. Before defining these nodes and constructing $p(x, y)$, we simplify the problem by transforming it to an equivalent interpolation problem over a standard triangle.

Let the vertices of Δ be denoted by $\{v_1, v_2, v_3\}$, $v_j \equiv (x_j, y_j)$. Introduce the *unit simplex*

$$\sigma = \{(s, t) \mid s, t \geq 0, s + t \leq 1\}$$

Define $T : \sigma \overset{1-1}{\underset{onto}{\to}} \Delta$ by

$$(x, y) = T(s, t) \equiv u v_1 + t v_2 + s v_3, \quad u = 1 - s - t \tag{5.1.10}$$

The coordinates (s, t, u) are called the *barycentric coordinates* of a point $(x, y) \in \Delta$. This type of mapping is called an *affine* mapping, being a translation of a linear mapping. Given an $(x, y) \in \Delta$, the inverse of this mapping is given by solving for (s, t) in the equation $(x, y) = T(s, t)$, or equivalently, by solving the linear system

$$
\begin{aligned}
s(x_3 - x_1) + t(x_2 - x_1) &= x - x_1 \\
s(y_3 - y_1) + t(y_2 - y_1) &= y - y_1
\end{aligned}
\tag{5.1.11}
$$

for (s, t). This system is nonsingular since the vertices $\{v_1, v_2, v_3\}$ have implicitly been assumed to not be collinear. Denote the inverse mapping by $(s, t) = Q(x, y)$, and it too is an affine mapping. We assert that if $p(x, y)$ is a polynomial of degree r in (x, y), then $P(s, t) \equiv p(T(s, t))$ is a polynomial of degree r in (s, t). Conversely, if $P(s, t)$ is a polynomial of degree r in (s, t), then $p(x, y) \equiv P(Q(x, y))$ is a polynomial of degree r in (x, y). The proof of this is left as an exercise for the reader.

We define interpolation over σ, and then we use the affine mapping $T(s, t)$ to define a corresponding interpolation polynomial over Δ. From what was just said in the last paragraph concerning the mappings T and Q, the degrees of the corresponding polynomials over σ and Δ will be the same.

For $r \geq 1$, define $\delta = 1/r$ and

$$(s_i, t_j) = (i\delta, j\delta), \quad i, j \geq 0, \quad i + j \leq r \tag{5.1.12}$$

These f_r nodes form a uniform grid over σ, and we will use them to define interpolation polynomials over σ. To make the development a bit more intuitive, we consider first the cases of degrees $r = 1$ and $r = 2$.

Example: Linear interpolation. Denote the above grid $\{(s_i, t_j)\}$ by

$$q_1 = (0, 0), \qquad q_2 = (0, 1), \qquad q_3 = (1, 0) \tag{5.1.13}$$

which are also the three vertices of σ. Introduce the Lagrange basis functions

$$\ell_1(s, t) = u, \qquad \ell_2(s, t) = t, \qquad \ell_3(s, t) = s \tag{5.1.14}$$

with

$$u = 1 - s - t$$

Then for a given $G \in C(\sigma)$, the formula

$$P_1(s, t) = \sum_{i=1}^{3} G(q_i)\ell_i(s, t) \tag{5.1.15}$$

is the unique linear polynomial that interpolates $G(s, t)$ at the nodes $\{q_1, q_2, q_3\}$. We leave the proof of the uniqueness of $P_1(s, t)$ to the reader.

Example: Quadratic interpolation. There are now six nodes $\{(s_i, t_j)\}$ in the earlier grid, and we denote them by $\{q_1, \ldots, q_6\}$, using (5.1.13) and

$$q_4 = (0, 0.5), \qquad q_5 = (0.5, 0.5), \qquad q_6 = (0.5, 0) \tag{5.1.16}$$

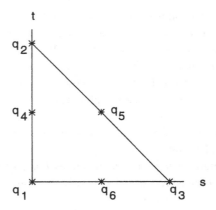

Figure 5.1. Unit simplex and quadratic interpolation nodes.

These are shown in Figure 5.1. Define

$$\begin{array}{lll} \ell_1 = u(2u - 1), & \ell_2 = t(2t - 1), & \ell_3 = s(2s - 1) \\ \ell_4 = 4tu, & \ell_5 = 4st, & \ell_6 = 4su \end{array} \qquad (5.1.17)$$

The polynomial

$$P_2(s, t) = \sum_{i=1}^{6} G(q_i)\ell_i(s, t) \qquad (5.1.18)$$

is the unique quadratic polynomial that interpolates $G(s, t)$ at the nodes $\{q_1, \ldots, q_6\}$. We leave the proof of the uniqueness of $P_2(s, t)$ to the reader.

We can continue this manner for interpolation of any degree $r \geq 1$, obtaining a formula

$$P_r(s, t) = \sum_{i=1}^{f_r} G(q_i)\ell_i(s, t), \quad (s, t) \in \sigma \qquad (5.1.19)$$

This is the Lagrange formula for the unique polynomial of degree $\leq r$ which interpolates $G(s, t)$ on the grid $\{(s_i, t_j)\}$ defined in (5.1.12). We use a sequential ordering of the nodes, $\{q_1, \ldots, q_{f_r}\}$, to simplify the notation and to lead to formulas more readily adaptable to implementation in computer languages such as Fortran.

Returning to our original triangle $\triangle$, we introduce a set of interpolation nodes on $\triangle$ by mapping $\{q_i\}$ into $\triangle$ by means of the mapping $T(s, t)$ of (5.1.10):

$$v_i = T(q_i), \quad i = 1, \ldots, f_r \qquad (5.1.20)$$

Given a function $g \in C(\Delta)$, the unique polynomial of degree $\leq r$ that interpolates g at the nodes $\{v_1, \ldots, v_{f_r}\}$ is given by

$$p_r(x, y) = \sum_{i=1}^{f_r} g(T(q_i))\ell_i(s, t), \ (x, y) = T(s, t)$$

$$= \sum_{i=1}^{f_r} g(v_i)\ell_{\Delta,i}(x, y) \qquad (5.1.21)$$

with

$$\ell_{\Delta,i}(x, y) = \ell_i(s, t), \quad (x, y) = T(s, t) \qquad (5.1.22)$$

In computer programs this is the usual form in which multivariable interpolation over triangles is defined and computed.

Piecewise polynomial interpolation

Given a polygonal region R in the plane $\mathbf{R}^2$, let $\mathcal{T}_n = \{\Delta_1, \ldots, \Delta_n\}$ denote a triangulation of R. We assume triangles Δ_j and Δ_k can intersect only at vertices or along all of a common edge. Later we assume additional properties for the triangulation. Also for later use, we denote the three vertices of Δ_k by $\{v_{k,1}, v_{k,2}, v_{k,3}\}$, with

$$v_{k,i} = T_k(q_i), \quad i = 1, 2, 3$$

with the $\{q_i\}$ of (5.1.13).

For each triangle Δ_k, we have a mapping $T_k : \sigma \xrightarrow[\text{onto}]{1-1} \Delta_k$ defined as in (5.1.10). For a given $g \in C(R)$, define $\mathcal{P}_n g$ by

$$\mathcal{P}_n g(T_k(s, t)) = \sum_{i=1}^{f_r} g(T_k(q_i))\ell_i(s, t), \quad (s, t) \in \sigma, \ k = 1, \ldots, n$$

$$(5.1.23)$$

We claim that $\mathcal{P}_n g(x, y)$ is a continuous function on R. To prove this requires more information about the formula (5.1.19) and the functions $\ell_j(s, t)$. We consider only the case of the quadratic interpolation formula (5.1.18), as a similar argument is valid for other degrees $r \geq 1$.

Assume Δ_j and Δ_k have a common edge; and without loss of generality, assume it corresponds to the edge $\{(0, t) \mid 0 \leq t \leq 1\}$ of σ, for both triangles. Consider the formulas (5.1.17) for (s, t) on that common edge of σ. The formula (5.1.18) becomes

$$P_2(0, t) = u(2u - 1)g(0, 0) + t(2t - 1)g(0, 1) + 4tug(0, 0.5)$$

which is a quadratic interpolation polynomial in t. This formula carries over to both triangles Δ_j and Δ_k, via the mappings T_j and T_k, and the uniqueness of quadratic interpolation will imply that the functions $\mathcal{P}_n g(T_k(0, t))$ and $\mathcal{P}_n g(T_k(0, t))$ must agree for all t. As stated earlier, this proof generalizes to other degrees $r \geq 1$, but it requires more information on the form of the basis functions $\ell_k(s, t)$ than has been included here.

The formula (5.1.23) defines a bounded projection on $C(R)$, with

$$\|\mathcal{P}_n\| = \max_{(s,t)\in\sigma} \sum_{j=1}^{f_r} |\ell_j(s, t)| \tag{5.1.24}$$

For the cases of linear and quadratic interpolation, considered earlier in (5.1.15) and (5.1.18),

$$\|\mathcal{P}_n\| = \begin{cases} 1, & \text{linear interpolation} \\ \dfrac{5}{3}, & \text{quadratic interpolation} \end{cases} \tag{5.1.25}$$

Let

$$\mathcal{X}_n = \text{range}(\mathcal{P}_n) \tag{5.1.26}$$

This subspace of $C(R)$ has dimension f_r, and it is often used as an approximating subspace when solving integral equations defined over the region R.

We have not yet considered the case of piecewise constant interpolation over a triangular mesh. For interpolation over σ, use

$$G(s, t) \approx G\left(\frac{1}{3}, \frac{1}{3}\right), \quad (s, t) \in \sigma \tag{5.1.27}$$

The point $(\frac{1}{3}, \frac{1}{3})$ is the centroid of σ. The analog of formula (5.1.23) is

$$\mathcal{P}_n g(T_k(s, t)) = g(c_k), \quad (s, t) \in \sigma, \quad k = 1, \dots, n \tag{5.1.28}$$

with

$$c_k = \frac{1}{3}(v_{k,1} + v_{k,2} + v_{k,3})$$

the centroid of Δ_k. The function $\mathcal{P}_n g$ is no longer continuous over R, but it can be regarded as a bounded projection on the larger space $L^\infty(R)$, with $\|\mathcal{P}_n\| = 1$. See [62] for details of how to extend $\mathcal{P}_n$ from $C(R)$ to $L^\infty(R)$.

Interpolation error formulas over triangles

Our first error formula shows that the above form of interpolation converges uniformly for all $g \in C(R)$.

Lemma 5.1.1. Let $\mathcal{T}_n$ be a triangulation of the polygonal region R. Let $g \in C(R)$, let $r \geq 0$ be an integer, and let $\mathcal{P}_n g$ be defined by (5.1.23) or (5.1.28). Then

$$\|g - \mathcal{P}_n g\|_\infty \leq \|\mathcal{P}_n\| \omega(\delta_n, g) \tag{5.1.29}$$

with $\omega(\delta, g)$ the modulus of continuity of g,

$$\omega(\delta, g) = \sup_{\substack{v, w \in R \\ |v-w| \leq \delta}} |g(v) - g(w)|$$

and δ_n the mesh size of the triangulation of R,

$$\delta_n = \max_{1 \leq k \leq n} \text{diameter}(\Delta_k)$$

Proof. Consider the interpolation error over a particular triangle $\Delta_k \in \mathcal{T}_n$. Use the identity

$$g(T_k(s, t)) - (\mathcal{P}_n g)(T_k(s, t))$$

$$= \sum_{j=1}^{f_r} [g(T_k(s, t)) - g(v_{k,j})] \ell_j(s, t), \quad (s, t) \in \sigma$$

Take bounds and use the result (5.1.24). □

The derivation of a more practical error formula for the interpolation of (5.1.21) is based on Taylor's formula with integral remainder. With it, we can obtain the following.

Lemma 5.1.2. Let Δ be a planar triangle, let $r \geq 0$ be an integer, and assume $g \in C^{r+1}(\Delta)$. Then for the interpolation polynomial $p_r(x, y)$ of (5.1.21),

$$\max_{(x,y) \in \Delta} |g(x, y) - p_r(x, y)| \leq c\delta^{r+1} \max_{\substack{i,j \geq 0 \\ i+j=r+1}} \max_{(\xi,\eta) \in \Delta} \left| \frac{\partial^{r+1} g(\xi, \eta)}{\partial \xi^i \partial \eta^j} \right| \tag{5.1.30}$$

with $\delta \equiv \text{diameter}(\Delta)$. The constant c depends on r, but it is independent of both g and Δ.

Proof. Introduce

$$G(s, t) = g(T(s, t))$$

Expand this as a Taylor polynomial about $(0, 0)$:

$$G(s, t) = H_r(s, t) + K_r(s, t)$$

with $H_r(s, t)$ the Taylor polynomial of degree r for $G(s, t)$ and

$$K_r(s, t) = \frac{1}{r!} \int_0^1 (1 - v)^r \frac{d^{r+1}G(vs, vt)}{dv^{r+1}} \, dv$$

Note that the interpolation of $H_r(s, t)$ by the method of (5.1.21) is exact because it is a polynomial of degree $\leq r$. Therefore,

$$g(x, y) - p_r(x, y) = K_r(s, t) - \sum_{j=1}^{f_r} K_r(q_j)\ell_j(s, t)$$

$$= \frac{1}{r!} \int_0^1 (1 - v)^r E_r(v; s, t) \, dv \qquad (5.1.31)$$

$$E_r(v; s, t) \equiv \frac{d^{r+1}G(vs, vt)}{dv^{r+1}} - \sum_{j=1}^{f_r} \ell_j(s, t) \frac{d^{r+1}G(vs_j, vt_j)}{dv^{r+1}}$$

with $q_j \equiv (s_j, t_j)$.

Consider the first derivative of $G(vs, vt)$:

$$\frac{dG(vs, vt)}{dv} = s \frac{\partial G}{\partial s} + t \frac{\partial G}{\partial t}$$

To relate these derivatives to those of $g(x, y)$, write

$$G(s, t) = g(x_1 + s(x_3 - x_1) + t(x_2 - x_1), y_1 + s(y_3 - y_1) + t(y_2 - y_1))$$

Then

$$\frac{\partial G(s, t)}{\partial s} = (x_3 - x_1)\frac{\partial g}{\partial x} + (y_3 - y_1)\frac{\partial g}{\partial y}$$

$$\frac{\partial G(s, t)}{\partial t} = (x_2 - x_1)\frac{\partial g}{\partial x} + (y_2 - y_1)\frac{\partial g}{\partial y} \qquad (5.1.32)$$

where the partial derivatives of g are evaluated at $(x, y) = T(s, t)$ and $v_i = (x_i, y_i)$, $i = 1, 2, 3$, are the vertices of Δ. With these formulas, we have

$$\max_{\substack{0 \leq v \leq 1 \\ (s,t)\in\sigma}} \left| \frac{dG(vs, vt)}{dv} \right| \leq c\,\delta \max \left\{ \left\| \frac{\partial g}{\partial x} \right\|_\infty, \left\| \frac{\partial g}{\partial y} \right\|_\infty \right\}$$

with

$$\left\| \frac{\partial g}{\partial x} \right\|_\infty = \max_{(x,y) \in \Delta} \left| \frac{\partial g(x,y)}{\partial x} \right|$$

and analogously for $\partial g / \partial y$. The constant c is independent of Δ and g.

Continuing in a similar fashion with the higher-order derivatives of G, and then combining the result with (5.1.31), we obtain the result (5.1.30).   □

Theorem 5.1.2. *Let R be a polygonal region in the plane* $\mathbf{R}^2$, *and let* $r \geq 0$ *be an integer.*

(a) *For all* $g \in C(R)$, *the interpolant* $\mathcal{P}_n g$ *converges uniformly to g on R.*
(b) *Assume* $g \in C^{r+1}(R)$. *Let* $\mathcal{T}_n = \{\Delta_1, \ldots, \Delta_n\}$ *be a triangulation of R, and define*

$$\delta_n = \max_{1 \leq k \leq n} \ diameter(\Delta_k)$$

Then

$$\|g - \mathcal{P}_n g\|_\infty \leq c \delta_n^{r+1} \|g\|_{r+1,\infty} \tag{5.1.33}$$

with

$$\|g\|_{r+1,\infty} \equiv \max_{\substack{i+j=r+1 \\ i,j \geq 0}} \max_{(x,y) \in R} \left| \frac{\partial^{r+1} g(x,y)}{\partial x^i \partial y^j} \right| \tag{5.1.34}$$

and c independent of n and g.

Proof. This is a straightforward application of the preceding two lemmas.   □

5.1.2. Numerical integration over triangles

Numerical integration over triangular regions is used in the numerical solution of both partial differential equations (in the finite element method) and multivariable integral equations. There is a wealth of literature on numerical integration over triangles, and we merely introduce the subject in this section. Much more extensive discussions are given in Lyness and Jespersen [357] and Stroud [537, pp. 306–315]. Our formulas are chosen with an eye toward their application in solving integral equations.

In analogy with interpolation over triangles, we first develop integration formulas over σ, and then these are extended to a general triangle Δ by the affine change of variables $(x, y) = T(s, t)$. In fact,

$$\int_\Delta g(x, y)\, dA = 2\,\text{Area}(\Delta) \int_\sigma g(T(s, t))\, d\sigma \qquad (5.1.35)$$

in which the area of Δ is computed by

$$\text{Area}(\Delta) = \frac{1}{2}\left| \det \begin{bmatrix} x_2 - x_1 & x_3 - x_1 \\ y_2 - y_1 & y_3 - y_1 \end{bmatrix} \right|$$

where $v_i = (x_i, y_i)$, $i = 1, 2, 3$, are the vertices of Δ. As a consequence, we consider the problem of approximating

$$I(g) \equiv \int_\sigma G(s, t)\, d\sigma$$

There are two major ways in which formulas are developed to approximate $I(g)$.

1. **Interpolation.** Replace $g(s, t)$ by a polynomial $P(s, t)$ that interpolates $G(s, t)$ at points $\{\mu_1, \ldots, \mu_f\}$ within σ. Then use

$$\int_\sigma G(s, t)\, d\sigma \approx \int_\sigma P(s, t)\, d\sigma = \sum_{j=1}^f w_j G(\mu_j) \qquad (5.1.36)$$

with the weights $\{w_j\}$ obtained by integrating the Lagrange basis functions in the Lagrange representation formula for $P(s, t)$. For example, consider integrating the interpolation formula (5.1.19). If the polynomial interpolation $P(s, t)$ is exact when $G(s, t)$ is a polynomial of degree $\leq r$, then the degree of precision d_f of the quadrature formula (5.1.36) is at least r, and it is possibly greater.

2. **Undetermined coefficients.** Choose a set of points $\{\mu_1, \ldots, \mu_f\} \subset \sigma$, and write

$$\int_\sigma G(s, t)\, d\sigma \approx \sum_{j=1}^f w_j G(\mu_j) \qquad (5.1.37)$$

Choose the weights $\{w_j\}$ so as to maximize the degree of precision of this formula, that is, choose the weights to make the integration formula exact for all polynomials of degree $\leq d_f$, with d_f chosen as large as possible. This leads to a linear system for $\{w_j\}$. Sometimes, the points $\{\mu_j\}$ are also considered as variables, thus generalizing the approach that leads to Gaussian quadrature formulas for functions of one variable.

Most of the formulas in Lyness and Jespersen [357] are developed by using the method of undetermined coefficients, with the addition of other constraints that correspond to other desirable properties for a quadrature rule (for example, positivity of the weights $\{w_j\}$). Stroud [537] uses both of the above approaches, as well as others.

With such formulas we can numerically approximate integrals over polygonal regions R:

$$I(g) = \int_R g(x, y) \, dA$$

Let $T_n = \{\Delta_1, \ldots, \Delta_n\}$ be a triangulation of R, and then write

$$I(g) = \sum_{k=1}^n \int_{\Delta_k} g(x, y) \, dA = 2 \sum_{k=1}^n \text{Area}(\Delta_k) \int_\sigma g(T_k(s, t)) \, d\sigma \qquad (5.1.38)$$

Apply an integration formula of the form (5.1.37) to obtain

$$I(g) \approx 2 \sum_{k=1}^n \text{Area}(\Delta_k) \sum_{j=1}^f w_j g(T_k(\mu_j)) \equiv I_n(g) \qquad (5.1.39)$$

where we have used the mapping $T_k : \sigma \overset{1-1}{\underset{onto}{\to}} \Delta_k$ defined as in (5.1.10). For the error,

$$E_n(g) \equiv I(g) - I_n(g)$$

$$= 2 \sum_{k=1}^n \text{Area}(\Delta_k) \left[\int_\sigma g(T_k(s, t)) \, d\sigma - \sum_{j=1}^f w_j g(T_k(\mu_j)) \right] \qquad (5.1.40)$$

and later we analyze the error by first analyzing the error in the quadrature formula (5.1.37) over σ.

Some quadrature formulas based on interpolation

Let $r \geq 1$ be an integer. Use the uniform grid $\{\mu_j\} \equiv \{q_1, \ldots, q_{f_r}\}$ of (5.1.12) that was used earlier in defining the interpolation polynomial $P_r(s, t)$ of (5.1.19). Integrate $P_r(s, t)$ to obtain the numerical integration formula

$$\int_\sigma G(s, t) \approx \sum_{j=1}^{f_r} w_j G(q_j), \qquad w_j \equiv \int_\sigma \ell_j(s, t) \, d\sigma \qquad (5.1.41)$$

For the case of $r = 0$, use (5.1.27) to obtain the approximation

$$\int_\sigma G(s, t) \approx \frac{1}{2} G\left(\frac{1}{3}, \frac{1}{3}\right) \qquad (5.1.42)$$

which is called the *centroid rule*.

The degree of precision d_r of these formulas is at least $d_r = r$. Moreover, when some such formulas are used in defining a composite rule [as in (5.1.39)] with some triangulations $\mathcal{T}_n$, the degree of precision is effectively $r + 1$ as regards the rate of convergence that is obtained.

Example. Take $r = 1$. The formula (5.1.41) yields the numerical integration formula

$$\int_\sigma G(s, t) \approx \frac{1}{6}[G(0, 0) + G(0, 1) + G(1, 0)] \tag{5.1.43}$$

It has degree of precision $d_1 = 1$, and it corresponds to the trapezoidal rule for functions of one variable.

Example. Take $r = 2$. Integrate $P_2(s, t)$ from (5.1.18) to obtain the numerical integration formula

$$\int_\sigma G(s, t) \approx \frac{1}{6}\left[G\left(0, \frac{1}{2}\right) + G\left(\frac{1}{2}, \frac{1}{2}\right) + G\left(\frac{1}{2}, 0\right)\right] \tag{5.1.44}$$

The integrals of $\ell_i(s, t)$, $i = 1, 2, 3$, over σ are all zero, thus giving a simpler formula than might have been expected. The degree of precision is $d_2 = 2$. When (5.1.44) is used in defining a composite formula for some triangulations $\mathcal{T}_n$ of a polygonal region R, it is effectively $d_2 = 3$, as we discuss later.

Numerical integration formulas with any desired degree of precision are obtained by choosing r appropriately. We can also use interpolation polynomials $P_r(s, t)$ based on other sets of interpolation node points $\{\mu_j\}$, but we do not consider such formulas here.

Other quadrature formulas

Following are two additional formulas that we have found useful in applications; both are 7-point formulas. First,

$$\int_\sigma G(s, t)\, d\sigma \approx \frac{9}{40}G\left(\frac{1}{3}, \frac{1}{3}\right) + \frac{1}{40}[G(0, 0) + G(0, 1) + G(1, 0)]$$
$$+ \frac{1}{15}\left[G\left(0, \frac{1}{2}\right) + G\left(\frac{1}{2}, \frac{1}{2}\right) + G\left(\frac{1}{2}, 0\right)\right] \tag{5.1.45}$$

This has degree of precision $d = 3$, and it can be derived using the method of undetermined coefficients. When used to form a composite formula, the nodes from adjacent triangles overlap. Consequently, the total number of integration nodes will be much less than $7n$, even though there are seven nodes per triangle.

The second formula is

$$\int_\sigma G(s,t)\,d\sigma \approx \frac{9}{80}G\left(\frac{1}{3},\frac{1}{3}\right) + B[G(\alpha,\alpha) + G(\alpha,\beta) + G(\beta,\alpha)]$$
$$+ C[G(\gamma,\gamma) + G(\gamma,\delta) + G(\delta,\gamma)] \qquad (5.1.46)$$

with

$$\alpha = \frac{6 - \sqrt{15}}{21} \qquad\qquad \beta = \frac{9 + 2\sqrt{15}}{21}$$

$$\gamma = \frac{6 + \sqrt{15}}{21} \qquad\qquad \delta = \frac{9 - 2\sqrt{15}}{21}$$

$$B = \frac{155 - \sqrt{15}}{2400} \qquad\qquad C = \frac{155 + \sqrt{15}}{2400}$$

This quadrature formula has degree of precision $d = 5$. It is the formula T2:5.1 from Stroud [537, p. 314], and it can be derived using tools based on orthogonal polynomials in two variables.

Error formulas for composite numerical integration formulas

From Lemma 5.1.1, we have convergence of the composite formula (5.1.39) for all continuous functions $g \in C(R)$,

$$|I(g) - I_n(g)| \le \text{Area}(R)\|g - \mathcal{P}_n g\|_\infty$$
$$\le \text{Area}(R)\|\mathcal{P}_n\|\omega(\delta_n, g) \qquad (5.1.47)$$

with $\omega(\delta_n, g)$ the modulus of continuity of g over R. To obtain more useful formulas on the rate of convergence, we proceed much as in Lemma 5.1.1 for polynomial interpolation error formulas, using Taylor's theorem to analyze the error.

Theorem 5.1.3. *Assume the formula*

$$\int_\sigma G(s,t)\,d\sigma \approx \sum_{j=1}^{f} w_j G(\mu_j) \qquad (5.1.48)$$

has degree of precision $d \ge 0$. Let R be a polygonal region in the plane $\mathbf{R}^2$, and let $\mathcal{T}_n = \{\Delta_1, \ldots, \Delta_n\}$ be a triangulation of R. Assume $g \in C^{d+1}(R)$. Then for the error $E_n(g)$ in the composite numerical integration formula (5.1.39),

$$|E_n(g)| \le c\,\delta_n^{d+1}\|g\|_{d+1,\infty} \qquad (5.1.49)$$

with $\|g\|_{d+1,\infty}$ as defined in (5.1.34) and

$$\delta_n = \max_{1 \le k \le n} \text{diameter}(\Delta_k)$$

Proof. For a generic triangle $\Delta \equiv \Delta_k$, introduce

$$G(s, t) = g(T(s, t))$$

Expand this as a Taylor polynomial about $(0, 0)$:

$$G(s, t) = H_d(s, t) + K_d(s, t) \tag{5.1.50}$$

with $H_d(s, t)$ the Taylor polynomial of degree d for $G(s, t)$ and

$$K_d(s, t) = \frac{1}{d!} \int_0^1 (1 - v)^d \frac{d^{d+1} G(vs, vt)}{dv^{d+1}} \, dv$$

For the error in the integration of $G(s, t)$, note that the numerical integration of $H_d(s, t)$ is exact because H_d is, by definition, a polynomial of degree $\leq d$. Then

$$E \equiv \int_\sigma G(s, t) \, d\sigma - \sum_{j=1}^f w_j G(\mu_j)$$

$$= \int_\sigma K_d(s, t) \, d\sigma - \sum_{j=1}^f w_j K_d(\mu_j)$$

$$= \frac{1}{d!} \int_0^1 (1 - v)^d \left\{ \int_\sigma \frac{d^{d+1} G(vs, vt)}{dv^{d+1}} \, d\sigma - \sum_{j=1}^f w_j \frac{d^{d+1} G(vs_j, vt_j)}{dv^{d+1}} \right\} dv \tag{5.1.51}$$

with $\mu_j \equiv (s_j, t_j)$.

As in the proof of Lemma 5.1.1, we can bound the derivatives of $G(vs, vt)$ to obtain

$$\max_{\substack{0 \leq v \leq 1 \\ (s,t) \in \sigma}} \left| \frac{d^{d+1} G(vs, vt)}{dv^{d+1}} \right| \leq c \delta^{d+1} \max_{\substack{i+j=d+1 \\ i,j \geq 0}} \left\| \frac{\partial^{d+1} g}{\partial^i x \partial^j y} \right\|_\infty$$

with the norm defined over Δ, $\delta = \text{diameter}(\Delta)$, and some constant $c > 0$. Combining this with (5.1.51),

$$|E| \leq c \, \delta^{d+1} \max_{\substack{i+j=d+1 \\ i,j \geq 0}} \left\| \frac{\partial^{d+1} g}{\partial^i x \partial^j y} \right\|_\infty$$

for a suitable constant $c > 0$.

If we use the preceding results in (5.1.40), we obtain the result (5.1.49). $\quad\square$

It is possible to construct asymptotic error formulas for the composite rules analyzed in the preceding theorem. In Lyness and Puri [358], the Euler-MacLaurin expansion is generalized from the one variable case to the multivariable case, to handle composite formulas (5.1.39), and of course, such an expansion furnishes an asymptotic error estimate. We do not use such error expansions in this book, but they can be quite useful.

How to refine a triangulation

Given a triangulation T_n of a polygonal region R, how does one refine it to a new triangulation with a smaller grid size δ_n? One might first think that it does not matter, so long as $\delta_n \to 0$ as $n \to \infty$, and this is usually the case with finite element methods for solving partial differential equations. However, when integration is involved, the "right" type of triangulation can lead to a "fortuitous cancellation" of errors.

This should not be too surprising, as this also occurs with numerical integration for functions of one variable. For example, let $g(x)$ be defined on $[a, b]$, and let $a < c < b$. Define $p(x)$ to be the quadratic polynomial interpolating $g(x)$ at $x = a, b, c$. The numerical integration formula

$$\int_a^b g(x)\,dx \approx \int_a^b p(x)\,dx$$

has degree of precision 2 for most choices of the node point c. But if $c = \frac{1}{2}(a + b)$, then the degree of precision is 3.

By imposing some symmetry on our triangulations T_n, we will sometimes obtain an increase in the degree of precision of a quadrature formula and thus of the rate of convergence of the resulting composite numerical integration formula (5.1.40).

Given a triangle $\Delta \in T_n$, refine it into smaller triangles by using straight line segments to connect the midpoints of the three sides of Δ. The four new triangles will be congruent, and they will be similar to the original triangle Δ. After such a refinement of all triangles in T_n, we will have a new triangulation T_{4n} with four times the number of triangles in T_n. Also,

$$\delta_{4n} = \frac{1}{2}\delta_n \tag{5.1.52}$$

Henceforth, we will refer to triangulations with this form of refinement as *symmetric triangulations*. For an integration method with degree of precision d, the ratio

$$\frac{E_n(g)}{E_{4n}(g)}$$

of the errors (5.1.49) should equal approximately $2^{-(d+1)}$.

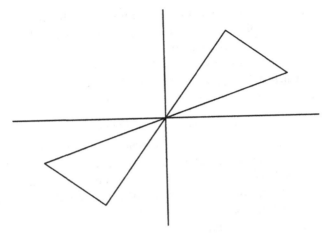

Figure 5.2. Symmetric triangles.

With a symmetric triangulation scheme, the degree of precision of certain composite formulas is increased. In particular, if the initial degree of precision d from integrating over σ is an even number, and if the integration formula has certain symmetry as regards its nodes and weights, then the degree of precision is increased effectively to $d + 1$. The formulas (5.1.41)–(5.1.46) all possess the requisite symmetry. Thus, those with degree of precision d an even number will have an effective degree of precision of $d + 1$ when used in the composite formula (5.1.40). For example, the 3-point rule (5.1.44) will lead to composite formulas with order of convergence $O(\delta_n^4)$. This latter rule can be looked on as the two-dimensional analog of Simpson's rule.

To understand this increase in the degree of precision, first consider the pair of triangles in Figure 5.2, which are symmetric with respect to their common vertex. With the use of symmetric triangulations $\mathcal{T}_n$ as described above, essentially all the triangles in $\mathcal{T}_n$ can be partitioned into pairs of such symmetric triangles. There will be at most $O(\sqrt{n}) = O(\delta_n^{-1})$ triangles not included in such pairs of triangles. Proceeding much as in the proof of Theorem 5.1.3, the total numerical integration error (5.1.40) over the set of all such symmetric pairs of triangles is $O(\delta_n^{d+2})$, where d is the degree of precision when integrating over a single triangle. The remaining triangles, of number $O(\delta_n^{-1})$, will have a composite error of

$$O\left(\delta_n^{-1}\right)\mathrm{Area}(\Delta)O\left(\delta_n^{d+1}\right) = O\left(\delta_n^{d+2}\right) \qquad (5.1.53)$$

To prove the above assertion on the increased accuracy when using symmetric pairs, consider the composite numerical integration rule based on one of the rules (5.1.41)–(5.1.42), which are based on polynomial interpolation of degree d. If d is even, then the composite rule over such a pair of symmetric triangles will have degree of precision $d + 1$. As an example of the proof of this, consider the formula (5.1.44) with degree of precision $d = 2$; and without loss of generality, let the common vertex in Figure 5.2 be the origin. To verify the degree of precision is 3, consider any monomial of degree 3. It will have the form

$$g(x, y) = x^i y^j$$

with $i, j \geq 0$ and $i + j = 3$. Thus one of the exponents, i or j, will be odd, and the other exponent will be even. If a point (x, y) is in one of the triangles, then $(-x, -y)$ is a point in the other triangle. Evaluating g,

$$g(-x, -y) = -g(x, y) \tag{5.1.54}$$

With this, the integral of such a monomial over the sum of these two triangles is zero. By the definition of the quadrature node points, if (a, b) is a node point in the upper triangle in Figure 5.2, then $(-a, -b)$ is a node point in the lower triangle. Using (5.1.54), it is straightforward to verify that the numerical integral based on (5.1.44) is also zero. Combining the results that both the integral and the numerical integral are zero over the sum of these two triangles, the error in the numerical integration over the union of the two triangles will be zero. This shows the degree of precision is 3, as desired. As a consequence, the composite form of (5.1.44) over all such symmetric pairs of triangles will have a rate of convergence of $O(\delta_n^4)$. A similar argument holds for other even degrees d.

5.2. Solving integral equations on polygonal regions

Consider the integral equation

$$\lambda \rho(x, y) - \int_R K(x, y, \xi, \eta) \rho(\xi, \eta) \, d\xi \, d\eta = \psi(x, y), \quad (x, y) \in R \tag{5.2.55}$$

We assume R is a polygonal region in the plane $\mathbf{R}^2$, and we assume the integral operator is compact on $C(R)$ into $C(R)$. Integral equations over nonpolygonal regions are considered in later sections. In this section we apply projection and Nyström methods to the above equation, with the discretizations based on approximations and numerical integration over triangulations $\mathcal{T}_n$ of R. The error analysis will be a straightforward application of the earlier theoretical frameworks developed in Chapters 3 and 4. Practical implementation of the methods is not as straightforward. We need to keep track of the triangulations and node points, to perform refinements, and to perform interpolation and numerical integration over these triangulations. In addition, the order of the linear systems to be solved can become very large with only a few refinements of the triangulation, and this necessitates the use of iteration methods for solving the system, a topic taken up in Chapter 6.

5.2.1. Collocation methods

To define the collocation method for solving (5.2.55), proceed as follows. Let $\mathcal{T}_n = \{\Delta_1, \ldots, \Delta_n\}$ be a triangulation of R. Let a polynomial interpolation method over σ be given by

$$G(s, t) \approx \sum_{j=1}^{f} G(q_j)\ell_j(s, t), \quad (s, t) \in \sigma, \quad G \in C(\sigma) \qquad (5.2.56)$$

and define the interpolatory operator $\mathcal{P}_n$ on $C(R)$ by

$$\mathcal{P}_n g(T_k(s, t)) = \sum_{j=1}^{f} g(T_k(q_j))\ell_j(s, t), \quad (s, t) \in \sigma, \quad k = 1, \ldots, n$$

$$(5.2.57)$$

for all $g \in C(R)$. This uses the affine mapping $T_k : \sigma \to \Delta_k$ of (5.1.10). Let $\mathcal{X}_n = \text{Range}(\mathcal{P}_n)$. If $\mathcal{P}_n g(x, y)$ is continuous over R, for all $g \in C(R)$, then $\mathcal{P}_n$ is a projection on $C(R)$ to $\mathcal{X}_n$. If $\mathcal{P}_n g(x, y)$ is not continuous, then the definition (5.2.57) must be extended to become a bounded projection on $L^\infty(R)$ to $\mathcal{X}_n$. The latter can be done by using the tools developed in Ref. [62], and a short discussion of this is given in Ref. [43, p. 31]. Using this extension of $\mathcal{P}_n$ to $L^\infty(R)$, the results of Theorem 5.2.1, given below, are still valid in the context of $L^\infty(R)$. We do not pursue this further, as all of our examples are based on schemes with $\mathcal{P}_n g(x, y)$ continuous on R.

Recall that the collocation method for solving (5.2.55) can be written symbolically as

$$(\lambda - \mathcal{P}_n \mathcal{K})\rho_n = \mathcal{P}_n \psi \qquad (5.2.58)$$

To find ρ_n, let

$$\rho_n(x, y) = \sum_{j=1}^{f} \rho_n(v_{k,j})\ell_j(s, t), \quad (x, y) = T_k(s, t) \in \Delta_k, \quad k = 1, \ldots, n$$

$$(5.2.59)$$

with

$$v_{k,j} = T_k(q_j), \quad j = 1, \ldots, f, \ k = 1, \ldots, n$$

We also refer collectively to the node points $\{v_{k,j} \mid 1 \le j \le f, \ 1 \le k \le n\}$ as

$$\mathcal{V}_n = \{v_1, \ldots, v_{n_v}\} \qquad (5.2.60)$$

Substituting (5.2.59) into (5.2.55) and then collocating at the node points in $\mathcal{V}_n$, we obtain the linear system

$$\lambda \rho_n(v_i) - 2 \sum_{k=1}^{n} \text{Area}(\Delta_k) \sum_{j=1}^{f} \rho_n(v_{k,j}) \int_\sigma K(v_i, T_k(s,t)) \ell_j(s,t) \, d\sigma$$

$$= \psi(v_i), \quad i = 1, \ldots, n_v \qquad (5.2.61)$$

This has order n_v, and for large n_v, it must be solved by iteration. Most of the integrals in this system must be evaluated numerically, thus producing a discrete collocation method. Discrete Galerkin methods are discussed briefly near the end of this section, and similar considerations apply to discrete collocation methods. A brief discussion of discrete collocation methods for integral equations on general surfaces is given at the end of §5.4.2. We also discuss the numerical integration of such collocation integrals for boundary integral equations in §9.2.1 of Chapter 9. For the error analysis of (5.2.58), we have the following.

Theorem 5.2.1. *Let R be a polygonal region in* $\mathbf{R}^2$, *and let* $\{\mathcal{T}_n\}$ *be a sequence of triangulations of R. Assume*

$$\delta_n \equiv \max_{k=1,\ldots,n} \ diameter(\Delta_k) \to 0 \quad as \ n \to \infty$$

Assume the integral equation $(\lambda - \mathcal{K})\rho = \psi$ *of (5.2.55) is uniquely solvable, with* $\mathcal{K}$ *a compact operator on* $C(R)$. *Then for all sufficiently large n, say* $n \ge N$, *the approximating equation (5.2.58) is uniquely solvable, and the inverses* $(\lambda - \mathcal{P}_n\mathcal{K})^{-1}$ *are uniformly bounded. For the error in* ρ_n,

$$\|\rho - \rho_n\|_\infty \le |\lambda| \, \|(\lambda - \mathcal{P}_n\mathcal{K})^{-1}\| \, \|\rho - \mathcal{P}_n\rho\|_\infty, \quad n \ge N \qquad (5.2.62)$$

Proof. Note first that the interpolatory projection $\mathcal{P}_n$ of (5.2.57) is pointwise convergent on $C(R)$:

$$\mathcal{P}_n g \to g \quad as \ n \to \infty, \ g \in C(R)$$

This follows from Lemma 5.1.1. The remainder of the proof is immediate from Theorem 3.1.1 and Lemma 3.1.2 of §3.1 in Chapter 3. Note that this theorem does not require a symmetric triangulation.   □

Rates of convergence for ρ_n can be obtained from (5.1.33) of Theorem 5.1.1. For the interpolation scheme of (5.1.21), with polynomial interpolation of degree r, (5.2.62) implies

$$\|\rho - \rho_n\|_\infty \le c\delta_n^{r+1}, \quad n \ge N, \quad \rho \in C^{r+1}(R) \qquad (5.2.63)$$

The iterated collocation method and superconvergence

The above rate of convergence can sometimes be improved at the interpolation node points by considering the iterated collocation solution

$$\hat{\rho}_n = \frac{1}{\lambda}[\psi + \mathcal{K}\rho_n] \qquad (5.2.64)$$

We also need to assume that the kernel function $K(x, y, \xi, \eta)$ is differentiable with respect to (ξ, η), with the derivatives uniformly bounded with respect to both (x, y) and (ξ, η).

Recall the discussion of iterated projection solutions in §3.4 of Chapter 3. From that discussion,

$$\mathcal{P}_n \hat{\rho}_n = \rho_n \qquad (5.2.65)$$

$$(\lambda - \mathcal{K}\mathcal{P}_n)\hat{\rho}_n = \psi \qquad (5.2.66)$$

$$\rho - \hat{\rho}_n = (\lambda - \mathcal{K}\mathcal{P}_n)^{-1}\mathcal{K}(I - \mathcal{P}_n)\rho \qquad (5.2.67)$$

To show improved rates of convergence for $\hat{\rho}_n$, we examine the term $\mathcal{K}(I - \mathcal{P}_n)\rho$.

Assume we are using symmetric triangulations $\mathcal{T}_n$. Let the interpolation scheme be that of (5.1.19), based on polynomial interpolation of degree r over σ and using the uniformly distributed nodes

$$\{q_i \equiv (s_i, t_i) \mid i = 1, \ldots, f_r\} \equiv \left\{ \left(\frac{j}{r}, \frac{k}{r} \right) \,\middle|\, j, k \ge 0, \ j + k \le r \right\}$$

In addition, let r be an even integer. For the error in (5.2.67),

$$\mathcal{K}(I - \mathcal{P}_n)\rho(x, y) = 2\sum_{k=1}^{n} \text{Area}(\Delta_k) \int_\sigma K(x, y, T_k(s, t))$$

$$\times \left\{ \rho(T_k(s, t)) - \sum_{j=1}^{f_r} \rho(v_{k,j})\ell_j(s, t) \right\} d\sigma \qquad (5.2.68)$$

Let $G_k(s, t) = \rho(T_k(s, t))$. Using Taylor's theorem, write

$$G_k(s, t) = H_{r,k}(s, t) + J_{r,k}(s, t) + L_{r,k}(s, t) \tag{5.2.69}$$

In this, $H_{r,k}(s, t)$ is the Taylor polynomial of degree r for $G_k(s, t)$, and $H_{r,k}(s, t) + J_{r,k}(s, t)$ is the Taylor polynomial of degree $r + 1$ for $G_k(s, t)$, so that

$$J_{r,k}(s, t) = \left(s\frac{\partial}{\partial \xi} + t\frac{\partial}{\partial \eta} \right)^{r+1} G_k(\xi, \eta)\Big|_{\substack{\xi=0 \\ \eta=0}} \tag{5.2.70}$$

The Taylor polynomial remainder term is

$$L_{r,k}(s, t) = \frac{1}{(r+1)!} \int_0^1 (1-v)^{r+1} \frac{d^{r+2} G_k(vs, vt)}{dv^{r+2}}\, dv$$

By an examination of the kind done earlier in the proofs of Lemma 5.1.1 and Theorem 5.1.3, we can show

$$\begin{aligned} J_{r,k}(s, t) &= O\left(\delta_n^{r+1}\right) \\ L_{r,k}(s, t) &= O\left(\delta_n^{r+2}\right) \end{aligned} \tag{5.2.71}$$

uniformly in (s, t). The interpolation error in (5.2.68) becomes

$$\rho(T_k(s, t)) - \sum_{j=1}^{f_r} \rho(v_{k,j})\ell_j(s, t) = J_{r,k}(s, t) - \sum_{j=1}^{f_r} J_{r,k}(s_j, t_j)\ell_j(s, t)$$

$$+ L_{r,k}(s, t) - \sum_{j=1}^{f_r} L_{r,k}(s_j, t_j)\ell_j(s, t)$$

We also expand the kernel function $K(x, y, T_k(s, t))$, considered as a function of (s, t):

$$K(x, y, T_k(s, t)) = K(x, y, T_k(0, 0)) + R_k \tag{5.2.72}$$

It is straightforward to show

$$R_k = O(\delta_n) \tag{5.2.73}$$

much as was done following (5.1.32). Again, this is uniform in (s, t), k, and (x, y).

To obtain an improved speed of convergence, we must examine the term $J_{r,k}(s, t)$ more carefully. Let Δ_k and Δ_l be a symmetric pair of triangles.

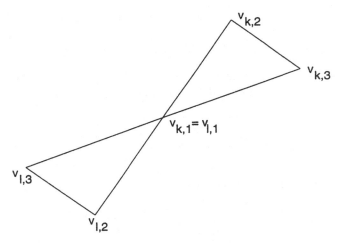

Figure 5.3. Symmetric triangles for showing (5.2.74).

These are pictured in Figure 5.3, with $v_{k,1} = v_{l,1}$ as the common vertex. Note that

$$v_{k,2} - v_{k,1} = -(v_{l,2} - v_{l,1}), \qquad v_{k,3} - v_{k,1} = -(v_{l,3} - v_{l,1})$$

Using this, and using an analysis of the type embodied in (5.1.32), we can show

$$J_{r,k}(s, t) = -J_{r,l}(s, t) \qquad (5.2.74)$$

for all $(s, t) \in \sigma$. We omit the proof.

Return to the integration of (5.2.68) over the symmetric pair of triangles Δ_k and Δ_l. By combining the above results, we obtain that over a symmetric pair of triangles, the integrals of the errors of the terms of size $O(\delta_n^{r+1})$ will cancel. The remaining terms are of size $O(\delta_n^{r+2})$, and we have

$$\|\mathcal{K}(I - \mathcal{P}_n)\rho\|_\infty = O(\delta_n^{r+2})$$

Thus from (5.2.67),

$$\|\rho - \hat{\rho}_n\|_\infty = O(\delta_n^{r+2})$$

Recalling (5.2.65), the solutions ρ_n and $\hat{\rho}_n$ agree at the collocation node points, and therefore

$$\max_{1 \le i \le n_v} |\rho(v_i) - \rho_n(v_i)| = O(\delta_n^{r+2}) \qquad (5.2.75)$$

Comparing with (5.2.63), we find that the collocation solution is superconvergent at the collocation node points. For example, the use of interpolation of degree $r = 2$ leads to a collocation method with order of convergence δ_n^4 at the collocation node points $\{v_i\}$.

5.2.2. Galerkin methods

Recall the earlier material on Galerkin methods, from §§3.1, 3.3, and 3.4 of Chapter 3. With Galerkin methods it is more convenient not to impose continuity conditions on the approximation functions. Let $r \geq 0$ be an integer. Define $\mathcal{X}_n$ to be the set of all $\phi \in L^\infty(R)$ for which $\phi \mid \Delta_k$ is a polynomial of degree $\leq r$, for $k = 1, \ldots, n$. The dimension of $\mathcal{X}_n$ is nf_r, with $f_r = \frac{1}{2}(r+1)$ $(r+2)$ the number of independent polynomials of degree $\leq r$ over a single triangle Δ.

Let $\{\phi_1, \ldots, \phi_{f_r}\}$ be a basis for the polynomials of degree $\leq r$ over σ, and then introduce

$$\phi_{k,j}(x,y) = \begin{cases} \phi_j(s,t), & (x,y) \in \Delta_k, \quad (x,y) = T_k(s,t) \\ 0, & (x,y) \notin \Delta_k \end{cases} \quad (5.2.76)$$

for $j = 1, \ldots, f_r$, $k = 1, \ldots, n$. These comprise a standardized set of basis functions for the piecewise polynomial functions of degree $\leq r$ over R, using the triangulation $\mathcal{T}_n = \{\Delta_1, \ldots, \Delta_n\}$ of R.

Let $\mathcal{P}_n$ be the orthogonal projection of $L^2(R)$ onto $\mathcal{X}_n$, which we will also regard as a projection operator on $L^\infty(R)$ to $\mathcal{X}_n$. The Galerkin method for solving (5.2.55) is given by

$$(\lambda - \mathcal{P}_n \mathcal{K})\rho_n = \mathcal{P}_n \psi \quad (5.2.77)$$

which has the same abstract form as the collocation method. We can analyze it in the context of both $\mathcal{X} = L^2(R)$ and $\mathcal{X} = L^\infty(R)$.

For the practical implementation of the Galerkin method of solving (5.2.55), let

$$\rho_n(x,y) = \sum_{k=1}^{n} \sum_{j=1}^{f_r} \alpha_{k,j} \phi_{k,j}(x,y) \quad (5.2.78)$$

Substitute into (5.2.55), form the inner product with $\phi_{l,i}$ on both sides of the equation, and force equality:

$$\sum_{k=1}^{n}\sum_{j=1}^{f_r}\alpha_{k,j}\left\{\lambda\int_R\phi_{l,i}(x,y)\phi_{k,j}(x,y)\,dx\,dy\right.$$

$$\left.-\int_R\phi_{l,i}(x,y)\int_R K(x,y,\xi,\eta)\phi_{k,j}(\xi,\eta)\,d\xi\,d\eta\,dx\,dy\right\}$$

$$=\int_R\phi_{l,i}(x,y)\psi(x,y)\,dx\,dy \qquad (5.2.79)$$

for $i = 1, \ldots, f_r$, $l = 1, \ldots, n$. Changing to integration over the standard triangle σ, we have

$$\lambda\sum_{j=1}^{f_r}\alpha_{l,j}\int_\sigma\phi_i(s,t)\phi_j(s,t)\,ds\,dt - \sum_{k=1}^{n}\text{Area}(\Delta_k)\sum_{j=1}^{f_r}\alpha_{k,j}\int_\sigma\phi_i(s,t)$$

$$\times\left\{\int_\sigma K(T_l(s,t),T_k(\zeta,\tau))\phi_j(\zeta,\tau)\,d\zeta\,d\tau\right\}\,ds\,dt$$

$$=\int_\sigma\phi_i(s,t)\psi(T_l(s,t))\,ds\,dt, \quad i = 1, \ldots, f_r, \ l = 1, \ldots, n$$
$$(5.2.80)$$

Most of the integrals in this system must be evaluated numerically, but we defer consideration of this until later in the section, following the discussion of Nyström methods.

In addition to the Galerkin solution ρ_n, also consider the iterated Galerkin solution $\hat{\rho}_n$, defined abstractly in (5.2.64) and satisfying (5.2.65)–(5.2.67). Using (5.2.78),

$$\hat{\rho}_n(x,y) = \frac{1}{\lambda}\psi(x,y) + \frac{1}{\lambda}\sum_{k=1}^{n}\text{Area}(\Delta_k)\sum_{j=1}^{f_r}\alpha_{k,j}$$

$$\times\int_\sigma K(x,y,T_k(s,t))\phi_j(s,t)\,ds\,dt, \quad (x,y)\in R \qquad (5.2.81)$$

An error analysis for ρ_n and $\hat{\rho}_n$ is given by the following.

Theorem 5.2.2. *Let R be a polygonal region in $\mathbf{R}^2$; and let $\{T_n\}$ be a sequence of triangulations of R. Assume*

$$\delta_n \equiv \max_{k=1,\ldots,n}\ diameter(\Delta_k) \to 0 \quad as\ n \to \infty$$

Assume the integral equation $(\lambda - \mathcal{K})\rho = \psi$ of (5.2.55) is uniquely solvable for $\psi \in L^2(R)$, with $\mathcal{K}$ a compact operator on $L^2(R)$. Then for all sufficiently large *n, say n $\geq$ N, the approximating equation (5.2.77) is uniquely solvable, and the inverses $(\lambda - \mathcal{P}_n\mathcal{K})^{-1}$ are uniformly bounded on $L^2(R)$. For the error in ρ_n,*

$$\|\rho - \rho_n\| \leq |\lambda| \, \|(\lambda - \mathcal{P}_n\mathcal{K})^{-1}\| \, \|\rho - \mathcal{P}_n\rho\|, \quad n \geq N \qquad (5.2.82)$$

For the iterated Galerkin solution the inverses $(\lambda - \mathcal{K}\mathcal{P}_n)^{-1}$ are uniformly bounded for n $\geq$ N, and

$$\|\rho - \hat{\rho}_n\| \leq \|(\lambda - \mathcal{K}\mathcal{P}_n)^{-1}\| \, \|(I - \mathcal{P}_n)\mathcal{K}^*\| \, \|\rho - \mathcal{P}_n\rho\|, \quad n \geq N \qquad (5.2.83)$$

Proof. Note first that

$$\mathcal{P}_n\phi \to \phi$$

for all $\phi \in L^2(R)$. The proof of this is essentially the same as that given for the use of piecewise linear functions following (3.3.65) in §3.3 of Chapter 3; we omit the proof. The remainder of the proof of the theorem is a repeat of earlier derivations in Chapter 3. The proof of the convergence of the Galerkin method is an application of Theorem 3.1.1 and Lemma 3.1.2, in §3.1. The error analysis for the iterated Galerkin solution follows from the discussion of (3.4.89) and what follows it. □

Corollary 5.2.1. *Assume $\rho \in C^{r+1}(R)$. Then*

$$\|\rho - \rho_n\| \leq c\delta_n^{r+1}, \quad n \geq N \qquad (5.2.84)$$

Assume further that $K(x, y, \xi, \eta)$ is $(r + 1)$-times differentiable with respect to (ξ, η), with the derivatives of order $r + 1$ uniformly bounded with respect to $(x, y), (\xi, \eta) \in R$. Then

$$\|\rho - \hat{\rho}_n\| \leq c\delta_n^{2r+2}, \quad n \geq N \qquad (5.2.85)$$

Proof. The proof of (5.2.84) is a straightforward application of (5.2.82) and Lemma 5.1.1. The proof of (5.2.85) is an application of (5.2.83), (5.2.84), and the application of Lemma 5.1.1 to $K(x, y, \xi, \eta)$, considered as a function of (ξ, η). □

Uniform convergence

The earlier result (3.4.99) of §3.4 in Chapter 3 states that

$$\|\rho - \hat{\rho}_n\|_\infty \leq \frac{1}{|\lambda|}\Big\{ \|\rho - \mathcal{P}_n\rho\| \max_{(x,y)\in R} \|(I - \mathcal{P}_n)K_{(x,y)}\|$$

$$+ \|\rho - \hat{\rho}_n\| \max_{(x,y)\in R} \|K_{(x,y)}\| \Big\} \qquad (5.2.86)$$

with $K_{(x,y)}(\xi, \eta) \equiv K(x, y, \xi, \eta)$. All the norms on the right side of the inequality are the inner product norm of $L^2(R)$. With this bound, the results for $L^2(R)$ of Theorem 5.2.2 will transfer to convergence results in the norm $\| \cdot \|_\infty$. In particular, the result (5.2.85) of Corollary 5.2.1 is valid for $\|\rho - \hat{\rho}_n\|_\infty$. Uniform convergence results for ρ_n follow from

$$\rho - \rho_n = \rho - \mathcal{P}_n \hat{\rho}_n = [\rho - \mathcal{P}_n \rho] + \mathcal{P}_n [\rho - \hat{\rho}_n]$$

$$\|\rho - \rho_n\|_\infty \leq \|\rho - \mathcal{P}_n \rho\|_\infty + \|\mathcal{P}_n\| \|\rho - \hat{\rho}_n\|_\infty \qquad (5.2.87)$$

where $\|\mathcal{P}_n\|$ denotes the norm of $\mathcal{P}_n$ when it is regarded as an operator on $L^\infty(R)$ to $\mathcal{X}_n$. The bound in (5.2.87) will imply the bound of (5.2.84) for $\|\rho - \rho_n\|_\infty$. The assumptions needed on $\rho(x, y)$ and $K(x, y, \xi, \eta)$ are the same as were assumed for Theorem 5.2.2.

5.2.3. The Nyström method

Let $K(x, y, \xi, \eta)$ be continuous over $R \times R$. Let a composite numerical integration rule be based on

$$\int_\sigma G(s, t) \, d\sigma \approx \sum_{i=1}^{f} w_i G(\mu_i) \qquad (5.2.88)$$

as in (5.1.39). We assume the quadrature nodes $\{\mu_1, \ldots, \mu_f\}$ are contained in σ. For specific formulas see the discussion following (5.1.36). Let

$$v_{k,i} = T_k(\mu_i), \quad i = 1, \ldots, f, \quad k = 1, \ldots, n$$

and collectively refer to these integration nodes as $\mathcal{V}_n = \{v_1, \ldots, v_{n_v}\}$.

Approximate the integral operator $\mathcal{K}$ by

$$\mathcal{K}_n \rho(x, y) = 2 \sum_{k=1}^{n} \text{Area}(\Delta_k) \sum_{i=1}^{f} w_i K(x, y, T_k(\mu_i)) \rho(T_k(\mu_i)), \quad (x, y) \in R$$

$$= \sum_{j=1}^{n_v} \omega_j K(x, y, \xi_j, \eta_j) \rho(\xi_j, \eta_j) \qquad (5.2.89)$$

with $v_j = (\xi_j, \eta_j)$ and $\rho \in C(R)$. For the weights $\{\omega_j\}$,

$$\omega_j = \sum_{\substack{k,i \\ T_k(\mu_i) = v_j}} 2 \, \text{Area}(\Delta_k) w_i$$

The integral equation (5.2.55) is approximated by $(\lambda - \mathcal{K}_n)\rho_n = \psi$, or equivalently,

$$\lambda\rho_n(x, y) - \sum_{j=1}^{n_v} \omega_j K(x, y, \xi_j, \eta_j)\, \rho_n(\xi_j, \eta_j) = \psi(x, y), \quad (x, y) \in R$$

$$(5.2.90)$$

Solve for ρ_n by first solving the linear system

$$\lambda\rho_n(\xi_i, \eta_i) - \sum_{j=1}^{n_v} \omega_j K(\xi_i, \eta_i, \xi_j, \eta_j)\, \rho_n(\xi_j, \eta_j) = \psi(\xi_i, \eta_i),$$

$$i = 1, \ldots, n_v \quad (5.2.91)$$

The solution ρ_n is obtained at the remaining points $(x, y) \in R$ by using the Nyström interpolation formula

$$\rho_n(x, y) = \frac{1}{\lambda}\left[\psi(x, y) + \sum_{j=1}^{n_v} \omega_j K(x, y, \xi_j, \eta_j)\, \rho_n(\xi_j, \eta_j) \right]$$

The error analysis for this approximation is a straightforward application of the general theory of Nyström methods from §4.1 of Chapter 4 in combination with the error analysis for quadrature methods from §5.1.

Theorem 5.2.3. *Let R be a polygonal region in $\mathbf{R}^2$, and let $\{\mathcal{T}_n\}$ be a sequence of triangulations of R. Assume*

$$\delta_n \equiv \max_{k=1,\ldots,n}\, diameter(\Delta_k) \to 0 \quad as\ n \to \infty$$

Assume the integral equation $(\lambda - \mathcal{K})\rho = \psi$ of (5.2.55) is uniquely solvable for $\psi \in C(R)$, with $\mathcal{K}$ a compact operator on $C(R)$. Assume the integration formula (5.2.88) has degree of precision $d \geq 0$. Then

(a) *For all sufficiently large n, say $n \geq N$, the approximating equation (5.2.90) is uniquely solvable, and the inverses $(\lambda - \mathcal{K}_n)^{-1}$ are uniformly bounded on $C(R)$. For the error in ρ_n,*

$$\rho - \rho_n = (\lambda - \mathcal{K}_n)^{-1}(\mathcal{K}\rho - \mathcal{K}_n\rho)$$

and $\rho_n \to \rho$ as $n \to \infty$.

(b) *Assume $K(x, y, \cdot, \cdot) \in C^{d+1}(R)$, for all $(x, y) \in R$, and $\rho \in C^{d+1}(R)$. Then*

$$\|\rho - \rho_n\|_\infty \leq c\, \delta_n^{d+1}, \quad n \geq N \quad (5.2.92)$$

Proof. The general error analysis is based on Theorem 4.1.2. With the assumption $\delta_n \to 0$ as $n \to \infty$ and the assumption that $K(x, y, \xi, \eta)$ is continuous, it is straightforward to prove $\{\mathcal{K}_n\}$ is collectively compact and pointwise convergent on $C(R)$. Use Theorem 4.1.2 to complete the proof of (a). For (b) apply the integration error result Theorem 5.1.3.   □

Corollary 5.2.2. *Assume the triangulation $\mathcal{T}_n$ used in the preceding theorem is a symmetric triangulation. Moreover, let the integration scheme (5.2.88) be that of (5.1.41), based on integrating the polynomial of degree r that interpolates the integrand at the evenly spaced nodes of (5.1.12). Let $r \geq 2$ be an even integer, and assume $K(x, y, \cdot, \cdot) \in C^{r+2}(R)$, for all $(x, y) \in R$, and $\rho \in C^{r+2}(R)$. Then*

$$\|\rho - \rho_n\|_\infty \leq c \, \delta_n^{r+2}, \quad n \geq N \tag{5.2.93}$$

Proof. This uses the discussion of integration error given at the conclusion of §5.1.   □

This result generalizes to other integration schemes (5.2.88), provided they have an improved degree of precision when used over pairs of symmetric triangles, as discussed at the end of §5.1.

Discrete Galerkin methods

The integrals in (5.2.79) or (5.2.80) must be evaluated numerically. We would like the integration to have such accuracy as to preserve the rates of convergence for ρ_n and $\hat{\rho}_n$ that were given in (5.2.84) and (5.2.85) of Corollary 5.2.1.

Recall the development in §4.4 of Chapter 4 of a general framework for discrete Galerkin methods. We must give a way for numerically integrating the inner products over R and the integral operator $\mathcal{K}$ defined by integrating over R. We assume the kernel function $K(x, y, \xi, \eta)$ is a smooth function, and we then use the same integration scheme to approximate both the inner product and the integral operator. It is a composite formula based on a basic formula

$$\int_\sigma g(s, t) \, d\sigma \approx \sum_{i=1}^{M} \omega_i g(\mu_i) \tag{5.2.94}$$

with degree of precision $d \geq 0$. We apply it to each of the integrals in (5.2.80) and (5.2.81), for calculating approximations to ρ_n and $\hat{\rho}_n$, respectively.

As was shown in §4.4 of Chapter 4, this leads to functions z_n and $\hat{z}_n$. Abstractly, they satisfy $(\lambda - \mathcal{Q}_n\mathcal{K}_n)z_n = \mathcal{Q}_n\psi$ and $(\lambda - \mathcal{K}_n\mathcal{Q}_n)\hat{z}_n = \psi$.

Table 5.1. *Quadrature formulas (5.2.94)*

d	M	Reference	Comment
1	1	Centroid rule	See (5.1.42)
2	3	Stroud [537, T_2:2-2]	See (5.1.44)
3	4	Stroud [537, T_2:3-1]	Some negative weights
3	6	Stroud [537, $^*T_2$:3-1]	
5	7	Stroud [537, $^*T_2$:5-1]	See (5.1.46)
7	16	Stroud [537, T_2:7-1]	

The operator $\mathcal{Q}_n$ is the discrete orthogonal projection associated with the discrete inner product. The error analysis for z_n and $\hat{z}_n$ is given in Theorems 4.4.2 and 4.4.3. The first is for the case $M = f_r$, and the second is for $M > f_r$.

When $M = f_r \equiv \frac{1}{2}(r + 1)(r + 2)$, the iterated discrete Galerkin solution $\hat{z}_n$ is simply the solution of the Nyström equation $(\lambda - \mathcal{K}_n)\hat{z}_n = \psi$. Then Theorem 5.2.3 and (5.2.92) tell us the rate of convergence is

$$\|\rho - \hat{z}_n\|_\infty = O\left(\delta_n^{\min\{d+1, 2r+2\}}\right) \qquad (5.2.95)$$

provided K and ρ are sufficiently differentiable. Comparing with (5.2.85) of Corollary 5.2.1, we need $d \geq 2r + 1$ if $\hat{z}_n$ is to converge as rapidly as $\hat{\rho}_n$. In Table 5.1, for several integers d we reference formulas with a relatively small number M of node points for a quadrature formula (5.2.94) with the degree of precision d. These are taken from the book of Stroud [537, pp. 307–315], and additional formulas can be found in Lyness and Jespersen [357].

As an example, consider taking $r = 2$. Then $f_r = 6$, and we would need a formula (5.2.94) with $M = 6$ nodes and a degree of precision of $d \geq 5$. We know of no such formula. A similarly negative result holds for other integers $r \geq 1$. Only for $r = 0$ are we able to have a suitably accurate formula, the centroid rule, with $M = f_r = 1$ and $d = 1$. The resulting order of convergence for $\hat{z}_n$ is then $O(\delta_n^2)$.

For $r \geq 1$ we must consider the case $M > f_r$. Then with K and ρ sufficiently differentiable we have (5.2.95) as before. But now we are free to choose M large enough as to have $d \geq 2r + 1$. As an example, for $r = 2$, we have $d \geq 5$ if we choose $M = 7$ and the formula T_2:5-1 of Stroud [537, p. 314] (cf. (5.1.46) in §5.1). The resulting rate of convergence will be $O(\delta_n^6)$. Thus the iterated discrete Galerkin solution $\hat{z}_n$ can be made to converge as rapidly as the iterated Galerkin solution $\hat{\rho}_n$, and the needed numerical integrations need not be very expensive.

5.3. Interpolation and numerical integration on surfaces

An important category of integral equations is that defining integral equations on surfaces in $\mathbf{R}^3$. In this section we generalize the results of §5.1 to such surfaces, and we also consider the approximation of such surfaces by simpler ones. The results of this section are applied to integral equations in the following section and in Chapter 9.

We assume S is a connected piecewise smooth surface in $\mathbf{R}^3$. By this, we mean S can be written as

$$S = S_1 \cup \cdots \cup S_J \tag{5.3.96}$$

with each S_j the continuous image of a polygonal region in the plane:

$$F_j : R_j \overset{1-1}{\underset{onto}{\to}} S_j, \quad j = 1, \ldots, J \tag{5.3.97}$$

Generally, the mappings F_j are assumed to be several times continuously differentiable. Other than these assumptions, we are allowing our surfaces to be quite general. Thus S can be either an "open" or "closed" surface.

We create triangulations for S by first triangulating each R_j and then mapping this triangulation onto S_j. Let $\{\hat{\Delta}_{j,k} \mid k = 1, \ldots, n_j\}$ be a triangulation of R_j, and then define

$$\Delta_{j,k} = F_j(\hat{\Delta}_{j,k}), \quad k = 1, \ldots, n_j, \quad j = 1, \ldots, J \tag{5.3.98}$$

This yields a triangulation of S, which we refer to collectively as $\mathcal{T}_n = \{\Delta_1, \ldots, \Delta_n\}$. We make the following assumptions concerning this triangulation.

T1. The set of all vertices of the surface S is a subset of the set of all vertices of the triangulation $\mathcal{T}_n$.

T2. The union of all edges of S is contained in the union of all edges of all triangles in $\mathcal{T}_n$.

T3. If two triangles in $\mathcal{T}_n$ have a nonempty intersection, then that intersection consists either of (*i*) a single common vertex, or (*ii*) all of a common edge.

The assumptions **T1** and **T2** are needed as a practical matter, as they reflect the likely lack of differentiability of solutions of many integral equations when crossing edges on the surface S. The third assumption can often be weakened, a point we discuss later in the section. We refer to triangulations satisfying **T1**–**T3** as *conforming triangulations*, while noting this definition may vary from that found elsewhere in the literature. We include **T3** because it simplifies some arguments by allowing them to be done in the framework of $C(S)$.

Let Δ_k be some element from $\mathcal{T}_n$, and let it correspond to some $\hat{\Delta}_k$, say $\hat{\Delta}_k \subset R_j$ and

$$\Delta_k = F_j(\hat{\Delta}_k)$$

Let $\{\hat{v}_{k,1}, \hat{v}_{k,2}, \hat{v}_{k,3}\}$ denote the vertices of $\hat{\Delta}_k$. Define $m_k : \sigma \overset{1-1}{\underset{onto}{\to}} \Delta_k$ by

$$m_k(s, t) = F_j(u\hat{v}_{k,1} + t\hat{v}_{k,2} + s\hat{v}_{k,3}), \quad (s, t) \in \sigma, \quad u = 1 - s - t$$
(5.3.99)

As in §5.1, we define interpolation and numerical integration over a triangular surface element Δ by means of a similar formula over σ. If F_j is the identity mapping, then m_k is the affine mapping T_k of (5.1.10) in §5.1.

5.3.1. Interpolation over a surface

Let $r \geq 1$ be an integer. From (5.1.12), recall the grid

$$(s_i, t_j) = (i\delta, j\delta), \quad i, j \geq 0, \quad i + j \leq r, \quad \delta = \frac{1}{r}$$
(5.3.100)

which is a uniform grid over σ. Collectively, we refer to it as $\{q_1, \ldots, q_{f_r}\}$. Recall the interpolation formula (5.1.19),

$$G(s, t) \approx \sum_{i=1}^{f_r} G(q_i)\ell_i(s, t)$$
(5.3.101)

which is a polynomial of degree r and interpolates G at the nodes $\{q_i\}$. For $g \in C(S)$, restrict g to some $\Delta \in \mathcal{T}_n$, and then identify $G(s, t)$ with $g(m_k(s, t))$. Define

$$(\mathcal{P}_n g)(m_k(s, t)) = \sum_{i=1}^{f_r} g(m_k(q_i))\ell_i(s, t), \quad (s, t) \in \sigma, \quad k = 1, \ldots, n$$
(5.3.102)

This is a polynomial in the parametrization variables (s, t), but it need not be a polynomial in the spatial variables (x, y, z). By an argument analogous to that used earlier in §5.1, following (5.1.23), we can show $(\mathcal{P}_n g)(x, y, z)$ is continuous over S. This also requires the assumptions **T1–T3**. We leave the proof to the reader.

The use of piecewise constant interpolation over S can be defined in a way analogous to (5.1.28):

$$(\mathcal{P}_n g)(m_k(s, t)) = g\left(m_k\left(\frac{1}{3}, \frac{1}{3}\right)\right), \quad (s, t) \in \sigma, \quad k = 1, \ldots, n$$

The function $\mathcal{P}_n g$ is no longer continuous in most cases, but by using the techniques found in Ref. [62], $\mathcal{P}_n$ can be extended to $L^\infty(S)$ as a projection, with preservation of norm. The use of piecewise constant interpolation over triangulations of surfaces is very popular in many engineering applications.

The operator $\mathcal{P}_n$ is a bounded projection operator on $C(S)$, with

$$\|\mathcal{P}_n\| = \max_{(s,t)\in\sigma} \sum_{j=1}^{f_r} |\ell_j(s,t)|$$

In particular,

$$\|\mathcal{P}_n\| = \begin{cases} 1, & r = 0, 1 \\ \dfrac{5}{3}, & r = 2 \end{cases}$$

The derivation of bounds for $\|g - \mathcal{P}_n g\|_\infty$ is similar to that done in Theorem 5.1.2.

Theorem 5.3.1. *Let $g \in C(S)$. Assume the functions $F_j \in C^{r+1}(R_j)$, $j = 1, \ldots, J$. Define*

$$\hat{g}_j(x, y) = g(F_j(x, y)), \quad (x, y) \in R_j \qquad (5.3.103)$$

and assume $\hat{g}_j \in C^{r+1}(R_j)$, $j = 1, \ldots, J$. [We will refer to this by saying $g \in C(S) \cap C^{r+1}(S_j)$, $j = 1, \ldots, J$; and this will implicitly assume the functions $F_j \in C^{r+1}(R_j)$.] Then

$$\|g - \mathcal{P}_n g\|_\infty \le c \hat{\delta}_n^{r+1} \max_{j=1,\ldots,J} N_{r+1}(\hat{g}_j) \qquad (5.3.104)$$

with

$$N_{r+1}(\hat{g}_j) = \max_{\substack{i+j=r+1 \\ i,j\ge 0}} \left[\max_{(x,y)\in R_j} \left| \frac{\partial^{r+1}\hat{g}_j(x, y)}{\partial^i x \partial^j y} \right| \right]$$

and

$$\hat{\delta}_n = \max_{j=1,\ldots,J} \left[\max_{k=1,\ldots,n_j} diameter(\hat{\Delta}_{j,k}) \right]$$

The constant c is independent of both g and n.

Proof. The proof is essentially the same as that given for Lemma 5.1.2 and Theorem 5.1.2. Merely identify each of the functions $\hat{g}_j$ with the function g in Lemma 5.1.2. □

5.3.2. Numerical integration over a surface

Recall the discussion in §5.1 of numerical integration formulas over σ. As a generic formula, we use

$$\int_\sigma G(s,t)\, d\sigma \approx \sum_{i=1}^{f} w_i G(\mu_i) \qquad (5.3.105)$$

with $\{\mu_1, \ldots, \mu_f\}$ the quadrature nodes within σ. For specific formulas, see the discussion following (5.1.36).

For a general $\Delta_k \in \mathcal{T}_n$, we convert an integral over Δ_k to an integral over σ.

$$\int_{\Delta_k} g(x,y,z)\, dS = \int_\sigma g(m_k(s,t))|(D_s m_k \times D_t m_k)(s,t)|\, d\sigma \qquad (5.3.106)$$

In this, $D_s m_k = \partial m_k/\partial s$ and $D_t m_k = \partial m_k/\partial t$. If Δ_k is a planar triangle, then $|(D_s m_k \times D_t m_k)(s,t)|$ is simply twice the area of Δ_k. Now apply (5.3.105) to (5.3.106) to obtain a numerical integration formula over Δ_k.

For an integration formula over S, first decompose S using a triangulation $\mathcal{T}_n$. Then apply (5.3.105)–(5.3.106) to each integral over a triangular surface element:

$$
\begin{aligned}
I(g) &\equiv \int_S g(x,y,z)\, dS \\
&= \sum_{k=1}^{n} \int_{\Delta_k} g(x,y,z)\, dS \\
&\approx \sum_{k=1}^{n} \sum_{i=1}^{f} w_i g((m_k(\mu_i))|(D_s m_k \times D_t m_k)(\mu_i)| \\
&\equiv I_n(g) \qquad (5.3.107)
\end{aligned}
$$

The following result on the rate of convergence is completely analogous to Theorem 5.1.3. We omit the proof.

Theorem 5.3.2. *Assume the numerical integration formula (5.3.105) has degree of precision $d \geq 0$. Assume the functions $F_j \in C^{d+2}(R_j)$, $j = 1, \ldots, J$, for the parametrization functions of (5.3.97). Assume $g \in C(S) \cap C^{d+1}(S_j)$, $j = 1, \ldots, J$. Then*

$$|I(g) - I_n(g)| \leq c\hat{\delta}_n^{d+1} \max_{j=1,\ldots,J} N_{d+1}(\gamma_j) \qquad (5.3.108)$$

with $\gamma_j(x,y) \equiv g(F_j(x,y))|(D_x F_j \times D_y F_j)(x,y)|$, $(x,y) \in R_j$, $j = 1, \ldots, J$.

Recall the result on the possibly improved rate of convergence when using symmetric triangulations, discussed in and about (5.1.54). This generalizes to integration over surfaces. For symmetric integration formulas, such as (5.1.41) with $d = r$ and r even, the order of convergence in (5.3.108) will be $O(\hat{\delta}_n^{d+2})$. The proof is essentially the same as earlier.

5.3.3. Approximating the surface

In evaluating the integral (5.3.106) over Δ_k, we require the Jacobian

$$|(D_s m_k \times D_t m_k)(s, t)|$$

Recalling (5.3.99), the computation of the derivatives $D_s m_k$ and $D_t m_k$ requires a knowledge of the derivatives of the functions $F_j(x, y)$, $j = 1, \ldots, J$. With some surfaces S, the functions $F_j(x, y)$ are easily given and computed. And with some surfaces, knowing these functions and their derivatives can be avoided. For example, with S a polyhedral surface, the Jacobian will be twice the area of the triangle Δ_k, and this area is easily computed. However, with most surfaces, gaining knowledge of the derivatives of $F_j(x, y)$ is a major inconvenience, both to specify and to program. For this reason we consider surface approximations $\tilde{S}$ for which the Jacobians are more easily computed, and then we perform our integrations over $\tilde{S}$.

Recall the interpolation scheme (5.3.101) with the interpolation nodes $\{q_j\}$ of (5.3.100). Let $\Delta_k \in \mathcal{T}_n$. For $r \geq 1$, define

$$\bar{m}_k(s, t) = \sum_{i=1}^{f_r} m_k(q_i)\ell_i(s, t), \quad (s, t) \in \sigma$$

$$= \sum_{i=1}^{f_r} v_{k,i}\ell_i(s, t) \tag{5.3.109}$$

This interpolates $m_k(s, t)$ at the nodes $\{q_j\}$. Each component of $\bar{m}_k(s, t)$ is a polynomial of degree r in (s, t). Let $\tilde{\Delta}_k = \bar{m}_k(\sigma)$, and note that $\tilde{\Delta}_k$ agrees with Δ_k at the nodes $v_{k,i} = m_k(q_i)$, $i = 1, \ldots, f_r$. Define the approximating surface by

$$\tilde{S} \equiv \tilde{S}^{(n)} = \bigcup_{k=1}^{n} \tilde{\Delta}_k \tag{5.3.110}$$

With our assumptions **T1–T3** on the triangulation $\mathcal{T}_n$, it can be shown that $\tilde{S}$ is a continuous surface, the same as S. The proof is essentially the same as that discussed earlier to show $\mathcal{P}_n g \in C(S)$, following (5.3.102). The most commonly used case in applications is with $r = 1$, which means $\tilde{S}$ is piecewise planar with triangular faces.

Before considering numerical integration over $\tilde{S}$, with

$$\tilde{I}^{(n)}(g) \equiv \int_{\tilde{S}} g(x, y, z)\, dS \qquad (5.3.111)$$

we must consider the possibility of extending $g \in C(S)$ to a neighborhood of S, one that contains $\tilde{S}$ for all sufficiently large n. This is a well-studied problem, and for all well-behaved piecewise smooth surfaces such extensions are possible. Assume $g \in C(S) \cap C^k(S_j)$, $j = 1, \ldots, J$. This means $\hat{g}_j \in C^k(R_j)$, with $\hat{g}_j$ as in (5.3.103). Then for each j, there is a function $\tilde{g}_j \in C^k(\Omega_j)$ with Ω_j an open neighborhood of S_j and $\tilde{g}_j \mid S_j = g \mid S_j$. For smooth surfaces S, stronger results are possible. Henceforth, we assume that such extensions $\tilde{g}$ to our surfaces $\tilde{S}$ are known explicitly, as they generally are in practice. A partial justification of this is provided in Günter [246, Chap. 1, §3].

The integral $\tilde{I}^{(n)}(g)$ of (5.3.111) is calculated by using

$$\tilde{I}^{(n)}(g) = \sum_{k=1}^{n} \int_{\tilde{\Delta}_k} g(x, y, z)\, dS$$

$$= \sum_{k=1}^{n} \int_{\sigma} g(\tilde{m}_k(s, t)) |(D_s \tilde{m}_k \times D_t \tilde{m}_k)(s, t)|\, d\sigma \qquad (5.3.112)$$

This contains integrals that also must be evaluated numerically. To do so, we apply the method used in (5.3.107), which is based on (5.3.105). This yields

$$\tilde{I}_n^{(n)}(g) \equiv \sum_{k=1}^{n} \sum_{i=1}^{f_r} w_i\, g(\tilde{m}_k(\mu_i)) |(D_s \tilde{m}_k \times D_t \tilde{m}_k)(\mu_i)| \qquad (5.3.113)$$

Theorem 5.3.3.

(a) *Assume $\tilde{S}^{(n)}$ is based on interpolation of degree $r \geq 1$, as in (5.3.110). Define*

$$\tilde{r} = \begin{cases} r, & r \text{ odd} \\ r + 1, & r \text{ even} \end{cases}$$

Assume the parametrization functions $F_j \in C^{\tilde{r}+1}(R_j)$, $j = 1, \ldots, J$, for the functions of (5.3.97). Assume $g \in C(S) \cap C^{\tilde{r}}(S_j)$, $j = 1, \ldots, J$. Then

$$|I(g) - \tilde{I}^{(n)}(g)| \leq c \hat{\delta}_n^{\tilde{r}} \qquad (5.3.114)$$

(b) *Assume the numerical integration formula (5.3.105) has degree of precision*
$d \geq 0$. *Let* $\kappa = \min\{d + 1, \bar{r}\}$. *Assume the parametrization functions*
$F_j \in C^{\kappa+1}(R_j)$, $j = 1, \ldots, J$, *for the functions of (5.3.97). Assume*
$g \in C(S) \cap C^{\kappa}(S_j)$, $j = 1, \ldots, J$. *Then*

$$\left| I(g) - \bar{I}_n^{(n)}(g) \right| \leq c\hat{\delta}_n^{\kappa} \tag{5.3.115}$$

In both cases, c is dependent on g and the parametrization functions
$\{F_1, \ldots, F_J\}$, *and it is a multiple of the maximum of the norms of all*
derivatives of $g(F_j(x, y))$ *of order* $\leq \kappa$ *and derivatives of* $F_j(x, y)$ *of*
order $\leq \bar{r} + 1$, *including products of these norms.*

Proof. We omit the proof, as it is much the same as that for the following
theorem, which we consider more important.   □

For the numerical integration scheme (5.3.105), base it on integrating
(5.3.101) with the uniform nodes $\{q_i\}$ of (5.3.100). Then (5.3.113) becomes

$$\bar{I}_n^{(n)}(g) \equiv \sum_{k=1}^{n} \sum_{i=1}^{f_r} w_i g(v_{k,i}) |(D_s \tilde{m}_k \times D_t \tilde{m}_k)(q_i)| \tag{5.3.116}$$

because $\{v_{k,i}\}$ are the node points at which Δ_k and $\tilde{\Delta}_k$ agree. Thus (5.3.116)
requires a knowledge of $g(x, y, z)$ on only S. We have the following improve-
ment on Theorem 5.3.3.

Theorem 5.3.4. *Let* $r \geq 1$ *be an integer. Assume* $\tilde{S}^{(n)}$ *is based on interpolation*
of degree r, as in (5.3.110). Define

$$\kappa = \begin{cases} r + 1, & r \text{ odd} \\ r + 2, & r \text{ even} \end{cases}$$

Assume the parametrization functions $F_j \in C^{\kappa+1}(R_j)$, $j = 1, \ldots, J$, *for the*
functions of (5.3.97). Assume $g \in C(S) \cap C^{\kappa}(S_j)$, $j = 1, \ldots, J$. *Finally,*
assume that the triangulations of each of the polygonal regions R_j are symmetric
triangulations. Then

$$\left| I(g) - \bar{I}^{(n)}(g) \right| \leq c\hat{\delta}_n^{\kappa} \tag{5.3.117}$$

Moreover, for the numerical integration method (5.3.116),

$$\left| I(g) - \bar{I}_n^{(n)}(g) \right| \leq c\hat{\delta}_n^{\kappa} \tag{5.3.118}$$

In both cases, c is dependent on g and the parametrization functions $\{F_1, \ldots, F_J\}$, *and it is a multiple of the maximum of the norms of all derivatives of* $g(F_j(x, y))$ *of order* $\leq \kappa$ *and derivatives of* $F_j(x, y)$ *of order* $\leq \kappa + 1$, *including products of these norms.*

Proof. These results are due to Chien [119], [120], and many of the following derivations are taken directly from Ref. [120], including much of the notation. We begin with the proof of (5.3.117). To make the proof more intuitive, we consider in detail only the case $r = 2$. Write

$$I(g) - \tilde{I}^{(n)}(g) = \sum_{k=1}^{n} \int_{\sigma} g(m_k(s, t)) \, |D_s m_k \times D_t m_k(s, t)| \, d\sigma$$

$$- \sum_{k=1}^{n} \int_{\sigma} g(\tilde{m}_k(s, t)) \, |D_s \tilde{m}_k \times D_t \tilde{m}_k(s, t)| \, d\sigma$$

$$= E_1 + E_2 + E_3 \tag{5.3.119}$$

$$E_1 = \sum_{k=1}^{n} \int_{\sigma} g(m_k(s, t)) \{ |D_s m_k \times D_t m_k(s, t)|$$

$$- |D_s \tilde{m}_k \times D_t \tilde{m}_k(s, t)| \} \, d\sigma$$

$$E_2 = \sum_{k=1}^{n} \int_{\sigma} g(m_k(s, t)) - g(\tilde{m}_k(s, t)) \,] |D_s m_k \times D_t m_k(s, t)| \, d\sigma$$

$$E_3 = \sum_{k=1}^{n} \int_{\sigma} [g(m_k(s, t)) - g(\tilde{m}_k(s, t))] \cdot \{ |D_s \tilde{m}_k \times D_t \tilde{m}_k(s, t)|$$

$$- |D_s m_k \times D_t m_k(s, t)| \} \, d\sigma$$

We bound each of these quantities, showing

$$E_1, E_2 = O\left(\hat{\delta}_n^4\right), \qquad E_3 = O\left(\hat{\delta}_n^5\right) \tag{5.3.120}$$

and this will prove (5.3.117) for $r = 2$. But first we must introduce some notation and consider a number of preliminaries.

(a) Preliminaries. For a given Δ_k we can write the mapping m_k as

$$m_k(s, t) = F_j(x, y) = \begin{bmatrix} x^1(x, y) \\ x^2(x, y) \\ x^3(x, y) \end{bmatrix}, \quad (x, y) = T_k(s, t)$$

where we assume $\Delta_k = f_j(\hat{\Delta}_k)$, $\hat{\Delta}_k \subset R_j$. This uses the affine mapping $T_k(s, t)$

of (5.1.10). We can also write

$$m_k(s, t) = F_j(u\hat{v}_{k,1} + t\hat{v}_{k,2} + s\hat{v}_{k,3}) = \begin{bmatrix} x^1(u\hat{v}_{k,1} + t\hat{v}_{k,2} + s\hat{v}_{k,3}) \\ x^2(u\hat{v}_{k,1} + t\hat{v}_{k,2} + s\hat{v}_{k,3}) \\ x^3(u\hat{v}_{k,1} + t\hat{v}_{k,2} + s\hat{v}_{k,3}) \end{bmatrix}$$

The component functions x^i can be considered as either functions of $(x, y) \in R_j$ or as functions of $(s, t) \in \sigma$. We will use both $x^i(x, y)$ and $x^i(s, t)$, with the context indicating which of the above is intended.

As earlier in (5.1.32), we can show

$$\begin{bmatrix} \dfrac{\partial x^i}{\partial s} & \dfrac{\partial x^i}{\partial t} \end{bmatrix} = \begin{bmatrix} \dfrac{\partial x^i}{\partial x} & \dfrac{\partial x^i}{\partial y} \end{bmatrix} \begin{bmatrix} \hat{v}_{k,3} - \hat{v}_{k,1} & \hat{v}_{k,2} - \hat{v}_{k,1} \end{bmatrix} \qquad (5.3.121)$$

for $i = 1, 2, 3$. Notationally, the left side of this expression is a 2×1 row vector, and the right side is a multiplication of 2×1 and 2×2 arrays. Thus

$$|D_s m_k(s, t)|, \ |D_t m_k(s, t)| \leq c \ \text{diameter}(\hat{\Delta}_k) \cdot \max \left\{ \left\| \frac{\partial F_j}{\partial x} \right\|_\infty, \left\| \frac{\partial F_j}{\partial y} \right\|_\infty \right\}$$
$$(5.3.122)$$

with the norms being taken over R_j and c independent of j, k, n, s, t. Thus

$$\max_{k=1,\dots,n} \ \max_{(s,t)\in\sigma} |(D_s m_k \times D_t m_k)(s, t)| \leq c\delta_n^2 \qquad (5.3.123)$$

with c independent of n.

For the components of $\bar{m}_k(s, t)$, write

$$\tilde{x}^i(s, t) = \sum_{j=1}^{6} x^i(q_j)\ell_j(s, t), \qquad (s, t) \in \sigma, \ i = 1, 2, 3$$

Using Taylor's formula,

$$x^i(s, t) - \tilde{x}^i(s, t) = H^i(s, t) + G^i(s, t) + O(\hat{\delta}_n^5) \qquad (5.3.124)$$

with

$$H^i(s, t) = \frac{1}{3!} \left[\left(s\frac{\partial}{\partial s} + t\frac{\partial}{\partial t} \right)^3 x^i(0, 0) \right.$$

$$\left. - \sum_{j=1}^{6} \left(s_j\frac{\partial}{\partial s} + t_j\frac{\partial}{\partial t} \right)^3 x^i(0, 0)\ell_j(s, t) \right] \qquad (5.3.125)$$

$$G^i(s, t) = \frac{1}{4!} \left[\left(s\frac{\partial}{\partial s} + t\frac{\partial}{\partial t} \right)^4 x^i(0, 0) \right.$$

$$\left. - \sum_{j=1}^{6} \left(s_j\frac{\partial}{\partial s} + t_j\frac{\partial}{\partial t} \right)^4 x^i(0, 0)\ell_j(s, t) \right] \qquad (5.3.126)$$

The term $O(\hat{\delta}_n^5)$ comes from the fifth derivatives of $x^i(s, t)$, with the type of derivation given in (5.3.121)–(5.3.122). A similar expansion can be given for the errors in the partial derivatives of $x^i(s, t)$, amounting to a use of the derivative of (5.3.124):

$$x_s^i(s, t) - \bar{x}_s^i(s, t) = H_s^i(s, t) + G_s^i(s, t) + O(\hat{\delta}_n^4) \qquad (5.3.127)$$

with an analogous formula for $x_t^i(s, t) - \bar{x}_t^i(s, t)$. We use the subscripts s and t to denote the associated partial derivatives, to be less expansive in our formulas.

Using Taylor's theorem and expanding about $(s, t) = (0, 0)$,

$$|(D_s m_k \times D_t m_k)(s, t)| - |D_s \tilde{m}_k \times D_t \tilde{m}_k(s, t)|$$

$$= E_4(s, t; \hat{v}_{k,2} - \hat{v}_{k,1}, \hat{v}_{k,3} - \hat{v}_{k,1})$$

$$+ E_5(s, t; \hat{v}_{k,2} - \hat{v}_{k,1}, \hat{v}_{k,3} - \hat{v}_{k,1}) + O(\hat{\delta}_n^6) \qquad (5.3.128)$$

$$E_4(s, t; \hat{v}_{k,2} - \hat{v}_{k,1}, \hat{v}_{k,3} - \hat{v}_{k,1})$$

$$= \left\{ \left(x_s^2 x_t^3 - x_s^3 x_t^2 \right) \left[x_s^2 H_t^3 + x_t^3 H_s^2 - x_s^3 H_t^2 - x_t^2 H_s^3 \right] \right.$$

$$+ \left(x_s^3 x_t^1 - x_s^1 x_t^3 \right) \left[x_s^3 H_t^1 + x_t^1 H_s^3 - x_s^1 H_t^3 - x_t^3 H_s^1 \right] + \left(x_s^1 x_t^2 - x_s^2 x_t^1 \right)$$

$$\left. \times \left[x_s^1 H_t^2 + x_t^2 H_s^1 - x_s^2 H_t^1 - x_t^1 H_s^2 \right] \right\} / |(D_s m_k \times D_t m_k)(0, 0)|$$

$$(5.3.129)$$

The terms x_s^i and x_t^i are abbreviations for $x_s^i(0, 0)$ and $x_t^i(0, 0)$. The terms H_s^i and H_t^i denote $H_s^i(s, t)$ and $H_t^i(s, t)$, respectively. The term $E_5(s, t; \hat{v}_{k,2} - \hat{v}_{k,1}, \hat{v}_{k,3} - \hat{v}_{k,1})$ is yet more complicated, but it has the crucial property

$$E_5(s, t; -(\hat{v}_{k,2} - \hat{v}_{k,1}), -(\hat{v}_{k,3} - \hat{v}_{k,1}))$$

$$= -E_5(s, t; \hat{v}_{k,2} - \hat{v}_{k,1}, \hat{v}_{k,3} - \hat{v}_{k,1}) \qquad (5.3.130)$$

and this is sufficient for our later analysis. We call this the "odd function" property.

We will need a further lemma in connection with using (5.3.128).

Lemma 5.3.1. Let r be an even integer $r \geq 2$. Let $g(s, t)$ be a polynomial of degree $r + 1$ on σ, and let $g_r(s, t)$ be its interpolant of degree r, based on interpolation at the nodes $\{q_1, \ldots, q_{f_r}\}$ as in (5.3.100). Then

$$\int_\sigma \frac{\partial}{\partial s} [g(s, t) - g_r(s, t)] d\sigma = 0$$

$$\int_\sigma \frac{\partial}{\partial t} [g(s, t) - g_r(s, t)] d\sigma = 0$$

Proof. A proof is given in Chien [121]. $\qquad\square$

(b) Bounding E_1. Return to the decomposition (5.3.119) and consider the term E_1. In particular, consider the error over a single triangle Δ_k:

$$E_{1,k} = \int_\sigma g(m_k(s, t))\{|D_s m_k \times D_t m_k(s, t)| - |D_s \bar{m}_k \times D_t \bar{m}_k(s, t)|\} d\sigma$$

Using Taylor's theorem,

$$g(m_k(s, t)) = g(m_k(0, 0)) + \left(s\frac{\partial}{\partial s} + t\frac{\partial}{\partial t}\right) g(m_k(0, 0)) + O(\hat{\delta}_n^2)$$

$$(5.3.131)$$

In this,

$$g_s(m_k(s, t)) \equiv \frac{\partial}{\partial s} g(m_k(s, t)) = \sum_{i=1}^3 \frac{\partial}{\partial x^i} g(x^1, x^2, x^3) \frac{\partial x^i(s, t)}{\partial s}$$

and the partial derivatives $\partial x^i(s, t)/\partial s$ are given in (5.3.121). Consequently, this derivative shares the "odd function" property of (5.3.130).

Using the definition of E_1 and (5.3.131),

$$E_{1,k} = g(m_k(0, 0)) \int_\sigma [E_4(s, t) + E_5(s, t)] d\sigma$$

$$+ \int_\sigma E_4(s, t) [sg_s(m_k(0, 0)) + tg_t(m_k(0, 0))] d\sigma + O(\hat{\delta}_n^6)$$

$$(5.3.132)$$

Note first that

$$\int_\sigma E_4(s, t) d\sigma = 0$$

This is due to Lemma 5.3.1 and the use in $E_4(s, t)$ of partial derivatives of $H^i(s, t)$, which are errors when integrating cubic polynomials. This leads to

$$E_{1,k} = O(\hat{\delta}_n^5)$$

$$(5.3.133)$$

Now consider the error (5.3.132) over a pair of symmetric triangles, say Δ_k and Δ_l, of the type shown in Figure 5.3. The remaining integral terms in (5.3.132) will cancel due to the odd function property of

$$E_4(s, t) [sg_s(m_k(0, 0)) + tg_t(m_k(0, 0))]$$

and $E_5(s, t)$. Thus

$$E_{1,k} + E_{1,l} = O(\hat{\delta}_n^6) \qquad (5.3.134)$$

If we divide a polygonal region R_j into triangles $\{\hat{\Delta}_{j,k} \mid 1 \leq k \leq n_j\}$, then it is fairly straightforward to show that the number of symmetric pairs is of the order $O(n_j) = O(\hat{\delta}_n^{-2})$ and the number of remaining triangles is $O(\sqrt{n_j}) = O(\hat{\delta}_n^{-1})$. Combining (5.3.133) and (5.3.134) for these two types of triangles, we will have

$$E_1 = O(\hat{\delta}_n^4)$$

(c) Bounding E_2. Consider the contribution to E_2 from integration over a single triangle Δ_k:

$$E_{2,k} = \int_\sigma [g(m_k(s, t)) - g(\tilde{m}_k(s, t))] |(D_s m_k \times D_t m_k)(s, t)| \, d\sigma$$

$$(5.3.135)$$

Using a Taylor expansion,

$$g(m_k(s, t)) - g(\tilde{m}_k(s, t))$$

$$= \sum_{i=1}^{3} \frac{\partial g(m_k(s, t))}{\partial x^i} [x^i(s, t) - \tilde{x}^i(s, t)] + O(\|m_k - \tilde{m}_k\|_\infty^2)$$

We use the expansion (5.3.124) for the interpolation error. Then

$$g(m_k(s, t)) - g(\tilde{m}_k(s, t)) = \sum_{i=1}^{3} H^i(s, t) \frac{\partial g(m_k(s, t))}{\partial x^i} + O(\hat{\delta}_n^6)$$

$$= \sum_{i=1}^{3} H^i(s, t) \frac{\partial g(m_k(0, 0))}{\partial x^i} + O(\hat{\delta}_n^6)$$

$$(5.3.136)$$

Using another Taylor expansion, and recalling (5.3.121), we can show

$$(D_s m_k \times D_t m_k)(s, t) = (D_s m_k \times D_t m_k)(0, 0) + O(\hat{\delta}_n^3)$$

Combining these results, and recalling that $H^i = O(\hat{\delta}_n^3)$, we have

$$
E_{2,k} = \int_\sigma \left[\sum_{i=1}^3 H^i(s,t) \frac{\partial g(m_k(0,0))}{\partial x^i} + O(\hat{\delta}_n^6) \right]
$$
$$
\times \left[|(D_s m_k \times D_t m_k)(0,0)| + O(\hat{\delta}_n^3) \right] d\sigma
$$
$$
= \sum_{i=1}^3 \frac{\partial g(m_k(0,0))}{\partial x^i} |(D_s m_k \times D_t m_k)(0,0)| \int_\sigma H^i(s,t)\, d\sigma + O(\hat{\delta}_n^6)
$$

Now recall that $H^i(s,t)$ has the odd function property. Consider the above error over a pair of symmetric triangles Δ_k and Δ_l. The constants outside the integral can be shown to be equal over the two triangles. The integrals are equal in magnitude, but opposite in sign, by the odd function property. Thus

$$
E_{2,k} + E_{2,l} = O(\hat{\delta}_n^6)
$$

Arguing as for E_1, we will obtain

$$
E_2 = O(\hat{\delta}_n^4)
$$

(d) Bounding E_3 and remaining topics. Showing that $E_3 = O(\hat{\delta}_n^5)$ is relatively straightforward based on the results and tools introduced above. As before, consider the contribution to E_3 from a single triangular element Δ_k:

$$
E_{3,k} = \int_\sigma [g(m_k(s,t)) - g(\tilde{m}_k(s,t))]
$$
$$
\times \{ |D_s \tilde{m}_k \times D_t \tilde{m}_k(s,t)| - |D_s m_k \times D_t m_k(s,t)| \}\, d\sigma
$$

From (5.3.129),

$$
|D_s \tilde{m}_k \times D_t \tilde{m}_k(s,t)| - |D_s m_k \times D_t m_k(s,t)| = O(\hat{\delta}_n^4)
$$

and from (5.3.136),

$$
g(m_k(s,t)) - g(\tilde{m}_k(s,t)) = O(\hat{\delta}_n^3)
$$

Combining these,

$$
E_{3,k} = O(\hat{\delta}_n^7)
$$

Arguing as for E_1, we will obtain

$$
E_2 = O(\hat{\delta}_n^5)
$$

as desired. This also completes the proof of (5.3.117) for $r = 2$.

The proof of (5.3.118) is fairly similar to the above, and we refer the reader to Chien [120, Theorem 2].

The proof for r an odd integer is slightly different. In this case the term E_4 of (5.3.129) has the odd function property, and thus we can show $E_{1,k} = O(\hat{\delta}_n^{r+3})$ and $E_{1,k} = O(\hat{\delta}_n^{r+1})$. The remaining part of the proof is similar to the above for the case of r an even integer. $\square$

Corollary 5.3.1. *Let $r \geq 1$ be an integer. Assume $\tilde{S}^{(n)}$ is based on interpolation of degree r, as in (5.3.110). Assume the parametrization functions $F_j \in C^{r+2}(R_j)$, $j = 1, \ldots, J$, for the functions of (5.3.97). Then for $g \in C(S)$,*

$$|I(g) - \tilde{I}^{(n)}(g)| \leq Area(S)\|g - \mathcal{P}_n g\|_\infty + c\|\mathcal{P}_n g\|\hat{\delta}_n^r \qquad (5.3.137)$$

Thus $\tilde{I}^{(n)}(g) \to I(g)$ as $n \to \infty$.

Proof. Use the identity

$$\int_\sigma g(m_k(s, t))|D_s m_k \times D_t m_k(s, t)| \, d\sigma$$

$$- \int_\sigma (\mathcal{P}_n g)(m_k(s, t))|D_s \tilde{m}_k \times D_t \tilde{m}_k(s, t)| \, d\sigma$$

$$= \int_\sigma [g(m_k(s, t)) - (\mathcal{P}_n g)(m_k(s, t))]|D_s m_k \times D_t m_k(s, t)| \, d\sigma$$

$$+ \int_\sigma (\mathcal{P}_n g)(m_k(s, t))[|D_s m_k \times D_t m_k(s, t)|$$

$$- |D_s \tilde{m}_k \times D_t \tilde{m}_k(s, t)|] \, d\sigma$$

Bounding this and applying the results of part (b) in the preceding proof, we obtain (5.3.137). $\square$

Example. Let S be the ellipsoidal surface given by

$$\left(\frac{x}{a}\right)^2 + \left(\frac{y}{b}\right)^2 + \left(\frac{z}{c}\right)^2 = 1$$

We will calculate approximations to

$$I = \int_S e^{-x+2y+\frac{1}{2}z} \, dS \qquad (5.3.138)$$

with the surface parameters $(a, b, c) = (2, 1, 0.5)$. The rule used is (5.3.116) with $r = 2$, and we let I_n denote the approximation to I with n faces.

Table 5.2. *Numerical integration*
of (5.3.138) on an ellipsoid

n	I_n	D_n	Ω_n
32	42.511471795	7.67E−1	
128	42.763926456	2.52E−1	1.60
512	42.799929963	3.60E−2	2.81
2048	42.802726376	2.80E−3	3.69
8192	42.802905921	1.80E−4	3.96

According to the preceding theorem, the rate of convergence should be $O(\hat{\delta}_n^4)$. We check this empirically. Table 5.2 contains the integrals I_n and the differences

$$D_n = I_n - I_{\frac{1}{4}n}$$

With these, we calculate

$$\Omega_n \equiv \log_2 \left[\frac{D_{\frac{1}{4}n}}{D_n} \right]$$

which is an estimate of the exponent in the order of convergence. Theoretically, the sequence $\{\Omega_n\}$ should converge to 4, and the values of Ω_n given in Table 5.2 agree with this. We emphasize that these results include the use of the approximate surface $\tilde{S}$ in the numerical integration.

Example. Let S be the elliptical paraboloid given by

$$z = \left(\frac{x}{a}\right)^2 + \left(\frac{y}{b}\right)^2, \quad 0 \le z \le c$$

together with its "cap"

$$\left(\frac{x}{a}\right)^2 + \left(\frac{y}{b}\right)^2 \le c$$

We calculate both the integral I of (5.3.138) and the integral

$$I = \int_S \frac{\partial}{\partial \mathbf{n}_P} (e^z)\, dS = \pi ab[(c-1)e^c + 1] \tag{5.3.139}$$

with $\mathbf{n}_P$ the external normal to S at $P = (x, y, z)$. The surface parameters used are $(a, b, c) = (2, 1, 0.5)$. This last integrand $g(P)$ is discontinuous along the edge of the paraboloid at $z = c$. We evaluate the numerical integral in the

Table 5.3. *Numerical integration of (5.3.138)*
on a paraboloid

n	I_n	D_n	Ω_n
32	13.966970109	−6.92E−1	
128	13.945419499	−2.16E−2	5.00
512	13.943158305	−2.26E−3	3.25
2048	13.942952696	−2.06E−4	3.46
8192	13.942936547	−1.62E−5	3.67
32768	13.942935361	−1.19E−6	3.77

Table 5.4. *Numerical integration of (5.3.139)*
on a paraboloid

n	I_n	E_n	Ω_n
8	1.1630059263	−5.94E−2	
32	1.1137998496	−1.02E−2	2.54
128	1.1052130807	−1.64E−3	2.64
512	1.1037575702	−1.83E−4	3.16
2048	1.1035917672	−1.71E−5	3.42
8192	1.1035761147	−1.44E−6	3.57
32768	1.1035747889	−1.14E−7	3.66

form (5.3.116), because g is continuous over each triangle Δ_k. The normal derivative in (5.3.139) is calculated exactly for the true surface S. The numerical integration rule is (5.3.116) with $r = 2$, with S being approximated by $\tilde{S}$.

The numerical results for approximating (5.3.138) are given in Table 5.3, and those for (5.3.139) are given in Table 5.4. For the latter case we give the true errors

$$E_n = I - I_n$$

and the empirical order is given by

$$\Omega_n \equiv \log_2 \left[\frac{E_{\frac{1}{4}n}}{E_n} \right]$$

In both cases the order of convergence is approaching the theoretically correct value of 4. The convergence is slower than for the case that S is an ellipsoid, but the trend is clear.

5.3.4. Nonconforming triangulations

When approximating or integrating functions $g(P)$ that have singular behavior at isolated parts of S, we sometimes use a graded mesh to obtain a better approximation. For functions of one variable such graded meshes were discussed in the concluding part of §4.2, with convergence results given in Theorem 4.2.3. We can also use graded meshes for multivariable problems, with triangular elements being relatively smaller near points of poor behavior of $g(P)$. In Figure 5.4 we show a graded mesh for a case with $g(P)$ having poor behavior at the vertex P_0. This triangulation is *nonconforming*, in the sense that the assumption T3 of §5.3 is no longer satisfied. We replace it with the following.

T3'. If two triangles in $\mathcal{T}_n$ have a nonempty intersection, then that intersection consists of (*i*) a single vertex of at least one of the triangles, or (*ii*) an entire edge of one of the triangles, with that edge either equal to or contained entirely within a single edge of the other triangle.

The interpolation of functions over such a triangulation will no longer yield a continuous function, and the use of such interpolation to approximate S will result in an approximating surface $\tilde{S}$ that is not continuous. Nonetheless, the types of error analyses given in this section can be generalized to such triangulation schemes, with the appropriate function space setting becoming $L^\infty(S)$. For a very brief discussion of interpolation over such *nonconforming* triangulations, see Ref. [43, p. 39]. For graded mesh numerical integration, see Yang and Atkinson [584]. In Chapter 9 we will discuss the use of graded mesh

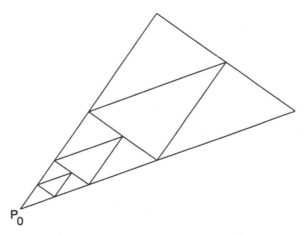

Figure 5.4. A nonconforming triangulation.

triangulations in solving boundary integral equations over piecewise smooth surfaces S.

5.4. Boundary element methods for solving integral equations

By the expression "boundary element method" we are referring to the combination of the tools of the preceding section with the projection methods and Nyström methods of Chapters 3 and 4. This name (often abbreviated to *BEM*) has been used primarily with the solution of boundary integral equation reformulations of partial differential equations; but since such methods can also be used for more general integral equations over surfaces, the name seems appropriate for such general integral equations. In Chapter 9 we modify the presentation and error analysis to the special case of boundary integral equation reformulations of Laplace's equation in $\mathbf{R}^3$.

The integral equation being solved is

$$\lambda\rho(P) - \int_S K(P, Q)\rho(Q)\, dS_Q = \psi(P), \quad P \in S \qquad (5.4.140)$$

with $\psi \in C(S)$. We begin with the Nyström method, as it is the most straightforward to explain and analyze. Following that, we present the collocation and Galerkin versions of the boundary element method. In all of the cases of this section, the integral operator is assumed to be compact on $C(S)$ to $C(S)$, and the integral equation is assumed to be uniquely solvable. Throughout the section we assume a sequence of triangulations $\mathcal{T}_n$ for S, of the kind discussed in (5.3.96)–(5.3.99). Moreover, we assume the mesh size $\hat{\delta}_n \to 0$ as $n \to \infty$.

5.4.1. The Nyström method

Consider the case that $K(P, Q)$ is continuous over S. Let a composite integration scheme be based on

$$\int_\sigma G(s, t)\, d\sigma \approx \sum_{i=1}^{f} w_i G(\mu_i) \qquad (5.4.141)$$

with $\{\mu_1, \ldots, \mu_f\}$ the quadrature nodes within σ. For specific formulas see the discussion following (5.1.36). Let

$$v_{k,i} = m_k(\mu_i), \quad i = 1, \ldots, f, \quad k = 1, \ldots, n \qquad (5.4.142)$$

and collectively refer to these integration nodes as $\mathcal{V}_n = \{v_1, \ldots, v_{n_v}\}$.

Approximate the integral operator $\mathcal{K}\rho(P)$ by

$$\mathcal{K}_n\rho(P) = \sum_{k=1}^{n} \sum_{i=1}^{f} w_i K(P, v_{k,i})\rho(v_{k,i})|(D_s m_k \times D_t m_k)(\mu_i)|, \quad P \in S$$

$$(5.4.143)$$

for all $\rho \in C(S)$. This can also be written in the simpler form

$$\mathcal{K}_n\rho(P) = \sum_{j=1}^{n_v} \omega_j K(P, v_j)\rho(v_j) \tag{5.4.144}$$

with

$$\omega_j = \sum_{\substack{i,k \\ v_j = m_k(\mu_i)}} w_i |(D_s m_k \times D_t m_k)(\mu_i)|$$

Approximate the integral equation $(\lambda - \mathcal{K})\rho = \psi$ by $(\lambda - \mathcal{K}_n)\rho_n = \psi$, or equivalently

$$\lambda \rho_n(P) - \sum_{j=1}^{n_v} \omega_j K(P, v_j)\rho_n(v_j) = \psi(P), \quad P \in S$$

Solve for ρ_n by first solving the linear system

$$\lambda \rho_n(v_i) - \sum_{j=1}^{n_v} \omega_j K(v_i, v_j)\rho_n(v_j) = \psi(v_i), \quad i = 1, \ldots, n_v \tag{5.4.145}$$

The solution ρ_n is obtained by Nyström interpolation:

$$\rho_n(P) = \frac{1}{\lambda}\left[\psi(P) + \sum_{j=1}^{n_v} \omega_j K(P, v_j)\rho_n(v_j)\right], \quad P \in C(S) \tag{5.4.146}$$

We give a theorem for the error analysis of this procedure. Later in this subsection, we will provide a numerical illustration.

Theorem 5.4.1. *Assume the integral equation (5.4.140) is uniquely solvable, with $K(P, Q)$ continuous on $S \times S$. Assume the numerical integration formula (5.4.141) has degree of precision $d \geq 0$. Assume the functions $F_j \in C^{d+2}(R_j)$, $j = 1, \ldots, J$, for the parametrization functions of (5.3.97). Then:*

(a) The family $\{\mathcal{K}_n\}$ is collectively compact and pointwise convergent on $C(S)$.

(b) *The inverse operators* $(\lambda - \mathcal{K}_n)^{-1}$ *exist and are uniformly bounded for all sufficiently large n, say* $n \geq N$.

(c) *The approximation* ρ_n *has the error*

$$\rho - \rho_n = (\lambda - \mathcal{K}_n)^{-1}(\mathcal{K}\rho - \mathcal{K}_n\rho) \qquad (5.4.147)$$

and thus $\rho_n \to \rho$ *as* $n \to \infty$.

(d) *Assume that* $K(P, \cdot) \in C(S) \cap C^{d+1}(S_j)$, $j = 1, \ldots, J$, *with the derivatives of order* $d + 1$ *uniformly bounded with respect to* $P \in S$. *Assume* $\rho \in C(S) \cap C^{d+1}(S_j)$, $j = 1, \ldots, J$. *Then*

$$\|\rho - \rho_n\|_\infty \leq c\hat{\delta}_n^{d+1}, \quad n \geq N \qquad (5.4.148)$$

Proof. Since $K(P, Q)$ is continuous, this is a straightforward application of the general Theorem 4.1.2. The error bound (5.4.148) comes from applying Theorem 5.3.2.   □

Using the approximate surface

The definition (5.4.143) of $\mathcal{K}_n$ requires the derivatives of the parametrization functions $\{F_j \mid j = 1, \ldots, J\}$. We now consider the use of the approximate surface $\tilde{S}$ of (5.3.109) and (5.3.110), based on interpolation of degree $r \geq 1$ in the parametrization variables. The integration rule is that of (5.3.116), with the integration nodes

$$(s_i, t_j) = (i\delta, j\delta), \quad i, j \geq 0, \quad i + j \leq r, \quad \delta = \frac{1}{r} \qquad (5.4.149)$$

which collectively are referred to as $\{q_1, \ldots, q_{f_r}\}$, with $f_r = \frac{1}{2}(r + 1)(r + 2)$. The basic numerical integration formula over σ,

$$\int_\sigma g(s, t)\, d\sigma \approx \sum_{j=1}^{f_r} w_j g(q_j) \qquad (5.4.150)$$

is based on integrating the polynomial of degree r that interpolates $g(s, t)$ at the nodes $\{q_j\}$. As such, its degree of precision is at least $d = r$.

Following (5.3.116) and (5.4.143), define the numerical integration operator $\mathcal{K}_n$ by

$$\mathcal{K}_n\rho(P) = \sum_{k=1}^{n} \sum_{i=1}^{f_r} w_i K(P, v_{k,i}) \rho(v_{k,i}) |(D_s\tilde{m}_k \times D_t\tilde{m}_k)(q_i)|, \quad P \in S$$
$$(5.4.151)$$

for $\rho \in C(S)$. This can then be used to define a Nyström method, just as was done above in (5.4.145)–(5.4.146). For the error analysis we have the following.

Theorem 5.4.2. *Assume the integral equation (5.4.140) is uniquely solvable, with $K(P, Q)$ continuous on $S \times S$. Let $r \geq 1$ be an integer. Let the basic numerical integration formula be that of (5.4.150), and let $\mathcal{K}_n$ be defined by (5.4.151). Assume the triangulations $\mathcal{T}_n$ are symmetric. Define*

$$\bar{r} = \begin{cases} r + 1, & r \text{ odd} \\ r + 2, & r \text{ even} \end{cases}$$

Assume the functions $F_j \in C^{\bar{r}+1}(R_j)$, $j = 1, \ldots, J$, for the parametrization functions of (5.3.97). Then:

(a) *The family $\{\mathcal{K}_n\}$ is collectively compact and pointwise convergent on $C(S)$.*
(b) *The inverse operators $(\lambda - \mathcal{K}_n)^{-1}$ exist and are uniformly bounded for all sufficiently large n, say $n \geq N$.*
(c) *The approximation ρ_n has the error*

$$\rho - \rho_n = (\lambda - \mathcal{K}_n)^{-1}(\mathcal{K}\rho - \mathcal{K}_n\rho)$$

and thus $\rho_n \to \rho$ as $n \to \infty$.
(d) *Assume that $K(P, \cdot) \in C(S) \cap C^{\bar{r}}(S_j)$, $j = 1, \ldots, J$, with the derivatives of order $\bar{r}$ uniformly bounded with respect to $P \in S$. Assume $\rho \in C(S) \cap C^{\bar{r}}(S_j)$, $j = 1, \ldots, J$. Then*

$$\|\rho - \rho_n\|_\infty \leq c\hat{\delta}_n^{\bar{r}}, \quad n \geq N \tag{5.4.152}$$

Proof. (a) Collective compact operator theory. From Theorem 5.3.3 it is straightforward to show that

$$\sum_{k=1}^n \sum_{j=1}^{f_r} w_i |(D_s \tilde{m}_k \times D_t \tilde{m}_k)(q_i)| \to \int_S dS = \text{Area}(S)$$

and this proves

$$\sup_n \sum_{k=1}^n \sum_{j=1}^{f_r} w_i |(D_s \tilde{m}_k \times D_t \tilde{m}_k)(q_i)| < \infty \tag{5.4.153}$$

From this and the boundedness of $K(P, Q)$,

$$\max_{P \in S} \sum_{k=1}^n \sum_{i=1}^{f_r} w_i |K(P, v_{k,i})| |(D_s \tilde{m}_k \times D_t \tilde{m}_k)(q_i)| < \infty \tag{5.4.154}$$

Thus $\mathcal{F} \equiv \{\mathcal{K}_n \mid n \geq 1 \text{ and } \|\rho\|_\infty \leq 1\}$ is uniformly bounded. The equicontinuity of $\mathcal{F}$ follows from (5.4.153) and the continuity of $K(P, Q)$. From Theorem 5.3.3, we have $\mathcal{K}_n \rho \to \mathcal{K}\rho$ for all ρ in a dense subspace of $C(S)$. Together with (5.4.153), this implies the convergence is valid for all $\rho \in C(S)$.

(b) Rates of convergence. Our analysis of the rate of convergence follows from extending the proof of (5.3.118) in Theorem 5.3.4, which we omitted. Write the error $\mathcal{K}\rho - \mathcal{K}_n\rho$ as

$$E(P) = \mathcal{K}\rho(P) - \mathcal{K}_n\rho(P)$$

$$= \int_S K(P, Q)\rho(Q)\, dS_Q$$

$$- \sum_{k=1}^{n} \sum_{i=1}^{f_r} w_i K(P, v_{k,i})\rho(v_{k,i}) \mid (D_s \tilde{m}_k \times D_t \tilde{m}_k)(q_i)\mid$$

$$= \sum_{k=1}^{n} E_{1,k} + \sum_{k=1}^{n} E_{2,k}$$

with

$$E_{1,k} = \int_\sigma K(P, m_k(s, t))\rho(m_k(s, t)) \mid (D_s m_k \times D_t m_k)(s, t)\mid d\sigma$$

$$- \sum_{i=1}^{f_r} w_i K(P, v_{k,i})\rho(v_{k,i}) \mid (D_s m_k \times D_t m_k)(q_i)\mid$$

$$E_{2,k} = \sum_{i=1}^{f_r} w_i K(P, v_{k,i})\rho(v_{k,i})\{\mid (D_s m_k \times D_t m_k)(q_i)\mid$$

$$- \mid (D_s \tilde{m}_k \times D_t \tilde{m}_k)(q_i)\mid\}$$

Recall the discussion of numerical integration in and following Theorem 5.3.2. From it, we have

$$\sum_{k=1}^{n} E_{1,k} = \begin{cases} O\left(\hat{\delta}_n^{r+1}\right), & r \text{ odd} \\ O\left(\hat{\delta}_n^{r+2}\right), & r \text{ even} \end{cases}$$

For E_2 we consider only the case $r = 2$, so that the basic integration formula over σ is given by

$$\int_\sigma g(s, t)\, d\sigma \approx \frac{1}{6} \sum_{j=4}^{6} g(q_j)$$

$$= \frac{1}{6}\left[g\left(\frac{1}{2}, 0\right) + g\left(0, \frac{1}{2}\right) + g\left(\frac{1}{2}, \frac{1}{2}\right) \right] \qquad (5.4.155)$$

We generalize Lemma 5.3.1 as follows.

Lemma 5.4.1. Let r be an even integer. Let $g(s, t)$ be a polynomial of degree $r + 1$ on σ, and let $g_r(s, t)$ be its interpolant of degree r, based on interpolation at the nodes $\{q_1, \ldots, q_{f_r}\}$ of (5.1.10). Then for the numerical integration formula (5.4.150),

$$\sum_{j=1}^{f_r} w_j \frac{\partial}{\partial s} [g(s, t) - g_r(s, t)]_{(s,t)=q_j} = 0$$

$$\sum_{j=1}^{f_r} w_j \frac{\partial}{\partial t} [g(s, t) - g_r(s, t)]_{(s,t)=q_j} = 0$$

Proof. See Chien [121].   □

Evaluate $E_{2,k}$ in a manner analogous to what was done with $E_{1,k}$ in the proof of Theorem 5.3.4; see (5.3.124)–(5.3.133). Expand

$$|(D_s m_k \times D_t m_k)(q_i)| - |(D_s \tilde{m}_k \times D_t \tilde{m}_k)(q_i)|$$

using (5.3.128)–(5.3.129). Expand

$$g(m_k(s, t)) \equiv K(P, m_k(s, t))\rho(m_k(s, t)))$$

about $(s, t) = (0, 0)$, as in (5.3.131). Then

$$E_{2,k} = \frac{1}{6} g(m_k(0, 0)) \sum_{j=4}^{6} [E_4(q_j) + E_5(q_j)] + \frac{1}{6} \sum_{j=4}^{6} E_4(q_j)[s_j g_s(m_k(0, 0))$$

$$+ t_j g_t(m_k(0, 0))] + O(\hat{\delta}_n^6) \tag{5.4.156}$$

By the preceding Lemma 5.4.1,

$$\sum_{j=4}^{6} E_4(q_j) = 0$$

The remaining terms are all of size $O(\hat{\delta}_n^5)$, which will prove $E_2 = O(\hat{\delta}_n^3)$.

Let Δ_k and Δ_l be symmetric triangles. Then use the odd function property of $E_5(s, t)$ and

$$E_4(s, t)[s g_s(m_k(0, 0)) + t g_t(m_k(0, 0))]$$

to show

$$E_{2,k} + E_{2,l} = O(\hat{\delta}_n^6)$$

By an argument analogous to that used at the end of part (b) of the proof of Theorem 5.3.4, we will have

$$E_2 = O\left(\hat{\delta}_n^4\right)$$

This also completes the proof, showing

$$\|\mathcal{K}\rho - \mathcal{K}_n\rho\|_\infty = O\left(\hat{\delta}_n^4\right)$$

and (5.4.152). □

Example. Let S be the ellipsoidal surface given by

$$\left(\frac{x}{a}\right)^2 + \left(\frac{y}{b}\right)^2 + \left(\frac{z}{c}\right)^2 = 1$$

We solve the integral equation

$$\lambda\rho(P) - \int_S \rho(Q)\frac{\partial}{\partial \mathbf{n}_Q}(|P - Q|^2)\, dS_Q = \psi(P), \quad P \in S \qquad (5.4.157)$$

with $\mathbf{n}_Q$ the inner normal to S at Q. In the evaluation of

$$K(P, Q) \equiv \frac{\partial}{\partial \mathbf{n}_Q}(|P - Q|^2)$$

we calculate the normal derivative with respect to the true surface S, not the approximate surface $\tilde{S}$. We choose

$$\rho(x, y, z) = e^z$$

and define $\psi(P)$ accordingly.

In defining the numerical integration operators $\mathcal{K}_n$, the basic integration rule being used is that of (5.4.155), which corresponds to $r = 2$ in Theorem 5.4.2. We also use the approximate surface $\tilde{S}$ in calculating the quadrature weights, as in (5.4.151). The Nyström method amounts to solving the linear system

$$\lambda\rho_n(v_i) - \sum_{j=1}^{n_v} \omega_j K(v_i, v_j)\rho_n(v_j) = \psi(v_i), \quad i = 1, \ldots, n_v \qquad (5.4.158)$$

with

$$\omega_j = \sum_{\substack{i,k \\ v_j = m_k(\mu_i)}} w_i |(D_s\tilde{m}_k \times D_t\tilde{m}_k)(q_i)| \qquad (5.4.159)$$

Table 5.5. *Nyström solution of (5.4.157) on an ellipsoid*

n	n_s	E_n	Ω_n	n	n_s	E_n	Ω_n
8	12	5.46E−2		20	30	1.61E−2	
32	48	7.81E−3	2.81	80	120	1.48E−3	3.45
128	192	6.61E−4	3.56	320	480	1.02E−4	3.86
512	768	4.47E−5	3.89				

The solution ρ_n is obtained by Nyström interpolation:

$$\rho_n(P) = \frac{1}{\lambda}\left[\psi(P) + \sum_{j=1}^{n_v} \omega_j K(P, v_j)\rho_n(v_j)\right], \quad P \in C(S) \qquad (5.4.160)$$

Note that the quadrature weights in (5.4.158) are zero for those nodes v_j that are vertices of a triangular element Δ_k. This is due to the form of (5.4.155), in which there is no contribution from the vertices of σ. As a consequence, we can solve (5.4.158) at only the remaining nodes, which are those corresponding to a "midpoint" of a side of a triangular element. We denote by n_s the number of such nodes, and we solve a system (5.4.158) of order n_s. The value of $\rho_n(v_i)$ with v_i a vertex of a triangular element is obtained by using (5.4.160).

Numerical results for the case $(a, b, c) = (1, 0.75, 0.5)$ and $\lambda = 30$ are given in Table 5.5. The error columns are

$$E_n \equiv \max_{1 \leq i \leq n_v} |\rho(v_i) - \rho_n(v_i)|$$

and the order of convergence is computed using

$$\Omega_n = \log_2 \left[\frac{E_{\frac{1}{4}n}}{E_n}\right]$$

From Theorem 5.4.2 the rate of convergence should be $O(\delta_n^4)$, and this appears to be confirmed empirically in Table 5.5.

Example. Let S be the elliptical paraboloid given by

$$z = \left(\frac{x}{a}\right)^2 + \left(\frac{y}{b}\right)^2, \quad 0 \leq z \leq c$$

together with its "cap"

$$\left(\frac{x}{a}\right)^2 + \left(\frac{y}{b}\right)^2 \leq c$$

Table 5.6. *Nyström solution*
of (5.4.157) on a paraboloid

n	n_s	E_n	Ω_n
8	12	2.14E−3	
32	48	4.33E−4	2.31
128	192	6.52E−5	2.73
512	768	7.28E−6	3.16

We again solve the integral equation (5.4.157). Note now that $K(P, Q)$ is discontinuous at the edges of S. In defining the numerical integration operator $\mathcal{K}_n$, we use the form (5.4.151) and integrate separately over each triangular element Δ_k. The preceding theory extends without difficulty to this case.

Again we define the numerical integration operator $\mathcal{K}_n$ with the basic integration method (5.4.155), which corresponds to $r = 2$ in Theorem 5.4.2. Numerical results for $(a, b, c) = (0.75, 0.6, 0.5)$ and $\lambda = 20$ are given in Table 5.6. The empirical rate of convergence is much less than the theoretical rate of $O(\hat{\delta}_n^4)$. However, if we recall the example of integration over an elliptical paraboloid, following (5.3.139), the orders Ω_n are consistent with those in Tables 5.3 and 5.4. Therefore, we expect the orders Ω_n to converge to 4.0 as n increases further.

5.4.2. Collocation methods

Approximate the integral equation (5.4.140) by using the collocation method with the interpolatory projection $\mathcal{P}_n$ of (5.3.102). As in Chapter 3, we write the collocation method in abstract form as

$$(\lambda - \mathcal{P}_n \mathcal{K})\rho_n = \mathcal{P}_n \psi \tag{5.4.161}$$

For the actual solution, write

$$\rho_n(m_k(s, t)) = \sum_{j=1}^{f_r} \rho_n(v_{k,j})\ell_j(s, t), \quad (s, t) \in \sigma, \quad k = 1, \ldots, n \tag{5.4.162}$$

The set of nodal values $\{\rho_n(v_{k,j})\}$ is determined by solving the linear system

$$\lambda \rho_n(v_i) - \sum_{k=1}^{n} \sum_{j=1}^{f_r} \rho_n(v_{k,j}) \int_\sigma K(v_i, m_k(s, t))\ell_j(s, t)$$
$$\times |(D_s m_k \times D_t m_k)(s, t)| \, d\sigma = \psi(v_i), \quad i = 1, \ldots, n_v \tag{5.4.163}$$

for $i = 1, \ldots, n_v$. The integrals must be evaluated numerically, thus leading
to a discrete collocation method. We discuss this later.

As in §3.4, we can also consider the iterated collocation solution

$$\hat{\rho}_n(P) = \frac{1}{\lambda}[\psi(P) + (\mathcal{K}\rho_n)(P)]$$

$$= \frac{1}{\lambda}\left[\psi(P) + \int_S K(P, Q)\rho_n(Q)\, dS\right]$$

$$= \frac{1}{\lambda}\psi(P) + \frac{1}{\lambda}\sum_{k=1}^{n}\sum_{j=1}^{f_r} \rho_n(v_{k,j})\int_\sigma K(P, m_k(s, t))\ell_j(s, t)$$

$$\times |(D_s m_k \times D_t m_k)(s, t)|\, d\sigma \qquad (5.4.164)$$

It satisfies the equation

$$(\lambda - \mathcal{K}\mathcal{P}_n)\hat{\rho}_n = \psi$$

Also, recall that

$$\hat{\rho}_n = \mathcal{P}_n\rho_n \qquad (5.4.165)$$

The error analysis for both ρ_n and $\hat{\rho}_n$ is a straightforward application of results
from Chapter 3.

Theorem 5.4.3. *Assume the integral equation (5.4.140) is uniquely solvable,
with $K(P, Q)$ continuous on $S \times S$. Let the interpolation of (5.3.102) used
in defining the collocation method (5.4.161) be of degree $r \geq 0$. Assume the
functions $F_j \in C^{r+2}(R_j)$, $j = 1, \ldots, J$, for the parametrization functions of
(5.3.97). Assume the triangulations T_n are such that $\hat{\delta}_n \to 0$ as $n \to \infty$. Then:*

(a) *The inverse operators $(\lambda - \mathcal{P}_n\mathcal{K})^{-1}$ exist and are uniformly bounded for
all sufficiently large n, say $n \geq N$.*

(b) *The approximation ρ_n has the error*

$$\rho - \rho_n = \lambda(\lambda - \mathcal{P}_n\mathcal{K})^{-1}(I - \mathcal{P}_n)\rho \qquad (5.4.166)$$

and thus $\rho_n \to \rho$ as $n \to \infty$.

(c) *Assume $\rho \in C(S) \cap C^{r+1}(S_j)$, $j = 1, \ldots, J$. Then*

$$\|\rho - \rho_n\|_\infty \leq c\hat{\delta}_n^{r+1}, \quad n \geq N \qquad (5.4.167)$$

(d) *Let the triangulations T_n be symmetric, and let r be an even integer. Assume
$F_j \in C^{r+3}(R_j)$, $\rho \in C(S) \cap C^{r+2}(S_j)$, and $K(P, \cdot) \in C(S) \cap C^1(S_j)$,*

$j = 1, \ldots, J$. *Then*

$$\|\rho - \hat{\rho}_n\|_\infty \le c\hat{\delta}_n^{r+2}, \quad n \ge N \tag{5.4.168}$$

and thus

$$\max_{1 \le i \le n_v} |\rho(v_i) - \rho_n(v_i)| \le c\hat{\delta}_n^{r+2}, \quad n \ge N \tag{5.4.169}$$

Proof. Combining Lemma 5.1.1 and the assumption $\hat{\delta}_n \to 0$, we have

$$\mathcal{P}_n g \to g \quad \text{as } n \to \infty \tag{5.4.170}$$

for all $g \in C(S)$. Together with the compactness of $\mathcal{K}$ and Lemma 3.1.2, this implies

$$\|\mathcal{K} - \mathcal{P}_n\mathcal{K}\| \to 0 \quad \text{as } n \to \infty$$

The existence and stability of $(\lambda - \mathcal{P}_n\mathcal{K})^{-1}$ is based on Theorem 3.1.1, and the error formula (5.4.166) is simply (3.1.30) of that theorem. The formula (5.4.167) is simply an application of Theorem 5.3.1.

The stability of the iterated collocation method comes from the identity

$$(\lambda - \mathcal{K}\mathcal{P}_n)^{-1} = \frac{1}{\lambda}[I + \mathcal{K}(\lambda - \mathcal{P}_n\mathcal{K})^{-1}\mathcal{P}_n]$$

and the uniform boundedness of $(\lambda - \mathcal{P}_n\mathcal{K})^{-1}$ for $n \ge N$. The error in the iterated collocation solution satisfies

$$\rho - \hat{\rho}_n = (\lambda - \mathcal{K}\mathcal{P}_n)^{-1}(\mathcal{K}\rho - \mathcal{K}\mathcal{P}_n\rho)$$

Applying the remark following Theorem 5.3.2, we have

$$\|\mathcal{K}\rho - \mathcal{K}\mathcal{P}_n\rho\|_\infty \le c\hat{\delta}_n^{r+2}$$

and this proves (5.4.168). The superconvergence result (5.4.169) for ρ_n follows from (5.4.168) and (5.4.165). □

Using the approximate surface

The use of the approximate surface $\tilde{S}$ in solving the integral equation (5.4.140) leads to the following. Let

$$\tilde{\rho}_n(m_k(s, t)) = \sum_{j=1}^{f_r} \tilde{\rho}_n(v_{k,j})\ell_j(s, t), \quad (s, t) \in \sigma, \quad k = 1, \ldots, n \tag{5.4.171}$$

Solve for the nodal values $\{\bar{\rho}_n(v_{k,j})\}$ by solving the linear system

$$
\lambda \bar{\rho}_n(v_i) - \sum_{k=1}^{n} \sum_{j=1}^{f_r} \bar{\rho}_n(v_{k,j}) \int_\sigma K(v_i, \tilde{m}_k(s,t)) \ell_j(s,t)
$$

$$
\times |(D_s \tilde{m}_k \times D_t \tilde{m}_k)(s,t)| \, d\sigma = \psi(v_i), \quad i = 1, \ldots, n_v \qquad (5.4.172)
$$

We can associate this with a Nyström method based on the following numerical integral operator:

$$
\mathcal{K}_n g(P) = \int_{\tilde{S}} K(P, Q) \, (\mathcal{P}_n g)(Q) \, dS
$$

$$
= \sum_{k=1}^{n} \int_{\tilde{\Delta}_k} K(P, Q) \, (\mathcal{P}_n g)(Q) \, dS
$$

$$
= \sum_{k=1}^{n} \sum_{j=1}^{f_r} g(v_{k,j}) \int_\sigma K(v_i, \tilde{m}_k(s,t)) \ell_j(s,t)
$$

$$
\times |(D_s \tilde{m}_k \times D_t \tilde{m}_k)(s,t)| \, d\sigma \qquad (5.4.173)
$$

for $P \in S$ and $g \in C(S)$. In this, we use the following as a simplifying notation:

$$
(\mathcal{P}_n g)(\tilde{m}_k(s,t)) \equiv (\mathcal{P}_n g)(m_k(s,t)),
$$

$$
(s,t) \in \sigma, \quad k = 1, \ldots, n, \quad g \in C(S)
$$

An error analysis for $\{\bar{\rho}_n\}$ can now be based on the theory of collectively compact operator approximations.

Theorem 5.4.4. *Assume the integral equation (5.4.140) is uniquely solvable, with $K(P, Q)$ continuous on $S \times S$. Let the interpolation of (5.3.102) used in defining the collocation method (5.4.161) be of degree $r \geq 1$. Define*

$$
\bar{r} = \begin{cases} r, & r \text{ odd} \\ r + 1, & r \text{ even} \end{cases}
$$

Assume the functions $F_j \in C^{\bar{r}+1}(R_j)$, $j = 1, \ldots, J$, for the parametrization functions of (5.3.97). Assume the triangulations $\mathcal{T}_n$ are such that $\hat{\delta}_n \to 0$ as $n \to \infty$. Then:

(a) *The family $\{\mathcal{K}_n\}$ are collectively compact and pointwise convergent to $\mathcal{K}$ on $C(S)$. Moreover, the inverse operators $(\lambda - \mathcal{K}_n)^{-1}$ exist and are uniformly bounded for all sufficiently large n, say $n \geq N$.*

(b) *The approximations* $\bar{\rho}_n$ *satisfy*

$$\rho - \bar{\rho}_n = (\lambda - \mathcal{K}_n)^{-1}(\mathcal{K}\rho - \mathcal{K}_n\rho) \qquad (5.4.174)$$

and thus $\bar{\rho}_n \to \rho$ *as* $n \to \infty$.
(c) *Assume* $\rho \in C(S) \cap C^{r+1}(S_j)$, $j = 1, \ldots, J$. *Then*

$$\|\rho - \bar{\rho}_n\|_\infty \le c\hat{\delta}_n^{\bar{r}}, \quad n \ge N \qquad (5.4.175)$$

(d) *Let the triangulations* $\mathcal{T}_n$ *be symmetric. Assume* $F_j \in C^{\bar{r}+2}(R_j)$, $\rho \in C(S) \cap C^{\bar{r}+1}(S_j)$, *and* $K(P, \cdot) \in C(S) \cap C^1(S_j)$, $j = 1, \ldots, J$. *Then*

$$\|\rho - \bar{\rho}_n\|_\infty \le c\hat{\delta}_n^{\bar{r}+1}, \quad n \ge N \qquad (5.4.176)$$

Proof. The family $\{\mathcal{K}_n\}$ can be shown to be uniformly bounded and point-wise convergent by using Corollary 5.3.1 of §5.3. The equicontinuity of $\{\mathcal{K}_n g \mid \|g\|_\infty \le 1$ and $n \ge 1\}$ can be shown much as in proof of Theorem 5.4.2, and we omit it here. The remainder of the proof of (a) and (b) follows from the general Theorem 4.1.2. The result (5.4.142) follows from Theorem 5.3.3, and that of (5.4.176) follows from Theorem 5.3.4.   □

Discrete collocation methods

The collocation methods described above, based on polynomial interpolation of degree r, lead to a numerical solution ρ_n with error of size $O(\hat{\delta}_n^{r+1})$. In addition, with r an even integer and $\mathcal{T}_n$ a symmetric triangulation, there is superconvergence with an error of size $O(\hat{\delta}_n^{r+2})$ at the nodes of the triangulation. These results remain valid when the surface S is replaced by $\tilde{S}$ based on interpolation of degree r. With a discrete collocation method, the integrals of (5.4.163) and (5.4.172) must be numerically evaluated so as to preserve this accuracy.

We do not develop a general theory of the needed discrete collocation methods here, because the Nyström methods developed in the preceding subsection of this section have the desired accuracy. In fact, the Nyström method of (5.4.143) or that based on using the approximate surface, as in (5.4.151), leads to exactly the desired rates of convergence given above. These Nyström methods can also be interpreted as iterated discrete collocation methods, with the integration for the discrete collocation method based on the formula

$$\int_\sigma g(s, t)\, d\sigma \approx \sum_{j=1}^{f_r} w_j g(q_j), \quad w_j \equiv \int_\sigma \ell_j(s, t)\, d\sigma$$

For a general discussion of the equivalence of such iterated discrete collocation methods and Nyström methods, see Theorem 4.3.1 in §4.3.

5.4.3. Galerkin methods

Define the approximating subspace $\mathcal{X}_n$ as the set of all functions that are piecewise polynomial of degree r or less in the parametrization variables, with $r \geq 0$ a given integer. A function $g \in \mathcal{X}_n$ if on each Δ_k, g can be written in the form

$$g(m_k(s, t)) = \sum_{j=1}^{f_r} \alpha_{k,j} \ell_j(s, t), \quad (s, t) \in \sigma \tag{5.4.177}$$

for some set of constants $\{\alpha_{k,1}, \ldots, \alpha_{k,f_r}\}$, $k = 1, \ldots, n$. The set of functions $\{\ell_1, \ldots, \ell_{f_r}\}$ is to be a basis for the polynomials in (s, t) of degree $\leq r$. For example, we might use the basis

$$\{s^i t^j \mid i, j \geq 0, \ i + j \leq r\}$$

Another choice is the set of Lagrange basis functions for collocation on the nodes $\{(ih, jh) \mid i, j \geq 0, \ i + j \leq r\}$ with $h = 1/r$; for example, see (5.1.14), (5.1.17), and (5.1.19). The dimension of $\mathcal{X}_n$ is

$$nf_r = n \frac{(r+1)(r+2)}{2}$$

As additional useful notation, introduce basis functions defined on Δ_k, as follows:

$$\ell_{k,j}(P) = \ell_j\big(m_k^{-1}(P)\big), \quad P \in \Delta_k, \ k = 1, \ldots, n$$

with $m_k^{-1} : \Delta_k \to \sigma$ the inverse of the mapping m_k.

Let $\mathcal{P}_n$ be the orthogonal projection of $L^2(S)$ onto $\mathcal{X}_n$. As an operator on $L^2(S)$, $\mathcal{P}_n$ is self-adjoint and has norm 1. As before, the Galerkin method for solving (5.4.140) is

$$(\lambda - \mathcal{P}_n \mathcal{K})\rho_n = \mathcal{P}_n \psi \tag{5.4.178}$$

The theory is first developed in the context of $L^2(S)$, and then it is extended to $L^\infty(S)$. As a review of earlier material on the Galerkin method, see §3.1 following (3.1.12), and see §3.3, both in Chapter 3. Also see the earlier material on Galerkin's method in §5.2 of this chapter, following (5.2.76).

The Galerkin method (5.4.178) for solving (5.4.140) amounts to writing

$$\rho_n(P) = \sum_{j=1}^{f_r} \alpha_{k,j} \ell_{k,j}(P), \quad P \in \Delta_k, \quad k = 1, \ldots, n \qquad (5.4.179)$$

and determining the constants $\{\alpha_{k,j} \mid j = 1, \ldots, f_r, \ k = 1, \ldots, n\}$ by solving the following linear system.

$$\lambda \sum_{j=1}^{f_r} \alpha_{k,j}(\ell_{k,j}, \ell_{k,i})_{\Delta_k} - \sum_{l=1}^{n} \sum_{j=1}^{f_r} \alpha_{l,j}(\mathcal{K}\ell_{l,j}, \ell_{k,i})_{\Delta_k} = (\psi, \ell_{k,i})_{\Delta_k}$$

$$\qquad (5.4.180)$$

for $i = 1, \ldots, f_r, \ k = 1, \ldots, n$. In this formula,

$$(g, h)_{\Delta_k} = \int_{\Delta_k} g(Q)h(Q)\,dS = \int_{\sigma} g(m_k(s,t))h(m_k(s,t))$$

$$\times |(D_s m_k \times D_t m_k)(s,t)|\,d\sigma \qquad (5.4.181)$$

and

$$\mathcal{K}\ell_{l,j}(P) = \int_{\Delta_l} K(P,Q)\ell_{l,j}(Q)\,dS$$

$$= \int_{\sigma} K(P, m_k(s,t))\ell_j(s,t)|(D_s m_k \times D_t m_k)(s,t)|\,d\sigma$$

$$\qquad (5.4.182)$$

Substituting these into (5.4.180), we have the linear system

$$\lambda \sum_{j=1}^{f_r} \alpha_{k,j} \int_{\sigma} \ell_j(s,t)\ell_i(s,t)|(D_s m_k \times D_t m_k)(s,t)|\,ds\,dt$$

$$- \sum_{l=1}^{n} \sum_{j=1}^{f_r} \int_{\sigma} \ell_i(s,t)\left[\int_{\sigma} K(m_k(s,t), m_k(\xi,\eta))\ell_j(\xi,\eta)\right.$$

$$\left. \times |(D_s m_k \times D_t m_k)(\xi,\eta)|\,d\xi\,d\eta\right]|(D_s m_k \times D_t m_k)(s,t)|\,ds\,dt$$

$$= \int_{\sigma} \psi(m_k(s,t))\ell_i(s,t))|(D_s m_k \times D_t m_k)(s,t)|\,ds\,dt \qquad (5.4.183)$$

The integrals over σ in this system must be evaluated numerically, and we return to this later.

Introduce the iterated Galerkin solution

$$\hat{\rho}_n = \frac{1}{\lambda}[\psi + \mathcal{K}\rho_n] \qquad (5.4.184)$$

It satisfies

$$\mathcal{P}_n \hat{\rho}_n = \rho_n$$

and

$$(\lambda - \mathcal{K}\mathcal{P}_n)\hat{\rho}_n = \psi$$

For the general error analysis of the iterated Galerkin solution, see §4.4. Given ρ_n in (5.4.179), evaluated $\hat{\rho}_n$ with

$$\hat{\rho}_n(P) = \frac{1}{\lambda}\psi(P) + \frac{1}{\lambda}\sum_{k=1}^{n}\sum_{j=1}^{f_r}\alpha_{k,j}\int_{\Delta_k} K(P, Q)\ell_{k,j}(Q)\,dS(Q), \quad P \in S$$

$$(5.4.185)$$

For many values of P, these integrals will have been evaluated in the process of setting up the linear system (5.4.183).

Theorem 5.4.5. *Assume the integral equation (5.4.140) is uniquely solvable, with $K(P, Q)$ so defined that $\mathcal{K}$ is compact from $L^2(S)$ into $L^2(S)$. Let the approximating subspace $\mathcal{X}_n$ be as defined above preceding (5.4.177), using piecewise polynomial functions of degree at most r, for some $r \geq 0$. Assume the functions $F_j \in C^{r+2}(R_j)$, $j = 1, \ldots, J$, for the parametrization functions of (5.3.97). Assume the triangulations $\mathcal{T}_n$ are such that $\hat{\delta}_n \to 0$ as $n \to \infty$. Then:*

(a) The inverse operators $(\lambda - \mathcal{P}_n\mathcal{K})^{-1}$ exist for all sufficiently large n, say $n \geq N$, and

$$\rho - \rho_n = \lambda(\lambda - \mathcal{P}_n\mathcal{K})^{-1}(\rho - \mathcal{P}_n\rho) \qquad (5.4.186)$$

which implies $\rho_n \to \rho$.

(b) Assume $\rho \in C(S) \cap C^{r+1}(S_j)$, $j = 1, \ldots, J$. Then

$$\|\rho - \rho_n\| \leq c\hat{\delta}_n^{r+1}, \quad n \geq N \qquad (5.4.187)$$

(c) For some integer $\kappa \leq r + 1$, assume $K(P, Q)$ satisfies the following: For each $P \in S$, the function $K(P, \cdot) \in C(S) \cap C^{\kappa}(S_j)$, $j = 1, \ldots, J$, with all derivatives of order κ continuous with respect to P. Then

$$\|\rho - \hat{\rho}_n\| \leq c\hat{\delta}_n^{r+1+\kappa}, \quad n \geq N \qquad (5.4.188)$$

Proof. The proof of (a) is immediate from Theorem 3.1.1, the general error analysis for projection methods, together with Lemma 3.1.2. The formula (5.4.186) is simply (3.1.30) from Theorem 3.1.1.

Apply Theorem 5.3.1 to prove the error bound (5.4.187). Finally, (5.4.188) comes from (3.4.93), together with Theorem 5.3.1.   □

With $\kappa = r + 1$, the order of convergence of $\hat{\rho}_n$ is $2r + 2$, which is double the order of $r + 1$ of the Galerkin solution ρ_n. But even with more poorly behaved kernel functions $K(P, Q)$, there will be some gain in the speed of convergence, as can be seen from (5.4.188).

Discrete Galerkin methods

The numerical integrations of the integrals in (5.4.183) and (5.4.185) must be performed with sufficient accuracy to preserve the rates of convergence given in (5.5.243) and (5.4.188). For a review of the framework for discrete Galerkin methods used in this book, see §4.4 in Chapter 4.

For the present discussion we will base the discrete inner product integration of $(g, h)_\Delta$ and the numerical integral operator $\mathcal{K}_n$ on the same integration rule. However, for cases with $K(P, Q)$ poorly behaved, different integration rules should be used in these two cases. All of our quadratures over triangular elements Δ will be based on quadratures over the unit simplex σ:

$$\int_\sigma g(s, t)\, d\sigma \approx \sum_{j=1}^{M} \omega_j g(\mu_j) \qquad (5.4.189)$$

with the nodes $\{\mu_1, \ldots, \mu_M\}$ contained in σ. If this integration rule has degree of precision $d \geq 0$, then the resulting composite integration rule over S will have an error of $O(\hat{\delta}_n^{d+1})$; see the quadrature formula (5.3.107) and the error formula in Theorem 5.3.2. For symmetric triangulations with suitable quadrature formulas, and with d even, this order of convergence can often be improved to $O(\hat{\delta}_n^{d+2})$; but we will not consider such a development here.

Recalling the development of §4.4, the discrete Galerkin method can be written as

$$(\lambda - \mathcal{Q}_n \mathcal{K}_n) z_n = \mathcal{Q}_n \psi$$

The iterated discrete Galerkin solution is

$$\hat{z}_n = \frac{1}{\lambda}[\psi + \mathcal{K}_n z_n]$$

It satisfies

$$Q_n \hat{z}_n = z_n$$

$$(\lambda - K_n Q_n)\hat{z}_n = \psi$$

For the errors in z_n and $\hat{z}_n$, see Theorems 4.4.2 and 4.4.3. From the latter theorem, we have

$$\|\rho - z_n\|_\infty = O\left(\hat{\delta}_n^{\min\{d+1,r+1\}}\right) \qquad (5.4.190)$$

$$\|\rho - \hat{z}_n\|_\infty = O\left(\hat{\delta}_n^{\min\{d+1,2r+2\}}\right) \qquad (5.4.191)$$

provided ρ and $K(P, Q)$ are sufficiently differentiable. Thus to preserve the order of convergence of ρ_n of (5.4.187), we must have $d \geq r$, and to preserve the order of convergence of $\hat{\rho}_n$ we must have $d \geq 2r + 1$. A similar discussion was given at the end of §5.2, and rather than repeating it here, we refer the reader to that earlier discussion as it also applies here.

5.5. Global approximation methods on smooth surfaces

Numerical methods that are based on using functions that are globally smooth (over the entire domain of integration) are often more rapidly convergent than are methods based on using piecewise polynomial functions. For integral equations defined on the smooth boundary of a planar region, many of the most effective numerical methods are based on representing the solution as a trigonometric polynomial. Such methods were discussed in §3.2, following (3.2.50) [see (3.2.57)–(3.2.58)], and in §3.3 [see (3.3.75)]. Kress's book [325] makes extensive use of approximations based on trigonometric polynomials for solving planar boundary integral equations. In this section we consider an important generalization to three dimensions of the use of trigonometric polynomials. We begin by showing that many integral equations on surfaces can be transformed to a standard form on the unit sphere in $\mathbf{R}^3$.

Consider the integral equation

$$\lambda \rho(P) - \int_S K(P, Q)\rho(Q)\, dS_Q = \psi(P), \quad P \in S \qquad (5.5.192)$$

Let S be a smooth surface in $\mathbf{R}^3$, and let U denote the unit sphere in $\mathbf{R}^3$. Assume there is a smooth mapping

$$M : U \xrightarrow[\text{onto}]{1-1} S \qquad (5.5.193)$$

with a smooth inverse M^{-1}. For simplicity, we assume M and M^{-1} are C^∞ mappings, although that is stronger than is generally needed. The integral equation (5.5.192) can be transformed to a new integral equation over U by means of the transformation M.

As a simple example, the ellipsoid

$$\left(\frac{x}{a}\right)^2 + \left(\frac{y}{b}\right)^2 + \left(\frac{z}{c}\right)^2 = 1$$

corresponds to the mapping function

$$M(x, y, z) = (ax, by, cz), \quad (x, y, z) \in U \tag{5.5.194}$$

Some examples for nonconvex but star-like regions are given in Ref. [45], but in general, the above assumptions do not require the interior of S to even be star-like. (By a region D being *star-like*, we mean there is a point $P_0 \in D$ such that for any other point $P \in D$, the straight line segment joining P_0 and P also is contained in D.)

For any function ρ defined on S, introduce the function

$$\hat{\rho}(P) = \rho(M(P)), \quad P \in U \tag{5.5.195}$$

For the function spaces $C(S)$ and $C(U)$, this defines an isometric isomorphism $\mathcal{M} : C(S) \overset{1-1}{\underset{onto}{\to}} C(U)$:

$$\mathcal{M}\rho = \hat{\rho} \tag{5.5.196}$$

Trivially,

$$\|\mathcal{M}\rho\|_\infty = \|\rho\|_\infty, \quad \rho \in C(S) \tag{5.5.197}$$

Function space properties of operators on $C(S)$ transfer easily to analogous properties on $C(U)$. The same definitions can be used to link $L^2(S)$ and $L^2(U)$, with much the same consequences.

Using the transformation M of (5.5.193), we can transform (5.5.192) to an equivalent integral equation over the surface U:

$$\lambda\hat{\rho}(P) - \int_U \hat{K}(P, Q)\hat{\rho}(Q)\, dS_Q = \hat{\psi}(P), \quad P \in U \tag{5.5.198}$$

with

$$\hat{K}(P, Q) = K(M(P), M(Q)) J_M(Q)$$

and $J_M(Q)$ the Jacobian of the transformation (5.5.193). As an example, the mapping (5.5.194) for an ellipsoid has the Jacobian

$$J_M(x, y, z) = \{(bcx)^2 + (acy)^2 + (abz)^2\}^{\frac{1}{2}}, \quad (x, y, z) \in U \tag{5.5.199}$$

The kernel function $\hat{K}(P, Q)$ on $U \times U$ has the same degree of smoothness as does the original kernel $K(P, Q)$ on $S \times S$, and analogous statements hold when comparing the smoothness of $\hat{\rho}$ and $\hat{\psi}$ with respect to ρ and ψ, respectively. In this chapter we restrict our study to the case of $K(P, Q)$ being continuous and smooth, and we leave until Chapter 9 the study of cases such as

$$K(P, Q) = \frac{1}{|P - Q|}$$

which arise in the study of boundary integral equations.

Because of the transformation of (5.5.192) to (5.5.198), we consider in the remainder of this section only the case of integral equations defined over the unit sphere U. We dispense with the notation of (5.5.195)–(5.5.198) and study the integral equation (5.5.192) with $S = U$. But we will create numerical examples based on the transformation of (5.5.192) to (5.5.198), using the ellipsoid (5.5.194).

5.5.1. *Spherical polynomials and spherical harmonics*

Consider an arbitrary polynomial in (x, y, z) of degree N, say

$$p(x, y, z) = \sum_{\substack{i,j,k \geq 0 \\ i+j+k \leq N}} a_{i,j,k} x^i y^j z^k \qquad (5.5.200)$$

and restrict (x, y, z) to lay on the unit sphere U. The resulting function is called a *spherical polynomial* of degree $\leq N$. [These are the analogs of the trigonometric polynomials, which can be obtained by replacing (x, y) in

$$\sum_{\substack{i,j \geq 0 \\ i+j \leq N}} a_{i,j} x^i y^j$$

with $(\cos \theta, \sin \theta)$.] Note that a polynomial $p(x, y, z)$ may reduce to an expression of lower degree. For example, $p(x, y, z) = x^2 + y^2 + z^2$ reduces to 1 when $(x, y, z) \in U$. Call the set of such functions S_N.

An alternative way of obtaining polynomials on U is to begin with *homogeneous harmonic polynomials*. Let $p(x, y, z)$ be a polynomial of degree n that satisfies Laplace's equation,

$$\Delta p(x, y, z) \equiv \frac{\partial^2 p}{\partial x^2} + \frac{\partial^2 p}{\partial y^2} + \frac{\partial^2 p}{\partial z^2} = 0, \quad (x, y, z) \in \mathbf{R}^3$$

and further, let p be homogeneous of degree n:

$$p(tx, ty, tz) = t^n p(x, y, z), \quad -\infty < t < \infty, \quad (x, y, z) \in \mathbf{R}^3$$

Restrict all such polynomials to U. Such functions are called *spherical harmonics of degree n*. As examples of spherical harmonics, we have

$n = 0$

$$p = 1$$

$n = 1$

$$p = x, y, z$$

$n = 2$

$$p = xy, \ xz, \ yz, \ x^2 + y^2 - 2z^2, \ x^2 + z^2 - 2y^2$$

where in all cases, we use $(x, y, z) = (\cos\phi\sin\theta, \sin\phi\sin\theta, \cos\theta)$. Nontrivial linear combinations of spherical harmonics of a given degree are again spherical harmonics of that same degree. For example, $p = x + y + z$ is also a spherical harmonic of degree 1. The number of linearly independent spherical harmonics of degree n is $2n + 1$, and thus the above sets are maximal independent sets for each of the given degrees $n = 0, 1, 2$.

Define $\hat{S}_N$ to be the smallest vector space to contain all of the spherical harmonics of degree $n \leq N$. Alternatively, $\hat{S}_N$ is the set of all finite linear combinations of spherical harmonics of all possible degrees $n \leq N$. Then it can be shown that

$$S_N = \hat{S}_N \tag{5.5.201}$$

and

$$\dim S_N = (N + 1)^2 \tag{5.5.202}$$

Below, we give a basis for S_N. See MacRobert [360, Chap. 7] for a proof of these results.

There are well-known formulas for spherical harmonics, and we will make use of some of them in working with spherical polynomials. The study of spherical harmonics is a very large one, and we can only touch on a few small parts of it. For further study, we recommend the classic book by T. MacRobert [360], although there are other fine books on this subject. The spherical harmonics of degree n are the analogs of the trigonometric functions $\cos n\theta$ and $\sin n\theta$, which are restrictions to the unit circle of the homogeneous harmonic polynomials

$$r^n \cos(n\theta), \quad r^n \sin(n\theta)$$

written in polar coordinates form.

The standard basis for spherical harmonics of degree n is

$$S_n^1(x, y, z) = c_n P_n(\cos\theta)$$

$$S_n^{2m}(x, y, z) = c_{n,m} P_n^m(\cos\theta)\cos(m\phi) \qquad (5.5.203)$$

$$S_n^{2m+1}(x, y, z) = c_{n,m} P_n^m(\cos\theta)\sin(m\phi), \qquad m = 1, \ldots, n$$

with $(x, y, z) = (\cos\phi\sin\theta, \sin\phi\sin\theta, \cos\theta)$. In this formula, $P_n(t)$ is a *Legendre polynomial* of degree n,

$$P_n(t) = \frac{1}{n!2^n}\frac{d^n}{dt^n}[(t^2 - 1)^n] \qquad (5.5.204)$$

and $P_n^m(t)$ is an *associated Legendre function*,

$$P_n^m(t) = (-1)^m(1 - t^2)^{\frac{m}{2}}\frac{d^m}{dt^m}P_n(t), \qquad 1 \le m \le n \qquad (5.5.205)$$

The constants in (5.5.203) are given by

$$c_n = \sqrt{\frac{2n + 1}{4\pi}}$$

$$c_{n,m} = \sqrt{\frac{2n + 1}{2\pi}\frac{(n - m)!}{(n + m)!}}$$

We will occasionally denote the Legendre polynomial P_n by P_n^0, to simplify referring to these Legendre functions. For more information on these functions, see Abramowitz and Stegun [1, Chap. 8] or MacRobert [360, Chap. 7]. The above function $P_n^m(t)$ is the function $T_n^m(t)$ of Ref. [360, p. 116, (14)]. A useful recursion relation for evaluating these functions is given in Ref. [1, p. 334, (8.5.3)], and a simple subroutine that uses it to evaluate $\{P_n^m(t) \mid 0 \le m \le n \le N\}$ for given N and t is included in the larger package given in Ref. [45].

The standard inner product on $L^2(U)$ is given by

$$(f, g) = \int_U f(Q)g(Q)\,dS_Q$$

Using this definition, the functions of (5.5.203) satisfy

$$(S_n^k, S_q^p) = \delta_{n,q}\delta_{k,p}$$

for $n, q = 0, 1, \ldots$, and $1 \le k \le 2n+1, 1 \le p \le 2q+1$. The set of functions

$$\{S_n^k \mid 1 \le k \le 2n + 1, \quad 0 \le n \le N\}$$

is an orthonormal basis for $\mathcal{S}_N$. To avoid some double summations, we will sometimes write this basis for $\mathcal{S}_N$ as

$$\{\Psi_1, \ldots, \Psi_{d_N}\} \tag{5.5.206}$$

with $d_N = (N+1)^2$ the dimension of the subspace.

The set $\{S_n^k \mid 1 \le k \le 2n+1, \ 0 \le n < \infty\}$ of spherical harmonics is an orthonormal basis for $L^2(U)$, and it leads to the expansion formula

$$g(Q) = \sum_{n=0}^{\infty} \sum_{k=1}^{2n+1} \left(g, S_n^k\right) S_n^k(Q), \quad g \in L^2(U) \tag{5.5.207}$$

This is called the *Laplace expansion* of the function g, and it is the generalization to $L^2(U)$ of the Fourier series on the unit circle in the plane. The function $g \in L^2(U)$ if and only if

$$\sum_{n=0}^{\infty} \sum_{k=1}^{2n+1} \left|\left(g, S_n^k\right)\right|^2 < \infty \tag{5.5.208}$$

This sum equals $\|g\|^2$ in $L^2(U)$. Other Sobolev spaces on U can be characterized by using this expansion, and we return to this in Chapter 9.

Of particular interest is the truncation of this series to terms of degree at most N, to obtain

$$\mathcal{P}_N g(Q) = \sum_{n=0}^{N} \sum_{k=1}^{2n+1} \left(g, S_n^k\right) S_n^k(Q) \tag{5.5.209}$$

This is the orthogonal projection of $L^2(U)$ onto $\mathcal{S}_N$, and of course, $\mathcal{P}_N g \to g$ as $N \to \infty$. Since it is an orthogonal projection on $L^2(U)$, we have $\|\mathcal{P}_N\| = 1$ as an operator from $L^2(U)$ in $L^2(U)$. However, we can also regard $\mathcal{S}_N \subset C(S)$, and then regarding $\mathcal{P}_N$ as a projection from $C(S)$ to $\mathcal{S}_N$, we have

$$\|\mathcal{P}_N\| = \left(\sqrt{\frac{8}{\pi}} + \delta_N\right) \sqrt{N} \tag{5.5.210}$$

with $\delta_N \to 0$ as $N \to \infty$. The proof of this is fairly involved, and we refer the reader to Gronwall [243] and Ragozin [442]. Later we use the projection $\mathcal{P}_N$ to define a Galerkin method for solving integral equations defined on U, with $\mathcal{S}_N$ as the approximating subspace.

Best approximations

Given $g \in C(U)$, define

$$\rho_N(g) = \inf_{p \in S_N} \|g - p\|_\infty \qquad (5.5.211)$$

This is called the *minimax error* for the approximation of g by spherical polynomials of degree $\leq N$. Using the Stone-Weierstraß theorem (e.g., see Ref. [376]), it can be shown that $\rho_N(g) \to 0$ as $N \to \infty$. In the error analysis of numerical methods that use spherical polynomials, it is important to have bounds on the rate at which $\rho_N(g)$ converges to zero. An initial partial result was given by Gronwall [243], and a much more complete theory was given many years later by Ragozin [441], a special case of which we give here. We first introduce some notation.

For given positive integer k, let $D^k g$ denote an arbitrary k^{th} order derivative of g on U, formed with respect to local surface coordinates on U. (One should consider a set of local patch coordinate systems over U, but what is intended is clear, and the present notation is simpler.) Let α be a real number, $0 < \alpha \leq 1$. Define $C^{k,\alpha}(U)$ to be the set of all functions $g \in C(U)$ for which all of its derivatives $D^k g \in C(U)$, with each of these derivatives satisfying a Hölder condition with exponent α:

$$|D^k g(P) - D^k g(Q)| \leq H_{k,\alpha}(g)|P - Q|^\alpha, \quad P, Q \in U \qquad (5.5.212)$$

The Hölder constant $H_{k,\alpha}(g)$ is to be uniform over all k^{th} order derivatives of g.

Lemma 5.5.1. Let $g \in C^{k,\alpha}(U)$. Then there is a sequence of spherical polynomials $\{p_N\}$ for which $\|g - p_N\|_\infty = \rho_N(g)$ and

$$\rho_N(g) \leq \frac{c_k H_{k,\alpha}(g)}{N^{k+\alpha}}, \quad N \geq 0 \qquad (5.5.213)$$

The constant c_k is dependent only on k. For later use, let $c_{k,\alpha}(g) \equiv c_k H_{k,\alpha}(g)$.

Proof. For the case $k = 0$, see Gronwall [243], and for the general case, see Ragozin [441, Theorem 3.3].   □

This result leads immediately to results on the rate of convergence of the Laplace series expansion of a function $g \in C^{k,\alpha}(U)$, given in (5.5.207). Using the norm of $L^2(U)$, and using the definition of the orthogonal projection $\mathcal{P}_N g$ being the best approximation in the inner product norm, we have

$$\|g - \mathcal{P}_N g\| \leq \|g - p_N\|$$

$$\leq 4\pi \|g - p_N\|_\infty$$

$$\leq \frac{4\pi c_{k,\alpha}(g)}{N^{k+\alpha}}, \quad N \geq 0 \qquad (5.5.214)$$

We can also consider the uniform convergence of the Laplace series. Write

$$\|g - \mathcal{P}_N g\|_\infty = \|g - p_N - \mathcal{P}_N(g - p_N)\|_\infty$$

$$\leq (1 + \|\mathcal{P}_N\|)\|g - p_N\|_\infty$$

$$\leq cN^{-(k+\alpha-\frac{1}{2})} \qquad (5.5.215)$$

with the last step using (5.5.210). In particular, if $g \in C^{0,\alpha}(U)$ with $\frac{1}{2} < \alpha \leq 1$, we have uniform convergence of $\mathcal{P}_N g$ to g on U. From (5.5.213), the constant c is a multiple of $H_{k,\alpha}(g)$. We make use of this in Chapter 9 with $k = 0$, when discussing the use of spherical harmonics in solving boundary integral equations.

5.5.2. *Numerical integration on the sphere*

For a function g defined on U, consider computing

$$I(g) = \int_U g(Q)\, dS_Q \qquad (5.5.216)$$

There is a fairly large literature on numerical methods for computing such integrals, and we refer the reader to Stroud [537, p. 40, 267] and Atkinson [42] for overviews on the subject. In the following, we consider only one numerical method for approximating $I(g)$, one based on Gaussian quadrature and the trapezoidal rule. The method is rapidly convergent and straightforward to implement, and it is related to the use of spherical polynomials.

Use spherical coordinates to rewrite $I(g)$ as

$$I(g) = \int_0^{2\pi} \int_0^\pi \tilde{g}(\theta, \phi) \sin\theta\, d\theta\, d\phi \qquad (5.5.217)$$

with

$$\tilde{g}(\theta, \phi) \equiv g(\cos\phi \sin\theta, \sin\phi \sin\theta, \cos\theta)$$

We approximate this by

$$I_m(g) \equiv \frac{\pi}{m} \sum_{j=1}^{2m} \sum_{i=1}^{m} w_i \bar{g}(\theta_i, \phi_j) \qquad (5.5.218)$$

The nodes $\{\theta_i\}$ are so chosen that $\{\cos \theta_i\}$ and $\{w_i\}$ are Gauss-Legendre nodes and weights on order m on $[-1, 1]$. The points $\{\phi_j\}$ are evenly spaced on $[0, 2\pi]$ with spacing π/m. Usually,

$$\phi_j = \frac{j\pi}{m}, \quad j = 1, \ldots, 2m$$

The method amounts to using the trapezoidal rule to integrate with respect to the variable ϕ in (5.5.217), and since the integrand $\bar{g}$ is periodic in ϕ with period 2π, the trapezoidal rule is very accurate. The Gauss-Legendre quadrature is with respect to the variable $z = \cos \theta$ in the equivalent integral formula

$$I(g) = \int_0^{2\pi} \int_{-1}^{1} g(\cos \phi \sqrt{1 - z^2}, \sin \phi \sqrt{1 - z^2}, z) \, dz \, d\phi$$

The use of this in (5.5.218) turns out to yield a very accurate integration formula.

Lemma 5.5.2. The integration formula (5.5.218) has degree of precision $2m - 1$.

Proof. See Stroud [537, p. 41] for a proof.   □

Lemma 5.5.3.

(a) Let $g \in C(U)$. Then

$$|I(g) - I_m(g)| \leq 8\pi \rho_{2m-1}(g) \qquad (5.5.219)$$

and as a consequence, $I_m(g) \rightarrow I(g)$.

(b) Let $g \in C^{k,\alpha}(U)$. Then

$$|I(g) - I_m(g)| \leq \frac{c}{(2m - 1)^{k+\alpha}} \qquad (5.5.220)$$

with c a suitable constant, depending on g, k, and α.

Table 5.7. *Numerical integration of (5.5.221)*
on an ellipsoid

m	$N(m)$	I_m	D_m
2	8	44.5199441881123	$-1.298E+00$
4	32	43.2223923740266	$-4.109E-01$
8	128	42.8115024378161	$-8.555E-03$
16	512	42.8029474311605	$-2.947E-05$
32	2048	42.8029179624289	$-1.899E-09$
64	8192	42.8029179605301	$<1.0E-15$

Proof. For $n \geq 0$, let p_{2m-1} be the minimax approximation to g from S_{2m-1}, as discussed in Lemma 5.5.1. Then using $I_m(p_{2m-1}) = I(p_{2m-1})$,

$$I(g) - I_m(g) = I(g - p_{2m-1}) - I_m(g - p_{2m-1})$$

$$|I(g) - I_m(g)| \leq \|g - p_{2m-1}\|_\infty \{I(1) + I_m(1)\}$$

$$= \rho_{2m-1}(g) \{4\pi + 4\pi\}$$

The latter equality uses the positivity of the quadrature weights and the exactness of $I_m(1) = I(1) = 4\pi$, from the degree of precision of (5.5.218) being $2m - 1 > 0$. This proves (5.5.219). The bound (5.5.220) is immediate from (5.5.213). □

For a further discussion of the efficiency of (5.5.218), see Ref. [42, p. 333].

Example. Let S be the ellipsoidal surface described by

$$\left(\frac{x}{a}\right)^2 + \left(\frac{y}{b}\right)^2 + \left(\frac{z}{c}\right)^2 = 1$$

We will use the formula (5.5.218) for I_m to estimate

$$\int_S e^{-x+2y-.5z} \, dS \doteq 42.8029179605301 \qquad (5.5.221)$$

in which $(a, b, c) = (2, 1, 0.5)$. The integral is first transformed to an equivalent integral over U, using the transformation (5.5.194) with Jacobian (5.5.199). The numerical results are given in Table 5.7. The column labeled $N(m)$ is equal to $2m^2$, the number of integration node points in the formula (5.5.218). The column labeled D_m denotes $I_{2m} - I_m$, which we take to be an approximation to the error $I - I_m$.

This same integral was calculated earlier, in §5.3, with results given in Table 5.2. In that table the number of integration node points is $2n + 2$, with n given in the table. Clearly, the formula (5.5.218) based on Gaussian quadrature is much more rapidly convergent. This is almost always the case when comparing methods based on piecewise polynomial interpolation with methods based on globally smooth approximations.

A discrete orthogonal projection operator

Use (5.5.218) to introduce a discrete inner product on the approximating subspace S_n:

$$(g_1, g_2)_n = \frac{\pi}{n+1} \sum_{j=1}^{2n+2} \sum_{i=1}^{n+1} w_i \tilde{g}_1(\theta_i, \phi_j) \tilde{g}_2(\theta_i, \phi_j) \qquad (5.5.222)$$

This is the formula (5.5.218) with $m = n + 1$. A corresponding discrete norm is defined by

$$\|g\|_n = \sqrt{(g, g)_n}$$

For some later work, let $\{\Theta_1, \ldots, \Theta_{q_n}\}$ denote some ordering of the node points $(\cos \phi_j \sin \theta_i, \sin \phi_j \sin \theta_i, \cos \theta_i)$ in the integration formula (5.5.222), with $q_n = 2(n + 1)^2$ the number of node points. Let $\{\omega_1, \ldots, \omega_{q_n}\}$ be the corresponding ordering of the weights $\{\frac{\pi}{n+1} w_i\}$.

When considered on S_n, this inner product satisfies the properties **H1–H4** of §4.4 of Chapter 4. For example, the dimension of S_n is $d_n = (n + 1)^2$, and $q_n > d_n$, so that **H1** is satisfied. The condition **H3** is trivial. We state the additional properties **H2** and **H4** as a lemma.

Lemma 5.5.4. Let $n \geq 0$, and consider the inner product (5.5.222).

(a) For all $g \in S_n$,

$$\|g\| = \|g\|_n \qquad (5.5.223)$$

Therefore, $\| \cdot \|_n$ is a norm on S_n.

(b) Recalling the basis $\{\Psi_1, \ldots, \Psi_{d_n}\}$ of (5.5.206), define the matrix Φ_n by

$$(\Phi_n)_{i,k} = \Psi_i(\Theta_k), \quad k = 1, \ldots, q_n, \ i = 1, \ldots, d_n$$

Then Φ_n has rank d_n.

Proof. Let $g_1, g_2 \in \mathcal{S}_n$. Then $g_1(P)g_2(P)$ defines a new spherical polynomial, and its degree is $\leq 2n$. Since the formula (5.5.222) has degree of precision $2n + 1$, we have

$$(g_1, g_2)_n = (g_1, g_2) \tag{5.5.224}$$

This implies (5.5.223).

Define a $q_n \times q_n$ diagonal matrix W by

$$W_{k,k} = \omega_k, \quad k = 1, \ldots, q_n$$

Then

$$\Phi_n W \Phi_n^{\mathrm{T}} = [(\Psi_i, \Psi_j)_n] = I \tag{5.5.225}$$

with the last equality using (5.5.224) and the orthogonality of $\{\Psi_i\}$. It is a standard consequence of linear algebra that (5.5.225) implies Φ_n has rank d_n. $\square$

As in §4.4, in and following (4.4.148), we introduce a *discrete orthogonal projection* operator $Q_n : C(U) \overset{onto}{\to} \mathcal{S}_n$. It satisfies

$$Q_n g(P) = \sum_{k=1}^{d_n} (g, \Psi_k)_n \Psi_k(P), \quad g \in C(U) \tag{5.5.226}$$

In Sloan [508] this is called a *hyperinterpolation operator*. Some of its properties are given in Lemma 4.4.1 of §4.4. We give additional properties in the following lemma.

Lemma 5.5.5. Let Q_n be the discrete orthogonal projection operator of (5.5.226). Then

(a) Regarding Q_n as an operator from $C(U)$ to $L^2(U)$,

$$\|Q_n g\| \leq 2\sqrt{\pi} \|g\|_\infty, \quad g \in C(U) \tag{5.5.227}$$

with $\|Q_n g\|$ the norm in $L^2(U)$. In addition,

$$\|g - Q_n g\| \leq (1 + 2\sqrt{\pi})\rho_n(g), \quad g \in C(U) \tag{5.5.228}$$

(b) Regarding Q_n as an operator from $C(U)$ to $C(U)$,

$$\|Q_n\| \leq c_1 n^2, \quad n \geq 1 \tag{5.5.229}$$

with c_1 a suitable constant. Therefore,

$$\|g - Q_n g\|_\infty \le c_2 n^2 \rho_n(g), \quad n \ge 1 \qquad (5.5.230)$$

with c_2 a suitable constant. Combined with (5.5.213),

$$\|g - Q_n g\|_\infty \le \frac{c_3}{n^{k+\alpha-2}}, \quad g \in C^{k,\alpha}(U) \qquad (5.5.231)$$

and $Q_n g \to g$ if $k + \alpha > 2$ (which means taking $k \ge 2$).

Proof.

(a) This result is taken from Sloan [508], and we repeat his argument. For $g \in C(U)$,

$$
\begin{aligned}
\|Q_n g\|^2 &= (Q_n g, Q_n g) \\
&= (Q_n g, Q_n g)_n \qquad \text{From (5.5.224)} \\
&\le (g, g)_n \qquad\qquad \text{Lemma 4.4.1(b)} \\
&= \sum_{j=1}^{q_n} \omega_j [g(\Theta_j)]^2 \\
&\le \sum_{j=1}^{q_n} \omega_j \|g\|_\infty^2 \\
&= 4\pi \|g\|_\infty^2
\end{aligned}
$$

This proves (5.5.227).

Let $g \in C(U)$. Then let p_n be the minimax approximation to g from $\mathcal{S}_n$. We know $Q_n p_n = p_n$, and therefore,

$$g - Q_n g = [g - p_n] - Q_n[g - p_n]$$
$$\|g - Q_n g\| \le (1 + 2\sqrt{\pi})\|g - p_n\|_\infty = (1 + 2\sqrt{\pi})\rho_n(g)$$

(b) The proof of (5.5.229) is given in Ganesh, Graham, and Sivaloganathan [209, Theorem 3.7], and this paper also contains significant extensions of the results of Ragozin [441] for approximation by spherical polynomials. The proofs of (5.5.230) and (5.5.231) are straightforward repetitions of earlier proofs, and we omit them. $\qquad\qquad\square$

5.5.3. Solution of integral equations on the unit sphere

Consider again the integral equation (5.5.192), with $S = U$:

$$\lambda \rho(P) - \int_U K(P, Q)\rho(Q)\, dS_Q = \psi(P), \quad P \in S \qquad (5.5.232)$$

As usual, we also write it in symbolic form as

$$(\lambda - \mathcal{K})\rho = \psi$$

An immediately accessible numerical method is to obtain a Nyström method by applying the integration scheme of (5.5.218) to the integral operator $\mathcal{K}$. Define

$$\mathcal{K}_n \rho(P) = \frac{\pi}{n} \sum_{j=1}^{2n} \sum_{i=1}^{n} w_j K(P, Q_{i,j})\rho(Q_{i,j}), \quad P \in U, \quad \rho \in C(U)$$

$$(5.5.233)$$

with $Q_{i,j} = (\sin\phi_j \cos\theta_i, \sin\phi_j \sin\theta_i, \cos\theta_i)$ the quadrature nodes on U for (5.5.218). The Nyström approximation is

$$(\lambda - \mathcal{K}_n)\rho_n = \psi$$

The error analysis is an immediate application of Theorem 4.1.2 from Chapter 4.

Theorem 5.5.1. *Assume* $\lambda - \mathcal{K} : C(U) \overset{1-1}{\underset{\text{onto}}{\to}} C(U)$.

(a) *Assume the kernel function $K(P, Q)$ is continuous for $P, Q \in U$. Then for all sufficiently large n, say $n \geq n_0$, $(\lambda - \mathcal{K}_n)^{-1}$ exists, and it is uniformly bounded in n. For the error in the approximate solution ρ_n when compared to the true solution ρ of (5.5.232),*

$$\|\rho - \rho_n\|_\infty \leq \|(\lambda - \mathcal{K}_n)^{-1}\| \|\mathcal{K}\rho - \mathcal{K}_n\rho\|_\infty, \quad n \geq n_0 \qquad (5.5.234)$$

(b) *Further assume that $g_P(Q) \equiv K(P, Q)\rho(Q)$ belongs to $C^{k,\alpha}(U)$ when considered as a function of Q, with the Hölder condition on the derivatives $D^k g_P(Q)$ being uniform in P:*

$$|D^k g_P(Q_1) - D^k g_P(Q_2)| \leq c_g |Q_1 - Q_2|^\alpha, \quad Q_1, Q_2, P \in U \qquad (5.5.235)$$

for all derivatives $D^k g_P$ of order k. The constant c_g is to be uniform over all such derivative $D^k g_P$. Then

$$\|\rho - \rho_n\|_\infty \leq \frac{c}{n^{k+\alpha}}, \quad n \geq n_0$$

Proof. Invoke the preceding Lemmas 5.5.1–5.5.3. The proof is then immediate from Theorem 4.1.2 of Chapter 4.   □

A Galerkin method

Apply the orthogonal projection $\mathcal{P}_n$ of (5.5.209) to the integral equation (5.5.192) to obtain the numerical approximation

$$(\lambda - \mathcal{P}_n \mathcal{K})\rho_n = \mathcal{P}_n \psi \qquad (5.5.236)$$

In the context of $L^2(U)$, the error analysis is a straightforward application of Theorem 3.1.1 of Chapter 3.

Theorem 5.5.2. *Assume* $\mathcal{K} : L^2(U) \to L^2(U)$ *is a compact operator. Then*

$$\|\mathcal{K} - \mathcal{P}_n \mathcal{K}\| \to 0 \quad \text{as } n \to \infty \qquad (5.5.237)$$

Assume $\lambda - \mathcal{K} : L^2(U) \overset{1-1}{\underset{\text{onto}}{\to}} L^2(U)$. *Then for all sufficiently large* n, *say* $n \geq n_0$, $(\lambda - \mathcal{P}_n \mathcal{K})^{-1}$ *exists, and it is uniformly bounded in* n. *For the error in the approximate solution* ρ_n,

$$\|\rho - \rho_n\| \leq |\lambda| \|(\lambda - \mathcal{P}_n \mathcal{K})^{-1}\| \|\rho - \mathcal{P}_n \rho\|$$

The function space norm being used is the inner product norm of $L^2(U)$.

Proof. This is an immediate consequence of Theorem 3.1.1 and Lemma 3.1.2.   □

We would like to give a similar theorem for (5.5.232) and (5.5.236) when regarded as equations in $C(U)$, but we must show the needed result (5.5.237) by other means. We cannot use Lemma 3.1.2, because (5.5.210) implies the family of projections $\{\mathcal{P}_n\}$ is no longer uniformly bounded when regarded as operators from $C(U)$ to $C(U)$. Thus (5.5.237) must be examined directly.

Theorem 5.5.3. *Assume the kernel function* $K(P, Q)$ *is Hölder continuous with respect to* P, *uniformly with respect to* Q:

$$|K(P_1, Q) - K(P_2, Q)| \leq c_K |P_1 - P_2|^\alpha, \quad P_1, P_2, Q \in U \qquad (5.5.238)$$

Assume $\frac{1}{2} < \alpha \leq 1$. *Then*

$$\|\mathcal{K} - \mathcal{P}_n \mathcal{K}\| \to 0 \quad \text{as } n \to \infty \qquad (5.5.239)$$

Assume $\lambda - \mathcal{K} : C(U) \overset{1-1}{\underset{\text{onto}}{\rightarrow}} C(U)$. Then for all sufficiently large n, say $n \geq n_0$, $(\lambda - \mathcal{P}_n\mathcal{K})^{-1}$ exists, and it is uniformly bounded in n. For the error in the approximate solution ρ_n,

$$\|\rho - \rho_n\|_\infty \leq |\lambda| \|(\lambda - \mathcal{P}_n\mathcal{K})^{-1}\| \|\rho - \mathcal{P}_n\rho\|_\infty \qquad (5.5.240)$$

[The quantity $\|(\lambda - \mathcal{P}_n\mathcal{K})^{-1}\|$ is now an operator norm on $C(U)$.]

Proof. Recall (5.5.16), Lemma 5.5.192, and (5.5.214). Let $K_Q(\cdot) = K(\cdot, Q)$, $Q \in U$. Then

$$\|K_Q - \mathcal{P}_n K_Q\|_\infty \leq \frac{c}{n^{\alpha - \frac{1}{2}}} \qquad (5.5.241)$$

with c a multiple of c_K. It is now straightforward to show (5.5.239). The remainder of the proof is a straightforward application of Theorem 3.1.2.

Also note that the lower bound in (3.1.31) of Theorem 3.1.2 implies that $\rho_n \to \rho$ if and only if $\mathcal{P}_n\rho \to \rho$. The lack of uniform boundedness for $\{\mathcal{P}_n\}$, from (5.5.210), implies that in $C(U)$, $\mathcal{P}_n\rho$ will fail to converge to ρ for some choices of ρ. Thus ρ_n will not converge to ρ for such choices of ρ. Convergence is guaranteed, however, for $\rho \in C^{0,\alpha}(U)$ with $\alpha > \frac{1}{2}$. □

A similar theorem and proof is given in Chapter 9, for integral equations of potential theory. It is shown that (5.5.238) can be replaced with a much weaker hypothesis, one allowing K to be a singular kernel.

A discrete Galerkin method

Using the discrete orthogonal projection operator $\mathcal{Q}_n$, replace the approximating equation (5.5.236) by

$$(\lambda - \mathcal{Q}_n\mathcal{K}_n)\zeta_n = \mathcal{Q}_n\psi \qquad (5.5.242)$$

In this, $\mathcal{K}_n$ is the numerical integration operator

$$\mathcal{K}_n g(P) = \frac{\pi}{n+1} \sum_{j=1}^{2n+2}\sum_{i=1}^{n+1} w_i \, K(P, Q_{i,j}) g(Q_{i,j}), \quad P \in U \qquad (5.5.243)$$

with $Q_{i,j} = (\sin\phi_j \cos\theta_i, \sin\phi_j \sin\theta_i, \cos\theta_i)$. Note this is only a minor modification on the definition (5.5.233) of $\mathcal{K}_n$ used with the Nyström method.

Theorem 5.5.4. *For all $Q \in U$, let $K_Q \equiv K(\cdot, Q)$ and assume $K_Q \in C^{2,\alpha}(U)$. Further, assume the Hölder constants for the second derivatives of K_Q are* *uniform with respect to Q, in the same manner as in (5.5.235) of Theorem 5.5.1. Assume $\lambda - \mathcal{K} : C(U) \xrightarrow[\text{onto}]{1-1} C(U)$. Then for all sufficiently large n, say $n \geq n_0$, $(\lambda - \mathcal{Q}_n\mathcal{K}_n)^{-1}$ exists, and it is uniformly bounded in n. For the error in the approximate solution ζ_n,*

$$\|\rho - \zeta_n\|_\infty \leq \|(\lambda - \mathcal{Q}_n\mathcal{K}_n)^{-1}\| \, [|\lambda| \|\psi - \mathcal{Q}_n\psi\|_\infty$$

$$+ \|\mathcal{Q}_n\| \|\mathcal{K}\rho - \mathcal{K}_n\rho\|_\infty] \tag{5.5.244}$$

Proof. This result is similar to Theorem 4.4.2, but we do not have the uniform boundedness of $\{\mathcal{Q}_n\}$ that is required in the earlier theorem. For the function $K_Q \equiv K(\cdot, Q)$, define

$$K_{Q,n} = \mathcal{Q}_n K_Q$$

and $K_n(P, Q) = K_{Q,n}(P)$, $P, Q \in U$. Then (5.5.231) and Lemma 5.5.1 imply

$$\sup_{P, Q \in} |K(P, Q) - K_n(P, Q)| \leq \frac{c}{n^\alpha}$$

(In this and the following, c denotes a generic constant.) Consequently, we can show

$$\|\mathcal{Q}_n\mathcal{K}_n - \mathcal{K}_n\| \leq \frac{c}{n^\alpha} \tag{5.5.245}$$

which converges to zero as $n \to \infty$.

From Theorem 5.5.1, $(\lambda - \mathcal{K}_n)^{-1}$ is uniformly bounded for all sufficiently large n, say $n \geq n_0$:

$$\|(\lambda - \mathcal{K}_n)^{-1}\| \leq B, \quad n \geq n_0$$

Using

$$\lambda - \mathcal{Q}_n\mathcal{K}_n = (\lambda - \mathcal{K}_n)[I + (\lambda - \mathcal{K}_n)^{-1}(\mathcal{K}_n - \mathcal{Q}_n\mathcal{K}_n)] \tag{5.5.246}$$

Choose $n_1 \geq n_0$ so that

$$\|\mathcal{Q}_n\mathcal{K}_n - \mathcal{K}_n\| \leq \frac{1}{2B}, \quad n \geq n_1$$

Then by applying the geometric series theorem (cf. Theorem A.1 in the Appendix) to the term in brackets on the right side of (5.5.242), we can show $(\lambda - \mathcal{Q}_n\mathcal{K}_n)^{-1}$ exists and is uniformly bounded for $n \geq n_1$.

For the error, apply $\mathcal{Q}_n$ to the original equation and then manipulate it to obtain

$$(\lambda - \mathcal{Q}_n\mathcal{K}_n)\rho = \mathcal{Q}_n\psi + \lambda(\rho - \mathcal{Q}_n\rho) + \mathcal{Q}_n(\mathcal{K}\rho - \mathcal{K}_n\rho)$$

Subtract the approximating equation (5.5.242) to obtain the error equation

$$(\lambda - \mathcal{Q}_n \mathcal{K}_n)(\rho - \zeta_n) = \lambda(\rho - \mathcal{Q}_n \rho) + \mathcal{Q}_n(\mathcal{K}\rho - \mathcal{K}_n \rho)$$

The bound (5.5.240) follows directly.

The result (5.5.215) can be applied to (5.5.244) to give rates of convergence for the approximate solution ζ_n. We omit the details. □

For continuous kernel functions K, the Nyström method of Theorem 5.5.1 is a far more practical method than is the discrete Galerkin method we have just discussed. But for problems of potential theory, taken up in Chapter 9, the above ideas on discrete Galerkin methods are of practical interest.

Discussion of the literature

The subject of multivariable interpolation is well-developed, but mostly in connection with the two subjects of computer graphics and the finite element method for solving partial differential equations. For examples in computer graphics of multivariable interpolation over planar regions and surfaces, see Farin [196], Lancaster and Šalkauskas [334], and Mortenson [389]. For examples from the study of finite element methods, see Babuška and Aziz [71], Johnson [294, Chaps. 3, 4], and Brenner and Scott [92, Chap. 3]. For examples from the numerical solution of integral equations, see Refs. [43], [46], Chen and Zhou [115, Chap. 5], Hackbusch [251, §9.2.3], and Wendland [560].

For numerical integration over planar triangular elements and over triangular-shaped surface elements, the basic reference is Stroud [537]. It contains many quadrature formulas, and it also discusses how to derive such formulas. Another important reference containing a number of new quadrature formulae is Lyness and Jespersen [357]. For additional information on the error in using quadrature formulas over triangular elements, see Lyness [350], [351], Lyness and Cools [355], Lyness and de Doncker [356], and Lyness and Puri [358]. It has also been recognized that in many cases, the approximation of surface integrals must also involve an approximation of the surface. For example, see Refs. [43], [46]; Chien [121]; and Georg and Tausch [214].

For the numerical analysis of general compact integral equations of the second kind over surfaces, see Refs. [43], [46]; Chien [120]–[122]; Georg and Tausch [214]; and Tausch [539]. An excellent and definitive research monograph on multivariate singular integral equations is that of Vainikko [550]. It examines the solvability of such equations, including giving the regularity of their solutions. The book also covers all aspects of the numerical analysis of such equations, and it has an excellent bibliography. There has been quite a bit of research on solving multivariate integral equations, but most of it has been for boundary integral equations defined on the boundary surfaces of three dimensional regions. We reserve to Chapter 9 a discussion of such boundary integral equations.

There are multivariate integral equations that are not boundary integral equations, and one of the more interesting ones is the *radiosity equation* from computer graphics. The radiosity equation is

$$u(P) - \frac{\rho(P)}{\pi} \int_S u(Q)G(P, Q)V(P, Q)\, dS_Q = E(P), \quad P \in S$$

with $u(P)$ the "brightness" or *radiosity* at P and $E(P)$ the *emissivity* at $P \in S$. The function $\rho(P)$ gives the *reflectivity* at $P \in S$, with $0 \le \rho(P) < 1$. In deriving this equation, the reflectivity is assumed to be independent of the angle at which the reflection takes place, which makes the surface a *Lambertian diffuse reflector*. The function G is given by

$$\begin{aligned} G(P, Q) &= \frac{\cos\theta_P \ \cos\theta_Q}{|P - Q|^2} \\ &= \frac{[(Q - P) \cdot \mathbf{n}_P][(P - Q) \cdot \mathbf{n}_Q]}{|P - Q|^4} \end{aligned}$$

In this, $\mathbf{n}_P$ is the inner normal to S at P, and θ_P is the angle between $\mathbf{n}_P$ and $Q - P$; $\mathbf{n}_Q$ and θ_Q are defined analogously. The function $V(P, Q)$ is a "line of sight" function. More precisely, if the points P and Q can "see each other" along a straight line segment that does not intersect S at any other point, then $V(P, Q) = 1$; otherwise, $V(P, Q) = 0$. An *unoccluded surface* is one for which $V \equiv 1$ on S. Note that S need not be connected, and it may be only piecewise smooth.

For an introduction to this topic, see Cohen and Wallace [124] and Sillion and Puech [500]; these books also include a discussion of some low-order numerical methods for solving the radiosity equation. Some higher-order numerical methods for solving the radiosity equation when the surface is *unoccluded* are defined and analyzed by Atkinson and Chandler [58]. When the surface S is smooth, the kernel of the radiosity equation is singular, but bounded, and this makes it better behaved than the kernels arising in solving boundary integral equations on surfaces. Nonetheless, many of the properties of the radiosity equation are similar to those of boundary integral equations, and the numerical methods of Chapter 9 seem most suitable for solving such radiosity integral equations.

6

Iteration methods

All the numerical methods of the preceding chapters involved the solution of systems of linear equations. When these systems are not too large, they can be solved by Gaussian elimination; for such systems, that is usually the simplest and most efficient approach to use. For larger linear systems, however, iteration is usually more efficient, and it is often the only practical means of solution. There is a large literature on general iteration methods for solving linear systems, but many of these general methods are often not efficient (or possibly, not even convergent) when used to solve the linear systems we have seen in the preceding chapters. In this chapter we define and analyze several iteration methods that seem especially suitable for solving the linear systems associated with the numerical solution of integral equations.

In §6.1 we give an iteration method for solving degenerate kernel integral equations and the associated linear systems. In §6.2 we define and analyze *two-grid iteration methods* for solving the systems associated with the Nyström method. And in §6.3 we consider related two-grid methods for projection methods. In our experience these are the most efficient numerical methods for solving the linear systems obtained when solving integral equations of the second kind. In §6.4 we define *multigrid iteration methods*, which are closely related to two-grid methods. Multigrid methods are among the most efficient methods for solving the linear systems associated with the numerical solution of elliptic partial differential equations, and they are also very efficient when solving Fredholm integral equations. Finally, in §6.5 we consider the *conjugate gradient method* for solving symmetric linear systems and variants of the conjugate gradient method for solving nonsymmetric linear systems.

6.1. Solving degenerate kernel integral equations by iteration

We use the framework for degenerate kernel methods that was developed in Chapter 2, and we begin by recalling some of that framework. To solve the integral equation

$$\lambda x(t) - \int_D K(t,s)x(s)\,ds = y(t), \quad t \in D \tag{6.1.1}$$

we use the solution of the approximating equation

$$\lambda x_n(t) - \int_D K_n(t,s)x_n(s)\,ds = y(t), \quad t \in D \tag{6.1.2}$$

In this equation, K_n is a degenerate kernel approximation of K. We write K_n as

$$K_n(t,s) = \sum_{j=1}^{d_n} \alpha_{j,n}(t)\beta_{j,n}(s) \tag{6.1.3}$$

with $d_n \geq n$, and we assume

$$\|K - K_n\| \to 0 \quad \text{as } n \to \infty \tag{6.1.4}$$

In this assumption, we are regarding the integral operators K and K_n of (6.1.1) and (6.1.2), respectively, as operators on $C(D)$ to $C(D)$. We also assume the original integral equation (6.1.1) is uniquely solvable on $C(D)$. With (6.1.4) it was shown earlier in Theorem 2.1.1 that (6.1.2) is a stable numerical method and that

$$\|x - x_n\|_\infty = O(\|K - K_n\|) \quad \text{as } n \to \infty$$

The solution x_n is given by

$$x_n(t) = \frac{1}{\lambda}\left[y(t) + \sum_{j=1}^{d_n} c_{j,n}\alpha_{j,n}(t)\right], \quad t \in D \tag{6.1.5}$$

with $\{c_{1,n}, \ldots, c_{d_n,n}\}$ the solution of the linear system

$$\lambda c_{i,n} - \sum_{j=1}^{d_n} c_{j,n}(\alpha_{j,n}, \beta_{i,n}) = (y, \beta_{i,n}), \quad i = 1, \ldots, d_n \tag{6.1.6}$$

This uses the inner product definition

$$(f,g) = \int_D f(s)g(s)\,ds$$

which we assume can be calculated for the coefficients $(\alpha_{j,n}, \beta_{i,n})$ and $(y, \beta_{i,n})$.

We define an iteration method for (6.1.6) by defining an iteration method for the associated functional equation

$$(\lambda - K_n)x_n = y \tag{6.1.7}$$

and then we look at the implication of it for solving (6.1.6). A philosophy throughout most of this chapter is that we iteratively solve $(\lambda - \mathcal{K}_n)x_n = y$ by assuming that we can solve directly the equation

$$(\lambda - \mathcal{K}_m)w = z \tag{6.1.8}$$

for some much smaller parametrization variable m, for arbitrary z. The equation (6.1.7) is often called the "fine grid" equation, and (6.1.8) is called the "coarse grid" equation. These names come from the associated discretizations.

Rewrite $(\lambda - \mathcal{K}_n)x_n = y$ as

$$(\lambda - \mathcal{K}_m)x_n = y + (\mathcal{K}_n - \mathcal{K}_m)x_n \tag{6.1.9}$$

with $m < n$. Define an iteration scheme by

$$(\lambda - \mathcal{K}_m)x_n^{(k+1)} = y + (\mathcal{K}_n - \mathcal{K}_m)x_n^{(k)}, \quad k = 0, 1, \ldots \tag{6.1.10}$$

with $x_n^{(0)}$ an initial guess at the solution. Often we will be solving $(\lambda - \mathcal{K}_n)x_n = y$ for a sequence of increasing values of n, say n_0, n_1, $\ldots$ Then we could define

$$x_{n_j}^{(0)} = x_{n_{j-1}} \tag{6.1.11}$$

using the numerical solution from the preceding value of n to define the initial guess for the new larger value of n.

We give an abstract error analysis for the convergence of the iteration (6.1.10), but we apply it only to the solution of the integral equation (6.1.1) in the function space $C(D)$. Other function spaces can be used, but $C(D)$ is the most commonly used one. Assume (6.1.4). Based on it and on the unique solvability of (6.1.1), assume that m has been chosen so large that for $n \geq m$,

$$\|\mathcal{K} - \mathcal{K}_n\| < \frac{1}{\|(\lambda - \mathcal{K}_n)^{-1}\|} \tag{6.1.12}$$

This implies the unique solvability of $(\lambda - \mathcal{K}_n)x_n = y$ for all $n \geq m$. Subtract (6.1.10) from (6.1.9), and then write it as

$$\begin{aligned} x_n - x_n^{(k+1)} &= (\lambda - \mathcal{K}_m)^{-1}(\mathcal{K}_n - \mathcal{K}_m)\left(x_n - x_n^{(k)}\right) \\ &\equiv \mathcal{M}_{m,n}\left(x_n - x_n^{(k)}\right) \end{aligned} \tag{6.1.13}$$

Taking bounds,

$$\left\|x_n - x_n^{(k+1)}\right\|_\infty \leq \|\mathcal{M}_{m,n}\| \left\|x_n - x_n^{(k)}\right\|_\infty, \quad k \geq 0 \tag{6.1.14}$$

For the multiplying factor $\|\mathcal{M}_{m,n}\|$, we have

$$\begin{aligned} \|\mathcal{M}_{m,n}\| &\leq \|(\lambda - \mathcal{K}_m)^{-1}\| \, \|\mathcal{K}_n - \mathcal{K}_m\| \\ &\leq \|(\lambda - \mathcal{K}_m)^{-1}\| \, [\|\mathcal{K} - \mathcal{K}_m\| + \|\mathcal{K} - \mathcal{K}_n\|] \end{aligned}$$

Thus,

$$\lim_{m,n\to\infty} \|\mathcal{M}_{m,n}\| = 0$$

It is also straightforward to show that if m is chosen sufficiently large, then

$$\lim_{n\to\infty} \|\mathcal{M}_{m,n}\| \leq \|(\lambda - \mathcal{K}_m)^{-1}\| \, \|\mathcal{K} - \mathcal{K}_m\| < 1 \qquad (6.1.15)$$

with the final inequality using (6.1.12). Thus for all sufficiently large n, $\|\mathcal{M}_{m,n}\| < 1$, and the iteration errors in (6.1.14) converge to zero as $k \to \infty$. The rate of convergence can be improved by increasing m. Moreover, (6.1.15) implies that the speed of convergence of the iteration is essentially independent of n, for all sufficiently large n.

6.1.1. *Implementation*

We now interpret (6.1.10) as an iteration method for solving (6.1.6). Earlier we assumed that the coarse grid equation $(\lambda - \mathcal{K}_m)w = z$ could be solved directly for all z. From Theorem 2.1.2 in Chapter 2, this implies that the associated linear system

$$\lambda c_i - \sum_{j=1}^{d_m} c_j(\alpha_{j,m}, \beta_{i,m}) = z_i, \quad i = 1, \ldots, d_m \qquad (6.1.16)$$

is uniquely solvable for all possible right-hand sides $[z_1, \ldots, z_{d_m}]^{\mathrm{T}}$. The equation (6.1.10) has the solution

$$
\begin{aligned}
x_n^{(k+1)}(t) &= \frac{1}{\lambda}\left[y(t) + (\mathcal{K}_n - \mathcal{K}_m)x_n^{(k)}(t) + \sum_{j=1}^{d_m} c_{j,m}^{(k+1)}\alpha_{j,m}(t) \right] \\
&= \frac{1}{\lambda}\left[y(t) + \sum_{j=1}^{d_n} (x_n^{(k)}, \beta_{j,n})\alpha_{j,n}(t) \right. \\
&\qquad \left. - \sum_{j=1}^{d_m} (x_n^{(k)}, \beta_{j,m})\alpha_{j,m}(t) + \sum_{j=1}^{d_m} c_{j,m}^{(k+1)}\alpha_{j,m}(t) \right] \\
&= \frac{1}{\lambda}\left[y(t) + \sum_{j=1}^{d_n} c_{j,n}^{(k)}\alpha_{j,n}(t) + \sum_{j=1}^{d_m} [c_{j,m}^{(k+1)} - c_{j,m}^{(k)}]\alpha_{j,m}(t) \right]
\end{aligned}
$$

$$(6.1.17)$$

in which we have defined

$$c_{j,m}^{(k)} = (x_n^{(k)}, \beta_{j,m}), \qquad c_{j,n}^{(k)} = (x_n^{(k)}, \beta_{j,n}), \quad k \geq 0$$

The values of $\{c_{j,m}^{(k+1)}\}$ are obtained by solving the linear system

$$\lambda c_{i,m}^{(k+1)} - \sum_{j=1}^{d_m} (\alpha_{j,m}, \beta_{i,m}) c_{j,m}^{(k+1)}$$

$$= (y, \beta_{i,m}) + \sum_{j=1}^{d_n} c_{j,n}^{(k)} (\alpha_{j,n}, \beta_{i,m}) - \sum_{j=1}^{d_m} c_{j,m}^{(k)} (\alpha_{j,m}, \beta_{i,m}) \qquad (6.1.18)$$

with $i = 1, \ldots, d_m$. By our earlier assumption regarding (6.1.16), this system is solvable. In practice, the LU-factorization of the matrix in (6.1.16) is likely to have been computed and saved; if so, the system (6.1.18) can be solved in only $O(d_m^2)$ arithmetic operations. The values of $\{c_{j,n}^{(k+1)}\}$ are obtained from

$$c_{j,n}^{(k+1)} = \frac{1}{\lambda} \Bigg[(y, \beta_{i,n}) + \sum_{j=1}^{d_m} c_{j,m}^{(k+1)} (\alpha_{j,m}, \beta_{i,n})$$

$$+ \sum_{j=1}^{d_n} c_{j,n}^{(k)} (\alpha_{j,n}, \beta_{i,n}) - \sum_{j=1}^{d_m} c_{j,m}^{(k)} (\alpha_{j,m}, \beta_{i,n}) \Bigg] \qquad (6.1.19)$$

Introduce

$$c_m^{(k)} = \left[c_{1,m}^{(k)}, \ldots, c_{d_m,m}^{(k)} \right]^{\mathrm{T}}, \qquad c_n^{(k)} = \left[c_{1,n}^{(k)}, \ldots, c_{d_n,n}^{(k)} \right]^{\mathrm{T}}$$

The use of (6.1.18) and (6.1.19) involves the inner products (y, β_i) and (α_j, β_i) for $\{\alpha_j\}$ and $\{\beta_j\}$ dependent on both m and n. These may need to be computed numerically, but we will not consider the effect of such numerical integrations. Rather, see the discussion following (2.3.59) in §2.3 of Chapter 2.

It is unnecessary to find $x_n^{(k+1)}$ at each step. Instead, we would usually iterate until

$$D_{k+1} \equiv \max \left[\left\| c_m^{(k+1)} - c_m^{(k)} \right\|_{\infty}, \left\| c_n^{(k+1)} - c_n^{(k)} \right\|_{\infty} \right]$$

is sufficiently small, and then we would compute $x_n^{(k)}$ for the final iterate $c_n^{(k)}$. To connect D_{k+1} to the error in $x_n^{(k+1)}$, use (6.1.17) to obtain

$$x_n^{(k+1)}(t) - x_n^{(k)}(t) = \frac{1}{\lambda} \Bigg[\sum_{j=1}^{d_n} \{ c_{j,n}^{(k)} - c_{j,n}^{(k-1)} \} \alpha_{j,n}(t)$$

$$+ \sum_{j=1}^{d_m} [\{ c_{j,m}^{(k+1)} - c_{j,m}^{(k)} \} - \{ c_{j,m}^{(k)} - c_{j,m}^{(k-1)} \}] \alpha_{j,m}(t) \Bigg]$$

Then

$$\left\| x_n^{(k+1)} - x_n^{(k)} \right\|_\infty \le \frac{1}{|\lambda|} \left[D_{k+1} \left\| \sum_{j=1}^{d_m} |\alpha_{j,n}(\cdot)| \right\|_\infty \right.$$

$$\left. + (D_{k+1} + D_k) \left\| \sum_{j=1}^{d_m} |\alpha_{j,m}(\cdot)| \right\|_\infty \right] \qquad (6.1.20)$$

We can force D_{k+1} to be so small as to have $\left\| x_n^{(k+1)} - x_n^{(k)} \right\|_\infty$ be as small as desired.

Note that from (6.1.14), the errors $\| x_n - x_n^{(k)} \|_\infty$ will decrease geometrically:

$$\left\| x_n - x_n^{(k+1)} \right\|_\infty \le c \left\| x_n - x_n^{(k)} \right\|_\infty, \quad k \ge 0$$

with $c \equiv \| \mathcal{M}_{m,n} \| < 1$. From this, it is a standard result that

$$\left\| x_n - x_n^{(k+1)} \right\|_\infty \le \frac{c}{1-c} \left\| x_n^{(k+1)} - x_n^{(k)} \right\|_\infty \qquad (6.1.21)$$

From (6.1.13), we can also show

$$\left\| x_n^{(k+1)} - x_n^{(k)} \right\|_\infty \le c \left\| x_n^{(k)} - x_n^{(k-1)} \right\|_\infty, \quad k \ge 1$$

for the same value of c. In practice, we usually estimate c by using the geometric mean of several recent values of the ratios

$$\nu_{k+1} \equiv \frac{\left\| x_n^{(k+1)} - x_n^{(k)} \right\|_\infty}{\left\| x_n^{(k)} - x_n^{(k-1)} \right\|_\infty}, \quad k \ge 1 \qquad (6.1.22)$$

For the initial guess $x_n^{(0)}$, we must compute

$$c_{j,m}^{(0)} = \left(x_n^{(0)}, \beta_{j,m} \right), \quad j = 1, \dots, d_m$$

$$c_{j,n}^{(0)} = \left(x_n^{(0)}, \beta_{j,n} \right), \quad j = 1, \dots, d_n$$

This could be done by numerical integration, but it is simpler to use $x_n^{(0)} = 0$, and thus have

$$c_m^{(0)} = 0, \qquad c_n^{(0)} = 0 \qquad (6.1.23)$$

This contradicts the earlier recommendation of (6.1.11), but it is often the most practical choice of $x_n^{(0)}$, even though it will result in the computation of more iterates than with (6.1.11).

The cost in arithmetic operations of the iteration method (6.1.18)–(6.1.19) can be shown to be $O(n^2)$ per iteration. We leave it as a problem for the reader to determine the exact details of the cost per iteration.

Another approach to the iteration method (6.1.10) is to regard it as a *residual correction method*. Given $x_n^{(k)}$, define the *residual*

$$r^{(k)} = y - (\lambda - \mathcal{K}_n)x_n^{(k)} \qquad (6.1.24)$$

Then

$$x_n - x_n^{(k)} = (\lambda - \mathcal{K}_n)^{-1}r^{(k)}$$

$$\approx (\lambda - \mathcal{K}_m)^{-1}r^{(k)}$$

Define

$$x_n^{(k+1)} = x_n^{(k)} + (\lambda - \mathcal{K}_m)^{-1}r^{(k)} \qquad (6.1.25)$$

We leave it to the reader to show the equivalence of (6.1.24)–(6.1.25) to the earlier iteration formula (6.1.10).

Example. We consider again the example (2.3.58) of §2.3 in Chapter 2:

$$\lambda x(s,t) - \int_0^{\sqrt{\pi}} \int_0^{\sqrt{\pi}} \cos(s\sigma)\cos(t\tau)\,x(\sigma,\tau)\,d\sigma\,d\tau = y(s,t),$$

$$0 \leq s, t \leq \sqrt{\pi}$$

with $\lambda = 5$ and $x(s,t) \equiv 1$. The degenerate kernel method being used is defined in (2.3.55), and it is based on piecewise bilinear interpolation of $K(s,t,\sigma,\tau)$ with respect to (s,t). Following the notation used earlier in §2.3, we define $h = \sqrt{\pi}/n$, $n \geq 1$. In finding x_n, the order of the linear system to be solved is $d_n = (n+1)^2$. For the error in this degenerate kernel approximation, it is shown in (2.3.56) that

$$\|\mathcal{K} - \mathcal{K}_n\| = O(h^2) \qquad (6.1.26)$$

and therefore,

$$\|x - x_n\|_\infty = O(h^2)$$

For examples showing this rate of convergence, see Table 2.6 in §2.3 of Chapter 2.

For presenting results about the iteration, define

$$\Delta_k = \left\|c_n^{(k)} - c_n^{(k-1)}\right\|_\infty, \quad k \geq 1$$

$$v_k = \frac{\Delta_k}{\Delta_{k-1}}, \quad k \geq 2$$

and let $\hat{v}$ denote the empirical limiting value of v_k as k increases. Table 6.1 contains numerical results on the behavior of the iteration for varying m and n.

Table 6.1. *Rate of convergence*
for iteration (6.1.10)

m	d_m	n	d_n	$\hat{v}$
1	4	2	9	.0694
2	9	4	25	.0179
2	9	8	81	.0225
2	9	16	289	.0237
4	25	8	81	.00493
4	25	16	289	.00617

In the table, we see that the rate of convergence improves as m increases. From the results (6.1.15) and (6.1.26) we expect the rate of convergence to decrease by a factor of approximately 4 when m is increased from 1 to 2 to 4. The values of $\hat{v}$ in the table give some empirical support for this.

6.2. Two-grid iteration for the Nyström method

We consider the iterative solution of the linear system associated with the Nyström method, which was defined and analyzed in Chapter 4. Again, the integral equation being solved is

$$\lambda x(t) - \int_D K(t,s)x(s)\,ds = y(t), \quad t \in D \qquad (6.2.27)$$

and initially, we assume $K(t,s)$ is continuous in s and t. To avoid constant references to Chapter 4, we repeat some definitions. Given a numerical integration scheme

$$\int_D g(t)\,dt \approx \sum_{j=1}^{q_n} w_{n,j}g(t_{n,j}), \quad n \geq 1 \qquad (6.2.28)$$

which is convergent for all $g \in C(D)$, define the numerical integration operator

$$\mathcal{K}_n x(t) = \sum_{j=1}^{q_n} w_{n,j}K(t,t_{n,j})x(t_{n,j}), \quad t \in D$$

for $x \in C(D)$ and $n \geq 1$. The Nyström approximation to $(\lambda - \mathcal{K})x = y$ is

$$\lambda x_n(t) - \sum_{j=1}^{q_n} w_{n,j}K(t,t_{n,j})x_n(t_{n,j}) = y(t), \quad t \in D \qquad (6.2.29)$$

which is denoted abstractly by $(\lambda - \mathcal{K}_n)x_n = y$. It is equivalent in solvability to the linear system

$$\lambda x_n(t_{n,i}) - \sum_{j=1}^{q_n} w_{n,j} K(t_{n,i}, t_{n,j}) x_n(t_{n,j}) = y(t_{n,i}), \quad i = 1, \ldots, q_n$$
(6.2.30)

by means of the Nyström interpolation formula

$$x_n(t) = \frac{1}{\lambda}\left[y(t) + \sum_{j=1}^{q_n} w_{n,j} K(t, t_{n,j}) x_n(t_{n,j}) \right], \quad t \in D$$
(6.2.31)

For a discussion and error analysis of (6.2.29)–(6.2.31), see §4.1 of Chapter 4.

As with the degenerate kernel method in §6.1, we first define an iteration method for solving the functional equation $(\lambda - \mathcal{K}_n)x_n = y$, and then we look at the implied iteration method for solving the linear system (6.2.30). Assume $x_n^{(0)}$ is an initial estimate of the solution x_n for (6.2.29). Define the residual

$$r^{(0)} = y - (\lambda - \mathcal{K}_n)x_n^{(0)}$$
(6.2.32)

This leads to

$$r^{(0)} = (\lambda - \mathcal{K}_n)\left(x_n - x_n^{(0)}\right)$$
$$x_n - x_n^{(0)} = (\lambda - \mathcal{K}_n)^{-1} r^{(0)}$$
(6.2.33)

By estimating $(\lambda - \mathcal{K}_n)^{-1} r^{(0)}$ in various ways, we can define various iteration methods. We give two such methods in this section, and we show others in § 6.4.

6.2.1. Iteration method 1 for Nyström's method

For some $m < n$, assume we can solve directly the approximating equation $(\lambda - \mathcal{K}_m)w = z$, for all $z \in C(D)$. Then consider the approximation

$$(\lambda - \mathcal{K}_n)^{-1} r^{(0)} \approx (\lambda - \mathcal{K}_m)^{-1} r^{(0)}$$
(6.2.34)

Using it in (6.2.33), define

$$x_n^{(1)} = x_n^{(0)} + (\lambda - \mathcal{K}_m)^{-1} r^{(0)}$$

The general iteration is defined by

$$r^{(k)} = y - (\lambda - \mathcal{K}_n)x_n^{(k)}$$
$$x_n^{(k+1)} = x_n^{(k)} + (\lambda - \mathcal{K}_m)^{-1} r^{(k)}, \quad k = 0, 1, 2, \ldots$$
(6.2.35)

This iteration turns out to be less than ideal, but it does converge when m is chosen sufficiently large.

Theorem 6.2.1. *Assume the integral equation* $(\lambda - \mathcal{K})x = y$ *of (6.2.27) is uniquely solvable for all* $y \in C(D)$, *and let* $K(t, s)$ *be continuous for* $t, s \in D$. *Assume the numerical integration scheme (6.2.28) is convergent for all* $g \in C(D)$. *Then if* m *is chosen sufficiently large, the iteration method (6.2.35) is convergent, that is,*

$$x_n^{(k)} \to x_n \quad as \ k \to \infty$$

for all $n > m$.

Proof. Introduce

$$\Psi = \{\mathcal{K}_n x \mid n \geq 1 \text{ and } \|x\|_\infty \leq 1\} \tag{6.2.36}$$

From the discussion following **A1–A3** in §4.1 of Chapter 4, the set Ψ has compact closure in $C(D)$. Define

$$B_K = \sup_{n \geq 1} \|\mathcal{K}_n\|$$

$$a_m = \sup_{n \geq m} \sup_{z \in \Psi} \|\mathcal{K}z - \mathcal{K}_n z\|_\infty$$

$$= \sup_{n \geq m} \sup_{l \geq 1} \|(\mathcal{K} - \mathcal{K}_n)\mathcal{K}_l\|$$

Since Ψ has compact closure, and since $\{\mathcal{K}_n\}$ is pointwise convergent on $C(D)$, it is relatively straightforward to show (a) the constant B_K is finite, and (b)

$$a_n \to 0 \quad as \ n \to \infty \tag{6.2.37}$$

monotonically. In fact, from the constructions used in Lemma 4.1.1 of Chapter 4,

$$\|(\mathcal{K} - \mathcal{K}_n)\mathcal{K}_l\| = \max_{t \in D} \sum_{j=1}^{q_l} |w_{l,j} E_n(t, t_{l,j})|$$

In this, $E_n(t, s)$ is the numerical integration error

$$E_n(t, s) \equiv \int_D K(t, u)K(u, s) \, du - \sum_{j=1}^{q_n} w_{n,j} K(t, t_{n,j}) K(t_{n,j}, s), \quad n \geq 1$$

and it was shown to converge to zero uniformly for $s, t \in D$.

Let $N(\lambda)$ be an integer for which

$$\|(\mathcal{K} - \mathcal{K}_n)\mathcal{K}_n\| < \frac{|\lambda|}{\|(\lambda - \mathcal{K})^{-1}\|}, \quad n \geq N(\lambda)$$

Then by Theorem 4.1.2, $(\lambda - \mathcal{K}_n)^{-1}$ exists and is uniformly bounded for $n \geq N(\lambda)$. Let

$$B_I(\lambda) = \sup_{n \geq N(\lambda)} \|(\lambda - \mathcal{K}_n)^{-1}\|$$

Trivially,

$$\|\mathcal{K}\| \leq B_K, \quad \|(\lambda - \mathcal{K})^{-1}\| \leq B_I(\lambda)$$

From (6.2.35),

$$
\begin{aligned}
r^{(k)} &= (\lambda - \mathcal{K}_n)\left(x_n - x_n^{(k)}\right) \\
x_n^{(k+1)} &= x_n^{(k)} + (\lambda - \mathcal{K}_m)^{-1}(\lambda - \mathcal{K}_n)\left(x_n - x_n^{(k)}\right) \\
x_n - x_n^{(k+1)} &= [I - (\lambda - \mathcal{K}_m)^{-1}(\lambda - \mathcal{K}_n)]\left(x_n - x_n^{(k)}\right) \qquad (6.2.38) \\
&= (\lambda - \mathcal{K}_m)^{-1}(\mathcal{K}_n - \mathcal{K}_m)\left(x_n - x_n^{(k)}\right) \\
&\equiv \mathcal{M}_{m,n}\left(x_n - x_n^{(k)}\right)
\end{aligned}
$$

We cannot show $\mathcal{M}_{m,n} \to 0$ for $n \geq m$ and $m \to \infty$, but we can show

$$\sup_{n \geq m} \left\|\mathcal{M}_{m,n}^2\right\| \to 0 \quad \text{as} \quad m \to \infty \qquad (6.2.39)$$

This turns out to be sufficient, since we can use (6.2.38) to write

$$x_n - x_n^{(k+1)} = \mathcal{M}_{m,n}^2\left(x_n - x_n^{(k-1)}\right), \quad k \geq 1 \qquad (6.2.40)$$

Then

$$\left\|x_n - x_n^{(k+1)}\right\|_\infty \leq \left\|\mathcal{M}_{m,n}^2\right\| \left\|x_n - x_n^{(k-1)}\right\|_\infty \qquad (6.2.41)$$

and we have that $\{x_n^{(k)}\}$ converges to x_n.

To show (6.2.39), write

$$
\begin{aligned}
\left\|\mathcal{M}_{m,n}^2\right\| &= \|[(\lambda - \mathcal{K}_m)^{-1}(\mathcal{K}_n - \mathcal{K}_m)]^2\| \\
&\leq \|(\lambda - \mathcal{K}_m)^{-1}\| \, [\|(\mathcal{K} - \mathcal{K}_m)(\lambda - \mathcal{K}_m)^{-1}(\mathcal{K}_n - \mathcal{K}_m)\| \\
&\quad + \|(\mathcal{K} - \mathcal{K}_n)(\lambda - \mathcal{K}_m)^{-1}(\mathcal{K}_n - \mathcal{K}_m)\|] \qquad (6.2.42)
\end{aligned}
$$

We need to bound

$$\|(\mathcal{K} - \mathcal{K}_r)(\lambda - \mathcal{K}_m)^{-1}\mathcal{K}_s\| = \sup_{\|x\|_\infty \leq 1} \|(\mathcal{K} - \mathcal{K}_r)(\lambda - \mathcal{K}_m)^{-1}\mathcal{K}_s x\|_\infty$$

for $r, s = m, n$. Write

$$u = (\lambda - \mathcal{K}_m)^{-1}\mathcal{K}_s x$$

Then

$$(\lambda - \mathcal{K}_m)u = \mathcal{K}_s x$$

$$u = \frac{1}{\lambda}[\mathcal{K}_s x + \mathcal{K}_m u]$$

Applying $\mathcal{K} - \mathcal{K}_r$ to u, we have

$$\|(\mathcal{K} - \mathcal{K}_r)u\|_\infty \leq \frac{1}{|\lambda|}[\|(\mathcal{K} - \mathcal{K}_r)\mathcal{K}_s x\|_\infty + \|(\mathcal{K} - \mathcal{K}_r)\mathcal{K}_m u\|_\infty]$$

$$\leq \frac{1}{|\lambda|}[a_r\|x\|_\infty + a_r\|u\|_\infty]$$

$$\leq \frac{a_r}{|\lambda|}[1 + B_K B_I(\lambda)]$$

This shows

$$\|(\mathcal{K} - \mathcal{K}_r)(\lambda - \mathcal{K}_m)^{-1}\mathcal{K}_s\| \leq \frac{a_r}{|\lambda|}[1 + B_K B_I(\lambda)] \qquad (6.2.43)$$

Applying this last inequality to (6.2.42), we have

$$\|\mathcal{M}_{m,n}^2\| \leq \frac{2B_I(\lambda)[1 + B_K B_I(\lambda)]}{|\lambda|}(a_m + a_n)$$

$$\leq \frac{4B_I(\lambda)[1 + B_K B_I(\lambda)]}{|\lambda|}a_m, \quad n \geq m \qquad (6.2.44)$$

Using (6.2.37), this proves (6.2.39) and completes the proof. $\qquad\qquad\square$

The results in (6.2.39)–(6.2.41) do not imply a predictable behavior for the convergence of $\{x_n^{(k)}\}$ to x_n, but only that the iteration error in $x_n^{(k+1)}$ will decrease when compared to that in $x_n^{(k-1)}$. We give a simple matrix example to emphasize this point.

Consider the iteration

$$x^{(k+1)} = y + Ax^{(k)}, \quad k = 0, 1, \ldots \qquad (6.2.45)$$

with

$$A = \begin{bmatrix} \lambda & c \\ 0 & -\lambda \end{bmatrix}$$

$0 < |\lambda| < 1$, x, $y \in \mathbf{R}^2$, and $x^{(0)} = 0$. Then

$$x - x^{(k)} = A^k x, \qquad x^{(k+1)} - x^{(k)} = A^k y$$

$$A^k = \lambda^k \begin{bmatrix} 1 & \dfrac{c}{2\lambda}[1 - (-1)^k] \\ 0 & -1 \end{bmatrix}$$

This leads to the two cases

$$x - x^{(k)} = \begin{cases} \lambda^k \begin{bmatrix} x_1 \\ -x_2 \end{bmatrix}, & k \text{ even} \\ \lambda^k \begin{bmatrix} x_1 + \dfrac{c}{\lambda}x_2 \\ -x_2 \end{bmatrix}, & k \text{ odd} \end{cases}$$

From this it is possible that the error in $x - x^{(k+1)}$ is larger than the error in $x - x^{(k)}$. For example, take

$$x_1 = x_2 = 1, \qquad \lambda = .1, \qquad c = 9.9$$

Then for k even,

$$x - x^{(k)} = 10^{-k} \begin{bmatrix} 1 \\ -1 \end{bmatrix}, \qquad x - x^{(k+1)} = 10^{-(k+1)} \begin{bmatrix} 100 \\ -1 \end{bmatrix}$$

$$\left\| x - x^{(k)} \right\|_\infty = 10^{-k}, \qquad \left\| x - x^{(k+1)} \right\|_\infty = 10^{-(k-1)} = 10 \left\| x - x^{(k)} \right\|_\infty$$

$$\text{(6.2.46)}$$

But when comparing errors over two iterations, we have

$$\left\| x - x^{(k+1)} \right\|_\infty = .01 \left\| x - x^{(k-1)} \right\|_\infty, \qquad k \geq 1$$

In practice, this type of behavior is fairly unusual for iteration methods for solving linear systems, but it does occur with some examples for our iteration method (6.2.35).

Implementation for solving the linear system

The linear system to be solved is (6.2.30), with unknown

$$\underline{x}_n = [x_n(t_{n,1}), \ldots, x_n(t_{n,q_n})]^{\mathrm{T}}$$

Turning to the iteration formula (6.2.35), we must first solve the system

$$\lambda e(t_{m,i}) - \sum_{j=1}^{q_m} w_{m,j} K(t_{m,i}, t_{m,j}) e(t_{m,j}) = r^{(k)}(t_{m,i}), \quad i = 1, \ldots, q_m$$

$$(6.2.47)$$

Then we use the Nyström interpolation formula to obtain

$$e(t_{n,i}) = \frac{1}{\lambda} \left[r^{(k)}(t_{n,i}) + \sum_{j=1}^{q_m} w_{m,j} K(t_{n,i}, t_{m,j}) e(t_{m,j}) \right], \quad i = 1, \ldots, q_n$$

$$(6.2.48)$$

Looking at these two formulas, we must evaluate the residual $r^{(k)}(t)$ at all points t in $\{t_{m,i}\}$ and $\{t_{n,i}\}$:

$$r^{(k)}(t) = y(t) - \lambda x^{(k)}(t) + \sum_{j=1}^{q_n} w_{n,j} K(t, t_{n,j}) x_n^{(k)}(t_{n,j}) \qquad (6.2.49)$$

The new iterate is defined by

$$x_n^{(k+1)}(t_{m,i}) = x_n^{(k)}(t_{m,i}) + e(t_{m,i}), \quad i = 1, \ldots, q_m$$
$$x_n^{(k+1)}(t_{n,i}) = x_n^{(k)}(t_{n,i}) + e(t_{n,i}), \quad i = 1, \ldots, q_n \qquad (6.2.50)$$

For the iteration we denote

$$\underline{x}_n^{(k)} = \left[x_n^{(k)}(t_{n,1}), \ldots, x_n^{(k)}(t_{n,q_n}) \right]^{\mathrm{T}}$$

Generally, we do not find $x_n^{(k)}(t)$ for other values of t, but rather we wait until our estimate of $\|\underline{x}_n - \underline{x}_n^{(k)}\|_\infty$ is sufficiently small and the iteration is completed. Then the Nyström interpolation formula can be used to evaluate $x_n^{(k)}(t)$. Because the iteration (6.2.35) may be ill-behaved in the manner with which it converges, one must be quite careful in bounding $\|\underline{x}_n - \underline{x}_n^{(k)}\|_\infty$ based on values of $\|\underline{x}_n^{(k)} - \underline{x}_n^{(k-1)}\|_\infty$, as was done in (6.1.21) with the iteration method for solving degenerate kernel integral equations. The possibly irregular behavior of (6.2.35) is illustrated in the example below.

Example. Consider solving the equation

$$\lambda x(t) - \int_0^1 K_\gamma(s+t) x(s) \, ds = y(t), \quad 0 \le t \le 1 \qquad (6.2.51)$$

with

$$K_\gamma(\tau) = \frac{1 - \gamma^2}{1 + \gamma^2 - 2\gamma \cos(2\pi\tau)} = 1 + 2 \sum_{j=1}^{\infty} \gamma^j \cos(2j\pi\tau)$$

and $0 \leq \gamma < 1$. The eigenvalues and eigenfunctions for the associated integral operator $\mathcal{K}$ are

$$\gamma^j, \quad \cos(2j\pi t), \quad j = 0, 1, 2, \ldots$$
$$-\gamma^j, \quad \sin(2j\pi t), \quad j = 1, 2, \ldots$$

(6.2.52)

This integral equation is obtained when reformulating the Dirichlet problem for Laplace's equation $\Delta u = 0$ on an elliptical region in the plane; see Kanwal [305, p. 119].

For the integration rule used to define the numerical integration operator $\mathcal{K}_n$, we use the midpoint rule:

$$\int_0^1 g(s)\, ds \approx \frac{1}{n} \sum_{j=1}^{n} g\left(\frac{2j-1}{2n}\right), \quad g \in C[0, 1]$$

(6.2.53)

For periodic integrands on $[0, 1]$, this method converges very rapidly; for example, see Ref. [48, p. 288]. We solve the integral equation (6.2.51) when $\gamma = 0.8$, and we have the unknown functions

$$x_1(t) \equiv 1, \qquad x_2(t) = \sin(2\pi t)$$

(6.2.54)

The approximate solutions will be referred to as $x_{1,n}$ and $x_{2,n}$, respectively.

In Table 6.2 we give numerical results from solving (6.2.51) with the Nyström method using (6.2.53). The initial guess was $\underline{x}_n^{(0)} = 0$, and the iteration was performed until $\|\underline{x}_n^{(k)} - \underline{x}_n^{(k-1)}\|_\infty$ was less than 10^{-13}. The column IT gives the number of iterates that were calculated, and the following column gives the final difference calculated. In the table,

$$\underline{x} = [x(t_{n,1}), \ldots, x(t_{n,q_n})]^{\mathrm{T}}$$

the true solution at the node points $\{t_{n,i}\}$ of the fine grid approximation $(\lambda - \mathcal{K}_n)x_n = y$. The column labeled $\|\underline{x} - \underline{x}_n^{(\kappa)}\|_\infty$ gives the error in the final computed iterate when compared with the true solution $\underline{x}$, with $\underline{x}_n^{(\kappa)}$ denoting that final iterate.

Note that in the table for $\lambda = -1$ and $x = x_1$ the rate of convergence of the iteration is improved when increasing the coarse mesh parameter from $m = 16$ to $m = 32$. The number of iterates to be computed is halved, approximately.

Table 6.2. *Nyström solution of (6.2.51) with iteration: Method 1*

Unknown	λ	m	n	IT	$\left\|\underline{x}_n^{(\kappa)} - \underline{x}_n^{(\kappa-1)}\right\|_\infty$	$\left\|\underline{x} - \underline{x}_n^{(\kappa)}\right\|_\infty$
x_1	-1.00	16	32	18	7.08E−14	7.92E−4
x_1	-1.00	16	64	19	4.15E−14	6.28E−7
x_1	-1.00	16	128	19	4.16E−14	3.92E−13
x_1	-1.00	32	64	10	1.22E−15	6.28E−7
x_1	-1.00	32	128	10	2.77E−15	3.94E−13
x_2	-1.00	32	64	14	1.08E−14	6.43E−6
x_2	-1.00	32	128	14	9.91E−15	4.03E−12
x_1	0.99	32	64	29	6.23E−16	1.26E−4
x_1	0.99	32	128	29	2.92E−14	7.83E−11

Also, note that for $\lambda = -1$ the rate of convergence varies between the two cases $x = x_1$ and $x = x_2$. In fact, for $\lambda = -1$ and $x = x_2$ the use of $m = 16$ leads to a divergent iteration.

The case of $\lambda = 0.99$ and $x = x_1$ is used to show the effect of letting λ be almost equal to an eigenvalue of the integral equation. For such λ, the size of $\|(\lambda - \mathcal{K})^{-1}\|$ is increased, and from (6.2.44), we would expect the rate of convergence to be made slower for such a choice of λ. This is confirmed experimentally in the table values, with nearly a tripling of the number of computed iterates when compared with $\lambda = -1$ and $x = x_1$ for the same value of $m = 32$.

For these iterations, we also computed the ratios

$$\nu_k = \frac{\left\|\underline{x}_n^{(k)} - \underline{x}_n^{(k-1)}\right\|_\infty}{\left\|\underline{x}_n^{(k-1)} - \underline{x}_n^{(k-2)}\right\|_\infty}, \quad k \geq 2$$

In line with our example (6.2.45), these turn out to vary widely in many cases. In Figures 6.1 and 6.2 we give graphs of the values of these ratios as k increases. In some cases, there seems to be a regularity to the behavior, as in Figure 6.1 for $m = 32$, $\lambda = -1$, and $x = x_1$. However, it is still sufficiently irregular as to make it difficult to predict iteration error by standard methods of the type used in (6.1.21) for degenerate kernel methods. In other cases in Table 6.2, the behavior of the ratios ν_k seems completely random, as is illustrated in Figure 6.2 for $m = 32$, $\lambda = 0.99$, and $x = x_1$. Note especially that some of the iteration ratios ν_k are greater than 1, which agrees with the type of behavior seen with (6.2.46) for the matrix iteration example (6.2.45).

Operations count

We will look at the number of arithmetic operations used in computing a single iteration of (6.2.47)–(6.2.50). In doing so, we assume such quantities as $\{y(t_{n,i})\}$

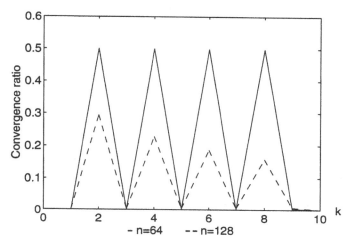

Figure 6.1. Ratios ν_k for $\lambda = -1$, $m = 32$, and $x = x_1$.

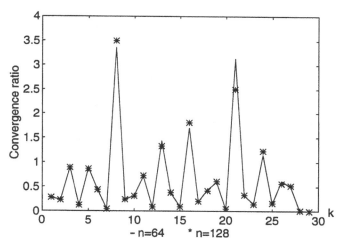

Figure 6.2. Ratios ν_k for $\lambda = 0.99$, $m = 32$, and $x = x_1$.

and $\{w_{n,i} K(t_{n,i}, t_{n,j})\}$ have already been calculated and saved for later use in the iteration.

1. To calculate the residuals $\{r^{(k)}(t_{n,i})\}$ and $\{r^{(k)}(t_{m,i})\}$, as in (6.2.49), requires approximately $2q_n(q_n + q_m)$ arithmetic operations (combining additions, subtractions, multiplications, and divisions).

2. To solve for $\{e(t_{m,i})\}$ requires approximately $(2/3)q_m^3$ operations to initially calculate an LU-factorization for the matrix in (6.2.47), but after it is done

258   6. Iteration methods

once, it can be reused in each succeeding iteration. For each iteration the number of arithmetic operations to solve (6.2.47) is approximately $2q_m^2$.

3. Finally, to calculate $\{e(t_{n,i})\}$ requires approximately $2q_m q_n$ operations.

The total cost in operations per iteration is approximately

$$2q_n(q_n + q_m) + 2q_m^2 + 2q_m q_n = 2(q_n + q_m)^2 \qquad (6.2.55)$$

We later compare this with operation counts for other iteration methods. Often $q_m \ll q_n$, and then the operations cost is approximately $2q_n^2$ operations per iteration. For perspective, this would also be the cost of solving the original system (6.2.30) associated with $(\lambda - \mathcal{K}_n)x_n = y$ if the LU-factorization was known for the linear system.

6.2.2. Iteration method 2 for Nyström's method

In some early work on iteration methods it was thought that the iteration method (6.2.35) would not converge in most cases. As shown in Theorem 6.2.1, this was not correct, but the method designed to replace it is a significantly better iteration method in most situations. Recall the equations (6.2.32)–(6.2.33) for relating the residual and the error in an initial guess $x_n^{(0)}$ for solving the approximating equation $(\lambda - \mathcal{K}_n)x_n = y$. Rather than approximating the error $e_n \equiv x_n - x_n^{(0)}$, which may be anywhere in some neighborhood of the zero element in the function space $C(D)$, we introduce a new unknown that can vary over only a much smaller (and compact) neighborhood of the origin.

Since

$$(\lambda - \mathcal{K}_n)e_n = r^{(0)} \equiv y - (\lambda - \mathcal{K}_n)x_n^{(0)} \qquad (6.2.56)$$

solve for e_n to get

$$e_n = \frac{1}{\lambda}\left[r^{(0)} + \mathcal{K}_n e_n\right]$$

Introduce the new unknown

$$\delta_n = \mathcal{K}_n e_n$$

so that

$$e_n = \frac{1}{\lambda}\left[r^{(0)} + \delta_n\right] \qquad (6.2.57)$$

Substitute this into (6.2.56) and simplify, obtaining

$$(\lambda - \mathcal{K}_n)\delta_n = \mathcal{K}_n r^{(0)} \qquad (6.2.58)$$

Estimate the unknown δ_n by solving the coarse mesh equation

$$(\lambda - \mathcal{K}_m)\delta_n^{(0)} = \mathcal{K}_n r^{(0)}$$

Then define

$$e_n^{(0)} = \frac{1}{\lambda}\left[r^{(0)} + \delta_n^{(0)}\right]$$

$$x_n^{(1)} = x_n^{(0)} + \frac{1}{\lambda}\left[r^{(0)} + \delta_n^{(0)}\right]$$

The general iteration is defined by

$$
\begin{aligned}
r^{(k)} &= y - (\lambda - \mathcal{K}_n)x_n^{(k)} \\
(\lambda - \mathcal{K}_m)\delta_n^{(k)} &= \mathcal{K}_n r^{(k)} \\
x_n^{(k+1)} &= x_n^{(k)} + \frac{1}{\lambda}\left[r^{(k)} + \delta_n^{(k)}\right], \quad k = 0, 1, \ldots
\end{aligned}
\tag{6.2.59}
$$

This method is usually superior to the iteration method developed in the preceding subsection. When the phrase "two-grid iteration for Nyström's method" is used, it generally refers to the iteration method (6.2.59).

Theorem 6.2.2. *Assume the integral equation $(\lambda - \mathcal{K})x = y$ of (6.2.27) is uniquely solvable for all $y \in C(D)$, and let $K(t, s)$ be continuous for $t, s \in D$. Assume the numerical integration scheme (6.2.28) is convergent for all $g \in C(D)$. Then if m is chosen sufficiently large, the iteration method (6.2.59) is convergent, that is,*

$$x_n^{(k)} \to x_n \quad \text{as } k \to \infty$$

for all $n > m$.

Proof. We use the notation given earlier in the proof of Theorem 6.2.1. From (6.2.36),

$$
\begin{aligned}
r^{(k)} &= (\lambda - \mathcal{K}_n)\left(x_n - x_n^{(k)}\right) \\
\delta_n^{(k)} &= (\lambda - \mathcal{K}_m)^{-1}\mathcal{K}_n r^{(k)} \\
x_n - x_n^{(k+1)} &= x_n - x_n^{(k)} - \frac{1}{\lambda}[I + (\lambda - \mathcal{K}_m)^{-1}\mathcal{K}_n]r^{(k)} \\
&= \left[I - \frac{1}{\lambda}[I + (\lambda - \mathcal{K}_m)^{-1}\mathcal{K}_n](\lambda - \mathcal{K}_n)\right]\left(x_n - x_n^{(k)}\right) \\
&= \frac{1}{\lambda}(\lambda - \mathcal{K}_m)^{-1}(\mathcal{K}_n - \mathcal{K}_m)\mathcal{K}_n\left(x_n - x_n^{(k)}\right)
\end{aligned}
$$

This gives the error equation

$$x_n - x_n^{(k+1)} = \mathcal{M}_{m,n}\big(x_n - x_n^{(k)}\big) \qquad (6.2.60)$$

with

$$\mathcal{M}_{m,n} = \frac{1}{\lambda}(\lambda - \mathcal{K}_m)^{-1}(\mathcal{K}_n - \mathcal{K}_m)\mathcal{K}_n \qquad (6.2.61)$$

As earlier in the proof of Theorem 6.2.1, let $N(\lambda)$ be an integer for which

$$\|(\mathcal{K} - \mathcal{K}_n)\mathcal{K}_n\| < \frac{|\lambda|}{\|(\lambda - \mathcal{K})^{-1}\|}, \qquad n \geq N(\lambda)$$

Then by Theorem 4.1.2, $(\lambda - \mathcal{K}_n)^{-1}$ exists and is uniformly bounded for $n \geq N(\lambda)$. As before, we denote

$$B_I(\lambda) = \sup_{n \geq N(\lambda)} \|(\lambda - \mathcal{K}_n)^{-1}\|$$

We restrict $m \geq N(\lambda)$. From (6.2.61),

$$\begin{aligned}
\|\mathcal{M}_{m,n}\| &\leq \frac{1}{|\lambda|}\|(\lambda - \mathcal{K}_m)^{-1}\|\,\|(\mathcal{K}_n - \mathcal{K}_m)\mathcal{K}_n\| \\
&\leq \frac{1}{|\lambda|}B_I(\lambda)[a_n + a_m] \\
&\leq \frac{2a_m}{|\lambda|}B_I(\lambda), \qquad n \geq m
\end{aligned} \qquad (6.2.62)$$

Since $a_m \to 0$ as $m \to \infty$, we have

$$\tau_m \equiv \sup_{n \geq m} \|\mathcal{M}_{m,n}\| < 1 \qquad (6.2.63)$$

for all sufficiently large values of m.

With (6.2.63) and (6.2.60) we have that $x_n^{(k)} \to x_n$ as $k \to \infty$, with a geometric rate of convergence τ_m that is bounded uniformly for $n \geq m$:

$$\left\|x_n - x_n^{(k+1)}\right\|_\infty \leq \tau_m\left\|x_n - x_n^{(k)}\right\|_\infty, \qquad k \geq 0 \qquad (6.2.64)$$

This completes the proof.   □

We can also show from (6.2.60) that the differences of the iterates also converge with this same bound on the geometric rate:

$$\left\|x_n^{(k+1)} - x_n^{(k)}\right\|_\infty \leq \tau_m\left\|x_n^{(k)} - x_n^{(k-1)}\right\|_\infty, \qquad k \geq 1 \qquad (6.2.65)$$

The pattern of convergence in this and (6.2.64) is much preferable to that seen with the earlier method 1 of (6.2.35), in which the errors and differences could increase as well as decrease with increasing values of k. From (6.2.64), it is straightforward to show

$$\left\| x_n - x_n^{(k+1)} \right\|_\infty \leq \frac{\tau_m}{1 - \tau_m} \left\| x_n^{(k+1)} - x_n^{(k)} \right\|_\infty \tag{6.2.66}$$

which can be used to estimate the iteration error, once an estimate or bound for τ_m is known.

Implementation for solving the linear system

The linear system to be solved is (6.2.30), with unknown

$$\underline{x}_n = [x_n(t_{n,1}), \dots, x_n(t_{n,q_n})]^T$$

Turning to the iteration formula (6.2.59), assume $\{x_n^{(k)}(t_{n,i})\}$ is known. First calculate the residual $r^{(k)}$ at the fine mesh node points:

$$r^{(k)}(t_{n,i}) = y(t_{n,i}) - \lambda x^{(k)}(t_{n,i}) + \sum_{j=1}^{q_n} w_{n,j} K(t_{n,i}, t_{n,j}) x_n^{(k)}(t_{n,j}),$$

$$i = 1, \dots, q_n \tag{6.2.67}$$

Second, calculate $\mathcal{K}_n r^{(k)}$ at both the coarse and fine mesh node points:

$$\mathcal{K}_n r^{(k)}(t) = \sum_{j=1}^{q_n} w_{n,j} K(t, t_{n,j}) r^{(k)}(t), \quad t \in \{t_{n,i}\} \cup \{t_{m,i}\} \tag{6.2.68}$$

Third, calculate the correction $\delta_n^{(k)}$ on the coarse mesh by solving the system

$$\lambda \delta_n^{(k)}(t_{m,i}) - \sum_{j=1}^{q_m} w_{m,j} K(t_{m,i}, t_{m,j}) \delta_n^{(k)}(t_{m,j}) = \mathcal{K}_n r^{(k)}(t_{m,i}),$$

$$i = 1, \dots, q_m \tag{6.2.69}$$

Fourth, extend this correction to the fine mesh with the Nyström interpolation formula:

$$\delta_n^{(k)}(t_{n,i}) = \frac{1}{\lambda} \left[\mathcal{K}_n r^{(k)}(t_{n,i}) + \sum_{j=1}^{q_m} w_{m,j} K(t_{n,i}, t_{m,j}) \delta_n^{(k)}(t_{m,j}) \right],$$

$$i = 1, \dots, q_n \tag{6.2.70}$$

Finally, define the new iterate $x_n^{(k+1)}$ on the fine mesh by

$$x_n^{(k+1)}(t_{n,i}) = x_n^{(k)}(t_{n,i}) + \frac{1}{\lambda}\left[r^{(k)}(t_{n,i}) + \delta_n^{(k)}(t_{n,i})\right], \quad i = 1, \ldots, q_n$$

$$(6.2.71)$$

For the iteration of the linear system, we denote

$$\underline{x}_n^{(k)} = \left[x_n^{(k)}(t_{n,1}), \ldots, x_n^{(k)}(t_{n,q_n})\right]^{\mathrm{T}}$$

We stop the iteration when $\|\underline{x}_n - \underline{x}_n^{(k+1)}\|_\infty$ is considered sufficiently small. Generally we define

$$\nu_k = \frac{\left\|\underline{x}_n^{(k)} - \underline{x}_n^{(k-1)}\right\|_\infty}{\left\|\underline{x}_n^{(k-1)} - \underline{x}_n^{(k-2)}\right\|_\infty}, \quad k = 2, 3, \ldots \qquad (6.2.72)$$

to estimate the convergence ratio τ_m of (6.2.63). Based on (6.2.64)–(6.2.66), we bound the error by using

$$\left\|\underline{x}_n - \underline{x}_n^{(k+1)}\right\|_\infty \leq \frac{\nu_{k+1}}{1 - \nu_{k+1}}\left\|\underline{x}_n^{(k+1)} - \underline{x}_n^{(k)}\right\|_\infty \qquad (6.2.73)$$

Sometimes in this formula we replace ν_{k+1} with the geometric mean of several successive values of ν_k to stabilize the ratios when necessary.

Example. We solve the integral equation (6.2.51), which was used to illustrate the earlier iteration method (6.2.35). As before, we use the Nyström method based on the rectangular rule of (6.2.53), and the true solutions x_i are given in (6.2.54), the same as used in the earlier example. The numerical results are given in Table 6.3. We include an additional column for the ratios of (6.2.72),

$$\bar{\nu} = \lim_{k \to \infty} \nu_k \qquad (6.2.74)$$

In all but one of the computations for Table 6.3, the sequence $\{\nu_k\}$ converged rapidly to an empirical limit $\bar{\nu}$, which we give in the table. For the case $m = 32$, $n = 128$, $\lambda = -1$, $x(t) \equiv 1$, (marked by a superscript "a" in the table) the convergence was so rapid that the value of $\bar{\nu} = .00130$ appears to be accurate to only two digits.

There are two main observations to make when comparing Tables 6.2 and 6.3. First, the iteration method 2 of (6.2.59) is much faster in its convergence, usually taking about one-half the number of iterations needed with the iteration method 1 of (6.2.35). Second, the ratios $\{\nu_k\}$ are quite well-behaved for method 2, as shown by the empirical existence of the limit $\bar{\nu}$ of (6.2.74). In contrast, for

Table 6.3. *Nyström solution of (6.2.51) with iteration: Method 2*

Unknown	λ	m	n	IT	$\left\|\underline{x}_n^{(\kappa)} - \underline{x}_n^{(\kappa-1)}\right\|_\infty$	$\left\|\underline{x} - \underline{x}_n^{(\kappa)}\right\|_\infty$	$\bar{\nu}$
x_1	−1.00	16	32	10	5.34E−14	7.92E−4	.0336
x_1	−1.00	16	64	11	4.57E−14	6.28E−7	.0452
x_1	−1.00	16	128	11	5.18E−14	3.96E−13	.0452
x_1	−1.00	32	64	6	7.77E−16	6.28E−7	.000797
x_1	−1.00	32	128	6	5.00E−15	3.94E−13	.00130[a]
x_2	−1.00	32	64	7	2.24E−14	6.43E−6	.00531
x_2	−1.00	32	128	7	4.77E−14	4.00E−12	.00599
x_1	0.99	32	64	17	2.00E−14	1.26E−4	.140
x_1	0.99	32	128	17	2.08E−14	7.87E−11	.139

[a] Rapid convergence forced accuracy to two digits only.

method 1 these ratios $\{\nu_k\}$ can be quite ill-behaved, as is illustrated in Figures 6.1 and 6.2.

With both methods there is a *mesh independence principle* working. This means that as n increases, the number of iterates to be computed to obtain a given reduction in the initial error is essentially independent of n. This is in contrast to what usually occurs with most iteration methods for solving discretizations of partial differential equations, where the number of iterates to be computed increases as the parametrization variable n increases. The present *mesh independence* can be supported theoretically. For method 1 the convergence operator $\mathcal{M}_{m,n}$ of (6.2.41) can be shown to satisfy

$$\lim_{n\to\infty} \mathcal{M}_{m,n}^2 = [(\lambda - \mathcal{K}_m)^{-1}(\mathcal{K} - \mathcal{K}_m)]^2 \qquad (6.2.75)$$

For method 2 the convergence operator $\mathcal{M}_{m,n}$ of (6.2.61) can be shown to satisfy

$$\lim_{n\to\infty} \mathcal{M}_{m,n} = \frac{1}{\lambda}(\lambda - \mathcal{K}_m)^{-1}(\mathcal{K} - \mathcal{K}_m)\mathcal{K} \qquad (6.2.76)$$

In both cases the norm of the limiting value is less than 1 if m is chosen sufficiently large, and thus the number of iterates to be computed can be shown to be independent of n, for n chosen sufficiently large. Also, with both methods, the rate of convergence is improved by increasing m.

The linear systems solved by iteration in the preceding examples were relatively small, and they could have been solved more simply by a direct method. In contrast, the systems studied in Chapter 5 are often so large that only iteration methods can be used. As an example of such, we give the following. Additional examples of the iterative solution of such large linear systems are given in Chapter 9.

Table 6.4. *Iterative solution of (6.2.77) on an ellipsoid*

m	m_v	n	n_v	IT	$\|\rho_n^{(\kappa)} - \rho_n^{(\kappa-1)}\|_\infty$	$\bar{\nu}$
8	18	32	66	19	3.61E$-$14	.181
8	18	128	258	21	2.53E$-$14	.211
8	18	512	1026	21	3.32E$-$14	.213
32	66	128	258	10	8.48E$-$14	.0352
32	66	512	1026	11	9.97E$-$15	.0383

Example. Recall the example of solving the integral equation (5.4.157),

$$\lambda\rho(P) - \int_S \rho(Q)\frac{\partial}{\partial\nu_Q}(|P - Q|^2)dS_Q = \psi(P), \quad P \in S \qquad (6.2.77)$$

given in §5.4 of Chapter 5. The function ψ was so chosen that the true solution was

$$\rho(x, y, z) = e^z$$

A Nyström method was used to approximate (6.2.77), as described in and following (5.4.158). We give results of using the above iteration method 2 to solve the linear system arising from the Nyström method, and we do so for the cases of S an ellipsoid and S an elliptical paraboloid.

For S the ellipsoid

$$\left(\frac{x}{a}\right)^2 + \left(\frac{y}{b}\right)^2 + \left(\frac{z}{c}\right)^2 = 1$$

we use the surface parameters $(a, b, c) = (1, 0.75, 0.5)$. In the integral equation, we choose $\lambda = 12$. [For this surface, the norm of the integral operator in (6.2.77) is $\|\mathcal{K}\| = 8\pi abc = 3\pi$.] The numerical results on the iteration method are given in Table 6.4.

For S the boundary of the elliptical paraboloid

$$\left(\frac{x}{a}\right)^2 + \left(\frac{y}{b}\right)^2 \leq z \leq c$$

we use the surface parameters $(a, b, c) = (0.75, 0.6, 0.5)$. In the integral equation, we choose $\lambda = 0.5$. [For this surface, the norm of the integral operator in (6.2.77) is $\|\mathcal{K}\| = \frac{1}{2}\pi abc^2 \doteq 1.06$.] The numerical results on the iteration method are given in Table 6.5.

In the tables, n denotes the number of fine mesh triangles in the triangulation of S, and n_v denotes the number of node points $[n_v = 2(n + 1)]$, which is also the order of the linear system being solved. The coarse mesh parameters m

Table 6.5. *Iterative solution of (6.2.77) on an elliptical paraboloid*

m	m_v	n	n_v	IT	$\|\underline{\rho}_n^{(\kappa)} - \underline{\rho}_n^{(\kappa-1)}\|_\infty$	$\bar{\nu}$
8	18	32	66	14	7.76E−14	.0930
8	18	128	258	16	5.08E−14	.124
8	18	512	1026	16	9.53E−14	.129
32	66	128	258	11	1.78E−14	.0368
32	66	512	1026	11	7.16E−14	.0435

and m_v are defined analogously. The norm of the difference in the successive iterates $\underline{\rho}_n^{(k)}$ is based on the differences at the node points:

$$\|\underline{\rho}_n^{(k)} - \underline{\rho}_n^{(k-1)}\|_\infty = \max_{1 \le i \le n_v} |\rho_n^{(k)}(v_{n,i}) - \rho_n^{(k-1)}(v_{n,i})|$$

with $\{v_{n,i} \mid 1 \le i \le n_v\}$ the nodes associated with the triangulation $\mathcal{T}_n = \{\Delta_{n,k} \mid 1 \le k \le n\}$. The initial guess is $\underline{\rho}_n^{(0)} = 0$, and the iteration was performed until $\|\underline{\rho}_n^{(k)} - \underline{\rho}_n^{(k-1)}\|_\infty$ was less than 10^{-13}. The column IT gives the number of iterates that were calculated, and the following column gives the final difference $\|\underline{\rho}_n^{(\kappa)} - \underline{\rho}_n^{(\kappa-1)}\|_\infty$ calculated, where $\kappa = IT$. The last column gives the empirical limiting ratio $\bar{\nu}$ of (6.2.74).

Operations count

We will look at the number of arithmetic operations used in computing a single iteration of (6.2.67)–(6.2.71). In doing so, we assume such quantities as $\{y(t_{n,i})\}$ and $\{w_{n,i}K(t_{n,i}, t_{n,j})\}$ have already been calculated and saved for later use in the iteration.

1. To calculate the residuals $\{r^{(k)}(t_{n,i})\}$ of (6.2.67) requires approximately $2q_n^2$ arithmetic operations (combining additions, subtractions, multiplications, and divisions).
2. To evaluate $\{\mathcal{K}_n r^{(k)}(t_{n,i})\}$ and $\{\mathcal{K}_n r^{(k)}(t_{m,i})\}$ requires approximately $2q_n(q_n + q_m)$ arithmetic operations.
3. To solve the linear system (6.2.69) for $\{\delta_n^{(k)}(t_{m,i})\}$ requires approximately $2q_m^2$ arithmetic operations provided an LU-factorization of the matrix of the linear system has already been calculated and saved.
4. To evaluate $\{\delta_n^{(k)}(t_{n,i})\}$ using the Nyström interpolation formula (6.2.70) requires approximately $2q_nq_m$ arithmetic operations.
5. The final step (6.2.71) requires only $3q_n$ arithmetic operations and is negligible in comparison to the other costs.

Combining these, we have a total cost of approximately

$$2q_n^2 + 2(q_n + q_m)^2 \tag{6.2.78}$$

arithmetic operations.

Recall that the cost of a single iteration of the earlier method (6.2.35) was approximately $2(q_n + q_m)^2$. Thus, for $q_n \gg q_m$, the new iteration method (6.2.59) is approximately twice as expensive per iteration when compared with the earlier iteration method (6.2.35). For the examples given in Tables 6.2 and 6.3 for these two methods, it can be seen that the total cost is nearly the same. The major benefit seems to be that the second method is much better behaved, which is important when attempting to predict the iteration error so as to know when to stop the iteration.

An algorithm with automatic error control

Consider the numerical solution of the integral equation

$$\lambda x(t) - \int_a^b K(t, s)x(s)\, ds = y(t), \quad a \leq t \leq b$$

It is possible to construct a program to estimate both the error $\|x - x_n\|_\infty$ and the iteration error $\|x_n - x_n^{(k)}\|_\infty$. This can be used to construct a program that will find the solution x to within any specified accuracy $\epsilon > 0$.

Two such programs are given in Ref. [38], and their structure is described in Ref. [37]. More examples of their use are given in Ref. [39, pp. 164–188]. The programs assume that the kernel function $K(t, s)$ is reasonably well-behaved. The Nyström method is used for the discretization of the integral equation, and the above iteration method 2 of (6.2.59) is used to solve the linear system when it becomes large. The error estimation for $\|\underline{x}_n - \underline{x}_n^{(k)}\|_\infty$ is based on the theory given above for the iteration method 2, and the error estimation for $\|\underline{x} - \underline{x}_n\|_\infty$ is based on the asymptotic error analysis for the Nyström method in (4.1.47)–(4.1.50) of §4.1 of Chapter 4.

Two programs are given in Ref. [38], one based on Simpson's rule and the other based on Gauss-Legendre quadrature. Each program begins by solving a discretization with parameter n_0 (and q_{n_0} equations), with an initial guess of $\underline{x}_{n_0}^{(0)} = 0$. Then the discretization parameter n is increased until an approximate solution with sufficient accuracy is obtained. Given $\epsilon > 0$, these programs compute an approximate solution $\underline{x}_\epsilon$ that satisfies

$$\|\underline{x} - \underline{x}_\epsilon\|_\infty \leq \epsilon \tag{6.2.79}$$

The total cost of this is $O(q_n^2)$, where n is the parameter for the final linear system being solved.

6.3. Two-grid iteration for collocation methods

The definitions and analyses of the collocation and iterated collocation methods for solving

$$\lambda x(t) - \int_D K(t,s)x(s)\,ds = y(t), \quad t \in D \tag{6.3.80}$$

are given in §§3.1, 3.2, and 3.4 of Chapter 3. This section begins with a short summary of those earlier results, followed by the introduction of some operators and other ideas needed in defining and analyzing the iteration method. The iteration method is defined by using an argument of the type that led to method 2 in the preceding section. An error analysis is given for the iteration method, and the section concludes with a numerical example for an integral equation with a singular kernel function.

Let $\mathcal{X} = C(D)$ with the uniform norm, and let $\mathcal{X}_n$ be a finite-dimensional approximating subspace, say of dimension d_n. Choose a set of collocation nodes $\{t_{n,1}, \ldots, t_{n,d_n}\} \subset D$. Letting $\{\phi_{n,1}, \ldots, \phi_{n,d_n}\}$ be a basis for $\mathcal{X}_n$, assume

$$\det [\phi_{n,i}(t_{n,j})] \neq 0$$

Let $\{\ell_{n,1}, \ldots, \ell_{n,d_n}\}$ be the basis of Lagrange functions for $\mathcal{X}_n$. This means $\ell_{n,i} \in \mathcal{X}_n$ and

$$\ell_{n,i}(t_{n,j}) = \delta_{i,j}, \quad i,j = 1, \ldots, d_n$$

The operator

$$\mathcal{P}_n x = \sum_{j=1}^{d_n} x(t_{n,j})\ell_{n,j} \tag{6.3.81}$$

is the interpolatory projection of $C(D)$ onto $\mathcal{X}_n$: $\mathcal{P}_n x$ is the unique element in $\mathcal{X}_n$ that interpolates x at the nodes $\{t_{n,i}\}$,

$$\mathcal{P}_n x(t_{n,i}) = x(t_{n,i}), \quad i = 1, \ldots, d_n$$

The collocation method for solving (6.3.80) is given abstractly by

$$(\lambda - \mathcal{P}_n \mathcal{K})x_n = \mathcal{P}_n y \tag{6.3.82}$$

Using the basis $\{\phi_{n,1}, \ldots, \phi_{n,d_n}\}$ for $\mathcal{X}_n$, write

$$x_n(t) = \sum_{j=1}^{d_n} c_j \phi_{n,j}(t)$$

The coefficients $\{c_1, \ldots, c_{d_n}\}$ are obtained by solving the linear system

$$\sum_{j=1}^{d_n} c_j \left\{ \lambda \phi_{n,j}(t_{n,i}) - \int_D K(t_{n,i}, s)\phi_{n,j}(s)\,ds \right\} = y(t_{n,i}), \quad i = 1, \ldots, d_n$$

$$(6.3.83)$$

Iterative methods for solving this system are the focus of this section.

One method for solving this system by iteration is to recognize that it can be seen as a Nyström method, as was noted at the end of §4.2 in Chapter 4. To this end, consider solving the iterated collocation equation

$$(\lambda - \mathcal{K}\mathcal{P}_n)\hat{x}_n = y \qquad (6.3.84)$$

in which

$$\hat{x}_n = \frac{1}{\lambda}[y + \mathcal{K}x_n]$$

Since $\mathcal{P}_n \hat{x}_n = x_n$,

$$\hat{x}_n(t_{n,i}) = x_n(t_{n,i}), \quad i = 1, \ldots, d_n$$

Thus finding $\hat{x}_n$ will yield $\{x_n(t_{n,1}), \ldots, x_n(t_{n,d_n})\}$. The equation (6.3.84) can be regarded as a Nyström method in which the numerical integral operator is defined by product integration:

$$\mathcal{K}_n x(t) \equiv \mathcal{K}\mathcal{P}_n x(t) = \sum_{j=1}^{d_n} x(t_{n,j}) \int_D K(t, s)\ell_{n,j}(s)\,ds, \quad x \in C(D)$$

$$(6.3.85)$$

The iteration methods developed in §6.2 can be applied to solve the linear system associated with (6.3.84).

The major difficulty with this approach is that often the kernel function $K(t, s)$ is singular and then the integrals

$$\int_D K(t_{n,i}, s)\ell_{n,j}(s)\,ds$$

are very time-consuming to evaluate. In addition, for multidimensional problems these integrals can be quite complicated to evaluate. This makes the Nyström interpolation formula expensive as a means to move from the coarse mesh space $\mathbf{R}^{d_m}$ to the fine mesh space $\mathbf{R}^{d_n}$, and in turn, this is motivation for finding less expensive iteration methods. One such method is due to Hackbusch [247]–[249], and it can lessen this difficulty somewhat.

For the collocation method the only case considered in this section is that for which the basis used is the set of Lagrange functions $\{\ell_{n,1}, \ldots, \ell_{n,d_n}\}$. The linear system (6.3.83) becomes

$$\lambda x_n(t_{n,i}) - \sum_{j=1}^{d_n} x_n(t_{n,j}) \int_D K(t_{n,i}, s)\ell_{n,j}(s)\, ds = y(t_{n,i}), \quad i = 1, \ldots, d_n$$

(6.3.86)

which is denoted in matrix form by

$$(\lambda - K_n)\mathbf{x}_n = \mathbf{y}_n \tag{6.3.87}$$

with $\mathbf{x}_n, \mathbf{y}_n \in \mathbf{R}^{d_n}$. In addition, we assume that the approximations in $\mathcal{X}_n$ are of a finite element type, based on decomposing D into small elements and using functions that are piecewise polynomial over these elements. In particular, let D be decomposed as $\mathcal{T}_n \equiv \{\Delta_1^{(n)}, \ldots, \Delta_n^{(n)}\}$.

For an interval $D = [a, b]$, this decomposition is usually based on a partition

$$a = \tau_0 < \tau_1 < \cdots < \tau_n = b$$

with

$$\Delta_i^{(n)} = [\tau_{i-1}, \tau_i], \quad i = 1, \ldots, n \tag{6.3.88}$$

For D a planar polygonal region or a surface in space, use the triangulations $\mathcal{T}_n$ introduced in Chapter 5. The function $\mathcal{P}_n x(t)$ is an interpolatory polynomial in t over each element $\Delta_i^{(n)}$, or for D a surface as in Chapter 5, it is an interpolatory polynomial in the parametrization variables. In what follows, let r denote the degree of this interpolation.

For the case (6.3.88),

$$\mathcal{X}_n = \{z \in C[a, b] \mid z|_{[\tau_{i-1}, \tau_i]} \in \pi_r, \ i = 1, \ldots, n\}$$

with π_r the polynomials of degree $\leq r$. The dimension of $\mathcal{X}_n$ equals $nr + 1$.

6.3.1. Prolongation and restriction operators

As in earlier sections, a coarse mesh discretization of (6.3.80) is used to define an iteration method for solving fine mesh discretizations. A coarse mesh parametrization

$$(\lambda - K_m)\mathbf{w} = \mathbf{z}, \quad \mathbf{w}, \mathbf{z} \in \mathbf{R}^{d_m} \tag{6.3.89}$$

with $m < n$, is used in the approximation of the fine mesh system $(\lambda - K_n)\mathbf{x}_n = \mathbf{y}_n$. As part of this process, we need to go between vectors in $\mathbf{R}^{d_m}$ and $\mathbf{R}^{d_n}$. *Prolongation* and *restriction* operators are used for this. For the development of the theory, we also need analogous operators between $C(D)$ and finite dimensional spaces $\mathbf{R}^{d_n}$.

The definition of the collocation node points must be made more precise. For each element $\Delta_k^{(n)}$, assume the number of independent polynomials of degree $\leq r$ over $\Delta_k^{(n)}$ is f_r. Let $\{t_{k,1}^{(n)}, \ldots, t_{k,f_r}^{(n)}\}$ denote nodes in $\Delta_k^{(n)}$ for the local definition of $\mathcal{P}_n x(t)$:

$$\mathcal{P}_n x(t) = \sum_{j=1}^{f_r} x\big(t_{k,j}^{(n)}\big)\ell_{k,j}^{(n)}(t), \quad t \in \Delta_k^{(n)} \tag{6.3.90}$$

for given $x \in C(D)$ and $k = 1, \ldots, n$. The functions $\{\ell_{k,j}^{(n)} \mid j = 1, \ldots, f_r\}$ are Lagrange functions defined locally on $\Delta_k^{(n)}$.

Usually, there is a mapping from a reference domain σ onto $\Delta_k^{(n)}$, say

$$\mu_k^{(n)} : \sigma \xrightarrow[onto]{1-1} \Delta_k^{(n)}$$

For subintervals (6.3.88), we use $\sigma = [0, 1]$, and for surfaces and planar polygonal regions, we let σ be the unit simplex, as in Chapter 5. Then a standard set of interpolation nodes $\{\xi_1, \ldots, \xi_{f_r}\} \subset \sigma$ is chosen, and we define

$$t_{k,j}^{(n)} = \mu_k^{(n)}(\xi_j), \quad j = 1, \ldots, f_r \tag{6.3.91}$$

for $k = 1, \ldots, d_n$. The Lagrange functions $\{\ell_{k,j}^{(n)}\}$ can also be defined in terms of an analogous family $\{\ell_1, \ldots, \ell_{f_r}\}$ over σ,

$$\ell_{k,j}^{(n)}\big(\mu_k^{(n)}(\xi)\big) = \ell_j(\xi), \quad \xi \in \sigma, \quad j = 1, \ldots, f_r \tag{6.3.92}$$

For example, see (5.1.19) in §5.1 of Chapter 5. Collectively, the interpolation nodes are referred to by $\{t_{n,1}, \ldots, t_{n,d_n}\}$. If the local definition (6.3.90) of the interpolation $\mathcal{P}_n x(t)$ does not lead to a function $\mathcal{P}_n x$ that is continuous over all of D, then the ideas in the remainder of this section can be generalized by using the ideas presented in Ref. [62]. For the remainder of the section, assume that $\mathcal{P}_n : C(D) \to C(D)$, although examples are discussed in which $\mathcal{P}_n x$ is only piecewise continuous.

To simplify the following definitions, regard vectors $\mathbf{w} \in \mathbf{R}^{d_n}$ as functions on $\{t_{n,i}\}$. Thus,

$$\mathbf{w}(t_{n,i}) = \mathbf{w}_i, \quad i = 1, \ldots, d_n$$

Define the *restriction* operator $\mathcal{R}_{\infty,n} : C(D) \to \mathbf{R}^{d_n}$ by

$$\mathcal{R}_{\infty,n}\rho(t_{n,i}) = \rho(t_{n,i}), \quad i = 1, \ldots, d_n, \quad \rho \in C(D) \tag{6.3.93}$$

Easily, $\mathcal{R}_{\infty,n}$ is a bounded mapping: $\|\mathcal{R}_{\infty,n}\rho\|_\infty \le \|\rho\|_\infty$, and easily

$$\|\mathcal{R}_{\infty,n}\| = 1 \tag{6.3.94}$$

where $\|\cdot\|_\infty$ denotes the norms on $\mathbf{R}^{d_n}$ and $C(D)$.

Define the *prolongation* operator $\mathcal{P}_{n,\infty} : \mathbf{R}^{d_n} \to C(D)$ as follows: for each element $\Delta_k^{(n)}$, use the interpolation formula (6.3.90) to write

$$\mathcal{P}_{n,\infty}\mathbf{v}(t) = \sum_{j=1}^{f_r} \mathbf{v}\big(t_{k,j}^{(n)}\big)\ell_{k,j}^{(n)}(t), \quad t \in \Delta_k^{(n)}, \quad \mathbf{v} \in \mathbf{R}^{d_n} \tag{6.3.95}$$

This defines a continuous function $\mathcal{P}_{n,\infty}\mathbf{v}$ over D. Moreover,

$$\|\mathcal{P}_{n,\infty}\| = \max_{k=1,\ldots,n} \max_{t \in \Delta_k^{(n)}} \sum_{j=1}^{f_r} \big|\ell_{k,j}^{(n)}(t)\big|$$

For the case referred to in (6.3.92), with the interpolation in (6.3.95) based on that over a standard region, it is straightforward that

$$\|\mathcal{P}_{n,\infty}\| = \max_{\xi \in \sigma} \sum_{j=1}^{f_r} |\ell_j(\xi)| \tag{6.3.96}$$

For the interpolatory projection operator $\mathcal{P}_n$ of (6.3.81),

$$\mathcal{P}_n = \mathcal{P}_{n,\infty}\mathcal{R}_{\infty,n} \tag{6.3.97}$$

Also, it is straightforward that

$$\mathcal{R}_{\infty,n}\mathcal{P}_{n,\infty} = I \tag{6.3.98}$$

the identity matrix of order d_n. Also, the collocation matrix K_n is given by

$$K_n = \mathcal{R}_{\infty,n}\mathcal{K}\mathcal{P}_{n,\infty} \tag{6.3.99}$$

The proof is left to the reader.

Assume $m < n$. Define the restriction operator $\mathcal{R}_{n,m} : \mathbf{R}^{d_n} \to \mathbf{R}^{d_m}$ by

$$\mathcal{R}_{n,m}\mathbf{v}(t_{m,i}) = \mathcal{P}_{n,\infty}\mathbf{v}(t_{m,i}), \quad i = 1, \ldots, d_m, \quad \mathbf{v} \in \mathbf{R}^{d_n} \tag{6.3.100}$$

In the case that $\{t_{m,i}\} \subset \{t_{n,i}\}$, it is immediate that

$$\mathcal{R}_{n,m}\mathbf{v}(t_{m,i}) = \mathbf{v}(t_{m,i}), \quad i = 1, \ldots, d_m, \quad \mathbf{v} \in \mathbf{R}^{d_n} \tag{6.3.101}$$

This assumption on the interpolation nodes is true for the interpolation scheme used in Chapter 5 with r even or $r = 1$, where the nodes were based on (6.3.91) with $\{\xi_1, \ldots, \xi_{f_r}\}$ evenly spaced in the unit simplex σ [see (5.1.12) in §5.1]. Define a prolongation operator $\mathcal{P}_{m,n} : \mathbf{R}^{d_m} \to \mathbf{R}^{d_n}$ by

$$\mathcal{P}_{m,n}\mathbf{v}(t_{n,i}) = \mathcal{P}_{m,\infty}\mathbf{v}(t_{n,i}), \quad i = 1, \ldots, d_n, \quad \mathbf{v} \in \mathbf{R}^{d_m} \tag{6.3.102}$$

For the most widely used methods for creating the decompositions $\mathcal{T}_n = \{\Delta_1^{(n)}, \ldots, \Delta_n^{(n)}\}$ of D, the following are usually true identities.

$$\mathcal{R}_{n,m}\mathcal{P}_{m,n} = I \tag{6.3.103}$$

$$\mathcal{P}_{n,\infty}\mathcal{P}_{m,n} = \mathcal{P}_{m,\infty} \tag{6.3.104}$$

A common feature of most methods of creating decompositions $\mathcal{T}_n$ of D is that each such subdivision is refined to obtain the next finer level of subdivision, say $\mathcal{T}_\nu$, $\nu > n$. This means that each element $\Delta_k^{(n)}$ is subdivided to obtain new smaller elements $\{\Delta_l^{(\nu)}\}$. We refer to such subdivision schemes $\{\mathcal{T}_n\}$ as *nested decompositions* of D. With this, and with the interpolation based on using a common reference domain σ, as in (6.3.91) and (6.3.92), we can prove the identities (6.3.103) and (6.3.104). In fact, (6.3.103) then follows from (6.3.104). The method used in Chapter 5 produced a sequence of nested decompositions $\{\mathcal{T}_n\}$.

If we make the stronger assumption that the nodes are nested, that is $\{t_{m,i}\} \subset \{t_{n,i}\}$, it is immediate that

$$\mathcal{R}_{n,m}\mathcal{R}_{\infty,n} = \mathcal{R}_{\infty,m} \tag{6.3.105}$$

This is true of the nested decomposition scheme used in Chapter 5, but there are many useful ways of choosing collocation nodes for which the nodes are not nested.

6.3.2. The two-grid iteration method

Assume $\mathbf{v}_n^{(0)}$ is an initial guess at the solution of the collocation linear system $(\lambda - K_n)\mathbf{x}_n = \mathbf{y}_n$ of (6.3.87). Calculate the residual

$$\mathbf{r}^{(0)} = \mathbf{y}_n - (\lambda - K_n)\mathbf{v}_n^{(0)}$$

which is a vector in $\mathbf{R}^{d_n}$. Then the error $\mathbf{e}_n^{(0)} = \mathbf{x}_n - \mathbf{v}_n^{(0)}$ satisfies

$$(\lambda - K_n)\mathbf{e}_n^{(0)} = \mathbf{r}^{(0)} \tag{6.3.106}$$

$$\mathbf{x}_n = \mathbf{v}_n^{(0)} + (\lambda - K_n)^{-1}\mathbf{r}^{(0)}$$

This could be approximated using

$$(\lambda - K_n)^{-1}\mathbf{r}^{(0)} \approx \mathcal{P}_{m,n}(\lambda - K_m)^{-1}\mathcal{R}_{n,m}\mathbf{r}^{(0)} \tag{6.3.107}$$

based on using $(\lambda - K_m)^{-1}$ for some $m < n$. It would lead to an analog of method 1 in §6.2, which was not a well-behaved iteration method. Instead, we follow the type of development that led to method 2 in §6.2. In (6.3.107) the prolongation and restriction operators permit us to move between finite dimensional spaces of different dimensions.

Writing $\mathbf{e}_n^{(0)} = \mathbf{x}_n - \mathbf{v}_n^{(0)}$, we rewrite (6.3.106) as

$$\mathbf{e}_n^{(0)} = \frac{1}{\lambda}\left[\mathbf{r}^{(0)} + K_n\mathbf{e}_n^{(0)}\right]$$

Introduce $\delta^{(0)} = K_n\mathbf{e}_n^{(0)}$ as a new unknown, so that

$$\mathbf{e}_n^{(0)} = \frac{1}{\lambda}\left[\mathbf{r}^{(0)} + \delta^{(0)}\right] \tag{6.3.108}$$

Substituting into (6.3.106), $\delta^{(0)}$ satisfies

$$(\lambda - K_n)\delta^{(0)} = K_n\mathbf{r}^{(0)}$$

We approximate the solution using

$$\delta^{(0)} \approx \mathcal{P}_{m,n}(\lambda - K_m)^{-1}\mathcal{R}_{n,m}K_n\mathbf{r}^{(0)}$$

An improved estimate of the solution $\mathbf{x}_n$ is obtained from (6.3.108),

$$\mathbf{x}_n \approx \mathbf{v}^{(0)} + \frac{1}{\lambda}\left[\mathbf{r}^{(0)} + \mathcal{P}_{m,n}(\lambda - K_m)^{-1}\mathcal{R}_{n,m}K_n\mathbf{r}^{(0)}\right]$$

The complete algorithm is as follows.

1. Given $\mathbf{v}_n^{(l)}$, compute the residual

$$\mathbf{r}^{(l)} = \mathbf{y}_n - (\lambda - K_n)\mathbf{v}_n^{(l)} \tag{6.3.109}$$

2. Calculate $K_n\mathbf{r}^{(l)}$

3. Calculate the correction

$$\mathbf{c}_n^{(l)} = \mathcal{P}_{m,n} (\lambda - K_m)^{-1} \mathcal{R}_{n,m} K_n \mathbf{r}^{(l)} \qquad (6.3.110)$$

This amounts to:

(a) Restrict $K_n \mathbf{r}^{(l)}$ to the vector $\mathbf{w}_m^{(l)} = \mathcal{R}_{n,m} K_n \mathbf{r}^{(l)}$ in $\mathbf{R}^{d_m}$.

(b) Solve the coarse mesh linear system

$$(\lambda - K_m) \mathbf{c}_m^{(l)} = \mathbf{w}_m^{(l)}$$

(c) Use interpolation to calculate the prolongation

$$\mathbf{c}_n^{(l)} = \mathcal{P}_{m,n} \mathbf{c}_m^{(l)}$$

which lies in $\mathbf{R}^{d_n}$.

4. Finally, calculate the new iterate

$$\mathbf{v}_n^{(l+1)} = \mathbf{v}_n^{(l)} + \frac{1}{\lambda} \left[\mathbf{r}^{(l)} + \mathbf{c}_n^{(l)} \right] \qquad (6.3.111)$$

This is essentially the two-grid algorithm of Hackbusch [247]. However, he is interested primarily in the case that n is the immediate successor of m, and then he uses his two-grid method to construct a multigrid method. We discuss multigrid methods in §6.4.

To provide an error analysis for this iteration, first construct a formula

$$\mathbf{x}_n - \mathbf{v}_n^{(l+1)} = M_{m,n} \left[\mathbf{x}_n - \mathbf{v}_n^{(l)} \right] \qquad (6.3.112)$$

and then examine the size of $M_{m,n}$ as m and n increase. Initially, we move from an analysis in the finite dimensional space $\mathbf{R}^{d_n}$ to the function space $C(D)$. Let

$$v_n^{(l)} = \mathcal{P}_{n,\infty} \mathbf{v}_n^{(l)}$$

Recall the collocation equation (6.3.82) and its matrix equivalent (6.3.87). Their solutions satisfy

$$\mathbf{x}_n = \mathcal{R}_{\infty,n} x_n$$

Using the definition (6.3.109) and the result (6.3.99) on K_n,

$$
\begin{aligned}
\mathcal{P}_{n,\infty} \mathbf{r}^{(l)} &= \mathcal{P}_{n,\infty} \mathbf{y}_n - \mathcal{P}_{n,\infty} (\lambda - K_n) \mathbf{v}_n^{(l)} \\
&= \mathcal{P}_n y - \lambda \mathcal{P}_{n,\infty} \mathbf{v}_n^{(l)} + \mathcal{P}_{n,\infty} \mathcal{R}_{\infty,n} \mathcal{K} \mathcal{P}_{n,\infty} \mathbf{v}_n^{(l)} \\
&= \mathcal{P}_n y - (\lambda - \mathcal{P}_n \mathcal{K}) v_n^{(l)} \\
&= (\lambda - \mathcal{P}_n \mathcal{K}) \left[x_n - v_n^{(l)} \right]
\end{aligned}
$$

Then

$$K_n \mathbf{r}^{(l)} = \mathcal{R}_{\infty,n} \mathcal{K} \mathcal{P}_{n,\infty} \mathbf{r}^{(l)}$$

$$= \mathcal{R}_{\infty,n} \mathcal{K} (\lambda - \mathcal{P}_n \mathcal{K}) \left[x_n - v_n^{(l)} \right]$$

$$= \mathcal{R}_{\infty,n} (\lambda - \mathcal{K} \mathcal{P}_n) \mathcal{K} \left[x_n - v_n^{(l)} \right]$$

For the correction $\mathbf{c}_n^{(l)}$,

$$\mathbf{c}_n^{(l)} = \mathcal{P}_{m,n} (\lambda - K_m)^{-1} \mathcal{R}_{n,m} K_n \mathbf{r}^{(l)}$$

$$= \mathcal{P}_{m,n} (\lambda - K_m)^{-1} \mathcal{R}_{n,m} \mathcal{R}_{\infty,n} (\lambda - \mathcal{K} \mathcal{P}_n) \mathcal{K} \left[x_n - v_n^{(l)} \right]$$

Combining these results with (6.3.111),

$$x_n - v_n^{(l+1)} = x_n - v_n^{(l)} - \frac{1}{\lambda} \mathcal{P}_{n,\infty} \left[\mathbf{r}^{(l)} + \mathbf{c}_n^{(l)} \right]$$

$$= \mathcal{M}_{m,n} \left[x_n - v_n^{(l)} \right]$$

$$\mathcal{M}_{m,n} = \frac{1}{\lambda} \{ \mathcal{P}_n - \mathcal{P}_{n,\infty} \mathcal{P}_{m,n} (\lambda - K_m)^{-1} \mathcal{R}_{n,m} \mathcal{R}_{\infty,n} (\lambda - \mathcal{K} \mathcal{P}_n) \} \mathcal{K}$$

$$\equiv \frac{1}{\lambda} \mathcal{C}_{m,n} \mathcal{K} \tag{6.3.113}$$

Using this formula, and under various reasonable assumptions on the operators $\{\mathcal{R}_{n,m}\}$, $\{\mathcal{P}_{m,n}\}$, $\{\mathcal{R}_{\infty,n}\}$, $\{\mathcal{P}_{n,\infty}\}$, and $\{\mathcal{P}_n\}$, it can be shown that

$$\lim_{m \to \infty} \sup_{n > m} \| \mathcal{M}_{m,n} \| = 0 \tag{6.3.114}$$

A general proof can be based on showing pointwise convergence of $\{\mathcal{C}_{m,n}\}$ to 0 on $C(D)$, and then the compactness of $\mathcal{K}$ will imply that the convergence of $\{\mathcal{C}_{m,n} \mathcal{K}\}$ is uniform. We will prove the result (6.3.114) under additional assumptions to simplify the proof and to provide some additional intuition on the behavior of $\{\mathcal{M}_{m,n}\}$.

Theorem 6.3.1. *Assume the integral equation $(\lambda - \mathcal{K})x = y$ of (6.3.80) is uniquely solvable for all $y \in C(D)$, and let $\mathcal{K}$ be a compact operator on $C(D)$ to $C(D)$. Assume the region D is decomposed into finite elements, $\mathcal{T}_n = \{\Delta_k^{(n)}\}$, using a nested decomposition scheme, and consistent with this, assume the prolongation and restriction operators satisfy the equations (6.3.103), (6.3.104). In addition, assume the nodes are nested: for each m, n with $m < n$, we have $\{t_{m,i}\} \subset \{t_{n,i}\}$. This implies the restriction operators satisfy (6.3.101) and (6.3.105). Finally, assume the interpolating projections $\{\mathcal{P}_n\}$ are pointwise convergent on $C(D)$:*

$$\mathcal{P}_n x \to x \quad as \ n \to \infty, \ x \in C(D)$$

Then

$$\mathcal{M}_{m,n} = (\lambda - \mathcal{P}_m \mathcal{K})^{-1}(\mathcal{P}_n - \mathcal{P}_m)\mathcal{K} \qquad (6.3.115)$$

and (6.3.114) is true. If the coarse mesh parameter m is chosen sufficiently large, we have that the iteration scheme (6.3.109)–(6.3.111) is convergent, for all $n \geq m$, and the rate of linear convergence will be bounded by a constant less than 1, uniformly for $n \geq m$.

Proof. Under the assumptions of this theorem, we can invoke Theorem 3.1.1 and Lemma 3.1.2 to give a complete convergence and stability theory for our collocation method. In particular,

$$\lim_{n \to \infty} \|(I - \mathcal{P}_n)\mathcal{K}\| = 0 \qquad (6.3.116)$$

In addition, there is an integer N such that for $n \geq N$, the inverse $(\lambda - \mathcal{P}_n \mathcal{K})^{-1}$ exists, and

$$\sup_{n \geq N} \|(\lambda - \mathcal{P}_n \mathcal{K})^{-1}\| < \infty$$

Using the assumptions on the restriction and prolongation operators, we can simplify $\mathcal{M}_{m,n}$:

$$\mathcal{M}_{m,n} = \frac{1}{\lambda}\{\mathcal{P}_n - \mathcal{P}_{m,\infty}(\lambda - K_m)^{-1}\mathcal{R}_{\infty,m}(\lambda - \mathcal{K}\mathcal{P}_n)\}\mathcal{K} \qquad (6.3.117)$$

Using $K_n = \mathcal{R}_{\infty,n}\mathcal{K}\mathcal{P}_{n,\infty}$ from (6.3.99),

$$\mathcal{P}_{m,\infty}(\lambda - K_m)^{-1}\mathcal{R}_{\infty,m} = \mathcal{P}_{m,\infty}(\lambda - \mathcal{R}_{\infty,m}\mathcal{K}\mathcal{P}_{m,\infty})^{-1}\mathcal{R}_{\infty,m}$$

We now use an identity based on the arguments given in Lemma 3.4.1 from §3.4 of Chapter 3. In particular,

$$(\lambda - \mathcal{R}_{\infty,m}\mathcal{K}\mathcal{P}_{m,\infty})^{-1} = \frac{1}{\lambda}[I + \mathcal{R}_{\infty,m}(\lambda - \mathcal{K}\mathcal{P}_{m,\infty}\mathcal{R}_{\infty,m})^{-1}\mathcal{K}\mathcal{P}_{m,\infty}]$$
$$(6.3.118)$$

from which we prove

$$(\lambda - \mathcal{R}_{\infty,m}\mathcal{K}\mathcal{P}_{m,\infty})^{-1}\mathcal{R}_{\infty,m} = \mathcal{R}_{\infty,m}(\lambda - \mathcal{K}\mathcal{P}_{m,\infty}\mathcal{R}_{\infty,m})^{-1}$$
$$= \mathcal{R}_{\infty,m}(\lambda - \mathcal{K}\mathcal{P}_m)^{-1}$$

Returning to (6.3.117), we have

$$\mathcal{M}_{m,n} = \frac{1}{\lambda}\{\mathcal{P}_n - \mathcal{P}_{m,\infty}\mathcal{R}_{\infty,m}(\lambda - \mathcal{K}\mathcal{P}_m)^{-1}(\lambda - \mathcal{K}\mathcal{P}_n)\}\mathcal{K}$$

$$= \frac{1}{\lambda}\{\mathcal{P}_n - \mathcal{P}_m(\lambda - \mathcal{K}\mathcal{P}_m)^{-1}(\lambda - \mathcal{K}\mathcal{P}_n)\}\mathcal{K}$$

$$= \frac{1}{\lambda}\{\mathcal{P}_n - (\lambda - \mathcal{P}_m\mathcal{K})^{-1}\mathcal{P}_m(\lambda - \mathcal{K}\mathcal{P}_n)\}\mathcal{K}$$

The last step used (3.4.86) from §3.4 of Chapter 3. Continuing with the simplification,

$$\mathcal{M}_{m,n} = \frac{1}{\lambda}(\lambda - \mathcal{P}_m\mathcal{K})^{-1}\{(\lambda - \mathcal{P}_m\mathcal{K})\mathcal{P}_n - \mathcal{P}_m(\lambda - \mathcal{K}\mathcal{P}_n)\}\mathcal{K}$$

$$= \frac{1}{\lambda}(\lambda - \mathcal{P}_m\mathcal{K})^{-1}[\lambda\mathcal{P}_n - \lambda\mathcal{P}_m]\mathcal{K}$$

$$= (\lambda - \mathcal{P}_m\mathcal{K})^{-1}[\mathcal{P}_n - \mathcal{P}_m]\mathcal{K}$$

which proves (6.3.115).

Use (6.3.116) to prove

$$\lim_{n\to\infty}\mathcal{M}_{m,n} = (\lambda - \mathcal{P}_m\mathcal{K})^{-1}[I - \mathcal{P}_m]\mathcal{K} \qquad (6.3.119)$$

By choosing m sufficiently large, we will have

$$\sup_{\acute{n}>m}\|\mathcal{M}_{m,n}\| < 1$$

Using the error formula

$$x_n - v_n^{(l+1)} = \mathcal{M}_{m,n}[x_n - v_n^{(l)}], \quad l = 0, 1, 2, \ldots \qquad (6.3.120)$$

we will have convergence of $v_n^{(l)}$ to x_n as $l \to \infty$. The rate of convergence is essentially

$$\lim_{n\to\infty}\|\mathcal{M}_{m,n}\| = \|(\lambda - \mathcal{P}_m\mathcal{K})^{-1}[I - \mathcal{P}_m]\mathcal{K}\| < 1 \qquad (6.3.121)$$

for larger values of n, and this can be used to show that the rate of convergence is uniform in n.

Referring back to the vectors $\mathbf{x}_n$ and $\mathbf{v}_n^{(l)}$, we have

$$\mathbf{x}_n - \mathbf{v}_n^{(l)} = \mathcal{R}_{\infty,n}[x_n - v_n^{(l)}]$$

$$\|\mathbf{x}_n - \mathbf{v}_n^{(l)}\|_\infty \le \|x_n - v_n^{(l)}\|_\infty$$

This shows the convergence of $\mathbf{v}_n^{(l)} \to \mathbf{x}_n$. ∎

One can carry the above analysis further, to show

$$\mathbf{x}_n - \mathbf{v}_n^{(l+1)} = M_{m,n}\big[\mathbf{x}_n - \mathbf{v}_n^{(l)}\big] \qquad (6.3.122)$$

with

$$M_{m,n} = \mathcal{R}_{\infty,n}\mathcal{M}_{m,n}\mathcal{P}_{n,\infty} \qquad (6.3.123)$$

More directly, from the definition of the iteration in (6.3.109)–(6.3.111) we have

$$M_{m,n} = \frac{1}{\lambda}[I - \mathcal{P}_{m,n}(\lambda - K_m)^{-1}\mathcal{R}_{n,m}(\lambda - K_n)]K_n \qquad (6.3.124)$$

This is used in a later chapter when we look at the effect of perturbing the elements of the coefficient matrix $\lambda - K_n$ in the collocation method, and it is used §6.4 in the analysis of the multigrid iteration method.

We can examine the convergence of the iteration method (6.3.109)–(6.3.111) under less restrictive assumptions than those of Theorem 6.3.1. This is based on using the formula (6.3.113) for $\mathcal{M}_{m,n}$. As indicated preceding the theorem, one can show pointwise convergence of the family $\mathcal{C}_{m,n}$ and then note that the convergence of $\mathcal{C}_{m,n}\mathcal{K}$ will be uniform because of the compactness of $\mathcal{K}$. Examples of this are left as exercises for the reader.

Example. Let D be a bounded, simply connected region in $\mathbf{R}^3$, and let it have a smooth boundary S. One of the ways of solving the exterior Neumann problem for Laplace's equation on D leads to the integral equation

$$2\pi u(P) + \int_S u(Q)\frac{\partial}{\partial \nu_Q}\left[\frac{1}{|P-Q|}\right]dS_Q = \int_S \frac{f(Q)}{|P-Q|}dS_Q, \quad P \in S$$
$$(6.3.125)$$

with $f(P)$ the given normal derivative of the unknown u on the boundary S. The quantity ν_Q denotes the normal to S at Q, directed into D, and the operator $\partial/\partial \nu_Q$ denotes the normal derivative. This equation is examined in detail in Chapter 9. For our purposes here, it is sufficient to note that when S is smooth and bounded, the integral operator

$$\mathcal{K}u(P) = \int_S u(Q)\frac{\partial}{\partial \nu_Q}\left[\frac{1}{|P-Q|}\right]dS_Q, \quad P \in S, \ u \in C(S)$$

is a compact operator on $C(S)$ to $C(S)$.

We use the collocation method to solve (6.3.125), with the approximating subspace based on the use of piecewise quadratic interpolation as defined in §5.4 of Chapter 5. The details of this method are discussed at length in Chapter 9;

Table 6.6. *Solving (6.3.125) using collocation and iteration*

(a, b, c)	m	m_v	n	n_v	$\tilde{v}$
(2,2.5,3)	8	18	32	66	.17
			128	258	.14
			512	1026	.14
	32	66	128	258	.089
			512	1026	.072
	20	42	80	162	.11
			320	642	.089
	80	162	320	642	.062
(1,2,3)	8	18	32	66	.31[a]
			128	258	.25
			512	1026	.27
	32	66	128	258	.19
			512	1026	.18

[a]Obtained by using the geometric mean of a few of the final iterates computed.

here we merely consider the iterative solution of the linear system associated with the collocation method.

The surface used is the ellipsoid

$$\left(\frac{x}{a}\right)^2 + \left(\frac{y}{b}\right)^2 + \left(\frac{z}{c}\right)^2 = 1$$

with various choices for (a, b, c). The function f in (6.3.125) is so chosen that the true solution is

$$u(x, y, x) = \frac{1}{\sqrt{x^2 + y^2 + z^2}}$$

The numerical results are given in Table 6.6. The notation is the same as that used earlier in Tables 6.4 and 6.5. Recall that $\tilde{v}$ is the empirically determined rate of convergence,

$$\tilde{v} = \lim_{l \to \infty} \frac{\left\|v_n^{(l+1)} - v_n^{(l)}\right\|_\infty}{\left\|v_n^{(l)} - v_n^{(l-1)}\right\|_\infty}$$

In one case, these ratios were still varying for large values of l. We used the geometric mean of a few of the final iterates computed, and this entry is indicated in the table by a superscript "a". The order of the fine mesh linear system being solved is n_v, the number of node points in the discretization of the surface

S. It can be seen that for a fixed coarse mesh parameter m, the convergence ratios usually do not vary by much as the fine mesh parameter n increases. Also, as m increases, the convergence ratio $\bar{\nu}$ becomes smaller, indicating more rapid convergence. These results are consistent with the theoretical results in (6.3.120) and (6.3.121).

An alternative formulation

The two-grid iteration can be written in an alternative way that is closer to the discussion of the multigrid method given in §6.4.

1. Given $\mathbf{v}_n^{(l)}$, compute its Picard or Sloan iterate:

$$\hat{\mathbf{v}}_n^{(l)} = \frac{1}{\lambda}\left[\mathbf{y}_n + K_n\mathbf{v}_n^{(l)}\right] \tag{6.3.126}$$

   This is often referred to as "smoothing" the iterate $\mathbf{v}_n^{(l)}$. Referring back to the material on iterated projection methods in §3.4, such smoothing often results in a more accurate approximation.
2. Calculate the "defect" in the approximation $\hat{\mathbf{v}}_n^{(l)}$ and project it into the coarse mesh space:

$$\mathbf{d}^{(l)} = \mathcal{R}_{n,m}\left[\mathbf{y}_n - (\lambda - K_n)\hat{\mathbf{v}}_n^{(l)}\right] \tag{6.3.127}$$

3. Calculate the coarse mesh correction:

$$\hat{\mathbf{c}}^{(l)} = (\lambda - K_m)^{-1}\mathbf{d}^{(l)} \tag{6.3.128}$$

4. Calculate the improved value $\mathbf{v}_n^{(l+1)}$:

$$\mathbf{v}_n^{(l+1)} = \hat{\mathbf{v}}_n^{(l)} + \mathcal{P}_{m,n}\hat{\mathbf{c}}^{(l)} \tag{6.3.129}$$

We leave it to the reader to show that the $\mathbf{v}_n^{(l+1)}$ is exactly the same as that obtained in the two-grid iteration (6.3.109)–(6.3.111).

Operations count

We count the number of arithmetic operations in calculating one iterate of the two-grid method. We assume that K_n is known and is of order q_n. Further, we assume that something such as the LU-decomposition of the matrix $\lambda - K_m$ is known, so that the solution of the coarse grid equation (6.3.128) will involve approximately $2q_m^2$ operations.

1. Smoothing the iterate $\mathbf{v}_n^{(l)}$, as in (6.3.126). This costs approximately $2q_n^2$ operations.
2. Calculating the defect $\mathbf{d}^{(l)}$, as in (6.3.127). This costs approximately $2q_nq_m$ operations, since it is needed only on the coarse mesh.

3. Calculate the coarse mesh correction $\hat{\mathbf{c}}^{(l)}$. This costs approximately $2q_m^2$ operations.
4. Calculate the new iterate $\mathbf{v}_n^{(l+1)}$. This costs $O(q_n)$ operations.

Combining these costs, we have a total operations count of

$$\text{Cost per iterate } = 2\left(q_n^2 + q_n q_m + q_m^2\right) + O(q_n) \text{ operations} \qquad (6.3.130)$$

Thus the cost per iterate is about one-half that of the two-grid iteration method of §6.2 for the Nyström method as given in (6.2.78). Nonetheless, our experience with the use of the Nyström interpolation formula as a means to move between the coarse and fine grid equations has generally shown it to be superior in efficiency to the use of the prolongation and restriction operator framework of the two-grid method of the present section. When both the methods of §§6.2, 6.3 are applicable to a problem, we have generally found the Nyström method requires the calculation of significantly fewer iterates.

6.4. Multigrid iteration for collocation methods

The numerical solution of a differential equation or integral equation leads usually to a sequence of approximating linear systems

$$A_n \mathbf{x}_n = \mathbf{y}_n$$

for an increasing sequence $n = n_0, n_1, \ldots, n_\ell, \ldots$, of the discretization parameter n. With two-grid iteration, we solve the linear system iteratively for $n = n_\ell$ by assuming it can be solved exactly for some smaller n, say $n = n_0$, and implicitly we use $A_{n_0}^{-1}$ to construct an approximate inverse for A_{n_ℓ}. With multigrid methods, we use information from all preceding levels of the discretization, for $n = n_0, n_1, \ldots, n_{\ell-1}$. There is a large literature on multigrid methods, with the major application being the solution of discretizations of elliptic partial differential equations. For an introduction and extensive analysis of multigrid methods, see Hackbusch [248]. For integral equations of the second kind, we will follow closely the development in Hackbusch [249, §5.5].

Since we will be working on multiple levels of discretization, we modify the notation for prolongation and restriction operators that was introduced in §6.3. Assume we are using discretization parameters $n = n_0, n_1, \ldots, n_\ell, \ldots$, with

$$n_0 < n_1 < \cdots < n_\ell < \cdots$$

When working with the discretization for the parameter $n = n_\ell$, we will refer to this as working on the level ℓ discretization. For the prolongation and restriction operators, use

$$\mathcal{R}_{\ell,\ell-1} \equiv \mathcal{R}_{n_\ell,n_{\ell-1}} \qquad \mathcal{R}_{\infty,\ell} \equiv \mathcal{R}_{\infty,n_\ell}$$

$$\mathcal{P}_{\ell-1,\ell} \equiv \mathcal{P}_{n_{\ell-1},n_\ell} \qquad \mathcal{P}_{\ell,\infty} \equiv \mathcal{P}_{n_\ell,\infty}$$

$$\mathcal{P}_\ell \equiv \mathcal{P}_{n_\ell}$$

We have restricted these definitions to movement between immediately adjacent levels at levels ℓ and $\ell - 1$ since the definition of the multigrid method requires only such prolongation and restriction operators. The linear system (6.3.87) at level ℓ will be written as

$$(\lambda - K_\ell)\mathbf{x}_\ell = \mathbf{y}_\ell \tag{6.4.131}$$

with $K_\ell \equiv K_{n_\ell}$ and similarly for $\mathbf{x}_\ell$ and $\mathbf{y}_\ell$.

The multigrid method is based on using the two-grid method of §6.3. When solving the linear system at level ℓ, we use a modification of that two-grid method with the coarse mesh being that at level $\ell - 1$. Since the multigrid method is best understood as a recursive process, we introduce a recursive algorithm called *multigrid*, taken from Hackbusch [249, p. 198].

PROGRAM *multigrid*$(\ell, \mathbf{x}, \mathbf{y})$
REMARK: Solve the linear system $(\lambda - K_\ell)\mathbf{x}_\ell = \mathbf{y}_\ell$
INTEGER ℓ; ARRAY $\mathbf{x}, \mathbf{y}$
IF $\ell = 0$ THEN $\mathbf{x} := (\lambda - K_0)^{-1}\mathbf{y}$ ELSE
   BEGIN
      INTEGER i; ARRAY $\mathbf{d}, \mathbf{c}$
      $\mathbf{x} := (1/\lambda)[\mathbf{y} + K_\ell\mathbf{x}]$ [The "smoothing" of $\mathbf{x}$]
      $\mathbf{d} := \mathcal{R}_{\ell,\ell-1}[\mathbf{y} - (\lambda - K_\ell)\mathbf{x}]$ [Calculating the defect]
      $\mathbf{c} := \mathbf{0}$
      FOR $i = 1, 2$ DO *multigrid*$(\ell - 1, \mathbf{c}, \mathbf{d})$ [Calculating the correction]
      $\mathbf{x} := \mathbf{x} + \mathcal{P}_{\ell-1,\ell}\mathbf{c}$ [Calculating the improved value of $\mathbf{x}$]
   END
END PROGRAM

The value of *multigrid*$(0, \mathbf{x}, \mathbf{y})$ is obtained by solving $(\lambda - K_0)\mathbf{x}_0 = \mathbf{y}_0$ exactly. The value of *multigrid*$(1, \mathbf{x}, \mathbf{y})$ amounts to solving $(\lambda - K_1)\mathbf{x}_1 = \mathbf{y}_1$ with one iteration of the two-grid method (6.3.109)–(6.3.111) of §6.3, with an initial guess of $\mathbf{x}$, and with the coarse mesh approximation that of level $\ell = 0$. To see this, refer back to (6.3.126)–(6.3.129), the alternative formulation of the two-grid iteration. For this case it would make sense to call *multigrid*$(0, \mathbf{c}, \mathbf{d})$ only once, since the same value of $\mathbf{c}$ is returned with both recursive calls to the procedure.

For general $\ell > 0$, we will denote the multigrid iterates for level ℓ by $\mathbf{v}_\ell^{(k)}$, $k \geq 0$, which is consistent with the notation of the preceding section. In the computation of $\mathbf{v}_\ell^{(k+1)}$ from $\mathbf{v}_\ell^{(k)}$, all operations are linear. Consequently, we can write

$$\mathbf{v}_\ell^{(k+1)} = M_\ell \mathbf{v}_\ell^{(k)} + \mathbf{g}_\ell, \quad k \geq 0 \qquad (6.4.132)$$

Also, note that from the definition of *multigrid* that if $\mathbf{v}_\ell^{(0)} = \mathbf{x}_\ell$, the true solution of $(\lambda - K_\ell)\mathbf{x}_\ell = \mathbf{y}_\ell$, then also $\mathbf{v}_\ell^{(1)} = \mathbf{x}_\ell$. In the above algorithm *multigrid*, $\mathbf{v}_\ell^{(0)} = \mathbf{x}_\ell$ implies $\mathbf{d} = \mathbf{0}$, and this will lead to the calling of *multigrid*$(\ell - 1, \mathbf{0}, \mathbf{0})$. Continuing inductively, we will have this same result at each level, until arriving at level $\ell = 0$, and at this coarsest level, the homogeneous system will be solved exactly for a solution of $\mathbf{c} = \mathbf{0}$. Thus $\mathbf{x}_\ell$ is a fixed point of (6.4.132). Using this,

$$\begin{aligned} \mathbf{g}_\ell &= (I - M_\ell)\mathbf{x}_\ell \\ &= (I - M_\ell)(\lambda - K_\ell)^{-1}\mathbf{y}_\ell \\ &\equiv N_\ell \mathbf{y}_\ell \end{aligned} \qquad (6.4.133)$$

Also, from $\mathbf{x}_\ell$ being a fixed point of (6.4.132), we can subtract (6.4.132) from $\mathbf{x}_\ell = M_\ell \mathbf{x}_\ell + \mathbf{g}_\ell$ to get

$$\mathbf{x}_\ell - \mathbf{v}_\ell^{(k+1)} = M_\ell \left[\mathbf{x}_\ell - \mathbf{v}_\ell^{(k)} \right], \quad k \geq 0 \qquad (6.4.134)$$

By bounding $\|M_\ell\|$, we can bound the rate of convergence of the multigrid iteration. This formula corresponds to (6.3.122) for the two-grid iteration methods of §6.3.

In keeping with the recursive definition of *multigrid*, we derive a recursion formula for the multiplying operator M_ℓ. First, for $\ell = 1$,

$$M_1 = M_{0,1} \qquad (6.4.135)$$

with $M_{0,1}$ given by (6.3.113) for $n = n_1$ and $m = n_0$.

For $\ell > 1$, we simplify the algebra by taking $\mathbf{y}_\ell = \mathbf{0}$. Then at level ℓ, $\mathbf{g}_\ell = \mathbf{0}$. When applying the algorithm *multigrid*$(\ell, \mathbf{v}_\ell^{(0)}, \mathbf{0})$, we first calculate the smoothing step:

$$\hat{\mathbf{v}}_\ell^{(0)} = \frac{1}{\lambda} K_\ell \mathbf{v}_\ell^{(0)} \qquad (6.4.136)$$

Since $\mathbf{g}_\ell = \mathbf{0}$, the defect is

$$\begin{aligned} \mathbf{d} &= -\mathcal{R}_{\ell,\ell-1}(\lambda - K_\ell)\hat{\mathbf{v}}_\ell^{(0)} \\ &= -\frac{1}{\lambda}\mathcal{R}_{\ell,\ell-1}(\lambda - K_\ell)K_\ell \mathbf{v}_\ell^{(0)} \end{aligned}$$

Letting $\mathbf{c} = \mathbf{0}$, apply *multigrid*$(\ell, \mathbf{c}, \mathbf{d})$. Apply the iteration formula (6.4.132) at level $\ell - 1$. This yields the new value for $\mathbf{c}$ of

$$\mathbf{c}^{(1)} = M_{\ell-1}\mathbf{c} + \mathbf{g}_{\ell-1}$$
$$= \mathbf{g}_{\ell-1}$$
$$= N_{\ell-1}\mathbf{d}$$

with the last line based on (6.4.133). Apply *multigrid*$(\ell, \mathbf{c}^{(1)}, \mathbf{d})$. Then $\mathbf{c}^{(1)}$ is replaced by

$$\mathbf{c}^{(2)} = M_{\ell-1}\mathbf{c}^{(1)} + \mathbf{g}_{\ell-1}$$
$$= M_{\ell-1}N_{\ell-1}\mathbf{d} + N_{\ell-1}\mathbf{d}$$
$$= (I + M_{\ell-1})N_{\ell-1}\mathbf{d}$$

Then

$$\mathbf{v}_\ell^{(1)} = \hat{\mathbf{v}}_\ell^{(0)} + \mathcal{P}_{\ell-1,\ell}\mathbf{c}^{(2)}$$
$$= \hat{\mathbf{v}}_\ell^{(0)} + \mathcal{P}_{\ell-1,\ell}(I + M_{\ell-1})N_{\ell-1}\mathbf{d}$$
$$= \frac{1}{\lambda}K_\ell\mathbf{v}_\ell^{(0)} - \frac{1}{\lambda}\mathcal{P}_{\ell-1,\ell}(I + M_{\ell-1})N_{\ell,\ell-1}\mathcal{R}_{\ell,\ell-1}(\lambda - K_\ell)K_\ell\mathbf{v}_\ell^{(0)}$$
$$= \frac{1}{\lambda}[I - \mathcal{P}_{\ell-1,\ell}(I + M_{\ell-1})N_{\ell,\ell-1}\mathcal{R}_{\ell,\ell-1}(\lambda - K_\ell)]K_\ell\mathbf{v}_\ell^{(0)}$$

$$(6.4.137)$$

From (6.4.132),

$$\mathbf{v}_\ell^{(1)} = M_\ell\mathbf{v}_\ell^{(0)} + \mathbf{g}_\ell = M_\ell\mathbf{v}_\ell^{(0)}$$

Identifying with (6.4.137) and noting that $\mathbf{v}_\ell^{(0)}$ is arbitrary, we have

$$M_\ell = \frac{1}{\lambda}[I - \mathcal{P}_{\ell-1,\ell}(I + M_{\ell-1})N_{\ell-1}\mathcal{R}_{\ell,\ell-1}(\lambda - K_\ell)]K_\ell$$
$$= \frac{1}{\lambda}[I - \mathcal{P}_{\ell-1,\ell}(I + M_{\ell-1})(I - M_{\ell-1})$$
$$\times (\lambda - K_{\ell-1})^{-1}\mathcal{R}_{\ell,\ell-1}(\lambda - K_\ell)]K_\ell$$
$$= \frac{1}{\lambda}\left[I - \mathcal{P}_{\ell-1,\ell}\left(I - M_{\ell-1}^2\right)(\lambda - K_{\ell-1})^{-1}\mathcal{R}_{\ell,\ell-1}(\lambda - K_\ell)\right]K_\ell$$

$$(6.4.138)$$

From (6.3.124), recall the result

$$M_{\ell-1,\ell} = \frac{1}{\lambda}[I - \mathcal{P}_{\ell-1,\ell}(\lambda - K_{\ell-1})^{-1}\mathcal{R}_{\ell,\ell-1}(\lambda - K_\ell)]K_\ell \qquad (6.4.139)$$

for the convergence matrix in the two-grid method (6.3.109)–(6.3.111) [see (6.3.122)]. Using this in the above (6.4.138), we have

$$M_\ell = M_{\ell-1,\ell} + \frac{1}{\lambda} \mathcal{P}_{\ell-1,\ell} M_{\ell-1}^2 (\lambda - K_{\ell-1})^{-1} \mathcal{R}_{\ell,\ell-1} (\lambda - K_\ell) K_\ell, \quad \ell > 1$$

$$(6.4.140)$$

Recalling (6.4.135), we also have $M_1 = M_{0,1}$.

Recall the identity (6.3.103),

$$\mathcal{R}_{\ell,\ell-1} \mathcal{P}_{\ell-1,\ell} = I \tag{6.4.141}$$

which is true for most families of prolongation and restriction operators. Using it, (6.4.140) takes the alternative form

$$M_\ell = M_{\ell-1,\ell} + \mathcal{P}_{\ell-1,\ell} M_{\ell-1}^2 \mathcal{R}_{\ell,\ell-1} \left[\frac{1}{\lambda} K_\ell - M_{\ell-1,\ell} \right], \quad \ell > 1$$

$$(6.4.142)$$

To see this, use (6.4.139) and the identity (6.4.141) to obtain

$$\mathcal{R}_{\ell,\ell-1} \left[\frac{1}{\lambda} K_\ell - M_{\ell-1,\ell} \right]$$

$$= \frac{1}{\lambda} \mathcal{R}_{\ell,\ell-1} [\mathcal{P}_{\ell-1,\ell} (\lambda - K_{\ell-1})^{-1} \mathcal{R}_{\ell,\ell-1} (\lambda - K_\ell) K_\ell]$$

$$= \frac{1}{\lambda} (\lambda - K_{\ell-1})^{-1} \mathcal{R}_{\ell,\ell-1} (\lambda - K_\ell) K_\ell$$

Using this in (6.4.140), we have (6.4.142). The formula (6.4.142) is used below to give a convergence analysis for the multigrid method.

We need some additional assumptions on the discretization method to carry out a convergence analysis. Associated with each level ℓ and the parameter n_ℓ, there is a mesh size

$$h_\ell = \phi(n_\ell), \quad \ell \geq 0$$

for some strictly decreasing function ϕ with $\phi(\infty) = 0$. For integral equations with an integration domain of dimension $d \geq 1$, we usually have $h_\ell = O(n_\ell^{-1/d})$. We assume

$$h_{\ell-1} \leq c_h h_\ell, \quad \ell \geq 0 \tag{6.4.143}$$

for some $c_h > 1$.

For the matrices $M_{\ell-1,\ell}$, we assume

$$\|M_{\ell-1,\ell}\| \leq c_{\text{TG}} h_\ell^q \tag{6.4.144}$$

for some $q > 0$ and some constant $c_{\mathrm{TG}} > 0$ (with the subscript "TG" used to denote "two-grid"). Note that all operator and matrix norms that we use are based on using the uniform norm $\| \cdot \|_\infty$ on the associated vector spaces. Under the assumptions of Theorem 6.3.1, for example, such a bound (6.4.144) can be obtained by using the formulas (6.3.115) and (6.3.123). Obtaining such a bound (6.4.144) is discussed at greater length in Ref. [249, §5.4]. We also assume that the following are finite:

$$c_P = \sup_{\ell \geq 1} \|\mathcal{P}_{\ell-1,\ell}\| \tag{6.4.145}$$

$$c_R = \sup_{\ell \geq 1} \|\mathcal{R}_{\ell,\ell-1}\| \tag{6.4.146}$$

$$c_K = \sup_{\ell \geq 0} \|K_\ell\| \tag{6.4.147}$$

Usually, $c_R = 1$ and $\|K_\ell\| \to \|\mathcal{K}\|$.

Theorem 6.4.1. *Assume the integral equation* $(\lambda - \mathcal{K})x = y$ *of (6.3.80) is uniquely solvable for all* $y \in C(D)$, *and let* $\mathcal{K}$ *be a compact operator on* $C(D)$ *to* $C(D)$. *Assume the two-grid convergence operators* $M_{\ell-1,\ell}$ *satisfy (6.4.144), and also assume the bounds (6.4.145)–(6.4.147) are finite. Then if* $h_0 < 1$ *is chosen sufficiently small, the multigrid operators* M_ℓ *satisfy*

$$\|M_\ell\| \leq \left(c_{\mathrm{TG}} + c^* h_\ell^q\right) h_\ell^q, \quad \ell \geq 1 \tag{6.4.148}$$

for a suitable constant $c^* > 0$. *In addition, for* h_0 *sufficiently small and* ℓ *sufficiently large, the multigrid iteration will converge, with* $\mathbf{v}_\ell^{(k)} \to \mathbf{x}_\ell$.

Proof. This is proven by induction. For $\ell = 1$,

$$\|M_\ell\| = \|M_{\ell-1,\ell}\| \leq c_{\mathrm{TG}} h_1^q$$

and the result (6.4.148) is true.

Assume the result (6.4.148) is true for $\ell = k - 1$, and prove it is then true for $\ell = k$. We use (6.4.142) to bound $\|M_\ell\|$, which assumes the relation (6.4.141), but a similar proof can be based on (6.4.140). From (6.4.142),

$$\|M_k\| \leq \|M_{k-1,k}\| + \|\mathcal{P}_{k-1,k}\| \|M_{k-1}\|^2 \|\mathcal{R}_{k,k-1}\| \left[\frac{1}{|\lambda|} \|K_k\| + \|M_{k-1,k}\|\right]$$

$$\leq c_{\mathrm{TG}} h_k^q + c_P c_R \|M_{k-1}\|^2 \left[\frac{c_K}{|\lambda|} + c_{\mathrm{TG}} h_k^q\right]$$

Apply the induction hypothesis that (6.4.148) is true for $\ell = k - 1$ and also use

(6.4.143). Then

$$
\begin{aligned}
\|M_k\| &\le c_{\mathrm{TG}} h_k^q + c_P c_R \left[\frac{c_K}{|\lambda|} + c_{\mathrm{TG}}\, h_k^q \right] \left[(c_{\mathrm{TG}} + c^* h_{k-1}^q) h_{k-1}^q \right]^2 \\
&\le c_{\mathrm{TG}} h_k^q + c_P c_R \left[\frac{c_K}{|\lambda|} + c_{\mathrm{TG}}\, c_h^q h_k^q \right] \left(c_{\mathrm{TG}} + c^* c_h^q h_k^q \right)^2 c_h^{2q} h_k^{2q} \\
&= \left\{ c_{\mathrm{TG}} + c_P c_R \left[\frac{c_K}{|\lambda|} + c_{\mathrm{TG}} c_h^q h_k^q \right] \left(c_{\mathrm{TG}} + c^* c_h^q h_k^q \right)^2 c_h^{2q} h_k^q \right\} h_k^q
\end{aligned}
$$

To have (6.4.148) be true for $\ell = k$, we must choose c^* so that

$$
c_P c_R \left[\frac{c_K}{|\lambda|} + c_{\mathrm{TG}} c_h^q h_k^q \right] \left(c_{\mathrm{TG}} + c^* c_h^q h_k^q \right)^2 c_h^{2q} \le c^*, \quad k \ge 1
$$

The left side is largest when $k = 1$, and thus we must choose c^* to satisfy

$$
c_P c_R \left[\frac{c_K}{|\lambda|} + c_{\mathrm{TG}} c_h^q h_1^q \right] \left(c_{\mathrm{TG}} + c^* c_h^q h_1^q \right)^2 c_h^{2q} \le c^* \tag{6.4.149}
$$

It is straightforward to show that there is such a c^* provided h_1 (or equivalently h_0) is sufficiently small. Since the left side converges to

$$
c_P c_R \frac{c_K}{|\lambda|} c_{\mathrm{TG}}^2 c_h^{2q}
$$

as $h_1 \to 0$, we can choose

$$
c^* = c_P c_R \frac{c_K}{|\lambda|} c_{\mathrm{TG}}^2 c_h^{2q} + 1
$$

The inequality (6.4.149) will then be true if h_1 is chosen sufficiently small.

When combined with (6.4.134), the bound (6.4.148) implies the convergence of the multigrid iteration. This completes the proof.   □

Compare (6.4.148) and (6.4.144). The result (6.4.148) says that the speed of convergence of the multigrid method is essentially the same as that of the two-grid method, which is based on solving iteratively at level ℓ by using the exact solvability of the discretization at level $\ell - 1$. As ℓ increases,

$$
\lim_{\ell \to \infty} \|M_\ell\| = 0 \tag{6.4.150}
$$

From (6.4.134),

$$
\left\| \mathbf{x}_\ell - \mathbf{v}_\ell^{(k+1)} \right\|_\infty \le \|M_\ell\| \left\| \mathbf{x}_\ell - \mathbf{v}_\ell^{(k)} \right\|_\infty, \quad k \ge 0 \tag{6.4.151}
$$

Thus we have convergence of $\mathbf{v}_\ell^{(k)} \to \mathbf{x}_\ell$ when $\|M_\ell\| < 1$, which is true for all sufficiently large ℓ.

6.4.1. Operations count

Before doing the operations count, we note that some savings can be made in the algorithm *multigrid*. When calling *multigrid*(0, **x**, **y**), only do so once since it always will return **x** $=(\lambda - K_0)^{-1}\mathbf{y}$. For each ℓ, note that when calling *multigrid*(ℓ − 1, **c**, **d**) from inside of *multigrid*(ℓ, **x**, **y**), the first call is with **c** = **0**. Therefore, the smoothing step inside of *multigrid*(ℓ − 1, **c**, **d**) will be greatly simplified, as the matrix-vector multiplication $K_{\ell-1}\mathbf{c}$ will not need to be calculated. This is implemented in the modified algorithm of Ref. [249, p. 207].

We will make the following assumptions regarding the number of arithmetic operations used in various parts of the iteration. Let $q_l \equiv q_{n_\ell}$, $\ell \geq 0$, be the order of the linear system being solved at level ℓ. Assume the solution of $(\lambda - K_0)\mathbf{x} = \mathbf{y}$ costs $2q_0^2$ operations to solve. This will be true if the LU-decomposition of $\lambda - K_0$ has been calculated previously and stored for later use. Assume the orders q_ℓ satisfy

$$q_{\ell-1} \leq c_N q_\ell, \quad \ell \geq 1 \tag{6.4.152}$$

Generally, the mesh size h_ℓ decreases by a factor of $1/2$ when ℓ increases by 1. If the integral equation being solved is defined on a region of dimension $d \geq 1$, then this rate of decrease in the mesh size h_ℓ implies

$$c_N = 2^{-d} \tag{6.4.153}$$

Assume that inside *multigrid*(ℓ, **x**, **y**) at level $\ell \geq 1$, the cost of one restriction operation (using $\mathcal{R}_{\ell,\ell-1}$), one prolongation operation (using $\mathcal{P}_{\ell-1,\ell}$), and the other operations of size $O(n_\ell)$ (such as multiplying by $\frac{1}{\lambda}$) is collectively $c'q_\ell$.

Theorem 6.4.2. *Assume (6.4.152) with*

$$c_N < \frac{1}{2} \tag{6.4.154}$$

Then the cost in arithmetic operations of multigrid(ℓ, 0, y) is bounded by

$$c_0 q_\ell^2 + c_2 q_\ell \tag{6.4.155}$$

and the cost of multigrid(ℓ, x, y) is bounded by

$$c_1 q_\ell^2 + c_2 q_\ell \tag{6.4.156}$$

with

$$c_0 = 2c_N \frac{1+c_N}{1-2c_N^2}, \quad c_1 = 2\frac{1+c_N-c_N^2}{1-2c_N^2}, \quad c_2 = \frac{c'}{1-2c_N} \tag{6.4.157}$$

Proof. The proof is by induction. For $\ell = 0$, *multigrid*$(0, \mathbf{x}, \mathbf{y})$ has a cost of $2q_0^2$ arithmetic operations, as assumed in the discussion preceding the theorem. Thus, we must require $c_0, c_1 \geq 2$.

Assume the results (6.4.155) and (6.4.156) for $\ell = k - 1$, and then prove it for $\ell = k$. First consider calling *multigrid*$(k, 0, \mathbf{y})$. The numbers of operations in the algorithm are as follows:

1. The smoothing step involves only $O(q_k)$ operations.
2. The calculation of the residual $\mathcal{R}_{k,k-1} [\mathbf{y} - (\lambda - K_k)\mathbf{x}]$ on the coarser mesh uses

$$2q_k q_{k-1} + O(q_k) \leq 2c_N q_k^2 + O(q_k)$$

   operations.
3. The combination of the two calls on *multigrid*$(k - 1, \mathbf{c}, \mathbf{y})$ has a bound on the number of operations of

$$\left[c_0 q_{k-1}^2 + c_2 q_{k-1} \right] + \left[c_1 q_{k-1}^2 + c_2 q_{k-1} \right] \leq c_N^2 (c_0 + c_1) q_k^2 + 2c_2 c_N q_k$$

4. The calculation of the new iterate in the step $\mathbf{x} := \mathbf{x} + \mathcal{P}_{k-1,k} \mathbf{c}$ requires $O(q_k)$ operations.

Thus the total number of operations is bounded by

$$\left[2c_N + c_N^2 (c_0 + c_1) \right] q_k^2 + (2c_2 c_N + c') q_k$$

Next, consider calling *multigrid*$(k, \mathbf{x}, \mathbf{y})$ for a general $\mathbf{x}$. The smoothing step now requires $2q_k^2 + O(q_k)$ operations, and the remaining numbers of operations are as above for *multigrid*$(k, 0, \mathbf{y})$. Thus the total number of operations for *multigrid*$(k, \mathbf{x}, \mathbf{y})$ is bounded by

$$2q_k^2 + \left[2c_N + c_N^2 (c_0 + c_1) \right] q_k^2 + (2c_2 c_N + c') q_k$$

To prove the induction step, we need to satisfy the inequalities

$$2c_N + c_N^2 (c_0 + c_1) \leq c_0$$

$$2 + 2c_N + c_N^2 (c_0 + c_1) \leq c_1$$

$$2c_2 c_N + c' \leq c_2$$

These will be satisfied with the quantities given in (6.4.157). $\qquad\square$

Consider one-variable problems, in which the domain of integration is an interval $[a, b]$. The mesh size h_ℓ is generally proportional to n_ℓ^{-1}, and the condition (6.4.154) requires the mesh size h_ℓ to satisfy

$$h_{\ell+1} \le c_N h_\ell$$

with $c_N < 1/2$. This a faster rate of decrease than is normally used for such problems, and according to Theorem 6.4.1, it is not necessary for convergence of the multigrid iteration. By using a slightly different argument when $c_N = 1/2$, we can obtain an operations count of

$$c_1 q_\ell^2 + O\left(q_\ell^{1+\epsilon}\right)$$

for *multigrid*$(\ell, \mathbf{x}, \mathbf{y})$. This will be true if $c_N < 1/\sqrt{2}$, and ϵ is a given positive real number that can be chosen arbitrarily small. Thus, the operations count will still be essentially $c_1 q_\ell^2$.

For problems in two-dimensions, using $c_N = 1/4$ [following (6.4.153)], we have

$$c_0 = \frac{5}{7}, \qquad c_1 = \frac{19}{7}$$

When solving an integral equation over a surface in three dimensions, the cost of the direct solution of $(\lambda - K_\ell)\mathbf{x}_\ell = \mathbf{y}_\ell$ by Gaussian elimination is approximately $(2/3)q_\ell^3$ arithmetic operations. In contrast, the cost of one call to *multigrid*$(\ell, \mathbf{x}, \mathbf{y})$ is

$$\frac{19}{7}q_\ell^2 + O(q_\ell) \text{ operations}$$

Also, because $\|M_\ell\| \to 0$ as $\ell \to \infty$, fewer iterations are needed as ℓ increases, if a good initial guess is chosen when beginning the iteration. When solving $(\lambda - K_\ell)\mathbf{x}_\ell = \mathbf{y}_\ell$, we generally use $\mathbf{x}_\ell^{(0)} = \mathcal{P}_{\ell-1,\ell}\mathbf{x}_{\ell-1}$. Then no more than one or two iterations are generally needed for larger values of ℓ.

The property that $\|M_\ell\| \to 0$ as $\ell \to \infty$ has been used to recommend multigrid iteration over two-grid iteration (as defined in §§6.2, 6.3). However, in practice two-grid iteration can also be used with a good initial guess. And by choosing $\|M_{m,n}\|$ sufficiently small, we can also get by with calculating only one or two iterations when solving each fine grid discretization of the original integral equation. This is done in the automatic programs of Ref. [37], in which $\|M_{m,n}\|$ is so chosen that only two iterations are needed for each value of the fine grid parameter n. These programs were discussed at the end of §6.2. With either program the total operations count for solving the integral equation to a desired accuracy $\delta > 0$ is $O(q_n^2)$, where q_n is the order of the final linear system being solved in the program.

6.5. The conjugate gradient method

The iteration methods of earlier sections are tied closely to the numerical solution of integral equations of the second kind. In this section we apply a general iteration method for solving simultaneous linear systems, one called the conjugate gradient method, to the solution of discretizations of integral equations of the second kind. The classical conjugate gradient method is restricted to solving linear systems $Av = b$ with A symmetric and positive definite. But recently, many generalizations have been given for solving linear systems that are not symmetric and positive definite. We first discuss the application of the classical conjugate gradient method, and later we discuss some generalizations to nonsymmetric linear systems. For an introduction to the conjugate gradient method, see Ref. [48, §8.9], Golub and Van Loan [224, §10.2], and Luenberger [349, Chap. 8].

We begin the section with a discussion of the direct application of the conjugate gradient method to solving integral equations, including an analysis of the speed of convergence. Next, we discuss the speed of convergence of the conjugate gradient method applied to the discretizations of the integral equation, where the discretization results in a symmetric linear system. We conclude the section with a discussion of one generalization for solving nonsymmetric problems.

6.5.1. *The conjugate gradient method for the undiscretized integral equation*

Consider the integral equation

$$x(t) - \int_D K(t,s)x(s)\,ds = y(t), \quad t \in D \tag{6.5.158}$$

and regard it as an equation in the space $L^2(D)$. Assume the kernel function is symmetric,

$$K(t,s) \equiv K(s,t), \quad s,t \in D$$

which implies the associated integral operator $\mathcal{K}$ is self-adjoint (note: we are assuming all functions are real). Introduce $\mathcal{A} \equiv I - \mathcal{K}$ and assume $\mathcal{A}$ is positive definite.

The conjugate gradient method for solving $\mathcal{A}x = y$ in $L^2(D)$ is defined as follows. Let x_0 be an initial guess for the solution $x^* = \mathcal{A}^{-1}y$. Define

$r_0 = y - \mathcal{A}x_0$ and $s_0 = r_0$. For $k \geq 0$, define

$$
\begin{aligned}
x_{k+1} &= x_k + \alpha_k s_k & \alpha_k &= \frac{\|r_k\|^2}{(\mathcal{A}s_k, s_k)} \\
r_{k+1} &= y - \mathcal{A}x_{k+1} & & \\
s_{k+1} &= r_{k+1} + \beta_k s_k & \beta_k &= \frac{\|r_{k+1}\|^2}{\|r_k\|^2}
\end{aligned}
\tag{6.5.159}
$$

The norm and inner product are those of $L^2(D)$. It can be shown that either $r_k = 0$ for some finite k (and thus $x_k = x^*$) or $\lim_{k \to \infty} r_k = 0$ (see [412, p. 159]). Here we consider only some results on the rate of convergence of x_k to x.

The discussion of the convergence requires results on the eigenvalues of $\mathcal{K}$. From Theorem 1.4.3 in §1.4 of Chapter 1, the eigenvalues of the symmetric integral operator $\mathcal{K}$ are real and the associated eigenfunctions form an orthonormal basis for $L^2(D)$:

$$
\mathcal{K}\phi_j = \lambda_j \phi_j, \quad j = 1, 2, \ldots
$$

with $(\phi_i, \phi_j) = \delta_{i,j}$. Without any loss of generality, let the eigenvalues be ordered as follows:

$$
|\lambda_1| \geq |\lambda_2| \geq |\lambda_3| \geq \cdots \geq 0
\tag{6.5.160}
$$

From Theorem 1.4.1 of §1.4 in Chapter 1,

$$
\lim_{j \to \infty} \lambda_j = 0
$$

The eigenvalues of $\mathcal{A}$ are $\{1 - \lambda_j\}$ with $\{\phi_j\}$ as the corresponding orthogonal eigenfunctions. The symmetric operator $\mathcal{A} = I - \mathcal{K}$ is positive definite if and only if

$$
\delta \equiv \inf_{j \geq 1}(1 - \lambda_j) = 1 - \sup_{j \geq 1} \lambda_j > 0
\tag{6.5.161}
$$

or equivalently, $\lambda_j < 1$ for all $j \geq 1$. For later use, also introduce

$$
\Delta \equiv \sup_{j \geq 1}(1 - \lambda_j)
$$

and we note that

$$
\|\mathcal{A}\| = \Delta, \qquad \|\mathcal{A}^{-1}\| = \frac{1}{\delta}
\tag{6.5.162}
$$

For error analyses of conjugate gradient methods, we often use the equivalent norm

$$
\|x\|_A = \sqrt{(\mathcal{A}x, x)}
$$

Easily,

$$\delta \|x\| \leq \|x\|_A \leq \Delta \|x\|, \quad x \in L^2(D) \tag{6.5.163}$$

From Patterson [412, p. 159], we have the result

$$\|x^* - x_{k+1}\|_A \leq \left(\frac{\Delta - \delta}{\Delta + \delta}\right) \|x^* - x_k\|_A, \quad k \geq 0 \tag{6.5.164}$$

This is valid for all bounded symmetric positive definite operators $\mathcal{A}$ that have a bounded inverse, with $\|\mathcal{A}\| = \Delta$ and $\|\mathcal{A}^{-1}\| = \delta^{-1}$ for some $\delta > 0$. The proof of (6.5.164) is similar to that used for corresponding finite dimensional results, and we omit it here.

It is possible to improve on this geometric rate of convergence when dealing with integral equations (6.5.158), to show a superlinear rate of convergence. The following result is due to Winther [573].

Theorem 6.5.1. *Let $\mathcal{K}$ be a self-adjoint compact operator on the Hilbert space $\mathcal{H}$. Assume $\mathcal{A} = I - \mathcal{K}$ is a symmetric positive definite operator (with the notation used above). Let $\{x_k\}$ be generated by the conjugate gradient iteration (6.5.159). Then $x_k \to x^*$ superlinearly:*

$$\|x^* - x_k\| \leq (c_k)^k \|x^* - x_0\|, \quad k \geq 0 \tag{6.5.165}$$

with $\lim_{k\to\infty} c_k = 0$.

Proof. It is a standard result that (6.5.159) implies that

$$x_k = x_0 + \bar{P}_{k-1}(\mathcal{A})r_0$$

with $\bar{P}_{k-1}(\lambda)$ a polynomial of degree $\leq k - 1$. Letting $\mathcal{A} = I - \mathcal{K}$, this can be rewritten in the equivalent form

$$x_k = x_0 + \hat{P}_{k-1}(\mathcal{K})r_0$$

for some other polynomial $\hat{P}_{k-1}(\lambda)$ of degree $\leq k - 1$. The conjugate gradient iterates satisfy an optimality property: if $\{y_k\}$ is another sequence of iterates, generated by another sequence of the form

$$y_k = y_0 + P_{k-1}(\mathcal{K})r_0, \quad k \geq 1, \quad y_0 = x_0 \tag{6.5.166}$$

for some sequence of polynomials $\{P_{k-1} : \deg(P_{k-1}) \leq k - 1, \ k \geq 1\}$, then

$$\|x^* - x_k\|_A \leq \|x^* - y_k\|_A, \quad k \geq 0 \tag{6.5.167}$$

For a proof, see Luenberger [349, p. 246].

Introduce

$$Q_k(\lambda) = \prod_{j=1}^{k} \frac{\lambda - \lambda_j}{1 - \lambda_j} \tag{6.5.168}$$

and note that $Q_k(1) = 1$. Define P_{k-1} implicitly by

$$Q_k(\lambda) = 1 - (1 - \lambda)P_{k-1}(\lambda)$$

and note that $\text{degree}(P_{k-1}) = k - 1$. Let $\{y_k\}$ be defined using (6.5.166). Define $\bar{e}_k = x^* - y_k$ and

$$\bar{r}_k = b - \mathcal{A}y_k = \mathcal{A}\bar{e}_k$$

We first bound $\bar{r}_k$, and then

$$\|\bar{e}_k\| \leq \|\mathcal{A}^{-1}\| \, \|\bar{r}_k\| = \frac{1}{\delta}\|\bar{r}_k\|$$

Moreover,

$$\|x^* - x_k\|_A \leq \|x^* - y_k\|_A$$
$$\leq \Delta\|x^* - y_k\|$$
$$\leq \frac{\Delta}{\delta}\|\bar{r}_k\|$$

$$\|x^* - x_k\| \leq \frac{1}{\delta}\|x^* - x_k\|_A \leq \frac{\Delta}{\delta^2}\|\bar{r}_k\| \tag{6.5.169}$$

From (6.5.166),

$$\bar{r}_k = b - \mathcal{A}[y_0 + P_{k-1}(\mathcal{K})r_0]$$
$$= [I - \mathcal{A}P_{k-1}(\mathcal{K})]r_0$$
$$= Q_k(\mathcal{A})r_0 \tag{6.5.170}$$

Expand r_0 using the eigenfunction basis $\{\phi_1, \phi_2, \dots\}$:

$$r_0 = \sum_{j=1}^{\infty} (r_0, \phi_j)\phi_j$$

Note that

$$Q_k(\mathcal{A})\phi_j = Q_k(\lambda_j)\phi_j, \quad j \geq 1$$

and thus $Q_k(\mathcal{A})\phi_j = 0$ for $j = 1, \dots, k$. Then (6.5.170) implies

$$\bar{r}_k = \sum_{j=1}^{\infty} (r_0, \phi_j)Q_k(\mathcal{A})\phi_j$$
$$= \sum_{j=k+1}^{\infty} (r_0, \phi_j)Q_k(\lambda_j)\phi_j$$

and

$$\|\tilde{r}_k\| \le \alpha_k \sqrt{\sum_{j=k+1}^{\infty} (r_0, \phi_j)^2} \le \alpha_k \|r_0\| \qquad (6.5.171)$$

with

$$\alpha_k = \sup_{j \ge k+1} |Q_k(\lambda_j)|$$

Examining $Q_k(\lambda_j)$ and using (6.5.160), we have

$$\alpha_k \le \prod_{j=1}^{k} \frac{|\lambda_{k+1}| + |\lambda_j|}{1 - \lambda_j}$$

$$\le \prod_{j=1}^{k} \frac{2|\lambda_j|}{1 - \lambda_j} \qquad (6.5.172)$$

Recall the well-known inequality

$$\left[\prod_{j=1}^{k} b_j \right]^{\frac{1}{k}} \le \frac{1}{k} \sum_{j=1}^{k} b_j$$

which relates the arithmetic and geometric means of k positive numbers $b_1, \ldots, b_k$. Applying this, we have

$$\alpha_k \le \left[\frac{2}{k} \sum_{j=1}^{k} \frac{|\lambda_j|}{1 - \lambda_j} \right]^k$$

Since $\lim_{j \to \infty} \lambda_j = 0$, it is a straightforward argument to show that

$$\lim_{k \to \infty} \frac{2}{k} \sum_{j=1}^{k} \frac{|\lambda_j|}{1 - \lambda_j} = 0 \qquad (6.5.173)$$

We leave the proof to the reader.

Returning to (6.5.171) and (6.5.169), we have

$$\|x^* - x_k\| \le \frac{\Delta}{\delta^2} \|\tilde{r}_k\|$$

$$\le \frac{\Delta}{\delta^2} \alpha_k \|r_0\|$$

$$\le \frac{\Delta^2}{\delta^2} \alpha_k \|x^* - x_0\|$$

To obtain (6.5.165), define

$$c_k = \left(\frac{\Delta}{\delta}\right)^{\frac{2}{k}} \left[\frac{2}{k} \sum_{j=1}^{k} \frac{|\lambda_j|}{1 - \lambda_j}\right] \tag{6.5.174}$$

From (6.5.173), $c_k \to 0$ as $k \to \infty$. $\square$

Bounds on c_k

It is of interest to know how rapidly c_k converges to zero and to know how this depends on the smoothness of the kernel function $K(t, s)$. This was examined in Flores [201], [202] and we give some of those results here. The speed of convergence of c_k is essentially the same as that of

$$\tau_k \equiv \frac{1}{k} \sum_{j=1}^{k} \frac{|\lambda_j|}{1 - \lambda_j} \tag{6.5.175}$$

In turn, the convergence of τ_k depends on the rate at which the eigenvalues λ_j converge to zero. We summarize some of the results of Flores [202] in the following theorem. In all cases, we also assume the operator $\mathcal{A} = I - \mathcal{K}$ is positive definite, which is equivalent to the assumption (6.5.161).

Theorem 6.5.2.

(a) *Assume the integral operator $\mathcal{K}$ of (6.5.158) is a self-adjoint Hilbert-Schmidt integral operator, that is,*

$$\|\mathcal{K}\|_{HS}^2 \equiv \int_D \int_D |K(t, s)|^2 \, ds \, dt < \infty$$

Then

$$\frac{1}{k} \cdot \frac{|\lambda_1|}{1 - \lambda_1} \le \tau_k \le \frac{1}{\sqrt{k}} \|\mathcal{K}\|_{HS} \|(I - \mathcal{K})^{-1}\| \tag{6.5.176}$$

(b) *Assume $K(t, s)$ is a symmetric kernel with continuous partial derivatives of order up to p for some $p \ge 1$. Then there is a constant $M \equiv M(p)$ with*

$$\tau_k \le \frac{M}{k} \zeta\left(p + \frac{1}{2}\right) \|(I - \mathcal{K})^{-1}\|, \quad k \ge 1 \tag{6.5.177}$$

with $\zeta(z)$ the Riemann zeta function.

Proof.

(a) It can be proven from Theorem 1.4.3 of §1.4 that

$$\sum_{j=1}^{\infty} \lambda_j^2 = \|\mathcal{K}\|_{HS}^2$$

From this, the eigenvalues λ_j can be shown to converge to zero with a certain speed. Namely,

$$j\lambda_j^2 \le \sum_{i=1}^{j} \lambda_i^2 \le \|\mathcal{K}\|_{HS}^2$$

$$\lambda_j \le \frac{1}{\sqrt{j}} \|\mathcal{K}\|_{HS}, \quad j \ge 1$$

This leads to

$$\tau_k = \frac{1}{k} \sum_{j=1}^{k} \frac{|\lambda_j|}{1 - \lambda_j}$$

$$\le \frac{1}{\delta} \frac{\|\mathcal{K}\|_{HS}}{k} \sum_{j=1}^{k} \frac{1}{\sqrt{j}}$$

$$\le \frac{1}{\sqrt{k}} \frac{\|\mathcal{K}\|_{HS}}{\delta}$$

Recalling that $\delta^{-1} = \|\mathcal{A}^{-1}\|$ proves the upper bound in (6.5.176). The lower bound in (6.5.176) is immediate from the definition of τ_k.

(b) From Fenyö and Stolle [198, §8.9], the eigenvalues $\{\lambda_j\}$ satisfy

$$\lim_{j \to \infty} j^{p+0.5} \lambda_j = 0$$

Let

$$M \equiv \sup j^{p+0.5} |\lambda_j|$$

so that

$$|\lambda_j| \le \frac{M}{j^{p+0.5}}, \quad j \ge 1 \tag{6.5.178}$$

With this bound on the eigenvalues,

$$\tau_k \le \frac{M}{k\delta} \sum_{j=1}^{k} \frac{1}{j^{p+0.5}} \le \frac{M}{k\delta} \zeta(p + 0.5)$$

This completes the proof of (6.5.177). □

We see that the speed of convergence of $\{\tau_k\}$ (or equivalently, $\{c_k\}$) is no better than $O(k^{-1})$, regardless of the differentiability of the kernel function K. Moreover, for virtually all cases of practical interest, it is no worse than $O(k^{-0.5})$. The result (6.5.165) was only a bound for the speed of convergence of the conjugate gradient method, although we believe the convergence speed is no better than this. For additional discussion of the convergence of $\{\tau_k\}$, see Flores [202].

6.5.2. The conjugate gradient iteration for Nyström's method

Let $K(t, s)$ in (6.5.158) be real and symmetric. Apply the Nyström method with a convergent numerical integration rule having positive weights. Then the approximating equation $(I - \mathcal{K}_n)x_n = y$ is equivalent to the linear system

$$x_n(t_{n,i}) - \sum_{j=1}^{q_n} w_{n,j} K(t_{n,i}, t_{n,j}) x_n(t_{n,j}) = y(t_{n,i}), \quad i = 1, \ldots, q_n$$

$$(6.5.179)$$

In general, this is not a symmetric linear system; but with the assumption that all weights $w_{n,i} > 0$, it can be converted to an equivalent symmetric system. Multiply equation i by $\sqrt{w_{n,i}}$ and introduce the new unknowns

$$z_i = \sqrt{w_{n,i}}\, x_n(t_{n,i}), \quad i = 1, \ldots, q_n$$

The system (6.5.179) is equivalent to

$$z_i - \sum_{j=1}^{q_n} \sqrt{w_{n,i} w_{n,j}} K(t_{n,i}, t_{n,j}) z_j = \sqrt{w_{n,i}}\, y(t_{n,i}), \quad i = 1, \ldots, q_n$$

$$(6.5.180)$$

which is a symmetric linear system.

Denote the matrix of coefficients in (6.5.179) by A_n and that of (6.5.180) by B_n. Then

$$B_n = W_n A_n W_n^{-1}$$

$$W_n = \mathrm{diag}[\sqrt{w_{n,1}}, \ldots, \sqrt{w_{n,q_n}}]$$

The matrices A_n and B_n are similar, and thus they have the same eigenvalues. Denote those eigenvalues by $1 - \lambda_{n,1}, \ldots, 1 - \lambda_{n,q_n}$, with

$$|\lambda_{n,1}| \geq \cdots \geq |\lambda_{n,q_n}| \geq 0 \qquad (6.5.181)$$

The numbers $\{\lambda_{n,j}\}$ are the nonzero eigenvalues of the numerical integral operator

$$\mathcal{K}_n x(t) \equiv \sum_{j=1}^{q_n} w_{n,j} K(t, t_{n,j}) x_n(t_{n,j}), \quad t \in D, \ x \in C(D)$$

Since B_n is symmetric, the eigenvalues $\{\lambda_{n,j}\}$ of $\mathcal{K}_n$ are real. The convergence of these eigenvalues $\{\lambda_{n,j}\}$ to the corresponding eigenvalues of $\mathcal{K}$ has been studied by many people. For a broad and extensive discussion of the numerical evaluation of the eigenvalues of $\mathcal{K}$, see the book of Chatelin [113].

The conjugate gradient method and its convergence

When the conjugate gradient method (6.5.159) is applied to the solution of (6.5.180), the method converges to the true answer in at most q_n iterations. Thus it is not appropriate to talk of the convergence of the method as $k \to \infty$ as was done with the conjugate gradient method when applied to the original integral equation (6.5.158). However, we can still go through the same algebraic manipulations, using the same optimality properties, to obtain

$$\left\| x_n - x_n^{(k)} \right\| \le (c_{n,k})^k \left\| x_n - x_n^{(0)} \right\|, \quad k = 0, 1, \ldots, q_n \tag{6.5.182}$$

$$c_{n,k} = \left(\frac{\Delta_n}{\delta_n} \right)^{\frac{2}{k}} \tau_{n,k} \tag{6.5.183}$$

with

$$\Delta_n = \sup_{1 \le j \le q_n} (1 - \lambda_{n,j}) \qquad \delta_n = \inf_{1 \le j \le q_n} (1 - \lambda_{n,j})$$

and

$$\tau_{n,k} = \frac{1}{k} \sum_{j=1}^{k} \frac{|\lambda_{n,j}|}{1 - \lambda_{n,j}} \tag{6.5.184}$$

The norm in (6.5.182) is the Euclidean vector norm on $\mathbf{R}^{q_n}$.

Using the convergence results of Ref. [30], it is straightforward to show that

$$\lim_{n \to \infty} \Delta_n = \Delta, \qquad \lim_{n \to \infty} \delta_n = \delta \tag{6.5.185}$$

Because we originally assumed in (6.5.161) that $\delta > 0$, it follows that $\delta_n > 0$ for all sufficiently large n, thus justifying the derivation leading to (6.5.183). Because of this convergence, the quantities

$$\left(\frac{\Delta}{\delta} \right)^{\frac{2}{k}} \quad \text{and} \quad \left(\frac{\Delta_n}{\delta_n} \right)^{\frac{2}{k}}, \quad k \ge 1$$

of (6.5.174) and (6.5.183), respectively, are uniformly bounded, say by a constant $\gamma > 1$. Then in the convergence statements (6.5.165) and (6.5.182),

$$c_k \leq \gamma \tau_k, \qquad c_{n,k} \leq \gamma \tau_{n,k}, \quad k \geq 1$$

for all n. We can compare $\{\tau_k\}$ and $\{\tau_{n,k}\}$ to compare the convergence speeds of the conjugate gradient method when applied to the original integral equation (6.5.158) and when applied to the approximating numerical integral equations of (6.5.179).

The results in Ref. [30] prove convergence of the approximate eigenvalues of collectively compact operator approximations. For our particular setting, the results in Ref. [30], [36] allow us to prove the following. For each eigenvalue λ_j of $\mathcal{K}$, the approximate eigenvalues $\lambda_{n,j}$ exist for all sufficiently large n, say $n \geq N_j$, and

$$|\lambda_j - \lambda_{n,j}| \leq \epsilon_j(n), \quad n \geq N_j \qquad (6.5.186)$$

with $\lim_{n \to \infty} \epsilon_j(n) = 0$. The error $\epsilon_j(n)$ is proportional to a numerical integration error; see Ref. [36].

Theorem 6.5.3. *Let p be any given positive integer. Then the constants $\{\tau_k\}$ and $\{\tau_{n,k}\}$ of (6.5.175) and (6.5.184), respectively, satisfy*

$$\max_{1 \leq k \leq p} |\tau_k - \tau_{n,k}| \to 0 \quad as \quad n \to \infty \qquad (6.5.187)$$

Proof. To simplify the notation used in this proof, we assume that all eigenvalues λ_j of $\mathcal{K}$ are positive. The essential nature of the proof can still be used for a more general distribution of real eigenvalues, but then we would need to modify the notation used here (with an additional modification needed when the number of nonzero eigenvalues of $\mathcal{K}$ is finite).

Using the assumption of positivity of $\lambda_1, \ldots, \lambda_p$, we can apply (6.5.186) to obtain the positivity of $\lambda_{n,1}, \ldots, \lambda_{n,p}$, $n \geq N_p$, provided N_p is chosen sufficiently large. Then from the definitions of τ_k and $\tau_{n,k}$,

$$\tau_k - \tau_{n,k} = \frac{1}{k} \sum_{j=1}^{k} \frac{\lambda_j - \lambda_{n,j}}{(1 - \lambda_j)(1 - \lambda_{n,j})}$$

Using (6.5.185) and (6.5.186), we have that for all sufficiently large n and for a suitable constant $\alpha > 0$,

$$|\tau_k - \tau_{n,k}| \leq \frac{1}{k\delta\delta_n} \sum_{j=1}^k |\lambda_j - \lambda_{n,j}|$$

$$\leq \frac{\alpha}{k\delta^2} \sum_{j=1}^k \epsilon_j(n)$$

$$\leq E_p(n)$$

with

$$E_p(n) = \frac{\alpha}{\delta^2} \max_{1 \leq k \leq p} \left[\frac{1}{k} \sum_{j=1}^k \epsilon_j(n) \right]$$

Since p is fixed and finite, it is easy to see that $E_p(n) \to 0$ as $n \to \infty$. This proves (6.5.187). $\qquad\square$

Using (6.5.187), the following form of *mesh independence principle* is valid for all sufficiently large values of n. If we desire a solution to (6.5.180) with a specified accuracy $\epsilon > 0$, then the needed number of iterations when using the conjugate gradient method will be approximately the same for all n, namely that needed when using the conjugate gradient method to solve the original integral equation with a right-hand function $y \in C(D)$ to an accuracy of the given ϵ. This is also observed experimentally in Ref. [202]. We note that as with earlier iteration methods, the operations cost is $O(q_n^2)$ per iteration, and by the preceding theorem, the total cost to solve the approximating equation to within a given accuracy is also $O(q_n^2)$.

The two-grid and multigrid methods of earlier sections also satisfied a similar kind of mesh independence principle. As to which of these methods is superior, it is probably necessary to carry out experiments for a carefully chosen variety of integral equations, both for functions of one variable and of several variables. We personally believe that the theory of two-grid and multigrid methods makes it easier to understand the behavior of such iteration methods than is true of the conjugate gradient iteration method.

6.5.3. Nonsymmetric integral equations

With no assumption of symmetry for $K(t, s)$ for the integral equation (6.5.158), consider again the linear system (6.5.179), and denote it by $A_n x_n = y_n$. In general, this system is nonsymmetric, and the conjugate gradient method cannot be used directly to solve it. During the late 1980s and early 1990s there was much research on developing generalizations of the conjugate gradient method to solve nonsymmetric linear systems. Much of the research dealt with what are called *Krylov subspace methods*, which are based on selecting the iterate $\mathbf{x}_n^{(k)}$ from the *Krylov subspace*

$$\left\{\mathbf{y}, A_n\mathbf{y}, A_n^2\mathbf{y}, \ldots, A_n^{(k-1)}\mathbf{y}\right\}$$

(assuming $\mathbf{x}_n^{(0)} = 0$) in some optimal manner, for $k = 1, 2, \ldots, q_n$. A good review of much of this work is given in Freund, Golub, Nachtigal [206]. The best known of such methods is called *GMRES*, an acronym for the way in which the method is defined. The method was introduced in Saad and Schultz [475], and a good theoretical perspective on it can be found in Ref. [206, §2.4]. We do not discuss it further here, but it is a popular iteration method for solving nonsymmetric linear systems.

Another approach to handling nonsymmetric linear systems is to convert them to symmetric linear systems and to then use the original conjugate gradient method. Multiplying $A_n\mathbf{x}_n = \mathbf{y}_n$ by A_n^T, we have the "normal equations"

$$A_n^\mathrm{T}A_n\mathbf{x}_n = A_n^\mathrm{T}\mathbf{y}_n \qquad (6.5.188)$$

The matrix of coefficients $A_n^\mathrm{T}A_n$ is symmetric and positive definite, and therefore, the conjugate gradient method is applicable.

The drawback with some important applications has been that the condition number of $A_n^\mathrm{T}A_n$ is the square of that of A_n alone,

$$\mathrm{cond}\left(A_n^\mathrm{T}A_n\right) = \mathrm{cond}(A_n)^2 \qquad (6.5.189)$$

In this, the condition number is based on the matrix norm that is induced by the Euclidean vector norm (cf. [48, p. 485]). When $A_n\mathbf{x}_n = \mathbf{y}_n$ is produced as a discretization of an elliptic partial differential equation, the condition number of A_n is large, and it becomes larger as the discretization becomes more accurate. With such systems, the larger the condition number, the slower is the convergence of the conjugate gradient iteration. Therefore, using (6.5.188) is usually unacceptable.

In contrast, with integral equations such as (6.5.158),

$$\mathrm{cond}(A_n) \approx c\|I - \mathcal{K}\|\,\|(I - \mathcal{K})^{-1}\|$$

with c a small constant. Well-conditioned integral equations $(I - \mathcal{K})x = y$ lead to well-conditioned linear systems $A_n\mathbf{x}_n = \mathbf{y}_n$; this was explored at length in Chapters 3 and 4. Therefore, solving $A_n\mathbf{x}_n = \mathbf{y}_n$ by applying the conjugate gradient to (6.5.188) seems a reasonable approach. We note one very important practical consideration. In actually implementing the conjugate gradient method as applied to (6.5.188), we never compute the matrix product $A_n^T A_n$. Instead, every matrix-vector multiplication is calculated using

$$\left(A_n^T A_n\right) \mathbf{x} = A_n^T \left(A_n \mathbf{x}\right), \quad \mathbf{x} \in \mathbf{R}^{q_n}$$

This results in an operations cost for solving (6.5.188) that is only double that of the original conjugate gradient method for solving $A_n \mathbf{x}_n = \mathbf{y}_n$. The extension of the earlier symmetric theory to this new situation is given in Flores [202, §4]. It is based on considering the system (6.5.188) to be just a Nyström-type of numerical solution of the integral equation

$$(I - \mathcal{K}^*)(I - \mathcal{K})x = (I - \mathcal{K}^*)y$$

with $\mathcal{K}^*$ the adjoint integral operator for $\mathcal{K}$. Numerical examples for nonsymmetric integral equations are also given in Ref. [202].

Discussion of the literature

The most widely used iteration method for solving $(\lambda - \mathcal{K})x = y$ is

$$x^{(k+1)} = \frac{1}{\lambda}\left(y + \mathcal{K}x^{(k)}\right), \quad k = 0, 1, 2, \ldots \tag{6.5.190}$$

This goes under a variety of names, including *successive approximations, fixed point iteration, Picard iteration, geometric series iteration*, and the *Neumann series iteration*. It converges if

$$\|\mathcal{K}\| < |\lambda|$$

as well as under slightly weaker assumptions, but it is not as flexible in its convergence when compared to the methods of this chapter. The iteration (6.5.190) is still widely used in engineering and other applications, but one of the methods of this chapter will usually be a better choice.

To the author's knowledge, very little has been written on the iterative solution of degenerate kernel integral equations, with the first general discussion of such methods appearing in Ref. [39, p. 128]. The method can also be considered as a special case of the *residual correction method*, and as such, an analysis is given in Kress [325, Chap. 14]. A more recent discussion of the iterative solution of degenerate kernel methods is given by Moret and Omari [388]. The paper of Dellwo [167] gives a somewhat related type of method, and his work also applies to projection and Nyström methods.

The Nyström iteration method 1 of §6.2.1 seems to have first been analyzed in Ref. [34], but it appears to have been used and understood earlier. It can also be considered as a residual correction method, although the type of analysis given in Theorem 6.2.1 is needed to analyze it, rather than the type of analysis that is often used with residual correction methods. The Nyström iteration method 2 of §6.2.2 was introduced in Brakhage [85], and it was subsequently generalized in Ref. [34]. Other analyses of the method can be found in Hackbusch [248, Chap. 16], [251, §5.4]. In Kelley [309] a modification is given to improve the Nyström iteration method 1. It yields a better-behaved and norm convergent iteration method, and it uses only approximately one-half the number of arithmetic operations of the Nyström iteration method 2. Programs to solve univariate Fredholm integral equations of the second kind, with automatic control of the error and using the Nyström iteration method 2, are given in Refs. [37], [38], and they are available from both the author's anonymous ftp site and World Wide Wib home page. For multivariable equations, analyses of the Nyström iteration method 2 are given in Atkinson [53] and Vainikko [550, p. 83], and programs implementing such iterative methods are contained in the boundary element package of Ref. [52].

Multigrid methods originated as a means to solve discretizations of elliptic partial differential equations, originating around 1970; they are also sometimes called *multilevel methods* and *multiresolution methods*.Later these methods were extended to the solution of integral equations of the second kind, principally by W. Hackbusch; much of this work is described in his books [248, Chap. 16], [251, §5.4]. Other treatments are given in Brandt and Lubrecht [88]; Hemker and Schippers [264]; Kress [325, Chap. 14]; Mandel [362]; and Schippers [489], [490], [491]. The thesis of Lee [338] contains a multigrid code, and numerical comparisons are given of his code and that of the two-grid code in Refs. [37], [38]. These comparisons show that the two-grid method is almost always more efficient, especially if the code is to estimate the error in its solution of the integral equation. The operations cost of our two-grid methods and the multigrid methods of Hackbusch are of the same order, $O(n^2)$ where n is the order of the linear system being solved. To a great extent, the method of choice depends on the type of method with which a user is most comfortable.

There are other iteration methods that can be applied to the iterative solution of the linear systems obtained in discretizing integral equations. A general presentation of iterative methods is given in Nevanlinna [399], focusing on many of the types of general iterative methods developed in the past 15–20 years for solving linear systems. Other general methods are discussed in Patterson [412] and Petryshyn [422].For linear systems that are symmetric, the conjugate gradient method has been a favorite, as was discussed in §6.5, and the literature for this topic is reviewed in Flores [201], [202]. For nonsymmetric systems, there are a number of possible methods. One can convert the linear system to a symmetric system, as described earlier in §6.5.3, or one can use a Krylov subspace method. For a general discussion of the latter, see Freund et al. [206], and for a discussion directed to integral equations, see Kelley and Xue [312].

An alternative to iterative methods is given in the thesis of Starr [527] for one-dimensional integral equations. In his thesis Starr uses wavelet-like bases to develop a discretization and solution technique that requires only $O(n \log^2 n)$ operations when using a discretization technique that usually would lead to solving a linear system of n equations. This type of approach is being used for solving discretizations of boundary integral equations on surfaces, and we will review the literature on such fast methods at the end of Chapter 9.

Boundary integral equations on a smooth planar boundary

A popular source of integral equations has been the study of elliptic partial differential equations. Given a boundary value problem for an elliptic partial differential equation over an open region D, the problem can often be reformulated as an equivalent integral equation over the boundary of D. Such a reformulation is called a *boundary integral equation*, and we will often refer to it as simply a *BIE*. As an example of such a reformulation, Carl Neumann [398] used what is now called the geometric series (cf. Theorem A.1) to investigate the solvability of some BIE reformulations of Laplace's equation $\Delta u \equiv 0$, thereby also obtaining solvability results for Laplace's equation. As a second example, the original work of Ivar Fredholm on the solvability of integral equations of the second kind (cf. §§1.3, 1.4 in Chapter 1) was also motivated by a desire to show the existence of solutions of Laplace's equation.

Various BIE reformulations have long been used as a means of numerically solving Laplace's equation, although this approach has been less popular than the use of finite difference and finite element methods. Since 1970 there has been a significant increase in the popularity of using boundary integral equations to solve Laplace's equation and many other elliptic equations, including the biharmonic equation, the Helmholtz equation, the equations of linear elasticity, and the equations for Stokes' fluid flow. Our presentation is limited to the study of Laplace's equation, but we note that many of our results generalize to the solution of BIE reformulations of these other elliptic equations.

In this and the following two chapters we consider the numerical solution of BIE reformulations of Laplace's equation, in both two and three dimensions. The present chapter considers the case that Laplace's equation is being solved on a region D that is planar with its boundary $S \equiv \partial D$ a smooth simple closed curve, and the following chapter considers the planar case with S only piecewise smooth. Chapter 9 investigates the case of $D \subset \mathbf{R}^3$ with S a surface in $\mathbf{R}^3$.

In this chapter, §7.1 contains a theoretical framework for BIE reformulations of Laplace's equation in $\mathbf{R}^2$, giving the most popular of such boundary integral equations. For much of the history of BIE, those of the second kind have been the most popular, and in §7.2 we discuss the numerical solution of such BIE. In §7.3 we introduce the study of BIE of the first kind, and we introduce the use of Fourier series as a means of studying these equations and numerical methods for their solution. We conclude in §7.4 with the framework for BIE which regards them as *pseudodifferential operator equations* and then solves them with generalizations of the finite element method.

7.1. Boundary integral equations

The main objective is to solve boundary value problems for Laplace's equation. The principal such problems are the *Dirichlet problem* and the *Neumann problem*; these are the only boundary value problems considered here. Let D be a bounded open simply connected region in the plane, and let its boundary S be a simple closed curve with a parametrization

$$\mathbf{r}(t) = (\xi(t), \eta(t)), \quad 0 \le t \le L \tag{7.1.1}$$

with $\mathbf{r} \in C^2[0, L]$ and $|\mathbf{r}'(t)| \ne 0$ for $0 \le t \le L$. We assume the parametrization traverses S in a counter-clockwise direction. Introduce the interior unit normal $\mathbf{n}(t)$ that is orthogonal to the curve S at $\mathbf{r}(t)$:

$$\mathbf{n}(t) = \frac{(-\eta'(t), \xi'(t))}{\sqrt{\xi'(t)^2 + \eta'(t)^2}}$$

In addition, introduce the exterior region $D_e \equiv \mathbf{R}^2 \backslash \bar{D}$. On occasion, we will use D_i in place of D to denote the region interior to S. The main boundary value problems for Laplace's equation are as follows.

The Dirichlet problem. Find $u \in C(\bar{D}) \cap C^2(D)$ that satisfies

$$\begin{aligned} \Delta u(P) &= 0, & P &\in D \\ u(P) &= f(P), & P &\in S \end{aligned} \tag{7.1.2}$$

with $f \in C(S)$ a given boundary function.

The Neumann problem. Find $u \in C^1(\bar{D}) \cap C^2(D)$ that satisfies

$$\begin{aligned} \Delta u(P) &= 0, & P &\in D \\ \frac{\partial u(P)}{\partial \mathbf{n}_P} &= f(P), & P &\in S \end{aligned} \tag{7.1.3}$$

with $f \in C(S)$ a given boundary function.

There is a corresponding set of problems on the exterior region D_e, and these are discussed later in the section. The above boundary value problems have been studied extensively, but we only give a very brief look at results on their solvability. The name *potential theory* is given to the study of Laplace's equation and to some closely related problems such as the Helmholtz equation. We also refer to functions satisfying Laplace's equation as *harmonic functions*.

The following theorem gives existence and uniqueness of solutions to the above problems.

Theorem 7.1.1. *Let the function* $f \in C(S)$, *and for the parametrization of* (7.1.1) *assume* $\mathbf{r} \in C^2[0, L]$. *Then:*

(a) *The Dirichlet problem* (7.1.2) *has a unique solution, and*
(b) *The Neumann problem* (7.1.3) *has a unique solution, up to the addition of an arbitrary constant, provided*

$$\int_S f(Q)\, dS = 0 \tag{7.1.4}$$

Proof. These results are contained in many textbooks, often in a stronger form than is given here. For example, see Colton [126, §5.3], Mikhlin [380, Chap. 18], or Pogorzelski [426, Chap. 12]. The proof is commonly based on a BIE reformulation of these problems. Also note that it can be shown that $u \in C^\infty(D)$ for both problems (7.1.2) and (7.1.3).   □

7.1.1. *Green's identities and representation formula*

A very important tool for studying elliptic partial differential equations is the *divergence theorem* or *Gauss's theorem*. Let Ω denote an open planar region. Let its boundary Γ consist of $m + 1$ distinct simple closed curves,

$$\Gamma = \Gamma_0 \cup \cdots \cup \Gamma_m$$

Assume $\Gamma_1, \ldots, \Gamma_m$ are contained in the interior of Γ_0, and for each $i = 1, \ldots, m$, let Γ_i be exterior to the remaining curves $\Gamma_1, \ldots, \Gamma_{i-1}, \Gamma_{i+1}, \ldots, \Gamma_m$. Further, assume each curve Γ_i is a piecewise smooth curve. We say a curve γ is piecewise smooth if (1) it can be broken into a finite set of curves $\gamma_1, \ldots, \gamma_k$ with each γ_j having a parametrization that is at least twice continuously differentiable, and (2) the curve γ does not contain any cusps (meaning that each pair of adjacent curves γ_i and γ_{i+1} join at an angle in the interval $(0, 2\pi)$). The region Ω is interior to Γ_0, but it is exterior to each of the curves $\Gamma_1, \ldots, \Gamma_m$.

The orientation of Γ_0 is to be counterclockwise, while the curves $\Gamma_1, \ldots, \Gamma_m$ are to be clockwise.

Theorem 7.1.2 (The divergence theorem). *Assume* $\mathbf{F}: \bar{\Omega} \to \mathbf{R}^2$ *with each component of F contained in* $C^1(\bar{\Omega})$. *Then*

$$\int_\Omega \nabla \cdot \mathbf{F}(Q)\, d\Omega = -\int_\Gamma \mathbf{F}(Q) \cdot \mathbf{n}(Q)\, d\Gamma \qquad (7.1.5)$$

Proof. This important result, which generalizes the fundamental theorem of calculus, is proven in most standard textbooks on advanced calculus. For example, see Apostol [23, p. 289] for a somewhat stronger form of this result. In the planar form given here, the result is also known as "Green's theorem." □

Using the divergence theorem, one can obtain *Green's identities* and *Green's representation formula*. Assuming $u \in C^1(\bar{\Omega})$ and $w \in C^2(\bar{\Omega})$, one can prove *Green's first identity* by letting $\mathbf{F} = u\nabla w$ in (7.1.5):

$$\int_\Omega u \Delta w\, d\Omega + \int_\Omega \nabla u \cdot \nabla w\, d\Omega = -\int_\Gamma u \frac{\partial w}{\partial \mathbf{n}}\, d\Gamma \qquad (7.1.6)$$

Next, assume $u, w \in C^2(\bar{\Omega})$. Interchanging the roles of u and w in (7.1.6) and then subtracting the two identities, one obtains *Green's second identity*:

$$\int_\Omega [u \Delta w - w \Delta u]\, d\Omega = \int_\Gamma \left[w \frac{\partial u}{\partial \mathbf{n}} - u \frac{\partial w}{\partial \mathbf{n}} \right] d\Gamma \qquad (7.1.7)$$

The identity (7.1.6) can be used to prove (*i*) if the Neumann problem (7.1.3) has a solution, then it is unique up to the addition of an arbitrary constant; and (*ii*) if the Neumann problem is to have a solution, then the condition (7.1.4) is necessary.

Return to the original domain D on which the problems (7.1.2) and (7.1.3) are posed, and assume $u \in C^2(\bar{D})$. Let $u(Q)$ be a solution of Laplace's equation, and let $w(Q) = \log|A - Q|$, with $A \in D$. Define Ω to be D after removing the small disk $B(A, \epsilon) \equiv \{Q \mid |A - Q| \le \epsilon\}$, with $\epsilon > 0$ so chosen that $B(A, \epsilon) \subset D$. Note that for the boundary Γ,

$$\Gamma = S \cup \{Q \mid |A - Q| = \epsilon\}$$

Apply (7.1.7) with this choice of Ω, and then let $\epsilon \to 0$. Doing so, and then carefully computing the various limits, we can obtain *Green's representation formula:*

$$u(A) = \frac{1}{2\pi} \int_S \left[\frac{\partial u(Q)}{\partial \mathbf{n}_Q} \log |A - Q| - u(Q) \frac{\partial}{\partial \mathbf{n}_Q} [\log |A - Q|] \right] dS_Q,$$

$$A \in D \qquad (7.1.8)$$

This expresses u over D in terms of the boundary values of u and its normal derivative on S. Assuming S has a parametrization (7.1.1) that is in C^2, we can take limits in (7.1.8) as A approaches a point on the boundary S. Let $P \in S$. Then after a careful calculation,

$$\lim_{A \to P} \int_S \frac{\partial u(Q)}{\partial \mathbf{n}_Q} \log |A - Q| \, dS_Q = \int_S \frac{\partial u(Q)}{\partial \mathbf{n}_Q} \log |P - Q| \, dS_Q$$

$$\lim_{\substack{A \to P \\ A \in D}} \int_S u(Q) \frac{\partial}{\partial \mathbf{n}_Q} [\log |A - Q|] \, dS_Q$$

$$= -\pi u(P) + \int_S u(Q) \frac{\partial}{\partial \mathbf{n}_Q} [\log |P - Q|] \, dS_Q \qquad (7.1.9)$$

A proof of (7.1.9), and of the associated limit in (7.1.25), can be found in Colton [126, pp. 197–202] or in many other texts on Laplace's equation.

Using these limits in (7.1.8) yields the relation

$$u(P) = \frac{1}{\pi} \int_S \left[\frac{\partial u(Q)}{\partial \mathbf{n}_Q} \log |P - Q| - u(Q) \frac{\partial}{\partial \mathbf{n}_Q} [\log |P - Q|] \right] dS_Q,$$

$$P \in S \qquad (7.1.10)$$

which gives a relationship between the values of u and its normal derivative on S. Finally, if $A \in D_e$, then the identity (7.1.7) implies

$$\frac{1}{2\pi} \int_S \left[\frac{\partial u(Q)}{\partial \mathbf{n}_Q} \log |A - Q| - u(Q) \frac{\partial}{\partial \mathbf{n}_Q} [\log |A - Q|] \right] dS_Q = 0,$$

$$A \in D_e \qquad (7.1.11)$$

The formula (7.1.10) is an example of a boundary integral equation, and it can be used to create other such boundary integral equations. First, however, we need to look at solving Laplace's equation on exterior regions D_e and to obtain formulas that correspond to (7.1.8)–(7.1.11) for such exterior regions.

7.1.2. The Kelvin transformation and exterior problems

Define a transformation $\mathcal{T} : \mathbf{R}^2 \backslash \{0\} \to \mathbf{R}^2 \backslash \{0\}$

$$\mathcal{T}(x, y) = (\xi, \eta) \equiv \frac{1}{r^2}(x, y), \quad r = \sqrt{x^2 + y^2} \qquad (7.1.12)$$

In polar coordinates,

$$T(r\cos\theta, r\sin\theta) = \frac{1}{r}(\cos\theta, \sin\theta)$$

Thus, a point (x, y) is mapped onto another point (ξ, η) on the same ray emanating from the origin, and we call (ξ, η) the inverse of (x, y) with respect to the unit circle. Note that $T(T(x, y)) = (x, y)$, so that $T^{-1} = T$. The Jacobian matrix for T is

$$J(T) = \begin{bmatrix} \dfrac{\partial\xi}{\partial x} & \dfrac{\partial\xi}{\partial y} \\[2mm] \dfrac{\partial\eta}{\partial x} & \dfrac{\partial\eta}{\partial y} \end{bmatrix} = \frac{1}{r^2} H$$

with

$$H = \begin{bmatrix} \dfrac{y^2 - x^2}{r^2} & \dfrac{-2xy}{r^2} \\[3mm] \dfrac{-2xy}{r^2} & \dfrac{x^2 - y^2}{r^2} \end{bmatrix}$$

The matrix H is orthogonal with determinant -1, and

$$\det[J(T(x, y))] = -\frac{1}{r^2}$$

Assume the bounded open region $D \equiv D_i$ contains the origin $\mathbf{0}$. For a function $u \in C(\bar{D}_e)$, define

$$\hat{u}(\xi, \eta) = u(x, y), \quad (\xi, \eta) = T(x, y), \quad (x, y) \in \bar{D}_e \qquad (7.1.13)$$

This is called the *Kelvin transformation*. Introduce the interior region $\hat{D} = T(D_e)$, and let $\hat{S}$ denote the boundary of $\hat{D}$. The boundaries S and $\hat{S}$ have the same degree of smoothness. In addition, the condition $(\xi, \eta) \to 0$ in $\hat{D}$ corresponds to $r \to \infty$ for points $(x, y) \in D_e$. For a function u satisfying Laplace's equation on D, it is a straightforward calculation to show

$$\Delta\hat{u}(\xi, \eta) = r^4 \Delta u(x, y) = 0, \quad (\xi, \eta) = T(x, y), \quad (x, y) \in D_e$$

thus showing $\hat{u}$ to be harmonic on $\hat{D}$. We can pass from the solution of Laplace's equation on the unbounded region D_e to the bounded open region $\hat{D}$.

If we were to impose the Dirichlet condition $u = f$ on the boundary S, this is equivalent to the Dirichlet condition

$$\hat{u}(\xi, \eta) = f(T^{-1}(\xi, \eta)), \quad (\xi, \eta) \in \hat{S}$$

From the existence and uniqueness result of Theorem 7.1.1, the interior Dirichlet problem on $\hat{D}$ will have a unique solution. This leads us to considering the following problem.

The exterior Dirichlet problem. Find $u \in C(\bar{D}_e) \cap C^2(D_e)$ that satisfies

$$
\begin{aligned}
\Delta u(P) &= 0, && P \in D_e \\
u(P) &= f(P), && P \in S
\end{aligned}
\tag{7.1.14}
$$
$$
\lim_{r \to \infty} \sup_{|P| \ge r} |u(P)| < \infty,
$$

with $f \in C(S)$ a given boundary function.

Using the above discussion on the Kelvin transform, this converts to the interior Dirichlet problem

$$
\begin{aligned}
\Delta \hat{u}(\xi, \eta) &= 0, && (\xi, \eta) \in \hat{D} \\
u(\xi, \eta) &= f(\mathcal{T}^{-1}(\xi, \eta)), && (\xi, \eta) \in \hat{S}
\end{aligned}
\tag{7.1.15}
$$

and Theorem 7.1.1 guarantees the unique solvability of this problem. The condition on $u(x, y)$ as $r \to \infty$ can be used to show that $\hat{u}(\xi, \eta)$ has a removable singularity at the origin, and $\hat{u}(0, 0)$ will be the value of $u(x, y)$ as $r \to \infty$. Thus, the above exterior Dirichlet problem has a unique solution.

For functions $u \in C^1(\bar{D}_e)$,

$$
\frac{\partial u(x, y)}{\partial \mathbf{n}(x, y)} = -\rho^2 \frac{\partial \hat{u}(\xi, \eta)}{\partial \hat{\mathbf{n}}(\xi, \eta)}, \qquad \rho = \frac{1}{r} = \sqrt{\xi^2 + \eta^2}
$$

with $\hat{\mathbf{n}}(\xi, \eta)$ the unit interior normal to $\hat{S}$ at (ξ, η). Thus the Neumann condition

$$
\frac{\partial u(x, y)}{\partial \mathbf{n}(x, y)} = f(x, y), \qquad (x, y) \in S
$$

is equivalent to

$$
\frac{\partial \hat{u}(\xi, \eta)}{\partial \hat{\mathbf{n}}(\xi, \eta)} = -\frac{1}{\rho^2} f(\mathcal{T}^{-1}(\xi, \eta)) \equiv \hat{f}(\xi, \eta), \qquad (\xi, \eta) \in \hat{S}
\tag{7.1.16}
$$

Also,

$$
\int_S \frac{\partial u}{\partial \mathbf{n}} \, dS = -\int_{\hat{S}} \frac{\partial \hat{u}}{\partial \hat{\mathbf{n}}} \, d\hat{S}
\tag{7.1.17}
$$

Using this information, consider the following problem.

The exterior Neumann problem. Find $u \in C^1(\bar{D}_e) \cap C^2(D_e)$ that satisfies

$$\Delta u(P) = 0, \qquad P \in D_e$$
$$\frac{\partial u(P)}{\partial \mathbf{n}_P} = f(P), \quad P \in S \tag{7.1.18}$$

$$u(r\cos\theta, r\sin\theta) = O\left(\frac{1}{r}\right), \quad \frac{\partial u(r\cos\theta, r\sin\theta)}{\partial r} = O\left(\frac{1}{r^2}\right) \tag{7.1.19}$$

as $r \to \infty$, uniformly in θ. The function $f \in C(S)$ is assumed to satisfy

$$\int_S f(Q)\, dS = 0 \tag{7.1.20}$$

just as in (7.1.4) for the interior Neumann problem.

Combining (7.1.17) with (7.1.20) yields

$$\int_{\hat{S}} \hat{f}(\xi, \eta)\, d\hat{S} = 0 \tag{7.1.21}$$

The problem (7.1.18) converts to the equivalent interior problem of finding $\hat{u}$ satisfying

$$\Delta\hat{u}(\xi, \eta) = 0, \qquad (\xi, \eta) \in \hat{D}$$
$$\frac{\partial\hat{u}(\xi, \eta)}{\partial\hat{\mathbf{n}}(\xi, \eta)} = \hat{f}(\xi, \eta), \quad (\xi, \eta) \in \hat{S} \tag{7.1.22}$$
$$\hat{u}(0, 0) = 0$$

By Theorem 7.1.1 and (7.1.21), this has a unique solution $\hat{u}$. This gives a complete solvability theory for the exterior Neumann problem.

The converted problems (7.1.15) and (7.1.22) can also be used for numerical purposes, and later we will return to these reformulations of exterior problems for Laplace's equation.

Green's representation formula on exterior regions. From the form of solutions to the interior Dirichlet problem, and using the Kelvin transform, we can assume the following form for potential functions u defined on D_e:

$$u(r\cos\theta, r\sin\theta) = u(\infty) + \frac{c(\theta)}{r} + O\left(\frac{1}{r^2}\right) \tag{7.1.23}$$

as $r \to \infty$ and with $c(\theta) = A\cos\theta + B\sin\theta$ for suitable constants A, B. The notation $u(\infty)$ denotes the limiting value of $u(r\cos\theta, r\sin\theta)$ as $r \to \infty$. From this, we can use the Green's representation formulas (7.1.8)–(7.1.11) for interior regions to obtain the following Green's representation formula for potential functions on exterior regions.

$$u(A) = u(\infty) - \frac{1}{2\pi} \int_S \frac{\partial u(Q)}{\partial n_Q} \log |A - Q| \, dS_Q$$

$$+ \frac{1}{2\pi} \int_S u(Q) \frac{\partial}{\partial n_Q} [\log |A - Q|] \, dS_Q, \quad A \in D_e \qquad (7.1.24)$$

To obtain a limiting value as $A \to P \in S$, we need the limit

$$\lim_{\substack{A \to P \\ A \in D_e}} \int_S u(Q) \frac{\partial}{\partial n_Q} [\log |A - Q|] \, dS_Q$$

$$= \pi u(P) + \int_S u(Q) \frac{\partial}{\partial n_Q} [\log |P - Q|] \, dS_Q \qquad (7.1.25)$$

Note the change of sign of $u(P)$ when compared to (7.1.9). Using this in (7.1.24), we obtain

$$u(P) = 2u(\infty) - \frac{1}{\pi} \int_S \frac{\partial u(Q)}{\partial n_Q} \log |P - Q| \, dS_Q$$

$$+ \frac{1}{\pi} \int_S u(Q) \frac{\partial}{\partial n_Q} [\log |P - Q|] \, dS_Q, \quad P \in S \qquad (7.1.26)$$

Finally, for points outside of D_e,

$$0 = u(\infty) - \frac{1}{2\pi} \int_S \frac{\partial u(Q)}{\partial n_Q} \log |A - Q| \, dS_Q$$

$$+ \frac{1}{2\pi} \int_S u(Q) \frac{\partial}{\partial n_Q} [\log |A - Q|] \, dS_Q, \quad A \in D_i \equiv D \qquad (7.1.27)$$

7.1.3. Boundary integral equations of direct type

The equations (7.1.8) and (7.1.24) give representations for functions harmonic in D_i and D_e, respectively, in terms of u and $\partial u / \partial n$ on the boundary S of these regions. When given one of these boundary functions, the equations (7.1.10) and (7.1.26) can often be used to obtain the remaining boundary function. We will illustrate some of the possibilities.

The interior Dirichlet problem (7.1.2)

The boundary condition is $u(P) = f(P)$ on S, and using it, (7.1.10) can be written as

$$\frac{1}{\pi} \int_S \rho(Q) \log |P - Q| \, dS_Q = g(P), \quad P \in S \qquad (7.1.28)$$

To emphasize the form of the equation, we have introduced

$$\rho(Q) \equiv \frac{\partial u(Q)}{\partial \mathbf{n}_Q}, \quad g(P) \equiv f(P) + \frac{1}{\pi} \int_S f(Q) \frac{\partial}{\partial \mathbf{n}_Q} [\log |P - Q|] \, dS_Q$$

The equation (7.1.28) is of the first kind, and it is often used as the prototype for studying boundary integral equations of the first kind. In §§7.3, 7.4, we examine the solution of (7.1.28) in some detail.

The interior Neumann problem (7.1.3)

The boundary condition is $\partial u / \partial \mathbf{n} = f$ on S, and using it, we write (7.1.10) as

$$u(P) + \frac{1}{\pi} \int_S u(Q) \frac{\partial}{\partial \mathbf{n}_Q} [\log |P - Q|] \, dS_Q$$

$$= \frac{1}{\pi} \int_S f(Q) \log |P - Q| \, dS_Q, \quad P \in S \qquad (7.1.29)$$

This is an integral equation of the second kind. Unfortunately, it is not uniquely solvable, this should not be surprising when given the lack of unique solvability for the Neumann problem itself. The homogeneous equation has $u \equiv 1$ as a solution, as can be seen by substituting the harmonic function $u \equiv 1$ into (7.1.10). The equation (7.1.29) is solvable if and only if the boundary function f satisfies the condition (7.1.4). The simplest way to deal with the lack of uniqueness in solving (7.1.29) is to introduce an additional condition such as

$$u(P^*) = 0$$

for some fixed point $P^* \in S$. This will lead to a unique solution for (7.1.29). Combine this with the discretization of the integral equation to obtain a suitable numerical approximation for u. There are other ways of converting (7.1.29) to a uniquely solvable equation, and some of these are explored in Ref. [29]. However, there are preferable ways to solve the interior Neumann problem, and some of these are discussed later.

The exterior Neumann problem (7.1.18)

The boundary condition is $\partial u/\partial \mathbf{n} = f$ on S, and u also satisfies $u(\infty) = 0$. Using this, (7.1.26) becomes

$$u(P) - \frac{1}{\pi} \int_S u(Q) \frac{\partial}{\partial \mathbf{n}_Q}[\log |P - Q|] \, dS_Q$$

$$= -\frac{1}{\pi} \int_S f(Q) \log |P - Q| \, dS_Q, \quad P \in S \qquad (7.1.30)$$

This equation is uniquely solvable, as will be discussed in greater detail below, following (7.2.48) in §7.2. This is considered a practical approach to solving the exterior Neumann problem, especially when one wants to find only the boundary data $u(P)$, $P \in S$. The numerical solution of the exterior Neumann problem using this approach is given following (7.2.66) in §7.2.

Using the representation $\mathbf{r}(t) = (\xi(t), \eta(t))$ of (7.1.1) for S and multiplying by $-\pi$, we can rewrite (7.1.30) as

$$-\pi u(t) + \int_0^L K(t, s)u(s) \, ds = g(t), \quad 0 \le t \le L \qquad (7.1.31)$$

$$K(t, s) = \frac{\eta'(s)[\xi(t) - \xi(s)] - \xi'(s)[\eta(t) - \eta(s)]}{[\xi(t) - \xi(s)]^2 + [\eta(t) - \eta(s)]^2}$$

$$= \frac{\eta'(s)\xi[s, s, t] - \xi'(s)\eta[s, s, t]}{|\mathbf{r}[s, t]|^2}, \quad s \ne t \qquad (7.1.32)$$

$$K(t, t) = \frac{\eta'(t)\xi''(t) - \xi'(t)\eta''(t)}{2\{\xi'(t)^2 + \eta'(t)^2\}} \qquad (7.1.33)$$

$$g(t) = \int_0^L f(\mathbf{r}(s))\sqrt{\xi'(s)^2 + \eta'(s)^2} \log |\mathbf{r}(t) - \mathbf{r}(s)| \, ds \qquad (7.1.34)$$

In (7.1.31) we have used $u(t) \equiv u(\mathbf{r}(t))$, for simplicity in notation. The second fraction in (7.1.32) uses first and second order Newton divided differences, to more easily obtain the limiting value $K(t, t)$ of (7.1.33). The value of $K(t, t)$ is one-half the curvature of S at $\mathbf{r}(t)$.

As in earlier chapters, we write (7.1.31) symbolically as

$$(-\pi + \mathcal{K})u = g \qquad (7.1.35)$$

By examining the formulas for $K(t, s)$, we have

$$\mathbf{r} \in C^\kappa[0, L] \implies K \in C^{\kappa-2}([0, L] \times [0, L]) \qquad (7.1.36)$$

The kernel function K is periodic in both variables, with period L, as are also the functions u and g. Introduce the space $C_p^{\ell}(L)$ as the set of all ℓ-times continuously differentiable functions h on $(-\infty, \infty)$ for which

$$h(t + L) \equiv h(t)$$

For a norm, use

$$\|h\|_{\ell} = \max \left\{ \|h\|_{\infty}, \|h'\|_{\infty}, \ldots, \left\|h^{(\ell)}\right\|_{\infty} \right\}$$

With this norm, $C_p^{\ell}(L)$ is a Banach space, and we identify $C_p(L) = C_p^0(L)$. We always assume for the parametrization that $\mathbf{r} \in C_p^{\kappa}(L)$, with $\kappa \geq 2$, and therefore the integral operator $\mathcal{K}$ is a compact operator from $C_p(L)$ to $C_p(L)$. Moreover, from (7.1.36), $\mathcal{K}$ maps $C_p(L)$ to $C_p^{\kappa-2}(L)$. The numerical solution of (7.1.31) is examined in detail in §7.2, along with related integral equations.

Using the Kelvin transform, the interior Neumann problem (7.1.2) can be converted to an equivalent exterior Neumann problem, as was done in passing between (7.1.18) and (7.1.22). Solving the exterior problem will correspond to finding that solution to the interior Neumann problem that is zero at the origin (where we assume $\mathbf{0} \in D_i$). This seems a very practical approach to solving interior Neumann problems, but it does not appear to have been used much in the past. It is used in the program given in Jeon and Atkinson [291].

The exterior Dirichlet problem (7.1.14)

The Kelvin transform can also be used to convert the exterior Dirichlet problem (7.1.14) to an equivalent interior Dirichlet problem. After doing so, there are many options for solving the interior problem, including using the first kind boundary integral equation (7.1.28). The value of $u(\infty)$ can be obtained as the value at $\mathbf{0}$ of the transformed problem.

7.1.4. Boundary integral equations of indirect type

Let $u_i \in C^1(\bar{D}_i)$ be harmonic in D_i, and let $u_e \in C^1(\bar{D}_e)$ be harmonic in D_e with

$$\sup_{P \in \bar{D}_e} |u_e(P)| < \infty$$

The function u_i satisfies (7.1.8), (7.1.10), and (7.1.11); u_e satisfies (7.1.24), (7.1.26), and (7.1.27). Introduce

$$[u(Q)] = u_i(Q) - u_e(Q), \quad Q \in S$$

$$\left[\frac{\partial u(Q)}{\partial \mathbf{n}_Q} \right] = \frac{\partial u_i(Q)}{\partial \mathbf{n}_Q} - \frac{\partial u_e(Q)}{\partial \mathbf{n}_Q}, \quad Q \in S$$

Adding (7.1.8) and (7.1.27),

$$u_i(A) = u_e(\infty) + \frac{1}{2\pi} \int_S \left[\frac{\partial u(Q)}{\partial \mathbf{n}_Q} \right] \log|A - Q| \, dS_Q$$

$$- \frac{1}{2\pi} \int_S [u(Q)] \frac{\partial}{\partial \mathbf{n}_Q} [\log|A - Q|] \, dS_Q, \quad A \in D_i \qquad (7.1.37)$$

Adding (7.1.11) and (7.1.24),

$$u_e(A) = u_e(\infty) - \frac{1}{2\pi} \int_S \left[\frac{\partial u(Q)}{\partial \mathbf{n}_Q} \right] \log|A - Q| \, dS_Q$$

$$+ \frac{1}{2\pi} \int_S [u(Q)] \frac{\partial}{\partial \mathbf{n}_Q} [\log|A - Q|] \, dS_Q, \quad A \in D_e \qquad (7.1.38)$$

And adding (7.1.10) and (7.1.26),

$$u_i(P) + u_e(P) = 2u_e(\infty) + \frac{1}{\pi} \int_S \left[\frac{\partial u(Q)}{\partial \mathbf{n}_Q} \right] \log|P - Q| \, dS_Q$$

$$- \frac{1}{\pi} \int_S [u(Q)] \frac{\partial}{\partial \mathbf{n}_Q} [\log|P - Q|] \, dS_Q, \quad P \in S$$

$$(7.1.39)$$

These formulas can be used to construct other boundary integral representations for harmonic functions.

Double layer potentials

Given a harmonic function $u_i \in C(\bar{D}_i)$, construct the above harmonic function u_e as the solution of the exterior Neumann problem with $u_e(\infty) = 0$ and

$$\frac{\partial u_e(Q)}{\partial \mathbf{n}_Q} = \frac{\partial u_i(Q)}{\partial \mathbf{n}_Q}, \quad Q \in S$$

Then from (7.1.37),

$$u_i(A) = -\frac{1}{2\pi} \int_S [u(Q)] \frac{\partial}{\partial \mathbf{n}_Q} [\log|A - Q|] \, dS_Q, \quad A \in D_i$$

We rewrite this as

$$u_i(A) = \int_S \rho(Q) \frac{\partial}{\partial \mathbf{n}_Q} [\log|A - Q|] \, dS_Q, \quad A \in D_i \qquad (7.1.40)$$

with

$$\rho(Q) \equiv -\frac{1}{2\pi} [u(Q)] \qquad (7.1.41)$$

Since u_i was an arbitrary function harmonic on D_i, formula (7.1.40) gives an integral representation for all such harmonic functions.

The integral in (7.1.40) is called a *double layer potential*, and the function ρ is called a *double layer density function* or a *dipole density function*. These names have their origin in the potential theory of physics, and a classical introduction from this viewpoint is given in Kellogg [307]. To use this integral representation, we solve directly for ρ. Suppose the function u_i is the solution of the interior Dirichlet problem with $u_i \equiv f$ on S. Then use (7.1.9) to take limits in (7.1.40) as $A \to P \in S$. This yields the boundary integral equation

$$-\pi\rho(P) + \int_S \rho(Q) \frac{\partial}{\partial \mathbf{n}_Q}[\log |P - Q|]\, dS_Q = f(P), \quad P \in S \qquad (7.1.42)$$

Note that the form of the left side of this equation is exactly that of (7.1.31) for the exterior Neumann problem. We discuss in detail the numerical solution of this and related equations in §7.2. Fredholm used (7.1.40) to show the solvability of the interior Dirichlet problem for Laplace's equation, and he did so by showing (7.1.42) is uniquely solvable for all $f \in C(S)$.

Single layer potentials

Given a harmonic function $u_i \in C(\bar{D}_i)$, construct the above harmonic function u_e as the solution of the exterior Dirichlet problem (7.1.14) with boundary data $f \equiv u_i|_S$. Then (7.1.37) implies

$$u_i(A) = u_e(\infty) + \frac{1}{2\pi} \int_S \left[\frac{\partial u(Q)}{\partial \mathbf{n}_Q}\right] \log |A - Q|\, dS_Q$$

We introduce

$$\rho(Q) \equiv \frac{1}{2\pi} \left[\frac{\partial u(Q)}{\partial \mathbf{n}_Q}\right] \qquad (7.1.43)$$

and rewrite the integral representation as

$$u_i(A) = u_e(\infty) + \int_S \rho(Q) \log |A - Q|\, dS_Q, \quad A \in D_i \qquad (7.1.44)$$

The integral operator is called a *single layer potential*, and the function ρ is called a *single layer density function*. We again refer to Kellogg [307] for the physical origins of this representation and its nomenclature.

The term $u_e(\infty)$ is a constant function, and it is therefore a harmonic function. In practice, it is usually omitted. This yields the representation

$$u_i(A) = \int_S \rho(Q) \log |A - Q|\, dS_Q, \quad A \in D_i \qquad (7.1.45)$$

When this representation is used to solve Neumann problems, the omission of $u_e(\infty)$ makes no difference, since the solution is unique only up to the addition of an arbitrary constant. The omission of $u_e(\infty)$ can make some difference in solving some Dirichlet problems, and this is discussed further in §7.3.

Other formulas for representing harmonic functions can be obtained from (7.1.37)–(7.1.39), for functions harmonic on interior or exterior regions. There are also additional representation formulas that can be obtained by other means. For example, one can form normal derivatives of (7.1.37) and (7.1.39) as the field point A approaches a point $P \in S$. For one such formula, which gives a way to solve mixed Dirichlet-Neumann boundary value problems for Laplace's equation, see Blue [81]. Representation formulas can also be obtained from the Cauchy integral formula for functions of a complex variable. All analytic functions $f(z)$ can be written in the form

$$f(z) = u(x, y) + i v(x, y)$$

Using the Cauchy-Riemann equations for u and v, it follows that both u and v are harmonic functions in the domain of analyticity for f. For results obtained from this approach, see Mikhlin [378]. Most of the representation formulas given above can also be obtained by using Cauchy's integral formula.

7.2. Boundary integral equations of the second kind

The original Fredholm theory for integral equations was developed to study the integral equations considered in this section, and these equations have also long been used as a means to solve boundary value problems for Laplace's equation. We begin with an indirect method for solving the interior Dirichlet problem for Laplace's equation, and then the results for this method are extended to integral equations for the interior and exterior Neumann problems.

Recall the double layer representation (7.1.40) for a function u harmonic on the interior region D_i:

$$u(A) = \int_S \rho(Q) \frac{\partial}{\partial \mathbf{n}_Q} [\log |A - Q|] \, dS_Q, \quad A \in D_i \qquad (7.2.46)$$

To solve the interior Dirichlet problem (7.1.2), the density ρ is obtained by solving the boundary integral equation given in (7.1.42), namely

$$-\pi \rho(P) + \int_S \rho(Q) \frac{\partial}{\partial \mathbf{n}_Q} [\log |P - Q|] \, dS_Q = f(P), \quad P \in S \qquad (7.2.47)$$

with f the given value of u on S. This is basically the same form of integral equation as in (7.1.30) for the exterior Neumann problem, with a different right-hand function. When the representation $\mathbf{r}(t) = (\xi(t), \eta(t))$ of (7.1.1) for S is applied, this integral equation becomes

$$-\pi\rho(t) + \int_0^L K(t,s)\rho(s)\,ds = f(t), \quad 0 \le t \le L \tag{7.2.48}$$

with $K(t,s)$ given in (7.1.32)–(7.1.33) and $f(t) \equiv f(\mathbf{r}(t))$. The smoothness and periodicity of K is discussed in and following (7.1.36), and the natural function space setting for studying (7.2.48) is $C_p(L)$ with the uniform norm. Symbolically, we write (7.2.48) as $(-\pi + \mathcal{K})\rho = f$.

The equation (7.2.47) has been very well studied, for well over 100 years; for example, see Colton [126, p. 216], Kress [325, p. 71] and Mikhlin [378, Chap. 4]. From this work, $(-\pi + \mathcal{K})^{-1}$ exists as a bounded operator from $C_p(L)$ to $C_p(L)$.

Because the kernel K and the functions f and ρ are periodic in both variables over $[0, L]$, and because K and ρ are usually smooth, the most efficient numerical method for solving the equation (7.2.48) is generally the Nyström method with the trapezoidal rule as the numerical integration rule. Because of the periodicity, the trapezoidal rule simplifies further, and the approximating equation takes the form

$$-\pi\rho_n(t) + h\sum_{j=1}^{n} K(t,t_j)\rho_n(t_j) = f(t), \quad 0 \le t \le L \tag{7.2.49}$$

with $h = L/n$, $t_j = jh$ for $j = 1, 2, \dots, n$. Symbolically, we write this as $(-\pi + \mathcal{K}_n)\rho_n = f$, with the numerical integration operator $\mathcal{K}_n$ defined implicitly by (7.2.49). Collocating at the node points, we obtain the linear system

$$-\pi\rho_n(t_i) + h\sum_{j=1}^{n} K(t_i,t_j)\rho_n(t_j) = f(t_i), \quad i = 1, \dots, n \tag{7.2.50}$$

whose solution is $[\rho_n(t_1), \dots, \rho_n(t_n)]^{\mathrm{T}}$. Then the Nyström interpolation formula can be used to obtain $\rho_n(t)$:

$$\rho_n(t) = \frac{1}{\pi}\left[-f(t) + h\sum_{j=1}^{n} K(t,t_j)\rho_n(t_j)\right], \quad 0 \le t \le L \tag{7.2.51}$$

This is a simple method to program, and usually the value of n is not too large, so that the linear system (7.2.50) can be solved directly, without iteration.

The error analysis for the above is straightforward from Theorem 4.1.1 of Chapter 4. This theorem shows that (7.2.49) is uniquely solvable for all

Table 7.1. *Errors in density function* ρ_n
for (7.2.54)

n	$(a, b) = (1, 2)$	$(a, b) = (1, 5)$
8	3.67E−3	4.42E−1
16	5.75E−5	1.13E−2
32	1.34E−14	1.74E−5
64		3.96E−11

sufficiently large values of n, say $n \geq N$; moreover,

$$\|\rho - \rho_n\|_\infty \leq \|(-\pi + \mathcal{K}_n)^{-1}\| \|\mathcal{K}\rho - \mathcal{K}_n\rho\|_\infty, \quad n \geq N \qquad (7.2.52)$$

It is well known that the trapezoidal rule is very rapidly convergent when the integrand is periodic and smooth. Consequently, $\rho_n \to \rho$ with a similarly rapid rate of convergence.

Example. Let the boundary S be the ellipse

$$\mathbf{r}(t) = (a \cos t, b \sin t), \quad 0 \leq t \leq 2\pi \qquad (7.2.53)$$

In this case, the kernel K of (7.1.32) can be reduced to

$$K(t, s) = \kappa \left(\frac{s + t}{2} \right), \qquad \kappa(\theta) = \frac{-ab}{2[a^2 \sin^2 \theta + b^2 \cos^2 \theta]}$$

and the integral equation (7.2.48) becomes

$$-\pi \rho(t) + \int_0^{2\pi} \kappa \left(\frac{s + t}{2} \right) \rho(s) \, ds = f(t), \quad 0 \leq t \leq 2\pi \qquad (7.2.54)$$

In Table 7.1, we give results for solving this equation with

$$f(x, y) = e^x \cos y, \quad (x, y) \in S \qquad (7.2.55)$$

The true solution ρ is not known explicitly; but we obtain a highly accurate solution by using a large value of n, and then this solution is used to calculate the errors shown in the table.

Results are given for $(a, b) = (1, 2)$ and $(1, 5)$. The latter ellipse is somewhat elongated, and this causes the kernel K to be more peaked. In particular,

$$p(a, b) \equiv \frac{\max |K(t, s)|}{\min |K(t, s)|} = \left[\frac{\max\{a, b\}}{\min\{a, b\}} \right]^2$$

and $p(1, 2) = 4$, $p(1, 5) = 25$. As the peaking factor becomes larger, it is

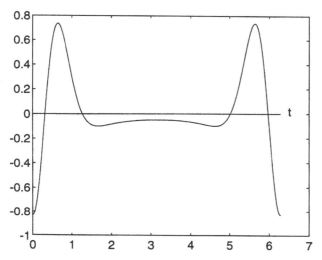

Figure 7.1. The density ρ for (7.2.54) with $(a, b) = (1, 5)$.

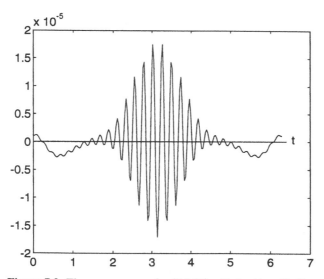

Figure 7.2. The error $\rho - \rho_{32}$ for (7.2.54) with $(a, b) = (1, 5)$.

necessary to increase n to retain comparable accuracy in approximating the integral $\mathcal{K}\rho$, and the consequences of this can be seen in the table.

A graph of ρ is given in Figure 7.1 for $(a, b) = (1, 5)$, and it shows a somewhat rapid change in the function around $t = 0$ or $(x, y) = (a, 0)$ on S. For the same curve S, a graph of the error $\rho(t) - \rho_n(t)$, $0 \le t \le 2\pi$, is given in Figure 7.2 for the case $n = 32$. Perhaps surprisingly in light of Figure 7.1, the error is largest around $t = \pi$ or $(x, y) = (-a, 0)$ on S, where ρ is better behaved.

7.2.1. Evaluation of the double layer potential

When using the representation $\mathbf{r}(s) = (\xi(s), \eta(s))$ of (7.1.1) for S, the double layer integral formula (7.2.46) takes the form

$$u(x, y) = \int_0^L M(x, y, s)\rho(s)\, ds, \quad (x, y) \in D_i \qquad (7.2.56)$$

$$M(x, y, s) = \frac{-\eta'(s)[\xi(s) - x] + \xi'(s)[\eta(s) - y]}{[\xi(s) - x]^2 + [\eta(s) - y]^2} \qquad (7.2.57)$$

This kernel is increasingly peaked as (x, y) approaches S. To see this more clearly, let S be the unit circle given by $\mathbf{r}(s) = (\cos s, \sin s)$, $0 \le s \le 2\pi$. Then

$$M(x, y, s) = \frac{-\cos s[\cos s - x] - \sin s[\sin s - y]}{[\cos s - x]^2 + [\sin s - y]^2}$$

To see the near-singular behavior more clearly, let (x, y) approach $(\cos s, \sin s)$ along the line

$$(x, y) = q(\cos s, \sin s), \quad 0 \le q < 1$$

Then after simplifying,

$$M(q \cos s, q \sin s, s) = \frac{1}{1 - q}$$

We use numerical integration to approximate (7.2.56), and since the integrand is periodic in s, the trapezoidal rule is an optimal choice. As (x, y) approaches S, the needed number of integration nodes will need to be increased to retain equivalent accuracy in the approximate values of $u(x, y)$. For (x, y) very close to S, other means should be used to approximate the integral (7.2.56), since the trapezoidal rule will be very expensive. Two such methods are explored in Miller [383, Chap. 4].

To solve the original Dirichlet problem (7.1.2), we first approximate the density ρ, obtaining ρ_n, and then we numerically integrate the double layer integral based on ρ_n. To aid in studying the resulting approximation of $u(x, y)$, introduce the following notation. Let $u_n(x, y)$ be the double layer potential using the approximate density ρ_n obtained by the Nyström method of (7.2.49):

$$u_n(x, y) = \int_0^L M(x, y, s)\rho_n(s)\, ds, \quad (x, y) \in D_i \qquad (7.2.58)$$

Let $u_{n,m}(x, y)$ denote the result of approximating $u_n(x, y)$ using the trapezoidal rule:

$$u_{n,m}(x, y) = h \sum_{i=1}^{m} M(x, y, t_i)\rho_n(t_i), \quad (x, y) \in D_i \qquad (7.2.59)$$

For the error in u_n, note that $u - u_n$ is a harmonic function, and therefore, by the maximum principle for such functions,

$$\max_{(x,y)\in \bar{D}_i} |u(x, y) - u_n(x, y)| = \max_{(x,y)\in S} |u(x, y) - u_n(x, y)| \qquad (7.2.60)$$

Since $u - u_n$ is also a double layer potential, the argument that led to the original integral equation (7.2.46) also implies

$$u(P) - u_n(P) = -\pi[\rho(P) - \rho_n(P)]$$
$$+ \int_S [\rho(Q) - \rho_n(Q)] \frac{\partial}{\partial \mathbf{n}_Q}[\log|P - Q|]\, dS_Q, \quad P \in S$$
$$\qquad (7.2.61)$$

Taking bounds,

$$|u(P) - u_n(P)| \leq [\pi + \|\mathcal{K}\|]\|\rho - \rho_n\|_\infty, \quad P \in S$$

Combined with (7.2.60),

$$\max_{(x,y)\in \bar{D}_i} |u(x, y) - u_n(x, y)| \leq [\pi + \|\mathcal{K}\|]\|\rho - \rho_n\|_\infty \qquad (7.2.62)$$

If the region D_i is convex, then the double layer kernel is strictly negative, and it can then be shown that

$$\|\mathcal{K}\| = \pi$$

For convex regions, therefore,

$$\max_{(x,y)\in \bar{D}_i} |u(x, y) - u_n(x, y)| \leq 2\pi\|\rho - \rho_n\|_\infty \qquad (7.2.63)$$

An algorithm for solving the interior Dirichlet problem (7.1.2) can be based on first solving for ρ_n to a prescribed accuracy. Then (7.2.62) says u_n has comparable accuracy uniformly on D_i. To complete the task of evaluating $u_n(x, y)$ for given values of (x, y), one can use the trapezoidal rule (7.2.59),

Table 7.2. *Errors* $u(\mathbf{c}(q)) - u_{n,m}(\mathbf{c}(q))$ *with* $n = 16$

q	$m = 16$	$m = 32$	$m = 64$	$m = 128$
0	−3.41E−1	−9.36E−3	3.51E−3	3.53E−3
.10	−2.31E−1	1.02E−2	1.92E−3	1.90E−3
.20	8.35E−2	1.54E−2	−3.40E−4	−4.11E−4
.30	3.08E−1	−3.32E−2	−2.17E−3	−1.99E−3
.40	3.96E−1	−9.03E−4	−2.05E−3	−2.07E−3
.50	4.46E−1	6.20E−2	1.62E−4	−8.19E−4
.60	4.52E−2	−7.80E−2	−2.12E−3	9.38E−4
.70	−7.95E−1	−8.37E−2	−1.13E−2	2.33E−3
.80	−1.25E+0	5.32E−1	2.64E−2	1.79E−3
.90	−1.07E+0	−1.13E+0	4.86E−1	3.49E−2
.98	2.72E+0	4.01E−1	−6.89E−1	−1.02E+0

varying m to obtain desired accuracy in $u_{n,m}(x, y)$. The total error is then given by

$$u(x, y) - u_{n,m}(x, y) = [u(x, y) - u_n(x, y)] + [u_n(x, y) - u_{n,m}(x, y)] \qquad (7.2.64)$$

Ideally, the two errors on the right side should be made comparable in size, to make the algorithm as efficient as possible. A Fortran program implementing these ideas is given in Jeon and Atkinson [291], and it also uses a slight improvement on (7.2.59) when (x, y) is near to S.

Example. We continue with the preceding example (7.2.53)–(7.2.55), noting that the true solution is also given by (7.2.55). For the case $(a, b) = (1, 5)$ and $n = 16$, we examine the error in the numerical solutions u_n and $u_{n,m}$ along the line

$$\mathbf{c}(q) = q\left(a \cos \frac{\pi}{4}, b \sin \frac{\pi}{4}\right), \qquad 0 \le q < 1 \qquad (7.2.65)$$

A graph of the error $u(\mathbf{c}(q)) - u_n(\mathbf{c}(q))$, $0 \le q \le .94$, is shown in Figure 7.3. Note that the size of the error is around 17 times smaller than is predicted from the error of $\|\rho - \rho_{16}\|_\infty = .0113$ of Table 7.1 and the bound (7.2.63). Table 7.2 contains the errors $u(\mathbf{c}(q)) - u_{n,m}(\mathbf{c}(q))$ for selected values of q and m, with $n = 16$. Graphs of these errors are given in Figure 7.4. Compare these graphs with that of Figure 7.3, noting the quite different vertical scales. It is clear that increasing m decreases the error, up to the point that the dominant error is that of $u(x, y) - u_n(x, y)$ in (7.2.64).

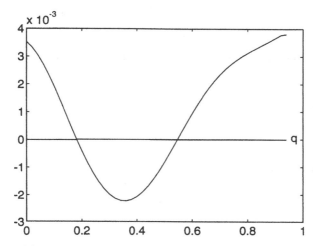

Figure 7.3. The errors $u(\mathbf{c}(q)) - u_n(\mathbf{c}(q))$ with $n = 16$.

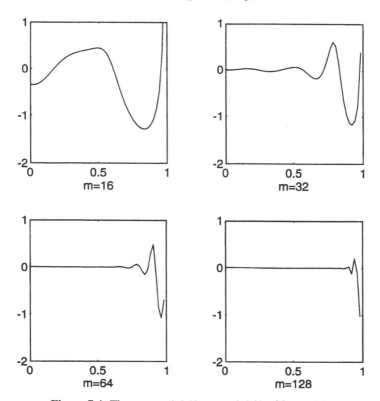

Figure 7.4. The errors $u(\mathbf{c}(q)) - u_{n,m}(\mathbf{c}(q))$ with $n = 16$.

7.2.2. The exterior Neumann problem

Recall the solving of the exterior Neumann problem (7.1.3) by means of the integral representation formula (7.1.24) and the boundary integral equation of (7.1.30). We rewrite the latter as

$$-\pi u(P) + \int_S u(Q) \frac{\partial}{\partial n_Q} [\log |P - Q|] \, dS_Q = \int_S f(Q) \log |P - Q| \, dS_Q,$$

$$P \in S \qquad (7.2.66)$$

The left side of this equation is the same as that of (7.2.47) for the interior Dirichlet problem, and it is therefore only the evaluation of the right side that concerns us here. Recalling (7.1.34), the right side is

$$g(t) = \int_0^L f(\mathbf{r}(s)) \sqrt{\xi'(s)^2 + \eta'(s)^2} \log |\mathbf{r}(t) - \mathbf{r}(s)| \, ds \qquad (7.2.67)$$

This could be approximated using the product integration techniques of §4.2 in Chapter 4, but we consider a more efficient method.

To simplify the notation, the parametrization $\mathbf{r}(t)$ of (7.2.46) is assumed to be defined on the standard interval $[0, 2\pi]$. Also, introduce

$$\varphi(s) = f(\mathbf{r}(s)) \sqrt{\xi'(s)^2 + \eta'(s)^2} \qquad (7.2.68)$$

The integral (7.2.67) becomes

$$g(t) = \int_0^{2\pi} \varphi(s) \log |\mathbf{r}(t) - \mathbf{r}(s)| \, ds, \quad 0 \le t \le 2\pi \qquad (7.2.69)$$

We write the kernel of this integral in the form

$$\log |\mathbf{r}(t) - \mathbf{r}(s)| = \log \left| 2e^{-\frac{1}{2}} \sin\left(\frac{t-s}{2}\right) \right| + B(t, s) \qquad (7.2.70)$$

$$B(t, s) = \begin{cases} \log \dfrac{\left| e^{\frac{1}{2}} [\mathbf{r}(t) - \mathbf{r}(s)] \right|}{\left| 2 \sin\left(\frac{t-s}{2}\right) \right|}, & t - s \ne 2m\pi \\ \log \left| e^{\frac{1}{2}} \mathbf{r}'(t) \right|, & t - s = 2m\pi \end{cases} \qquad (7.2.71)$$

The integral (7.2.69) becomes

$$g(t) = \int_0^{2\pi} \varphi(s) \log \left| 2e^{-\frac{1}{2}} \sin\left(\frac{t-s}{2}\right) \right| ds + \int_0^{2\pi} B(t, s)\varphi(s) \, ds$$

$$\equiv -\pi \mathcal{A}\varphi(t) + \mathcal{B}\varphi(t) \qquad (7.2.72)$$

Assuming $\mathbf{r} \in C_p^\kappa(2\pi)$, the kernel function $B \in C^{\kappa-1}([0, 2\pi] \times [0, 2\pi])$, and B is periodic in both variables t and s. Consequently, the second integral $B\varphi(t)$ in (7.2.72) can be accurately and efficiently approximated using the trapezoidal rule.

The first integral in (7.2.72) is a minor modification of the integral operator associated with the kernel $\log |P - Q|$ for S equal to the unit circle about the origin, where we have introduced

$$A\varphi(t) = -\frac{1}{\pi} \int_0^{2\pi} \varphi(s) \log \left| 2e^{-\frac{1}{2}} \sin \left(\frac{t-s}{2} \right) \right| ds, \quad 0 \le t \le 2\pi$$

(7.2.73)

To give some of its properties, write the Fourier expansion of an arbitrary $\varphi \in L^2(0, 2\pi)$:

$$\varphi(s) = \frac{1}{\sqrt{2\pi}} \sum_{m=-\infty}^\infty \hat{\varphi}(m) e^{ims}$$

$$\hat{\varphi}(m) = \frac{1}{\sqrt{2\pi}} \int_0^{2\pi} \varphi(s) e^{-ims} ds$$

Then it can be shown that

$$A\varphi(t) = \frac{1}{\sqrt{2\pi}} \left[\hat{\varphi}(0) + \sum_{|m|>0} \frac{\hat{\varphi}(m)}{|m|} e^{imt} \right]$$

(7.2.74)

This is an expansion of $A\varphi$ using the eigenfunctions $\psi_m(t) \equiv e^{imt}$ and the corresponding eigenvalues of A. For a proof of this result, and for a much more extensive discussion of the properties of A, see Yan and Sloan [580]. Extensive use is made of the decomposition (7.2.72) in §7.3.

Later, we need bounds for $A\varphi(t)$. When A is considered as an operator from $C_p(2\pi)$ to $C_p(2\pi)$, we can use

$$\|A\| = \max_t \int_0^{2\pi} \left| \log \left| 2e^{-\frac{1}{2}} \sin \left(\frac{t-s}{2} \right) \right| \right| ds$$

A straightforward bounding of this leads to

$$\|A\| \le \int_0^{2\pi} \log \left(2e^{-\frac{1}{2}} \right) ds + 2 \int_0^{2\pi} \log \sin \left(\frac{v}{2} \right) dv$$

$$= \pi[4 \log 2 - 1] \doteq 5.57$$

(7.2.75)

An improved bound comes from a direct examination of the formula (7.2.74). Applying the Cauchy-Schwarz inequality to (7.2.74),

$$|\mathcal{A}\varphi(t)|^2 \leq \frac{1}{2\pi}\left[1 + 2\sum_{|m|>0}\frac{1}{m^2}\right]\left[\sum_{m=-\infty}^{\infty}|\hat{\varphi}(m)|^2\right] \tag{7.2.76}$$

Using Parseval's equality,

$$\|\varphi\|_{L^2(0,2\pi)} = \sqrt{\sum_{m=-\infty}^{\infty}|\hat{\varphi}(m)|^2}$$

and using the well-known series

$$\sum_{m=1}^{\infty}\frac{1}{m^2} = \frac{\pi^2}{6}$$

the inequality (7.2.76) implies

$$|\mathcal{A}\varphi(t)| \leq \sqrt{\frac{1}{2\pi}\left(1 + \frac{\pi^2}{3}\right)}\|\varphi\|_{L^2(0,2\pi)} \tag{7.2.77}$$

Using

$$\|\varphi\|_{L^2(0,2\pi)} \leq \sqrt{2\pi}\|\varphi\|_{\infty}$$

the bound (7.2.77) implies

$$\|\mathcal{A}\varphi\|_{\infty} \leq \sqrt{1 + \frac{\pi^2}{3}}\|\varphi\|_{\infty}, \quad \varphi \in C_p(2\pi) \tag{7.2.78}$$

This also yields

$$\|\mathcal{A}\| \leq \sqrt{1 + \frac{\pi^2}{3}} \doteq 2.07 \tag{7.2.79}$$

which is an improvement over (7.2.75).

To approximate $\mathcal{A}\varphi$, we approximate φ using trigonometric interpolation, and then (7.2.74) is used to evaluate exactly the resulting approximation of $\mathcal{A}\varphi$. Let $n \geq 1$, $h = 2\pi/(2n+1)$, and

$$t_j = jh, \quad j = 0, \pm 1, \pm 2, \ldots \tag{7.2.80}$$

Let $\mathcal{Q}_n\varphi$ denote the trigonometric polynomial of degree $\leq n$ that interpolates $\varphi(t)$ at the nodes $\{t_0, t_1, \ldots, t_{2n}\}$, and by periodicity at all other nodes t_j. Also, let $T_k(\varphi)$ denote the trapezoidal rule on $[0, 2\pi]$ with k subdivisions:

$$T_k(\varphi) = \frac{2\pi}{k} \sum_{j=0}^{k-1} \varphi\left(\frac{2\pi j}{k}\right), \quad \varphi \in C_p(2\pi)$$

and introduce the discrete inner product

$$(f, g)_k = T_k(f(\cdot)\bar{g}(\cdot)) = \frac{2\pi}{k} \sum_{j=0}^{k-1} f\left(\frac{2\pi j}{k}\right) \bar{g}\left(\frac{2\pi j}{k}\right)$$

Introduce

$$\psi_m(s) = e^{ims}, \quad m = 0, \pm 1, \pm 2, \ldots$$

It is a relatively straightforward exercise to show that

$$T_k(\psi_m) = \begin{cases} 2\pi, & m = jk, \ j = 0, \pm 1, \pm 2, \ldots \\ 0, & \text{otherwise} \end{cases} \tag{7.2.81}$$

Using the discrete inner product $(\cdot, \cdot)_{2n+1}$, and applying (7.2.81), it follows that

$$(\psi_j, \psi_k)_{2n+1} = \begin{cases} 2\pi, & j = k \\ 0, & j \neq k \end{cases} \tag{7.2.82}$$

for $-n \leq j, k \leq n$. This says $\{\psi_{-n}, \ldots, \psi_n\}$ is an orthogonal basis for $\mathcal{X}_n \equiv \text{span}\{\psi_{-n}, \ldots, \psi_n\}$ when using the discrete inner product $(\cdot, \cdot)_{2n+1}$.

Given $\varphi \in C_p(2\pi)$, the interpolatory polynomial $\mathcal{Q}_n\varphi$ is also the discrete orthogonal projection into $\mathcal{X}_n$, as discussed in §4.4 of Chapter 4; see Lemma 4.4.1(d). Using (7.2.82), it will follow that

$$\mathcal{Q}_n\varphi(t) = \sum_{j=-n}^{n} \alpha_j \psi_j(t) \tag{7.2.83}$$

$$\alpha_j = \frac{1}{2\pi}(\varphi, \psi_j)_{2n+1} \tag{7.2.84}$$

For the error in $\mathcal{Q}_n\varphi$, recall the results of (3.2.56)–(3.2.58) in §3.2 of Chapter 3. Then

$$\|\varphi - \mathcal{Q}_n\varphi\|_\infty = O\left(\frac{\log n}{n^{\ell+\alpha}}\right), \quad \varphi \in C_p^{\ell,\alpha}(2\pi) \tag{7.2.85}$$

In this, φ is assumed to be ℓ-times continuously differentiable, and $\varphi^{(\ell)}$ is assumed to satisfy the Hölder condition

$$\left|\varphi^{(\ell)}(s) - \varphi^{(\ell)}(t)\right| \leq c\,|s - t|^{\alpha}, \quad -\infty < s, t < \infty$$

with c a finite constant.

We approximate $\mathcal{A}\varphi(t)$ using $\mathcal{A}\mathcal{Q}_n\varphi(t)$. From (7.2.74),

$$\mathcal{A}\psi_j = \begin{cases} \psi_0, & j = 0 \\ \dfrac{1}{|j|}\psi_j, & |j| > 0 \end{cases} \tag{7.2.86}$$

Applying this with (7.2.83),

$$\mathcal{A}\varphi(t) \approx \mathcal{A}\mathcal{Q}_n\varphi(t) = \alpha_0 + \sum_{\substack{j=-n \\ j \neq 0}}^{n} \frac{\alpha_j}{|j|}\psi_j(t), \tag{7.2.87}$$

where the coefficients $\{\alpha_j\}$ are given in (7.2.84).

To bound the error in $\mathcal{A}\mathcal{Q}_n\varphi$, we apply (7.2.78), yielding

$$\|\mathcal{A}\varphi - \mathcal{A}\mathcal{Q}_n\varphi\|_{\infty} \leq \|\mathcal{A}\| \|\varphi - \mathcal{Q}_n\varphi\|_{\infty}$$

Using (7.2.85), this bound implies

$$\|\mathcal{A}\varphi - \mathcal{A}\mathcal{Q}_n\varphi\|_{\infty} = O\left(\frac{\log n}{n^{\ell+\alpha}}\right), \quad \varphi \in C_p^{\ell,\alpha}(2\pi) \tag{7.2.88}$$

provided $\ell + \alpha > 0$. The approximation $\mathcal{A}\mathcal{Q}_n\varphi$ is rapidly convergent to $\mathcal{A}\varphi$.

To complete the approximation of the original integral (7.2.72), approximate $\mathcal{B}\varphi(t)$ using the trapezoidal rule with the nodes $\{t_j\}$ of (7.2.80):

$$\begin{aligned} \mathcal{B}\varphi(t) &\approx T_{2n+1}(B(t, \cdot)\varphi) \\ &= \frac{2\pi}{2n+1} \sum_{k=0}^{2n} B(t, t_k)\varphi(t_k) \\ &\equiv \mathcal{B}_n\varphi(t) \end{aligned} \tag{7.2.89}$$

To bound the error, we can use the standard Euler-MacLaurin error formula to show

$$|\mathcal{B}\varphi(t) - \mathcal{B}_n\varphi(t)| \leq O(n^{-\ell}), \quad \varphi \in C_p^{\ell}(2\pi) \tag{7.2.90}$$

This assumes that $\mathbf{r} \in C_p^{\kappa}(2\pi)$ with $\kappa \geq \ell + 1$. (Note that this result can be extended somewhat by using the alternative error formula given in Lemma 7.3.2 in §7.3.)

Table 7.3. *The error* $\|u - u_n\|_\infty$
for (7.2.91)

n	m	$\|u - u_n\|_\infty$
8	17	3.16E$-$2
16	33	3.42E$-$4
32	65	4.89E$-$8
64	129	1.44E$-$15

To solve the original integral equation (7.2.66), we use the Nyström method of (7.2.49)–(7.2.51) based on the trapezoidal numerical integration method with the $2n + 1$ nodes $\{t_0, \ldots, t_{2n}\}$ of (7.2.80). The right side g of (7.2.72) is approximated by using (7.2.87) and (7.2.89), yielding the approximation

$$(-\pi + \mathcal{K}_n)u_n = -\pi \mathcal{A}\mathcal{Q}_n\varphi + \mathcal{B}_n\varphi(t) \tag{7.2.91}$$

Error bounds can be produced by combining (7.2.88) and (7.2.90) with the earlier error analysis based on (7.2.52). We leave it to the reader to show that if $\varphi \in C_p^\ell(2\pi)$ for some $\ell \geq 1$, and if $\mathbf{r} \in C_p^\kappa(2\pi)$ with $\kappa \geq \ell + 1$, then the approximate Nyström solution u_n of (7.2.66) satisfies

$$\|u - u_n\|_\infty \leq O\left(\frac{\log n}{n^\ell}\right) \tag{7.2.92}$$

Example. We solve the exterior Neumann problem on the region outside the ellipse

$$\mathbf{r}(t) = (a\cos t, b\sin t), \quad 0 \leq t \leq 2\pi$$

For purposes of illustration, we use a known true solution,

$$u(x, y) = \frac{x}{x^2 + y^2}$$

This function is harmonic, and $u(x, y) \to 0$ as $x^2 + y^2 \to \infty$. The Neumann boundary data is generated from u. Numerical results for $(a, b) = (1, 2)$ are given in Table 7.3, and in it, $m = 2n + 1$ is the order of the linear system being solved by the Nyström method. A graph of the error $u(\mathbf{r}(t)) - u_n(\mathbf{r}(t))$ is given in Figure 7.5 for the case $n = 16$.

7.2.3. Other boundary value problems

The exterior Dirichlet problem of (7.1.14) is probably handled best by first re-formulating it as an interior Dirichlet problem by means of the Kelvin

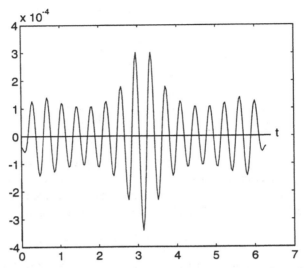

Figure 7.5. The error $u(\mathbf{r}(t)) - u_n(\mathbf{r}(t))$ for $n = 16$ and $(a, b) = (1, 2)$.

transformation, as in (7.1.15). Then apply any of the standard methods for the interior Dirichlet problem, for example, as in (7.2.46)–(7.2.47). When solved in this way, the exact asymptotic behavior of $u(x, y)$ as $x^2 + y^2 \to \infty$ can be obtained, as discussed following (7.1.15).

The interior Neumann problem (7.1.3) can be handled similarly. Using the Kelvin transformation (7.1.13), we can transform (7.1.3) to an equivalent exterior Neumann problem, carrying out the reverse of the transformation in (7.1.18)–(7.1.22). Then the exterior Neumann problem can be solved numerically, for example, as above in (7.2.91). This is implemented in Jeon and Atkinson [291]. Here we examine another popular approach to solving the interior Neumann problem.

For easier reference, the interior Neumann problem is to find $u \in C^1(\bar{D}_i) \cap C^2(D_i)$ that satisfies

$$
\begin{aligned}
\Delta u(P) &= 0, & P \in D_i \\
\frac{\partial u(P)}{\partial \mathbf{n}_P} &= f(P), & P \in S
\end{aligned}
\tag{7.2.93}
$$

with $f \in C(S)$ a given boundary function assumed to satisfy

$$
\int_S f(P)\, dS = 0
\tag{7.2.94}
$$

To solve this problem, we represent u as a single layer potential, as was done in (7.1.45):

$$u(A) = \int_S \rho(Q) \log |A - Q| \, dS_Q, \quad A \in D_i \tag{7.2.95}$$

Employing the Neumann boundary condition in (7.2.93), ρ must satisfy the integral equation

$$\pi \rho(P) + \int_S \rho(Q) \frac{\partial}{\partial \mathbf{n}_P} \log |P - Q| \, dS_Q = f(P), \quad P \in S \tag{7.2.96}$$

This is a well-studied integral equation; for example, see Kress [325, p. 72]. However, it has the serious disadvantage of not being uniquely solvable, and that causes difficulty when trying to solve it numerically. Writing (7.2.96) abstractly as

$$(\pi + \mathcal{K})\rho = f \tag{7.2.97}$$

we consider the adjoint operator $\pi + \mathcal{K}^*$. This adjoint is known to have a one-dimensional eigenfunction subspace:

$$(\pi + \mathcal{K}^*)\varphi = 0$$

with φ any constant function. To see this, return to Green's representation formula (7.1.26) and let u be a constant function. Recalling Theorem 1.4.2 of §1.4 in Chapter 1, the equation $(\pi + \mathcal{K})\rho = 0$ must also have a one-dimensional solution space, say all constant multiples of some $\psi \in C(S)$:

$$(\pi + \mathcal{K})\psi = 0 \tag{7.2.98}$$

Also, from Theorem 1.4.2,

$$\psi \perp \text{Range}(\pi + \mathcal{K}^*) \tag{7.2.99}$$

$$\varphi \perp \text{Range}(\pi + \mathcal{K}) \tag{7.2.100}$$

say with $\varphi \equiv 1$.

For most boundaries S, the function ψ can be obtained as follows. Find the single layer density function ψ for which

$$\int_S \psi(Q) \log |A - Q| \, dS_Q = 1, \quad A \in D_i \tag{7.2.101}$$

The eigenfunction ψ yields a constant potential field inside of D_i. Forming the normal derivative of the left side and letting $A \to P \in S$, we have (7.2.98). Such functions ψ do not exist for curves S called Γ-contours (see Jaswon and Symm [286, §4.3]). One way to avoid such Γ-contours S is to make D_i sufficiently small, for example, to have

$$\text{diameter}(D_i) < 1 \qquad (7.2.102)$$

For justification and greater detail, see Yan and Sloan [580]. Since harmonic functions remain harmonic under a constant rescaling in all coordinates, we assume the region D_i has been so chosen that S is not a Γ-contour.

Using the parametrization $\mathbf{r}(t)$ of (7.1.1), say with $0 \le t \le 2\pi$ as the parametrization interval, the integral equation (7.2.96) can be written as

$$\pi\rho(t) + \int_0^{2\pi} K(t,s)\rho(s)\,ds = f(t), \qquad 0 \le t \le 2\pi \qquad (7.2.103)$$

In this, we have used $\rho(s) \equiv \rho(\mathbf{r}(s))$ and $f(s) \equiv f(\mathbf{r}(s))$; and

$$K(t,s) = \frac{\eta'(t)[\xi(s) - \xi(t)] - \xi'(t)[\eta(s) - \eta(t)]}{[\xi(t) - \xi(s)]^2 + [\eta(t) - \eta(s)]^2} \sqrt{\frac{\xi'(s)^2 + \eta'(s)^2}{\xi'(t)^2 + \eta'(t)^2}}$$

$$= \frac{\eta'(t)\xi[t,t,s] - \xi'(t)\eta[t,t,s]}{|\mathbf{r}[s,t]|^2} \sqrt{\frac{\xi'(s)^2 + \eta'(s)^2}{\xi'(t)^2 + \eta'(t)^2}}, \qquad s \ne t$$

$$K(t,t) = \frac{\eta'(t)\xi''(t) - \xi'(t)\eta''(t)}{2\{\xi'(t)^2 + \eta'(t)^2\}}$$

As with the integral equation (7.2.46) for the interior Dirichlet problem, the Nyström method with the trapezoidal rule can be used to discretize (7.2.103). But before doing so, the nonunique solvability of (7.2.103) must be confronted.

One popular way of removing the nonuniqueness is to impose some added condition, for example,

$$\int_S \rho(Q)\,dS = 0$$

This must be implemented with some care, to produce discretized linear systems that are well-behaved. We use a slightly different approach, and we modify the integral equation itself, under the assumption that S is not a Γ-contour. Let $P^* = \mathbf{r}(s^*)$ be some fixed point on S. Referring back to the integral equation $(\pi + \mathcal{K})\rho = f$ of (7.2.97), introduce the integral equation

$$(\pi + \mathcal{K})R + R(P^*) = f \qquad (7.2.104)$$

We claim this is a uniquely solvable integral equation, and in addition, if the solvability requirement (7.2.94) is satisfied, then the solution R is one of the solutions to the original equation $(\pi + \mathcal{K})\rho = f$.

The equation (7.2.104) is of the form $(\pi + \mathcal{L})R = f$, with $\mathcal{L}$ the compact operator

$$\mathcal{L}g = \mathcal{K}g + g(P^*), \quad g \in C(S)$$

To show unique solvability, the Fredholm alternative theorem says we must only show

$$(\pi + \mathcal{L})g = 0 \Rightarrow g = 0$$

The condition $(\pi + \mathcal{L})g = 0$ can be rewritten to say

$$(\pi + \mathcal{K})g = -g(P^*)$$

in which the right side is a constant function. The result (7.2.100) then implies

$$(\pi + \mathcal{K})g = 0 \qquad g(P^*) = 0$$

From (7.2.98), g is some multiple of the eigenfunction ψ for $\mathcal{K}$, say $g = c\psi$ for some constant c. But since $g(P^*) = 0$, we have $c\psi(P^*) = 0$. It can be shown that the function ψ is nonzero on S, provided S is not a Γ-contour; for example, see Pogorzelski [426, p. 240]. From this, the constant $c = 0$ and thus $g = 0$. This shows $\pi + \mathcal{L}$ is one-to-one, and therefore that (7.2.104) is uniquely solvable for all $f \in C(S)$.

Assume now that $f \in C(S)$, and that it satisfies (7.2.94). Rewrite the equation (7.2.104) as

$$(\pi + \mathcal{K})R - f = -R(P^*)$$

By (7.2.100), $f \in \text{Range}(\pi + \mathcal{K})$, and therefore the left side of the equation is in $\text{Range}(\pi + \mathcal{K})$. Again by (7.2.100), this implies

$$(\pi + \mathcal{K})R - f = 0 \qquad R(P^*) = 0 \qquad (7.2.105)$$

This shows that the solution R of (7.2.104) is one of the solutions of the original integral equation $(\pi + \mathcal{K})\rho = f$.

The integral equation (7.2.104) is a very simple way to deal with the lack of unique solvability of $(\pi + \mathcal{K})\rho = f$. We can now apply the Nyström method to the solution of (7.2.104), again using the trapezoidal numerical integration method. The theory of the numerical method is almost identical to that given earlier for the numerical solution of (7.2.47), and therefore we dispense with it here. For more details on this approach to solving integral equations that lack unique solvability, see Ref. [29].

A final desirable property of the above formulation is that it provides a simple way to calculate ψ without directly solving an eigenvalue problem. Solve the uniquely solvable equation

$$(\pi + \mathcal{K})R + R(P^*) = 1 \qquad (7.2.106)$$

Its solution will be a nonzero multiple of ψ. This equation can be solved by using the same Nyström method as is used to solve (7.2.104) for the Neumann problem. The actual ψ satisfying (7.2.101) can then be found by means of a simple quadrature. We leave the details of this as a problem for the reader.

7.3. Boundary integral equations of the first kind

Most of the original theoretical work with boundary integral equations was for integral equations of the second kind, and consequently, these types of boundary integral equations came to be the principal type used in applications. In addition, some integral equations of the first kind can be quite ill-conditioned, and that has led some people to avoid such equations in general. Boundary integral equations of the first kind, however, are generally quite well-behaved, and recently, they have been an increasingly popular approach to solving various boundary value problems. In this section we look at the best studied of such boundary integral equations of the first kind, and we introduce some of the principal tools used in the error analysis of numerical methods for solving such integral equations.

Returning to §7.1, the interior Dirichlet problem (7.1.2) leads to the first kind equation (7.1.28). Introducing a change of sign (for later convenience), we write this integral equation as

$$-\frac{1}{\pi} \int_S \rho(Q) \log |P - Q| \, dS_Q = g(P) \qquad (7.3.107)$$

In the case of (7.1.28), the unknown density ρ is the value of the normal derivative on S of the unknown harmonic function u. In this section we consider various numerical methods for solving this integral equation, building on the ideas introduced in §7.2 following (7.2.67).

7.3.1. Sobolev spaces

For a function $\varphi \in L^2(0, 2\pi)$, we write its Fourier expansion as

$$\varphi(s) = \sum_{m=-\infty}^{\infty} a_m \psi_m(s), \quad \psi_m(s) = \frac{1}{\sqrt{2\pi}} e^{ims} \qquad (7.3.108)$$

$$a_m = \int_0^{2\pi} \varphi(s) \overline{\psi_m(s)} \, ds \qquad (7.3.109)$$

Note the slightly changed definition of ψ_m from that used in the preceding §7.2. From Parseval's equality,

$$\sqrt{\sum_{m=-\infty}^{\infty} |a_m|^2} = \sqrt{\int_0^{2\pi} |\varphi(s)|^2 \, ds} \equiv \|\varphi\|_{L^2}$$

the norm of φ in $L^2(0, 2\pi)$. If the function φ is k-times continuously differentiable, then we can differentiate (7.3.108) to obtain

$$\varphi^{(k)}(s) = \sum_{\substack{m=-\infty \\ m\neq 0}}^{\infty} (im)^k a_m \psi_m(s) \qquad (7.3.110)$$

which is convergent in $L^2(0, 2\pi)$. This implies $(im)^k a_m$ is the mth Fourier coefficient of $\varphi^{(k)}$, a fact that can also be proven directly from integration by parts in the formula

$$\int_0^{2\pi} \varphi^{(k)}(s) \overline{\psi_m(s)} \, ds$$

for the mth Fourier coefficient. From Parseval's equality applied to (7.3.110), we have

$$\sqrt{\sum_{\substack{m=-\infty \\ m\neq 0}}^{\infty} |m|^{2k} |a_m|^2} = \sqrt{\int_0^{2\pi} \left|\varphi^{(k)}(s)\right|^2 ds} = \left\|\varphi^{(k)}\right\|_{L^2} \qquad (7.3.111)$$

We use this as motivation for the following definition.

For any real number $q \geq 0$, define $H^q(2\pi)$ to be the set of all functions $\varphi \in L^2(0, 2\pi)$ for which

$$\|\varphi\|_q \equiv \left[|a_0|^2 + \sum_{\substack{m=-\infty \\ m\neq 0}}^{\infty} |m|^{2q} |a_m|^2 \right]^{\frac{1}{2}} < \infty \qquad (7.3.112)$$

where (7.3.108) is the Fourier series for φ. The space $H^q(2\pi)$ is a Hilbert space, and the inner product associated with it is given by

$$(\varphi, \psi)_q = a_0 \bar{b}_0 + \sum_{\substack{m=-\infty \\ m\neq 0}}^{\infty} |m|^{2q} a_m \bar{b}_m \qquad (7.3.113)$$

where

$$\varphi(s) = \sum_{m=-\infty}^{\infty} a_m \psi_m(s), \qquad \psi(s) = \sum_{m=-\infty}^{\infty} b_m \psi_m(s),$$

It can be shown that for $q = k$ a positive integer, $H^k(2\pi)$ is the set of all functions φ that are k-times differentiable with $\varphi^{(k)} \in L^2(0, 2\pi)$; the norm in $L^2(0, 2\pi)$ of $\varphi^{(k)}$ is given by (7.3.111). For q not an integer, write $q = k + \alpha$ with $0 < \alpha < 1$. Then $H^q(2\pi)$ is the set of all functions $\varphi \in H^k(2\pi)$ for which

$$\int_0^{2\pi} \int_0^{2\pi} \frac{\left|\varphi^{(k)}(s) - \varphi^{(k)}(\sigma)\right|^2}{\left| \sin \frac{s-\sigma}{2} \right|^{2\alpha+1}} \, ds \, d\sigma < \infty \qquad (7.3.114)$$

This can be seen as a type of Hölder condition with exponent α for $\varphi^{(k)}$. For proofs, see Kress [325, Theorems 8.4, 8.5]. The space $H^q(2\pi)$ has many important properties, but we can give only a few of them here. For a more complete introduction, see Kress [325, Chap. 8].

Lemma 7.3.1. Let $q > k + \frac{1}{2}$ for some integer $k \geq 0$, and let $\varphi \in H^q(2\pi)$. Then $\varphi \in C_p^k(2\pi)$.

Proof. We give a proof for only the case $k = 0$, as the general case is quite similar. We show the Fourier series (7.3.108) for φ is absolutely and uniformly convergent on $[0, 2\pi]$, and it then follows by standard arguments that φ is continuous and periodic. From the definition (7.3.108), and by using the Cauchy-Schwarz inequality,

$$|\varphi(s)| \leq \sum_{m=-\infty}^{\infty} |a_m|$$

$$= |a_0| + \sum_{\substack{m=-\infty \\ m \neq 0}}^{\infty} |m|^{-q} |m|^q |a_m|$$

$$\leq |a_0| + \sqrt{\sum_{\substack{m=-\infty \\ m \neq 0}}^{\infty} |m|^{-2q}} \sqrt{\sum_{\substack{m=-\infty \\ m \neq 0}}^{\infty} |m|^{2q} |a_m|^2}$$

$$\leq |a_0| + \sqrt{2\zeta(2q)} \|\varphi\|_q \qquad (7.3.115)$$

In this, $\zeta(r)$ denotes the *zeta* function:

$$\zeta(r) = \sum_{m=1}^{\infty} \frac{1}{m^r}, \quad r > 1$$

By standard arguments on the convergence of infinite series, (7.3.115) implies that the Fourier series (7.3.108) for φ is absolutely and uniformly convergent. In addition,

$$\|\varphi\|_\infty \le [1 + \sqrt{2\zeta(2q)}]\|\varphi\|_q, \quad \varphi \in H^q(2\pi) \tag{7.3.116}$$

which shows the identity mapping from $H^q(2\pi)$ into $C_p(2\pi)$ is bounded. $\quad\square$

Lemma 7.3.2. Let $q > p \ge 0$. Then $H^q(2\pi)$ is dense in $H^p(2\pi)$, and the identity mapping

$$I : H^q(2\pi) \to H^p(2\pi), \quad I(\varphi) = \varphi$$

is a compact operator.

Proof. See Kress [325, p. 111]. $\quad\square$

The trapezoidal rule and trigonometric interpolation

Consider again the trapezoidal rule

$$T_k(\varphi) = h \sum_{j=1}^{k} \varphi(jh), \quad h = \frac{2\pi}{k}$$

$$\approx \int_0^{2\pi} \varphi(s)\,ds \equiv I(\varphi)$$

with $k \ge 1$, $\varphi \in H^q(2\pi)$, and $q > \frac{1}{2}$. The latter assumption guarantees that evaluation of $\varphi(s)$ makes sense for all s. We give another error bound.

Lemma 7.3.3. Assume $q > \frac{1}{2}$, and let $\varphi \in H^q(2\pi)$. Then

$$|I(\varphi) - T_k(\varphi)| \le \frac{\sqrt{4\pi\zeta(2q)}}{k^q}\|\varphi\|_q, \quad k \ge 1 \tag{7.3.117}$$

Proof. Recall the result (7.2.81) for the trapezoidal rule applied to e^{ims}. Using it, and applying T_k to the Fourier series representation (7.3.108) of $\varphi(s)$, we have

$$I(\varphi) - T_k(\varphi) = -\sqrt{2\pi} \sum_{\substack{m=-\infty \\ m\neq 0}}^{\infty} a_{km} = -\sqrt{2\pi} \sum_{\substack{m=-\infty \\ m\neq 0}}^{\infty} a_{km}(km)^q (km)^{-q}$$

Applying the Cauchy-Schwarz inequality to the last sum,

$$|I(\varphi) - T_k(\varphi)| \leq \sqrt{2\pi} \left[\sum_{\substack{m=-\infty \\ m\neq 0}}^{\infty} |a_{km}|^2 (km)^{2q} \right]^{\frac{1}{2}} \left[\sum_{\substack{m=-\infty \\ m\neq 0}}^{\infty} (km)^{-2q} \right]^{\frac{1}{2}}$$

$$\leq \sqrt{2\pi} \|\varphi\|_q k^{-q} \sqrt{2\zeta(2q)}$$

This completes the proof.   □

Recall the discussion of the trigonometric interpolation polynomial $\mathcal{Q}_n\varphi$ of (7.2.83)–(7.2.85). We give another error bound for this interpolation. It is not clear who first propounded this result, but a complete derivation of it can be found in Kress and Sloan [328].

Lemma 7.3.4. Let $q > \frac{1}{2}$, and let $\varphi \in H^q(2\pi)$. Then for $0 \leq r \leq q$,

$$\|\varphi - \mathcal{Q}_n\varphi\|_r \leq \frac{c}{n^{q-r}} \|\varphi\|_q, \quad n \geq 1 \tag{7.3.118}$$

The constant c depends on only q and r.

Proof. The proof is based on using the Fourier series representation (7.3.108) of φ to obtain a Fourier series for $\mathcal{Q}_n\varphi$. This is then subtracted from that for φ, and the remaining terms are bounded to give the result (7.3.118). For details, see Ref. [328]. This is only a marginally better result than (7.2.85), but it is an important tool when doing error analyses of boundary integral equation methods in the Sobolev spaces $H^q(2\pi)$.   □

7.3.2. Some pseudodifferential equations

Recall the first kind boundary integral equation (7.3.107), and using the parametrization (7.1.1) for S, write this integral equation in the form

$$-\frac{1}{\pi} \int_0^{2\pi} \varphi(s) \log \left| 2e^{-\frac{1}{2}} \sin\left(\frac{t-s}{2}\right) \right| ds$$

$$-\frac{1}{\pi} \int_0^{2\pi} B(t,s)\varphi(s)\, ds = g(t), \quad 0 \leq t \leq 2\pi \tag{7.3.119}$$

with $\varphi(s) \equiv \rho(\mathbf{r}(s))|\mathbf{r}'(s)|$. The kernel $B(t,s)$ is given in (7.2.71), and for $\mathbf{r} \in C_p^1(2\pi)$, $B(t,s)$ is continuous. We write (7.3.119) in operator form as

$$\mathcal{A}\varphi + \mathcal{B}\varphi = g \tag{7.3.120}$$

[Note that this operator $\mathcal{B}$ is off by a factor of $-\frac{1}{\pi}$ from that used earlier in (7.2.72).]

Recall the discussion of $\mathcal{A}$ from the preceding section, in and following (7.2.73). We rewrite the formula (7.2.74) as

$$\mathcal{A}\varphi(t) = a_0\psi_0(t) + \sum_{|m|>0}\frac{a_m}{|m|}\psi_m(t) \qquad (7.3.121)$$

where $\varphi = \sum a_m\psi_m$. Then this implies that

$$\mathcal{A}: H^q(2\pi)\xrightarrow[onto]{1-1}H^{q+1}(2\pi), \quad q \geq 0 \qquad (7.3.122)$$

and

$$\|\mathcal{A}\| = 1$$

Because of the continuity of B, the operator $\mathcal{B}$ maps $H^q(2\pi)$ into $H^{q+2}(2\pi)$, at least. By Lemma 7.3.2, $\mathcal{B}$ is therefore a compact operator when considered as an operator from $H^q(2\pi)$ into $H^{q+1}(2\pi)$. Because of these mapping properties of $\mathcal{A}$ and $\mathcal{B}$, we consider the integral equation (7.3.120) with the assumption $g \in H^{q+1}(2\pi)$, and we seek a solution $\varphi \in H^q(2\pi)$ to the equation.

From (7.3.122), the equation (7.3.120) is equivalent to

$$\varphi + \mathcal{A}^{-1}\mathcal{B}\varphi = \mathcal{A}^{-1}g \qquad (7.3.123)$$

This is an integral equation of the second kind on $H^q(2\pi)$, and $\mathcal{A}^{-1}\mathcal{B}$ is a compact integral operator when regarded as an operator on $H^q(2\pi)$ into itself. Consequently, the standard Fredholm alternative theorem applies, and if the homogeneous equation $\varphi + \mathcal{A}^{-1}\mathcal{B}\varphi = 0$ has only the zero solution, then the original nonhomogeneous equation has a unique solution for all right sides $\mathcal{A}^{-1}g$. It is shown by Yan and Sloan in [580] that if S is not a Γ-contour, then the homogeneous version of the original integral equation (7.3.120) has only the zero solution, and thus by means of the Fredholm alternative theorem applied to (7.3.123), the integral equation (7.3.120) is uniquely solvable for all $g \in H^{q+1}(2\pi)$. Recalling the discussion following (7.2.101), we can ensure that S is not a Γ-contour by assuming

$$\text{diameter}(D) < 1 \qquad (7.3.124)$$

which was also imposed earlier in (7.2.102). This is discussed at length in Ref. [580].

Later in this section and in the following section, we look at numerical methods for solving (7.3.119). The error analysis is carried out within the context of Sobolev spaces, making use of the equivalent equation (7.3.123).

The Cauchy singular integral operator

Let D be a bounded simply connected region in the complex plane, and let S be its boundary. Let S have a parametrization

$$r(t) = \xi(t) + i\eta(t), \quad 0 \le t \le 2\pi \tag{7.3.125}$$

which we assume to be at least $(q + 2)$-times continuously differentiable, for some $q \ge 0$. Consider the integral operator

$$\mathcal{C}\rho(z) = \frac{1}{\pi i} \int_S \frac{\rho(\zeta)}{\zeta - z} \, d\zeta, \quad z \in S \tag{7.3.126}$$

The integral is to be interpreted as a Cauchy principal value integral, namely

$$\mathcal{C}\rho(z) = \frac{1}{\pi i} \lim_{\epsilon \to 0} \int_{S_\epsilon} \frac{\rho(\zeta)}{\zeta - z} \, d\zeta$$

where $S_\epsilon = \{\zeta \in S \mid |\zeta - z| \ge \epsilon\}$, with $\epsilon > 0$. As a consequence of the Cauchy representation theorem for functions ρ analytic in D, it follows that

$$\mathcal{C}\rho = \rho, \quad \text{for } \rho \text{ analytic in } D \tag{7.3.127}$$

It also can be shown that $\mathcal{C}^2 = I$ on $L^2(0, 2\pi)$, and therefore $\mathcal{C} : L^2(0, 2\pi) \overset{1-1}{\underset{onto}{\to}} L^2 (0, 2\pi)$; see Mikhlin [377, p. 22].

To obtain the function space properties of this operator $\mathcal{C}$, first consider the case in which $S = U$, the unit circle, with the parametrization

$$r_u(t) = \cos(t) + i \sin(t) \equiv e^{it} \tag{7.3.128}$$

We denote the Cauchy transform by $\mathcal{C}_u$ in this case. We write

$$\mathcal{C}_u\varphi(t) = \frac{1}{\pi} \int_0^{2\pi} \frac{\varphi(s)e^{is}}{e^{is} - e^{it}} \, ds, \quad 0 \le t \le 2\pi \tag{7.3.129}$$

with $\varphi(t) \equiv \rho(e^{it})$, and we can interpret $\mathcal{C}_u$ as an operator on $C_p(2\pi)$ or $H^q(2\pi)$. From Henrici [267, p. 109],

$$\mathcal{C}_u : e^{ikt} \to \text{sign}(k) \cdot e^{ikt}, \quad k = 0, \pm 1, \pm 2, \ldots \tag{7.3.130}$$

with sign(0) = 1. Writing $\varphi = \sum a_m \psi_m$,

$$C_u \varphi(t) = \sum_{m=0}^{\infty} a_m \psi_m(t) - \sum_{m=1}^{\infty} a_{-m} \psi_{-m}(t) \qquad (7.3.131)$$

This says

$$C_u : H^q(2\pi) \overset{1-1}{\underset{onto}{\longrightarrow}} H^q(2\pi), \quad q \geq 0 \qquad (7.3.132)$$

Consider now the case of a general region D with smooth boundary S. Using (7.3.125),

$$C\rho(r(t)) = \frac{1}{\pi i} \int_0^{2\pi} \frac{\rho(r(s))r'(s)}{r(s) - r(t)} ds \qquad (7.3.133)$$

Introduce a function φ defined on $[0, 2\pi]$, and implicitly on U, by

$$\varphi(t) = \rho(r(t)), \quad 0 \leq t \leq 2\pi$$

Using this, write (7.3.133) as

$$C\varphi(t) = \frac{1}{\pi i} \int_0^{2\pi} \frac{\varphi(s)ie^{is}}{e^{is} - e^{it}} \left[\frac{r'(s)[e^{is} - e^{it}]}{ie^{is}[r(s) - r(t)]} \right] ds, \quad 0 \leq t \leq 2\pi \qquad (7.3.134)$$

We examine in detail the quantity in brackets in this integral, and we are interested in the behavior as $s - t$ converges to 0. Note that the denominators can become zero when $s - t = \pm 2\pi$, but by periodicity we can always change these cases to $s = t = 0$.

Using standard Taylor series arguments, we have

$$r(t) = r(s) + (t - s)r'(s) + (t - s)^2 \int_0^1 (1 - v) r''(s + v(t - s)) dv$$

$$\equiv r(s) + (t - s)r'(s) + (t - s)^2 M_r(t, s)$$

$$e^u = 1 + u + u^2 \int_0^1 (1 - v)e^{uv} dv$$

$$\equiv 1 + u + u^2 M_e(u)$$

$$e^{i(t-s)} = 1 + i(t - s) - (t - s)^2 M_e(i(t - s))$$

Note that $M_r(t, s)$ and $M_e(i(t-s))$ are continuous functions of (t, s). Returning to the bracketed fraction in the integrand of (7.3.134),

$$
\frac{r'(s)[e^{is} - e^{it}]}{ie^{is}[r(s) - r(t)]} = \frac{r'(s)[e^{i(t-s)} - 1]}{i[r(t) - r(s)]},
$$

$$
= \frac{r'(s)[i(t-s) - (t-s)^2 M_e(i(t-s))]}{i[(t-s)r'(s) + (t-s)^2 M_r(t, s)]}
$$

$$
= \frac{r'(s)[1 + i(t-s)M_e(i(t-s))]}{r'(s) + (t-s)M_r(t, s)}
$$

$$
= 1 + (t-s)L(t, s)
$$

$$
L(t, s) = \frac{ir'(s)M_e(i(t-s)) - M_r(t, s)}{r'(s) + (t-s)M_r(t, s)}
$$

This function L is continuous as $t - s \to 0$.

Using these results in (7.3.134), we have

$$
C\varphi(t) = \frac{1}{\pi i} \int_0^{2\pi} \frac{\varphi(s)ie^{is}}{e^{is} - e^{it}} \, ds + \frac{1}{\pi i} \int_0^{2\pi} \frac{\varphi(s)ie^{is}}{e^{is} - e^{it}} (t-s)L(t, s) \, ds
$$

$$
= C_u\varphi(t) + \mathcal{G}\varphi(t) \tag{7.3.135}
$$

The integral operator $\mathcal{G}$ has a continuous smooth kernel function, although that does not follow completely from the form shown here. The only difficulty occurs when $(t, s) \to (0, 2\pi)$ or $(2\pi, 0)$, and those cases can be fixed with a modification of the above arguments. As a consequence, $\mathcal{G}: H^q(2\pi) \to H^{q+1}(2\pi)$, and by Lemma 7.3.2, this implies $\mathcal{G}$ is a compact operator on $H^q(2\pi)$ into $H^q(2\pi)$. Thus (7.3.135) implies that the essential properties of C, for example as to the mapping properties of C, follow from those of C_u, and

$$
C = C_u \left[I + C_u^{-1}\mathcal{G} \right] \tag{7.3.136}
$$

with $C_u^{-1}\mathcal{G}$ a compact operator on $H^q(2\pi)$ into $H^q(2\pi)$. Equations involving C can be converted using this decomposition into new forms to which the earlier result (7.3.132) and the Fredholm alternative theorem can be applied.

A hypersingular integral operator

Recall from (7.1.40) the definition of the double layer potential,

$$
u(A) = \int_S \rho(Q) \frac{\partial}{\partial \mathbf{n}_Q} [\log |A - Q|] \, dS_Q, \quad A \in D
$$

with D a bounded simply connected planar region with a smooth boundary S. Let $P \in S$, and consider forming the derivative of $u(A)$ in the direction $\mathbf{n}_P$, the inner normal to S at P:

$$\frac{\partial u(A)}{\partial \mathbf{n}_P} = \mathbf{n}_P \cdot \nabla_A \int_S \rho(Q) \frac{\partial}{\partial \mathbf{n}_Q} [\log |A - Q|] \, dS_Q$$

Take the limit as $A \to P$, thus obtaining the normal derivative

$$\frac{\partial u(P)}{\partial \mathbf{n}_P} = \lim_{A \to P} \mathbf{n}_P \cdot \nabla_A \int_S \rho(Q) \frac{\partial}{\partial \mathbf{n}_Q} [\log |A - Q|] \, dS_Q$$

$$\equiv \frac{\partial}{\partial \mathbf{n}_P} \int_S \rho(Q) \frac{\partial}{\partial \mathbf{n}_Q} [\log |P - Q|] \, dS_Q$$

$$\equiv \mathcal{H}\rho(P), \quad P \in S \tag{7.3.137}$$

If the integral and derivative operators are interchanged, then the resulting integral contains an integrand with a strongly nonintegrable singularity; this integral does not exist in any usual sense. Such integral operators $\mathcal{H}$ are often referred to as *hypersingular*.

To obtain some intuition for such integral operators, we consider the calculation of (7.3.137) when $S = U$, the unit circle, and we denote the operator by $\mathcal{H}_u$. We begin by calculating the harmonic functions

$$V_n(A) = \int_U \begin{bmatrix} \cos ns \\ \sin ns \end{bmatrix} \frac{\partial}{\partial \mathbf{n}_Q} [\log |A - Q|] \, dS_Q \tag{7.3.138}$$

with $Q = (\cos s, \sin s)$ in the integrand and $A = r (\cos t, \sin t)$, $0 \le r < 1$. We obtain boundary values for $V_n(A)$ by letting $A \to P = (\cos t, \sin t)$. Referring back to (7.1.9), and using (7.1.32) or (7.2.54) to calculate the kernel of the double layer integral, we have

$$V_n(P) = -\pi \begin{bmatrix} \cos nt \\ \sin nt \end{bmatrix} + \int_U \begin{bmatrix} \cos ns \\ \sin ns \end{bmatrix} \left(-\frac{1}{2} \right) dS_Q$$

Considering separately the cases of $n = 0$ and $n \ge 1$, we have the boundary values

$$V_n(P) = \begin{cases} -2\pi, & n = 0 \\ -\pi \begin{bmatrix} \cos nt \\ \sin nt \end{bmatrix}, & n \ge 1 \end{cases}$$

Solving Laplace's equation with these as boundary values is straightforward, yielding

$$
V_n(A) = \begin{cases} -2\pi, & n = 0 \\ -\pi \begin{bmatrix} r^n \cos nt \\ r^n \sin nt \end{bmatrix}, & n \geq 1 \end{cases} \tag{7.3.139}
$$

Returning to the definition (7.3.137), we note that the inner normal derivative on U coincides with the negative of the radial derivative with respect to r. Thus

$$
\frac{\partial V_n(P)}{\partial \mathbf{n}_P} = \begin{cases} 0, & n = 0 \\ \pi \begin{bmatrix} n \cos nt \\ n \sin nt \end{bmatrix}, & n \geq 1 \end{cases}
$$

with $P = (\cos t, \sin t)$; therefore

$$
\mathcal{H}_u : e^{int} \to \pi |n| e^{int}, \quad n = 0, \pm 1, \pm 2, \ldots \tag{7.3.140}
$$

We use this to calculate the value of $\mathcal{H}_u \rho$ in (7.3.137) for general ρ.

For ρ defined on U, introduce $\varphi(t) \equiv \rho(e^{it})$ and write $\varphi = \sum a_m \psi_m$. Assuming $\varphi \in H^q(2\pi)$ with $q \geq 1$, we have

$$
\mathcal{H}_u \rho = \pi \sum_{m \neq 0} |m| a_m \psi_m \tag{7.3.141}
$$

This shows

$$
\mathcal{H}_u : H^q(2\pi) \to H^{q-1}(2\pi) \tag{7.3.142}
$$

with $\mathcal{H}_u$ bounded,

$$
\|\mathcal{H}_u\| = \pi
$$

The range of $\mathcal{H}_u$ is the span of $\{\psi_m \mid m = \pm 1, \pm 2, \ldots\}$, a closed subspace of $H^{q-1}(2\pi)$ with codimension 1.

In some of its mapping properties, $\mathcal{H}_u$ is related closely to the derivative operator. For $\varphi \in H^1(2\pi)$ with $\varphi = \sum a_m \psi_m$,

$$
\mathcal{D}\varphi(t) \equiv \frac{d\varphi(t)}{dt} = i \sum_{m \neq 0} m a_m \psi_m \tag{7.3.143}
$$

Regarding the Cauchy singular integral operator $\mathcal{C}_u$ of (7.3.129) as an operator on $H^q(2\pi)$, and using the mapping property (7.3.130), we have

$$
\mathcal{H}_u \varphi = -\pi i \mathcal{D} \mathcal{C}_u \varphi = -\pi i \mathcal{C}_u \mathcal{D}\varphi \tag{7.3.144}
$$

Recall that $C_u : H^q(2\pi) \overset{1-1}{\underset{onto}{\rightarrow}} H^q(2\pi)$. Thus $\mathcal{H}_u \varphi$ is proportional to the derivative of a transformed version of φ. Extensions of this work to a general smooth boundary S are given in Chien and Atkinson [122], and it is analogous in form to the extension of the special Cauchy singular integral operator C_u to the general operator C, as in (7.3.136).

Pseudodifferential operators

We tie together the above operators $\mathcal{A}$, $\mathcal{A} + \mathcal{B}$, C_u, C, and $\mathcal{H}_u$ as follows. Each of them is a bounded mapping $\mathcal{L}$ with

$$\mathcal{L} : H^q(2\pi) \to H^{q-\gamma}(2\pi) \tag{7.3.145}$$

for some real γ and for all sufficiently large real q; with each such $\mathcal{L}$, Range($\mathcal{L}$) is a closed subspace of $H^{q-\gamma}(2\pi)$. We will call such an operator a *pseudodifferential operator*, and we will say it has order γ. Thus, the operators $\mathcal{A}$ and $\mathcal{A} + \mathcal{B}$ have order $\gamma = -1$, C_u and C have order $\gamma = 0$, and $\mathcal{H}_u$ has order $\gamma = 1$. Beginning in the late 1970s, and continuing today, a general numerical analysis has been developed for the solution of pseudodifferential operator equations, and in the next subsection, we consider one approach to developing such methods.

The concept of a pseudodifferential operator is usually introduced in other ways, using the concept of distributional derivative, but we chose the above approach to simplify our development. In §7.4 we give a further discussion of pseudodifferential operators, and we give numerical methods that generalize the finite element method, which is used in solving partial differential equations.

7.3.3. Two numerical methods

We give two related numerical methods for solving the first kind single layer equation (7.3.107) in the space $L^2(0, 2\pi)$. The first method is a Galerkin method using trigonometric polynomials as approximants, and the second is a further discretization of the Galerkin method. We assume that the integral equation (7.3.107) is uniquely solvable for all $g \in H^1(2\pi)$.

Introduce

$$\mathcal{X}_n = \text{span}\{\psi_{-n}, \ldots, \psi_0, \ldots, \psi_n\}$$

for a given $n \geq 0$, and let $\mathcal{P}_n$ denote the orthogonal projection of $L^2(0, 2\pi)$ onto $\mathcal{X}_n$. For $\varphi = \sum a_m \psi_m$, it is straightforward that

$$\mathcal{P}_n \varphi(s) = \sum_{m=-n}^{n} a_m \psi_m(s)$$

the truncation of the Fourier series for φ. See the earlier discussion following (3.3.69) in §3.3 of Chapter 3.

Recall the decomposition (7.3.119)–(7.3.120) of (7.3.107),

$$A\varphi + B\varphi = g \tag{7.3.146}$$

with $A\varphi$ given in (7.3.121). It is immediate that

$$\mathcal{P}_n A = A\mathcal{P}_n, \qquad \mathcal{P}_n A^{-1} = A^{-1}\mathcal{P}_n \tag{7.3.147}$$

Approximate (7.3.146) by the equation

$$\mathcal{P}_n(A\varphi_n + B\varphi_n) = \mathcal{P}_n g, \qquad \varphi_n \in \mathcal{X}_n \tag{7.3.148}$$

Letting

$$\varphi_n(s) = \sum_{m=-n}^{n} a_m^{(n)} \psi_m(s)$$

and recalling (7.2.80), the equation (7.3.148) implies that the coefficients $\{a_m^{(n)}\}$ are determined from the linear system

$$\frac{a_k^{(n)}}{\max\{1, |k|\}} + \sum_{m=-n}^{n} a_m^{(n)} \int_0^{2\pi} \int_0^{2\pi} B(t, s)\psi_m(s)\overline{\psi_k(t)}\,ds\,dt$$

$$= \int_0^{2\pi} g(t)\overline{\psi_k(t)}\,dt, \qquad k = -n, \ldots, n \tag{7.3.149}$$

Generally these integrals must be evaluated numerically. Later we give a numerically integrated version of this method.

The equation (7.3.146) is equivalent to

$$\varphi + A^{-1}B\varphi = A^{-1}g \tag{7.3.150}$$

The right side function $A^{-1}g \in L^2(0, 2\pi)$, by (7.3.122) and by the earlier assumption that $g \in H^1(2\pi)$. From the discussion following (7.3.123), $A^{-1}B$ is a compact mapping from $L^2(0, 2\pi)$ into $L^2(0, 2\pi)$, and thus (7.3.150) is a Fredholm integral equation of the second kind. By the earlier assumption on the unique solvability of (7.3.146), we have $(I + A^{-1}B)^{-1}$ exists on $L^2(0, 2\pi)$ to $L^2(0, 2\pi)$.

Using (7.3.147), the approximating equation (7.3.148) is equivalent to

$$\varphi_n + \mathcal{P}_n A^{-1}B\varphi_n = \mathcal{P}_n A^{-1}g \tag{7.3.151}$$

Equation (7.3.151) is simply a standard Galerkin method for solving (7.3.150), as discussed earlier in §3.3 of Chapter 3, following (3.3.76).

Since $\mathcal{P}_n \varphi \to \varphi$, for all $\varphi \in L^2(0, 2\pi)$, and since $\mathcal{A}^{-1}\mathcal{B}$ is a compact operator, we have

$$\|(I - \mathcal{P}_n)\mathcal{A}^{-1}\mathcal{B}\| \to 0 \quad \text{as } n \to \infty$$

from Lemma 3.1.2 in §3.1 of Chapter 3. Then by standard arguments, the existence of $\left(I + \mathcal{A}^{-1}\mathcal{B}\right)^{-1}$ implies that of $\left(I + \mathcal{P}_n\mathcal{A}^{-1}\mathcal{B}\right)^{-1}$, for all sufficiently large n. This is simply a repetition of the general argument given in Theorem 3.1.1, in §3.1 of Chapter 3. From (3.1.31) of that theorem,

$$\|\varphi - \varphi_n\|_0 \leq \|(I + \mathcal{P}_n\mathcal{A}^{-1}\mathcal{B})^{-1}\| \|\varphi - \mathcal{P}_n\varphi\|_0 \tag{7.3.152}$$

where $\|\cdot\|_0$ is the norm for $H^0(2\pi) \equiv L^2(0, 2\pi)$. For more detailed bounds on the rate of convergence, apply the result (3.3.74) of §3.3 from Chapter 3, obtaining

$$\|\varphi - \varphi_n\|_0 \leq \frac{c}{n^q} \|\varphi\|_q, \quad \varphi \in H^q(2\pi) \tag{7.3.153}$$

for any $q > 0$.

A discrete Galerkin method

We give a numerical method that amounts to using the trapezoidal rule to numerically integrate the integrals in (7.3.149). Introduce the discrete inner product

$$(f, g)_n = h \sum_{j=0}^{2n} f(t_j)\overline{g(t_j)}, \quad f, g \in C_p(2\pi) \tag{7.3.154}$$

with $h = 2\pi/(2n + 1)$, and $t_j = jh$, $j = 0, 1, \ldots, 2n$. This uses the trapezoidal rule with $2n + 1$ subdivisions of the integration interval $[0, 2\pi]$. Also, approximate the integral operator $\mathcal{B}$ of (7.3.119) by

$$\mathcal{B}_n\varphi(t) = -\frac{h}{\pi} \sum_{j=0}^{2n} B(t, t_j)\varphi(t_j), \quad \varphi \in C_p(2\pi) \tag{7.3.155}$$

Associated with the above discrete inner product (7.3.154) is the discrete orthogonal projection operator $\mathcal{Q}_n$ mapping $C_p(2\pi)$ into $\mathcal{X}_n$. This was explored at length in §4.4 from Chapter 4, following (4.4.147), and a number of properties of $\mathcal{Q}_n$ were given in Lemma 4.4.1. In particular,

$$(\mathcal{Q}_n\varphi, \psi)_n = (\varphi, \psi)_n, \quad \text{all } \psi \in \mathcal{X}_n \qquad (7.3.156)$$

$$\mathcal{Q}_n\varphi = \sum_{m=-n}^{n} (\varphi, \psi_m)_n \psi_m \qquad (7.3.157)$$

and

$$\mathcal{Q}_n\varphi(t_j) = \varphi(t_j), \quad j = 0, 1, \ldots, 2n \qquad (7.3.158)$$

We approximate (7.3.149) using

$$\sigma_n(s) = \sum_{m=-n}^{n} b_m^{(n)} \psi_m(s) \qquad (7.3.159)$$

with $\{b_m^{(n)}\}$ determined from the linear system

$$\frac{b_k^{(n)}}{\max\{1, |k|\}} + \sum_{m=-n}^{n} b_m^{(n)} (\mathcal{B}_n\psi_m, \psi_k)_n = (g, \psi_k)_n, \quad k = -n, \ldots, n \qquad (7.3.160)$$

Using the above properties of $\mathcal{Q}_n$, this can be written symbolically as

$$\mathcal{Q}_n(\mathcal{A}\sigma_n + \mathcal{B}_n\sigma_n) = \mathcal{Q}_n g, \quad \sigma_n \in \mathcal{X}_n \qquad (7.3.161)$$

Using (7.3.158), we can also determine $\{b_m^{(n)}\}$ from solving the linear system

$$\sum_{m=-n}^{n} b_m^{(n)} \left\{ \frac{1}{\max\{1, |m|\}} \psi_m(t_j) + \mathcal{B}_n\psi_m(t_j) \right\} = g(t_j), \quad j = 0, 1, \ldots, 2n \qquad (7.3.162)$$

This is less expensive to set up than is the system (7.3.160), and the two systems can be shown to have condition numbers of essentially the same size.

We construct an error analysis for (7.3.161), doing so in the space $\mathcal{X} = C_p(2\pi)$. Begin by noting that the equation (7.3.161) is equivalent to the equation

$$\mathcal{A}\sigma_n + \mathcal{Q}_n\mathcal{B}_n\sigma_n = \mathcal{Q}_n g, \quad \sigma_n \in \mathcal{X} \qquad (7.3.163)$$

To prove the equivalence, begin by assuming (7.3.163) is solvable. Then

$$\mathcal{A}\sigma_n = \mathcal{Q}_n g - \mathcal{Q}_n\mathcal{B}_n\sigma_n \in \mathcal{X}_n$$

Using the formula (7.3.121) for $\mathcal{A}$, this implies $\sigma_n \in \mathcal{X}_n$ and $\mathcal{Q}_n\sigma_n = \sigma_n$. Using this in (7.3.163) implies the equation (7.3.161). A similarly simple argument shows that (7.3.161) implies (7.3.163).

Equation (7.3.163) is equivalent to

$$\sigma_n + \mathcal{A}^{-1}\mathcal{Q}_n\mathcal{B}_n\sigma_n = \mathcal{A}^{-1}\mathcal{Q}_n g \qquad (7.3.164)$$

Introduce

$$\mathcal{C} = \mathcal{A}^{-1}\mathcal{B}, \qquad \mathcal{C}_n = \mathcal{A}^{-1}\mathcal{Q}_n\mathcal{B}_n$$

Then we write the original equation (7.3.146) and its approximation (7.3.164) as

$$\varphi + \mathcal{C}\varphi = \mathcal{A}^{-1}g \qquad (7.3.165)$$
$$\sigma_n + \mathcal{C}_n\sigma_n = \mathcal{A}^{-1}\mathcal{Q}_n g \qquad (7.3.166)$$

respectively. These equations become the vehicle for an error analysis of our discrete Galerkin method (7.3.160)–(7.3.162).

Theorem 7.3.3. *Assume the boundary S has a parametrization* $\mathbf{r} \in C_p^\infty(2\pi)$. *Assume the integral equation (7.3.160) is uniquely solvable, and further assume that the solution* $\varphi \in C_p(2\pi)$. *Then:*

(a) The family $\{\mathcal{C}_n\}$ *is collectively compact and pointwise convergent on* $C_p(2\pi)$;
(b) For all sufficiently large n, say $n \geq N$, *the inverses* $(I + \mathcal{C}_n)^{-1}$ *exist on* $C_p(2\pi)$ *to* $C_p(2\pi)$, *with*

$$\|(I + \mathcal{C}_n)^{-1}\| \leq M < \infty, \quad n \geq N \qquad (7.3.167)$$

The approximate solutions σ_n *satisfy*

$$\|\varphi - \sigma_n\|_\infty \leq M\{\|\mathcal{A}^{-1}(g - \mathcal{Q}_n g)\|_\infty + \|\mathcal{C}\varphi - \mathcal{C}_n\varphi\|_\infty\} \qquad (7.3.168)$$

Proof. **(a)** The operator $\mathcal{C}$ is an integral operator,

$$\mathcal{C}f(t) = \int_0^{2\pi} C(t, s) f(s)\, ds, \quad f \in C_p(2\pi)$$

with

$$C(\cdot, s) = -\frac{1}{\pi}\mathcal{A}^{-1}B_s, \qquad B_s(t) \equiv B(t, s) \qquad (7.3.169)$$

Use the smoothness of $\mathbf{r}(t)$ to imply that $B \in C_p^\infty(2\pi)$ in both variables. Then it is straightforward to show that the operator $\mathcal{C}$ is compact on $C_p(2\pi)$ into $C_p(2\pi)$.

Recall the formula (7.3.155) for $\mathcal{B}_n$, and write

$$\mathcal{Q}_n \mathcal{B}_n f(t) = -\frac{h}{\pi} \sum_{j=0}^{2n} B_n(t, t_j) f(t_j) \qquad (7.3.170)$$

with

$$B_n(\cdot, s) = \mathcal{Q}_n B_s$$

Next define

$$C(\cdot, s) = -\frac{1}{\pi} \mathcal{A}^{-1} B_{n,s}, \qquad B_{n,s}(t) \equiv B_n(t, s) \qquad (7.3.171)$$

Then

$$\mathcal{C}_n f(t) = h \sum_{j=0}^{2n} C_n(t, t_j) f(t_j), \qquad f \in C_p(2\pi) \qquad (7.3.172)$$

We will examine the functions $\{C_n\}$, showing:

1. $C_n(t, s) \to C(t, s)$ uniformly in (t, s);
2. $\{C_n\}$ is uniformly bounded and equicontinuous in (t, s).

From these properties, it is relatively straightforward to prove that $\{\mathcal{C}_n\}$ is a collectively compact and pointwise convergent family on $C_p(2\pi)$.

Begin the proof of property 1 by noting that Lemma 7.3.4 implies

$$\|B(\cdot, s) - B_n(\cdot, s)\|_r \leq \frac{c}{n^{q-r}} \|B(\cdot, s)\|_q, \qquad n \geq 1 \qquad (7.3.173)$$

with $q > \frac{1}{2}$, $0 \leq r \leq q$, and with c dependent on only q and r. From the assumption $\mathbf{r} \in C_p^\infty(2\pi)$, and from (7.2.48), the function $B(t, s)$ is infinitely differentiable in t, s; for (7.3.173), this implies

$$\max_s \|B(\cdot, s) - B_n(\cdot, s)\|_r \leq \frac{c_1}{n^{q-r}}, \qquad n \geq 1$$

for a suitable constant $c_1 > 0$, again dependent on only q and r. From (7.3.122), (7.3.169), and (7.3.171),

$$\|C(\cdot, s) - C_n(\cdot, s)\|_{r-1} \leq \frac{1}{\pi} \|\mathcal{A}^{-1}\| \|B(\cdot, s) - B_n(\cdot, s)\|_r$$

From Lemma 7.3.1 and (7.3.116),

$$\max_t |C(t, s) - C_n(t, s)| \leq c_2 \|C(\cdot, s) - C_n(\cdot, s)\|_1$$

for some $c_2 > 0$ [see (7.3.116)]. Combining these results, and choosing $r = 2$ and $q = 3$, we have

$$\max_{t,s} |C(t,s) - C_n(t,s)| \leq \frac{c_3}{n} \qquad (7.3.174)$$

This proves the above property 1. In fact, we have proven the more general result

$$\max_{t,s} |C(t,s) - C_n(t,s)| \leq \frac{c_3}{n^k} \qquad (7.3.175)$$

with c_3 dependent on k, for any $k \geq 1$.

It follows immediately from (7.3.174) and the continuity of $C(t,s)$ that $\{C_n(t,s)\}$ is uniformly bounded. For equicontinuity of $\{C_n(\cdot,s)\}$, begin with

$$|C_n(t,s) - C_n(\tau,s)| \leq |t - \tau| \max_t \left| \frac{\partial C_n(t,s)}{\partial t} \right| \qquad (7.3.176)$$

From the definition of C, and from the infinite differentiability of $B(t,s)$, it follows that $C(t,s)$ is also infinitely differentiable. As in (7.3.173) and following, we can show

$$\max_{t,s} \left| \frac{\partial^m C(t,s)}{\partial t^m} - \frac{\partial^m C_n(t,s)}{\partial t^m} \right| \leq \frac{c_4}{n}, \quad n \geq 1 \qquad (7.3.177)$$

for all $m \geq 1$, with c_4 a constant dependent on m. Consequently, the derivative in (7.3.176) is uniformly bounded, independent of n, and

$$|C_n(t,s) - C_n(\tau,s)| \leq c_5 |t - \tau|$$

for a suitable constant c_5. This completes the proof of the above property 2.

(b) From Theorem 4.1.1 in §4.1 of Chapter 4, the collective compactness and pointwise convergence of $\{C_n\}$ implies the existence on $C_p(2\pi)$ of $(I + C_n)^{-1}$ and their uniform boundedness, as in (7.3.167). For the error in σ_n, subtract (7.3.166) from (7.3.165), obtaining

$$(\varphi + C\varphi) - (\sigma_n + C_n\sigma_n) = \mathcal{A}^{-1}g - \mathcal{A}^{-1}\mathcal{Q}_n g$$
$$(I + C_n)(\varphi - \sigma_n) = (C_n\varphi - C\varphi) + \mathcal{A}^{-1}(g - \mathcal{Q}_n g)$$
$$\varphi - \sigma_n = (I + C_n)^{-1}[(C_n\varphi - C\varphi) + \mathcal{A}^{-1}(g - \mathcal{Q}_n g)] \qquad (7.3.178)$$

This leads immediately to (7.3.168). $\qquad\qquad\qquad\qquad\qquad\qquad\qquad\square$

To derive bounds showing the speed of convergence with respect to n, consider separately the terms $g - \mathcal{Q}_n g$ and $C_n \varphi - C \varphi$. From Lemma 7.3.4 and (7.3.118), we have

$$\|g - \mathcal{Q}_n g\|_r \leq \frac{c}{n^{q-r}} \|g\|_q, \quad g \in H^q(2\pi), \quad q > \frac{1}{2}$$

Then for any $\eta > \frac{1}{2}$,

$$\|\mathcal{A}^{-1}(g - \mathcal{Q}_n g)\|_\infty \leq d_1 \|\mathcal{A}^{-1}(g - \mathcal{Q}_n g)\|_\eta$$
$$\leq d_2 \|g - \mathcal{Q}_n g\|_{\eta+1}$$
$$\leq \frac{d_3}{n^{q-\eta-1}} \|g\|_q, \quad g \in H^q(2\pi) \qquad (7.3.179)$$

with suitable constants $d_1, d_2, d_3 > 0$. We have uniform convergence of $\mathcal{A}^{-1}$ $(g - \mathcal{Q}_n g)$ to zero if $g \in H^q(2\pi)$, with $q > \frac{3}{2}$, by choosing η suitably close to $\frac{1}{2}$.

For the term $C_n \varphi - C \varphi$, write

$$C \varphi(t) - C_n \varphi(t) = \left[\int_0^{2\pi} C(t, s) \varphi(s) \, ds - h \sum_{j=0}^{2n} C(t, t_j) \varphi(t_j) \right]$$
$$+ h \sum_{j=0}^{2n} [C(t, t_j) - C_n(t, t_j)] \varphi(t_j) \qquad (7.3.180)$$

The second term on the right side is bounded by

$$\max_t \left| h \sum_{j=0}^{2n} [C(t, t_j) - C_n(t, t_j)] \varphi(t_j) \right| \leq 2\pi \|\varphi\|_\infty \max_{t,s} |C(t, s) - C_n(t, s)|$$
$$(7.3.181)$$

and by (7.3.175), this converges to zero faster than any power of $1/n$. Using Lemma 7.3.3, it follows that the first term on the right side of (7.3.180) satisfies

$$\|C \varphi - C_n \varphi\|_\infty \leq \frac{c}{n^r} \max_t \|C(t, \cdot) \varphi\|_q \qquad (7.3.182)$$

for $\varphi \in H^q(2\pi)$ with $q > \frac{1}{2}$.

Combining (7.3.178), (7.3.179), (7.3.181), and (7.3.182), we have

$$\|\varphi - \sigma_n\|_\infty \leq \frac{c}{n^{q-1.5-\epsilon}} \qquad (7.3.183)$$

when $g \in H^q(2\pi)$ and $\varphi \in C_p(2\pi) \cap H^{q-1}(2\pi)$, for some $q > \frac{3}{2}$ and any small $\epsilon > 0$. The constant c depends on q, ϵ, and φ. Improved rates of convergence

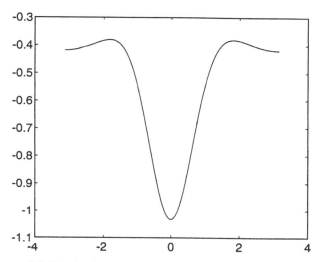

Figure 7.6. The density $\rho(\mathbf{r}(t))$ for (7.3.186) with $(a, b) = (1.0, 0.4)$.

can be obtained by using Sobolev norms to measure the error, and we leave their derivation to the reader. However, we will still need to assume $g \in C_p(2\pi)$ to have the right side in (7.3.162) be defined. A modified version of the method, one allowing more generality in the choice of g, is given in Saranen [481].

Example. Consider the interior Dirichlet problem (7.1.2) with boundary data

$$f(x, y) = e^x \cos(y) \qquad (7.3.184)$$

and an elliptical boundary S,

$$\mathbf{r}(t) = (a \cos t, b \sin t), \quad -\pi \le t \le \pi$$

Represent the solution of (7.1.2) by the single layer

$$u(P) = \int_S \rho(Q) \log |P - Q| \, dS_Q, \quad P \in D \qquad (7.3.185)$$

Letting $P \to S$, we must solve

$$\int_S \rho(Q) \log |P - Q| \, dS_Q = f(P), \quad P \in S \qquad (7.3.186)$$

We convert this to the form (7.3.107) by multiplying by $-\frac{1}{\pi}$, and then we solve it numerically with the method (7.3.159)–(7.3.162).

A graph of the true solution ρ is given in Figure 7.6. Table 7.4 contains the errors in $\|\rho - \sigma_n\|_\infty$ for varying n, and a graph of the error $\rho - \sigma_{10}$ is given

Table 7.4. *Errors in σ_n*
for (7.3.186) with
$(a, b) = (1.0, 0.4)$

n	$\|\rho - \sigma_n\|_\infty$
1	2.02E−1
2	7.42E−2
3	1.97E−2
4	3.64E−3
5	5.42E−4
6	6.47E−5
7	6.78E−6
8	6.20E−7
9	5.61E−8
10	8.84E−9

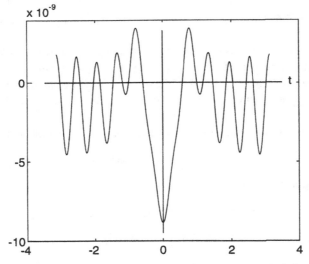

Figure 7.7. The error $\rho - \sigma_{10}$ for (7.3.186) with $(a, b) = (1.0, 0.4)$.

in Figure 7.7. We define a numerical approximation to the potential $u(P)$ by replacing ρ by σ_n in (7.3.185), followed by numerical integration using the trapezoidal rule with $2m + 1$ subdivisions:

$$u_{n,m}(P) = \frac{\pi}{m} \sum_{j=0}^{2m} \sigma_n(jh_m) \log |P - \mathbf{r}(jh_m)|, \quad h_m = \frac{\pi}{m} \qquad (7.3.187)$$

Most people think first of using this with $m = n$, since (7.3.162) involves

Table 7.5. *Errors in $u_{n,n}(P_j)$ for (7.3.185)*
with $(a, b) = (1.0, 0.4)$

n	$j = 1$	$j = 2$	$j = 3$	$j = 4$
	1.99E−1	5.21E−1	6.83E−1	7.78E−1
2	6.84E−4	−1.32E−1	2.40E−1	5.96E−1
3	1.21E−3	2.68E−2	−2.71E−1	−1.22E−1
4	−8.02E−5	−5.87E−4	5.40E−2	−1.49E−1
5	1.95E−4	4.59E−3	6.39E−2	2.06E−1
6	−6.38E−5	3.91E−3	−5.18E−2	1.88E−1
7	2.78E−5	−2.04E−3	−5.49E−3	−1.07E−1
8	−1.06E−5	7.00E−4	2.48E−2	−5.70E−2
9	4.23E−6	−7.26E−5	−6.72E−3	1.21E−1
10	−1.66E−6	−1.15E−5	−9.79E−3	1.06E−1

Table 7.6. *Errors in $u_{10,m}(P_j)$ for (7.3.185)*
with $(a, b) = (1.0, 0.4)$

m	$j = 1$	$j = 2$	$j = 3$	$j = 4$
10	−1.66E−6	−1.15E−4	−9.79E−3	1.06E−1
20	1.29E−9	−1.38E−7	−4.51E−4	−9.70E−3
40	1.42E−9	1.10E−9	−1.85E−6	2.04E−3
80	1.42E−9	1.10E−9	1.11E−9	3.86E−3
160	1.42E−9	1.10E−9	1.17E−9	1.35E−3
320	1.42E−9	1.10E−9	1.17E−9	−5.29E−5
640	1.42E−9	1.10E−9	1.17E−9	3.88E−7

collocation at the points $\{jh_n \mid j = 0, 1, \ldots, 2n\}$. Table 7.5 contains the errors $u(P) - u_{n,n}(P)$, for increasing n, at the selected points

$$P_j = \alpha_j \mathbf{r}\left(\frac{\pi}{4}\right), \quad j = 1, 2, 3, 4$$

with $\alpha = (0, .4, .8, .99)$. To show the effect of letting m be larger than n, Table 7.6 contains the errors $u(P) - u_{10,m}(P)$ for increasing values of m. The increasing accuracy is the analog of what was seen in the example involving (7.2.65) in §7.2; see (7.2.59), Figure 7.4, and Table 7.2.

7.4. Finite element methods

We consider numerical methods using approximations that are piecewise polynomial functions, using Galerkin's method and discrete variants of it. The techniques and theory from Chapter 4 can be applied to boundary integral equations of the second kind with a compact integral operator, but alternative approaches are required when dealing with other forms of boundary integral equations. In this section we describe an approach that generalizes the finite element method, a numerical method for solving partial differential equations. We regard a boundary integral equation as a pseudodifferential operator equation between Sobolev spaces. Then the abstract version of the finite element method is applied to this equation, yielding both a new numerical method and an error analysis for it.

We begin by extending to negative values of q the Sobolev spaces $H^q(2\pi)$ that were introduced in §7.3, and we extend earlier boundary integral equations to these new spaces. There are more general approaches to Sobolev spaces and to pseudodifferential operator equations defined on them (for example, see Aubin [70]), but we will only allude to these results, both for reasons of space and simplicity. When consistent with giving a relatively simple presentation, we derive finite element methods and do error analyses of them without specific use of properties of the planar BIE being studied.

7.4.1. Sobolev spaces – A further discussion

As before, define

$$\psi_m(t) = \frac{1}{\sqrt{2\pi}} e^{imt}, \quad -\infty < t < \infty, \quad m = 0, \pm 1, \pm 2, \ldots$$

For any real number q, define $H^q(2\pi)$ to be the space of all "functions"

$$\varphi(t) = \sum_{m=-\infty}^{\infty} a_m \psi_m(t) \tag{7.4.188}$$

for which

$$\|\varphi\|_q \equiv \left[|a_0|^2 + \sum_{|m|>0} |m|^{2q} |a_m|^2 \right]^{\frac{1}{2}} < \infty \tag{7.4.189}$$

This is a Hilbert space, with the inner product defined by

$$(\varphi, \rho)_q = a_0 \bar{b}_0 + \sum_{|m|>0} |m|^{2q} a_m \bar{b}_m \tag{7.4.190}$$

where $\varphi = \sum a_m \psi_m$ and $\rho = \sum b_m \psi_m$. The number q is called the *index* of the Sobolev space $H^q(2\pi)$.

For $q \geq 0$, this is the Sobolev space introduced in §7.3 [see (7.3.112)]. For $q < 0$, however, the space $H^q(2\pi)$ contains series that are divergent according to most usual definitions of convergence. These new "functions" (7.4.188) are referred to as both *generalized functions* and *distributions*. One way of giving meaning to these new functions is to introduce the concept of *distributional derivative*, which generalizes the ordinary sense of derivative. With the ordinary differentiation operator $\mathcal{D} \equiv d/dt$, we have

$$\mathcal{D}\varphi(t) \equiv \frac{d\varphi(t)}{dt} = i \sum_{m=-\infty}^{\infty} m a_m \psi_m(t) \qquad (7.4.191)$$

and $\mathcal{D}: H^q(2\pi) \to H^{q-1}(2\pi), q \geq 1$. The distributional derivative gives meaning to differentiation of periodic functions in $L^2(0, 2\pi)$ and also to repeated differentiation of generalized functions. To prove that there exists a unique such extension of the definition of $\mathcal{D}$, proceed as follows.

Introduce the space $\mathcal{Y}$ of all trigonometric polynomials,

$$\mathcal{Y} = \left\{ \varphi \equiv \sum_{m=-n}^{n} a_m \psi_m \right\} \qquad (7.4.192)$$

with n and $\{a_m\}$ arbitrary. It is straightforward to show this is a dense subspace of $H^q(2\pi)$ for arbitrary q, meaning that when using the norm (7.4.189), the closure of $\mathcal{Y}$ equals $H^q(2\pi)$. Considering $\mathcal{Y}$ as a subspace of $H^q(2\pi)$, define $\mathcal{D}: \mathcal{Y} \to H^{q-1}(2\pi)$ by

$$\mathcal{D}\varphi = \varphi', \quad \varphi \in \mathcal{Y} \qquad (7.4.193)$$

This is a bounded operator, and using the representation of φ in (7.4.192), it is straightforward that

$$\|\mathcal{D}\| = 1$$

Since $\mathcal{Y}$ is dense in $H^q(2\pi)$, and since $\mathcal{D}: \mathcal{Y} \subset H^q(2\pi) \to H^{q-1}(2\pi)$ is bounded, we have that there is a unique bounded extension of $\mathcal{D}$ to all of $H^q(2\pi)$; see Theorem A.6 in the Appendix. We will retain the notation $\mathcal{D}$ for the extension. Combining the representation of $\varphi \in \mathcal{Y}$ in (7.4.192) with the definition (7.4.193), and using the continuity of the extension $\mathcal{D}$, the formula (7.4.191) remains valid for all q.

As an example, define

$$\varphi(t) = \begin{cases} 1, & 2k\pi < t < (2k+1)\pi \\ 0, & (2k-1)\pi < t < 2k\pi \end{cases}$$

for all integers k. The Fourier series of this is given by

$$\varphi(t) = \frac{1}{2} - \frac{i}{\pi} \sum_{k=0}^{\infty} \frac{1}{2k+1} \left[e^{(2k+1)it} - e^{-(2k+1)it} \right], \quad -\infty < t < \infty$$

which converges almost everywhere. Regarding this as a function defined on $-\infty < t < \infty$, the distributional derivative of $\varphi(t)$ is

$$\varphi'(t) = \frac{1}{\pi} \sum_{k=0}^{\infty} \left[e^{(2k+1)it} + e^{-(2k+1)it} \right]$$

$$= \sum_{j=-\infty}^{\infty} (-1)^j \delta(t - \pi j)$$

The function $\delta(t)$ is the *Dirac delta function*, and it is a well-studied linear functional on elements of $H^q(2\pi)$ for $q > \frac{1}{2}$:

$$\delta[\varphi] \equiv \langle \varphi, \delta \rangle = \varphi(0), \quad \varphi \in H^q(2\pi), \quad q > \frac{1}{2}$$

with $\delta[\varphi]$ denoting the action of δ on φ.

This leads to another interpretation of $H^q(2\pi)$ for negative q. Let ℓ be a bounded linear functional on $H^p(2\pi)$ for some $p \geq 0$, bounded with respect to the norm $\|\cdot\|_p$. Then using the *Riesz representation theorem* (cf. Theorem A.4 in the Appendix), it can be shown that there is a unique element $\eta_\ell \in H^{-p}(2\pi)$, $\eta_\ell = \sum b_m \psi_m$, with

$$\ell[\varphi] \equiv \langle \varphi, \eta_\ell \rangle$$

$$= \sum_{m=-\infty}^{\infty} a_m \bar{b}_m, \quad \text{for } \varphi = \sum_{m=-\infty}^{\infty} a_m \psi_m \in H^p(2\pi) \qquad (7.4.194)$$

It is also straightforward to show that when given two such linear functionals, say ℓ_1 and ℓ_2, we have

$$\eta_{c_1 \ell_1 + c_2 \ell_2} = \bar{c}_1 \eta_{\ell_1} + \bar{c}_2 \eta_{\ell_2}$$

for all scalars c_1, c_2. Moreover,

$$|\langle \varphi, \eta_\ell \rangle| \leq \|\varphi\|_p \|\eta_\ell\|_{-p}$$

and

$$\|\ell\| = \|\eta_\ell\|_{-p}$$

The space $H^{-p}(2\pi)$ can be used to represent the space of bounded linear functionals for $H^p(2\pi)$, and it is usually called the *dual space* for $H^p(2\pi)$.

In this framework, we are regarding $H^0(2\pi) \equiv L^2(0, 2\pi)$ as self-dual. The evaluation of linear functionals on $H^p(2\pi)$, as in (7.4.194), can be considered as a bilinear function defined on $H^p(2\pi) \times H^{-p}(2\pi)$; in that case,

$$|\langle \varphi, \eta \rangle| \leq \|\varphi\|_p \|\eta\|_{-p}, \quad \varphi \in H^p(2\pi), \quad \eta \in H^{-p}(2\pi) \qquad (7.4.195)$$

Define $b : H^p(2\pi) \times L^2(0, 2\pi) \to \mathbf{C}$ by

$$b(\varphi, \psi) = (\varphi, \psi), \quad \varphi \in H^p(2\pi), \quad \psi \in L^2(0, 2\pi) \qquad (7.4.196)$$

using the usual inner product of $L^2(0, 2\pi)$. Then the bilinear *duality pairing* $\langle \cdot, \cdot \rangle$ is the unique bounded extension of b to $H^p(2\pi) \times H^{-p}(2\pi)$, when regarding $L^2(0, 2\pi)$ as a dense subspace of $H^{-p}(2\pi)$. For a more extensive discussion of this topic with much greater generality, see Aubin [70, Chapter 3].

We have considered $\langle \cdot, \cdot \rangle$ as defined on $H^p(2\pi) \times H^{-p}(2\pi)$ with $p \geq 0$. But we can easily extend this definition to allow $p < 0$. Using (7.4.194), we define

$$\langle \varphi, \eta \rangle = \sum_{m=-\infty}^{\infty} a_m \bar{b}_m$$

for $\varphi = \sum a_m \psi_m$ in $H^p(2\pi)$ and $\eta = \sum b_m \psi_m$ in $H^{-p}(2\pi)$, for any real number p. The bound (7.4.195) is also still valid. For the remainder of this section, we allow this extended meaning of $\langle \cdot, \cdot \rangle$.

Extensions of boundary integral operators

The boundary integral operators $\mathcal{L}$ of the preceding §7.3, (for example, $\mathcal{L}$ equal to $\mathcal{A}$, $\mathcal{C}$, and $\mathcal{H}_u$) were defined on Sobolev spaces of positive index. Using their Fourier series representation formulas, we can extend them in a unique way to operators on $H^q(2\pi)$ for all real numbers q. The original definitions as integral operators are no longer valid, but we can still consider solving equations

$$\mathcal{L}\varphi = \psi \qquad (7.4.197)$$

where $\mathcal{L} : H^p(2\pi) \to H^q(2\pi)$ and $\psi \in H^q(2\pi)$. Initially in using the theory of finite element methods, we will look at such pseudodifferential operator equations with

$$\mathcal{L} : H^\alpha(2\pi) \to H^{-\alpha}(2\pi) \qquad (7.4.198)$$

where 2α is the index of the pseudodifferential operator (see the discussion following (7.3.145) in §7.3).

7.4.2. An abstract framework

Consider a bounded operator $\mathcal{L}$ with

$$\mathcal{L}: H^\alpha(G) \xrightarrow[onto]{1-1} H^{-\alpha}(G) \tag{7.4.199}$$

We assume G is some reasonably well-behaved bounded domain or surface in some space $\mathbf{R}^m$, $m \geq 1$, with Sobolev spaces $H^\alpha(G)$ and $H^{-\alpha}(G)$ that are suitable generalizations of the spaces of periodic functions discussed above. We consider solving the equation

$$\mathcal{L}\varphi = \psi \tag{7.4.200}$$

with $\psi \in H^{-\alpha}(G)$. Denote its solution by φ^*.

An equivalent way of viewing the solving of (7.4.200) is in a *variational formulation*: find $\varphi^* \in H^\alpha(G)$ such that

$$\langle \eta, \mathcal{L}\varphi^* \rangle = \langle \eta, \psi \rangle, \quad \text{for all } \eta \in H^\alpha(G) \tag{7.4.201}$$

This reformulation of the problem uses the duality pairing referred to following (7.4.196). It is straightforward that any solution of this problem will also be a solution of (7.4.200), and vice versa.

To obtain a numerical method, introduce a sequence of finite dimensional approximating subspaces $\mathcal{X}_n \subset H^\alpha(G)$, $n \geq 1$. For each n, find a function $\varphi_n^* \in \mathcal{X}_n$ with

$$\langle \eta, \mathcal{L}\varphi_n^* \rangle = \langle \eta, \psi \rangle, \quad \text{for all } \eta \in \mathcal{X}_n \tag{7.4.202}$$

When the subspaces $\mathcal{X}_n$ are defined as piecewise polynomial functions of a suitable type, this numerical procedure is known as *Galerkin's method*. It is also sometimes referred to as the *finite element method*, although historically, that name is associated with the finite dimensional minimization problem given below in (7.4.214).

To look at the above numerical procedure in a more abstract framework, we introduce some additional notation. Define a bilinear operator

$$\mathcal{A}: H^\alpha(G) \times H^\alpha(G) \to \mathbf{C}$$

by

$$\mathcal{A}(\eta, \varphi) \equiv \langle \eta, \mathcal{L}\varphi \rangle, \quad \varphi, \eta \in H^\alpha(G) \tag{7.4.203}$$

We assume $\mathcal{L}$ is *symmetric*, in the sense that $\mathcal{A}$ satisfies

$$\mathcal{A}(\eta, \varphi) = \overline{\mathcal{A}(\varphi, \eta)}, \quad \varphi, \eta \in H^\alpha(G) \tag{7.4.204}$$

Moreover, we assume $\mathcal{L}$ is *strongly elliptic*, meaning that

$$|\mathcal{A}(\varphi, \varphi)| \geq c_e \|\varphi\|_\alpha^2, \qquad \varphi \in H^\alpha(G) \tag{7.4.205}$$

for some $c_e > 0$. Finally, from the boundedness of $\mathcal{L}$, we have boundedness of $\mathcal{A}$:

$$|\mathcal{A}(\eta, \varphi)| \leq \|\mathcal{L}\| \|\varphi\|_\alpha \|\eta\|_\alpha, \qquad \varphi, \eta \in H^\alpha(G) \tag{7.4.206}$$

In this, $\|\mathcal{L}\|$ refers to the norm when regarding $\mathcal{L}$ as an operator from $H^\alpha(G)$ to $H^{-\alpha}(G)$.

Define a linear functional $\ell : H^\alpha(G) \to \mathbf{C}$ by

$$\ell[\eta] \equiv \langle \eta, \psi \rangle, \qquad \eta \in H^\alpha(G) \tag{7.4.207}$$

We have boundedness of ℓ, with

$$|\ell[\eta]| \leq \|\ell\| \|\eta\|_\alpha, \qquad \|\ell\| = \|\psi\|_{-\alpha}$$

Using this notation, (7.4.201) can be rewritten as the problem of finding the solution $\varphi = \varphi^*$ to

$$\mathcal{A}(\eta, \varphi) = \ell[\eta], \quad \text{for all } \eta \in H^\alpha(G) \tag{7.4.208}$$

The variational problem (7.4.201), or (7.4.208), can be reformulated as an equivalent minimization problem. Introduce the nonlinear functional $\mathcal{F}$ on $H^\alpha(G)$ by

$$\mathcal{F}[\eta] \equiv \frac{1}{2} \mathcal{A}(\eta, \eta) - \text{Real}(\ell[\eta]), \qquad \eta \in H^\alpha(G) \tag{7.4.209}$$

and consider the problem of finding a function η^* for which

$$\mathcal{F}[\eta^*] = \min_{\eta \in H^\alpha(G)} \mathcal{F}[\eta] \tag{7.4.210}$$

If (7.4.208) has a unique solution $\varphi = \varphi^*$, then the function $\mathcal{F}[\eta]$ has $\eta = \varphi^*$ as a unique minimizer. To see this, replace η by $\varphi^* + \xi$ and expand $\mathcal{F}[\varphi^* + \xi]$ to get

$$\mathcal{F}[\varphi^* + \xi] = \mathcal{F}[\varphi^*] + \frac{1}{2} \mathcal{A}(\varphi^*, \xi) + \frac{1}{2} \mathcal{A}(\xi, \varphi^*) + \frac{1}{2} \mathcal{A}(\xi, \xi) - \text{Real}(\ell[\xi])$$

$$= \mathcal{F}[\varphi^*] + \text{Real}(\mathcal{A}(\xi, \varphi^*) - \ell[\xi]) + \frac{1}{2} \mathcal{A}(\xi, \xi)$$

$$= \mathcal{F}[\varphi^*] + \frac{1}{2} \mathcal{A}(\xi, \xi)$$

The last step uses the fact that φ^* satisfies (7.4.208), with η replaced by ξ. Using the strong ellipticity condition (7.4.205), we have

$$\mathcal{F}[\varphi^* + \xi] > \mathcal{F}[\varphi^*]$$

for all $\xi \neq 0$. Thus φ^* is the unique minimizer of $\mathcal{F}[\eta]$ as η ranges over $H^\alpha(G)$.

The converse statement is also true: if $\mathcal{F}[\eta]$ has a unique minimizer φ^*, then $\varphi = \varphi^*$ is the unique solution to (7.4.208). To see this, let ϵ be an arbitrary nonzero real number, and let ξ be an arbitrary element in $H^\alpha(G)$, $\xi \neq 0$. Then

$$\mathcal{F}[\varphi^*] < \mathcal{F}[\varphi^* + \epsilon \xi] \tag{7.4.211}$$

Expand the right side, obtaining

$$\mathcal{F}[\varphi^* + \epsilon \xi] = \mathcal{F}[\varphi^*] + \epsilon \, \mathrm{Real}\left(\mathcal{A}(\xi, \varphi^*) - \ell[\xi]\right) + \frac{\epsilon^2}{2} \mathcal{A}(\xi, \xi) \tag{7.4.212}$$

Using (7.4.211) to cancel $\mathcal{F}[\varphi^*]$, we have

$$0 < \epsilon \left[\mathrm{Real}(\mathcal{A}(\xi, \varphi^*) - \ell[\xi]) + \frac{\epsilon}{2}\mathcal{A}(\xi, \xi)\right]$$

Since ϵ can be negative or positive, we can let $\epsilon \to 0$ to imply that

$$\mathrm{Real}(\mathcal{A}(\xi, \varphi^*) - \ell[\xi]) = 0 \tag{7.4.213}$$

is necessary. Return to (7.4.211) and (7.4.212), and replace ϵ by $i\epsilon$. Using (7.4.204), the expansion (7.4.212) becomes

$$\mathcal{F}[\varphi^* + i\epsilon \xi] = \mathcal{F}[\varphi^*] - \epsilon \, \mathrm{Imag}(\mathcal{A}(\xi, \varphi^*) - \ell[\xi]) + \frac{\epsilon^2}{2}\mathcal{A}(\xi, \xi)$$

The earlier argument that led to (7.4.213) can be repeated, to obtain

$$\mathrm{Imag}(\mathcal{A}(\xi, \varphi^*) - \ell[\xi]) = 0$$

Combining this result with (7.4.213), we have shown φ^* is the solution to (7.4.208).

The *Ritz method* is the numerical method based on minimizing $\mathcal{F}[\eta]$ over an approximating subspace $\mathcal{X}_n$:

$$\mathcal{F}[\varphi_n^*] = \min_{\eta \in \mathcal{X}_n} \mathcal{F}[\eta] \tag{7.4.214}$$

When the approximating subspace $\mathcal{X}_n$ is based on using piecewise polynomial functions over some mesh defined on the domain G for the problem, as is described following (7.4.202), then the Ritz method is more commonly called a finite element method. Since all of our subspaces will be of the latter kind, we will refer to the solving of (7.4.214) as a finite element method. By adapting the argument showing the equivalence of the variational problem and the minimization problem over $H^\alpha(G)$, to the finite dimensional problem over $\mathcal{X}_n$, we can show the solution φ_n^* of this minimization is exactly the same as the solution of the earlier finite dimensional variational problem (7.4.202). In applications to partial differential equations, the function $\mathcal{F}[\eta]$ is often related to the *energy* in the physical problem under consideration, and minimizing $\mathcal{F}[\eta]$ is usually related to some basic physical principle. Using the finite element method, we also have

$$\mathcal{F}[\varphi^*] \leq \mathcal{F}[\varphi_n^*] \qquad (7.4.215)$$

Thus a solution φ_n^* will allow the computing of an upper bound for $\mathcal{F}[\varphi^*]$, without knowing φ^*.

A general existence theorem

The solvability theory for (7.4.208) or (7.4.210) is given in Theorem 7.4.1, given below. The theory is stated in an even more abstract form, as it has applicability that extends past our interest in solving boundary integral equations.

Repeating some earlier material from above, we assume the following. Let $\mathcal{X}$ be a complex Hilbert space, with norm $\|\cdot\|$ and inner product $(\cdot, \cdot)$. Assume $\mathcal{A} : \mathcal{X} \times \mathcal{X} \to \mathbf{C}$ is a function satisfying:

A1. For all $v_1, v_2, w \in \mathcal{X}$ and all complex numbers c_1, c_2,

$$\mathcal{A}(c_1 v_1 + c_2 v_2, w) = c_1 \mathcal{A}(v_1, w) + c_2 \mathcal{A}(v_2, w)$$

Also,

$$\mathcal{A}(v, w) = \overline{\mathcal{A}(w, v)}, \quad v, w \in \mathcal{X}$$

A2. $\mathcal{A}$ is bounded,

$$|\mathcal{A}(v, w)| \leq c_\mathcal{A} \|v\| \|w\|, \quad v, w \in \mathcal{X} \qquad (7.4.216)$$

for some finite constant $c_\mathcal{A}$.

A3. $\mathcal{A}$ is *strongly elliptic*:

$$|\mathcal{A}(v, v)| \geq c_e \|v\|^2, \quad v \in \mathcal{X} \qquad (7.4.217)$$

with some constant $c_e > 0$.

A4. Further, let ℓ be a bounded linear functional on $\mathcal{X}$.

As before, we are interested in solving the variational problem of finding $w \in \mathcal{X}$ such that

$$A(v, w) = \ell[v], \quad \text{for all } v \in \mathcal{X} \qquad (7.4.218)$$

A solution will be denoted by $w = u$. Also as before, introduce

$$\mathcal{F}[v] = \frac{1}{2} A(v, v) - \text{Real}(\ell[v]), \quad v \in \mathcal{X} \qquad (7.4.219)$$

which we seek to minimize as v ranges over $\mathcal{X}$. Again, these two problems are equivalent in their solvability, possessing the same solution u.

With these assumptions, we introduce a new inner product and a new norm on $\mathcal{X}$. Define

$$(v, w)_A = A(v, w), \quad v, w \in \mathcal{X} \qquad (7.4.220)$$

$$\|v\|_A = \sqrt{(v, v)_A}, \quad v \in \mathcal{X} \qquad (7.4.221)$$

With the above properties **A1–A3**, it is straightforward to show these define an inner product and its associated norm. This norm is sometimes called an *energy norm*, since in some applications it is proportional to the energy in the physical problem being studied. The norm $\|\cdot\|_A$ is equivalent to the original norm on $\mathcal{X}$, since

$$\sqrt{c_e}\|v\| \le \|v\|_A \le \sqrt{c_A}\|v\|, \quad v \in \mathcal{X} \qquad (7.4.222)$$

Using the energy norm, we can view the minimization of the nonlinear functional $\mathcal{F}$ from another perspective. Let u be a minimizer for $\mathcal{F}[v]$. Then straightforward expansion using (7.4.221) shows

$$\begin{aligned}
\mathcal{F}[v] - \frac{1}{2}\|v - u\|_A^2 &= \frac{1}{2} A(v, v) - \text{Real}(\ell[v]) - \frac{1}{2} A(v - u, v - u) \\
&= -\frac{1}{2} A(u, u) + \text{Real}\{A(v, u) - \ell[v]\} \\
&= -\frac{1}{2} A(u, u)
\end{aligned}$$

since $A(v, u) - \ell[v] = 0$ from (7.4.218). Thus

$$\mathcal{F}[v] = \frac{1}{2}\|v - u\|_A^2 - \frac{1}{2}\|u\|_A^2 \qquad (7.4.223)$$

The minimization of $\mathcal{F}[v]$ is equivalent to finding the element v for which the energy norm $\|v - u\|_A$ is minimized, namely $v = u$.

Theorem 7.4.1 (Lax-Milgram Theorem [336]). *Assume* **A1–A4** *from above. Then problem (7.4.218), or equivalently the minimization of* $\mathcal{F}[v]$ *in (7.4.219), has a unique solution* $w = u$, *and the solution satisfies*

$$\|u\| \le \frac{1}{c_e}\|\ell\| \tag{7.4.224}$$

Proof. (a) We begin by showing there is a unique bounded linear operator $\mathcal{B} : \mathcal{X} \to \mathcal{X}$ with

$$\mathcal{A}(v, \mathcal{B}z) = (v, z), \quad v, z \in \mathcal{X} \tag{7.4.225}$$

and

$$\|\mathcal{B}\| \le \frac{1}{c_e} \tag{7.4.226}$$

Begin by defining a linear functional $L_w : \mathcal{X} \to \mathbf{R}$,

$$L_w[v] = \mathcal{A}(v, w), \quad v \in \mathcal{X} \tag{7.4.227}$$

for an arbitrary $w \in \mathcal{X}$. This functional is bounded, with $\|L_w\| \le c_{\mathcal{A}}\|w\|$. By the Riesz representation theorem (cf. Theorem A.4 in the Appendix), there is a unique element $\mathcal{S}(w) \in \mathcal{X}$ with

$$(v, \mathcal{S}(w)) = L_w[v], \quad v \in \mathcal{X} \tag{7.4.228}$$

Moreover, $\|\mathcal{S}(w)\| = \|L_w\|$, and thus

$$\|\mathcal{S}(w)\| \le c_{\mathcal{A}}\|w\| \tag{7.4.229}$$

We can easily show S is linear.

L1. For $c \in \mathbf{C}$, by definition,

$$
\begin{aligned}
(v, \mathcal{S}(cw)) &= L_{cw}[v] \\
&= \mathcal{A}(v, cw) \\
&= \bar{c}\mathcal{A}(v, w), \quad v \in \mathcal{X}
\end{aligned}
$$

But also,

$$
\begin{aligned}
(v, c\mathcal{S}(w)) &= \bar{c}(v, \mathcal{S}(w)) \\
&= \bar{c}L_w[v] \\
&= \bar{c}\mathcal{A}(v, w), \quad v \in \mathcal{X}
\end{aligned}
$$

By the uniqueness part of the Riesz representation theorem, we have

$$\mathcal{S}(cw) = c\mathcal{S}(w), \quad w \in \mathcal{X}$$

L2. Let $w_1, w_2 \in \mathcal{X}$. Then

$$
\begin{aligned}
(v, \mathcal{S}(w_1 + w_2)) &= L_{w_1+w_2}[v] \\
&= \mathcal{A}(v, w_1 + w_2) \\
&= \mathcal{A}(v, w_1) + \mathcal{A}(v, w_2), \quad v \in \mathcal{X}
\end{aligned}
$$

But also,

$$
\begin{aligned}
(v, \mathcal{S}(w_1) + \mathcal{S}(w_2)) &= (v, \mathcal{S}(w_1)) + (v, \mathcal{S}(w_2)) \\
&= \mathcal{A}(v, w_1) + \mathcal{A}(v, w_2), \quad v \in \mathcal{X}
\end{aligned}
$$

Again, by the uniqueness part of the Riesz representation theorem, we have

$$\mathcal{S}(w_1 + w_2) = \mathcal{S}(w_1) + \mathcal{S}(w_2)$$

This shows $\mathcal{S}$ is linear. By (7.4.229),

$$\|\mathcal{S}\| \le c_{\mathcal{A}} \tag{7.4.230}$$

Moreover from the (7.4.227)–(7.4.228),

$$(v, \mathcal{S}w) = \mathcal{A}(v, w), \quad v, w \in \mathcal{X} \tag{7.4.231}$$

The operator $\mathcal{S}$ is one-to-one. To see this, use the strong ellipticity condition **A3** to obtain

$$
\begin{aligned}
c_e \|w\|^2 &\le \mathcal{A}(w, w) = (w, \mathcal{S}w) \le \|w\| \|\mathcal{S}w\| \\
c_e \|w\| &\le \|\mathcal{S}w\|, \quad w \in \mathcal{X}
\end{aligned}
\tag{7.4.232}
$$

Thus $\mathcal{S}$ is one-to-one, and $\mathcal{S}^{-1} : \text{Range}(\mathcal{S}) \to \mathcal{X}$ has the bound

$$\|\mathcal{S}^{-1}\| \le \frac{1}{c_e} \tag{7.4.233}$$

We next show that $\text{Range}(\mathcal{S}) = \mathcal{X}$.

We first show that $\text{Range}(\mathcal{S})$ is closed. Let $\{z_n\} \subset \mathcal{X}$ and $\mathcal{S}(z_n) \to w$ in $\mathcal{X}$. Note that this implies $\{\mathcal{S}z_n\}$ is a Cauchy sequence in $\mathcal{X}$. From (7.4.232),

$$\|z_n - z_m\| \le \frac{1}{c_e} \|\mathcal{S}z_n - \mathcal{S}z_m\| \to 0$$

as $n, m \to \infty$, since $\{Sz_n\}$ is Cauchy. This shows $\{z_n\}$ is also a Cauchy sequence in $\mathcal{X}$, and by the completeness of $\mathcal{X}$, $z_n \to z$ for some $z \in \mathcal{X}$. By the boundedness of S, shown in (7.4.230), $Sz_n \to Sz$. Therefore, $w = Sz \in \text{Range}(S)$. This proves $\text{Range}(S)$ is closed.

Since $\text{Range}(S)$ is a closed subspace, we can decompose $\mathcal{X}$ as

$$\mathcal{X} = \text{Range}(S) \oplus \text{Range}(S)^{\perp} \tag{7.4.234}$$

with $\text{Range}(S)^{\perp}$ the orthogonal complement of $\text{Range}(S)^{\perp}$. Assume $w \in \text{Range}(S)^{\perp}$. Then

$$0 = (w, Sz) = \mathcal{A}(w, z) \quad \text{for all } z \in \mathcal{X} \tag{7.4.235}$$

Let $z = w$ and use the strong ellipticity condition **A3** to obtain

$$\mathcal{A}(w, w) \geq c_e \|w\|^2 > 0$$

for $w \neq 0$. This contradicts (7.4.235) unless $w = 0$, and this proves $\text{Range}(S)^{\perp} = \{0\}$. Combined with (7.4.234), we have $\text{Range}(S) = \mathcal{X}$.

We have shown

$$S : \mathcal{X} \xrightarrow[\text{onto}]{1-1} \mathcal{X} \tag{7.4.236}$$

with a bound for S^{-1} given by (7.4.233). Define $\mathcal{B} = S^{-1}$. Then rewriting (7.4.231), and letting $z = Sw$,

$$(v, z) = \mathcal{A}(v, \mathcal{B}z), \quad v, z \in \mathcal{X}$$

This proves the assertions of (7.4.225)–(7.4.226).

To prove the uniqueness of $\mathcal{B}$, assume there is a second such operator, which we denote by $\hat{\mathcal{B}}$. Then it too satisfies (7.4.225), and therefore,

$$\mathcal{A}(v, \mathcal{B}z) = \mathcal{A}(v, \hat{\mathcal{B}}z)$$
$$\mathcal{A}(v, \mathcal{B}z - \hat{\mathcal{B}}z) = 0, \quad v, z \in \mathcal{X}$$

Let $v = \mathcal{B}z - \hat{\mathcal{B}}z$ and then use the strong ellipticity condition to obtain

$$c_e \|\mathcal{B}z - \hat{\mathcal{B}}z\|^2 \leq \mathcal{A}(\mathcal{B}z - \hat{\mathcal{B}}z, \mathcal{B}z - \hat{\mathcal{B}}z) = 0, \quad z \in \mathcal{X}$$

This proves $\mathcal{B}z = \hat{\mathcal{B}}z$ for all $z \in \mathcal{X}$, thus showing $\mathcal{B} = \hat{\mathcal{B}}$.

(b) We show the existence and uniqueness of a solution u to (7.4.218). By the Riesz representation theorem (cf. Theorem A.4 in the Appendix), there is a unique element $z_\ell \in \mathcal{X}$ for which

$$\ell[v] = (v, z_\ell), \quad v \in \mathcal{X}$$

with

$$\|\ell\| = \|z_\ell\| \qquad (7.4.237)$$

Thus we want to find $u \in \mathcal{X}$ for which

$$\mathcal{A}(v, u) = (v, z_\ell), \quad v \in \mathcal{X} \qquad (7.4.238)$$

Let $u = \mathcal{B} z_\ell$, since then (7.4.225) implies

$$\mathcal{A}(v, \mathcal{B} z_\ell) = (v, z_\ell), \quad v \in \mathcal{X}$$

as desired.

To prove the uniqueness of u, assume there is a second solution to (7.4.218), which we denote by $\hat{u}$. Using (7.4.218), we have

$$\mathcal{A}(v, u) = \mathcal{A}(v, \hat{u})$$
$$\mathcal{A}(v, u - \hat{u}) = 0, \quad v \in \mathcal{X}$$

Let $v = u - \hat{u}$ and use the strong ellipticity condition **A3** to obtain $u - \hat{u} = 0$.

To bound the solution u of (7.4.238),

$$\begin{aligned} \|u\| &= \|\mathcal{B} z_\ell\| \\ &\leq \|\mathcal{B}\| \|z_\ell\| = \|\mathcal{B}\| \|\ell\| \\ &\leq \frac{1}{c_e} \|\ell\| \end{aligned}$$

This uses (7.4.237) and (7.4.226). This proves (7.4.224) and completes the proof.   □

This theorem is used to obtain existence results for boundary value problems for elliptic partial differential equations. As an example of such results, see Renardy and Rogers [460, Chap. 8]. The above theorem can be generalized further, to handle a pseudodifferential operator that is the sum of an operator of the type treated above and a compact operator; see Hildebrandt and Wienholtz [270]. These results are sometimes needed when developing finite element methods for boundary integral equations other than Laplace's equation.

An abstract finite element theory

Consider the finite element method for solving (7.4.218). Let $\mathcal{X}_n$ be a finite dimensional subspace of the Hilbert space $\mathcal{X}$. Find $w \in \mathcal{X}_n$ for which

$$\mathcal{A}(v, w) = \ell[v], \quad v \in \mathcal{X}_n \qquad (7.4.239)$$

Since the assumptions **A1–A4** are still valid, we can apply the preceding Lax-Milgram theorem to this problem.

To find u_n, let $\mathcal{X}_n$ have a basis $\{\varphi_{n,1}, \ldots, \varphi_{n,d_n}\}$. Write

$$u_n = \sum_{j=1}^{d_n} \bar{\alpha}_j \varphi_{n,j} \tag{7.4.240}$$

in which we have used conjugated coefficients to simplify some formulas given later. Substitute this into (7.4.239) and let $v = \varphi_{n,i}, i = 1, \ldots, d_n$. This yields the linear system

$$\sum_{j=1}^{d_n} \alpha_j \mathcal{A}(\varphi_{n,i}, \varphi_{n,j}) = \ell[\varphi_{n,i}], \quad i = 1, \ldots, d_n \tag{7.4.241}$$

We write this linear system as

$$A_n \alpha = b \tag{7.4.242}$$

with $(A_n)_{i,j} = \mathcal{A}(\varphi_{n,i}, \varphi_{n,j})$ and $b_i = \ell[\varphi_{n,i}]$.

This system could also have been obtained from minimizing the function $\mathcal{F}[v]$ of (7.4.219) over $\mathcal{X}_n$. Represent $v \in \mathcal{X}_n$ as $v = \sum \bar{\eta}_j \varphi_{n,j}$ and substitute into the formula for $\mathcal{F}[v]$. This yields

$$\mathcal{F}[v] = \frac{1}{2} \eta^* A_n \eta - \text{Real}(\eta^* b) \tag{7.4.243}$$

The minimization of this leads to (7.4.242). The details are left to the reader.

Theorem 7.4.2 (Cea's lemma). *Assume* **A1–A4** *for the original space $\mathcal{X}$ and the original variational problem (7.4.218). Then the finite dimensional variational problem (7.4.239) has a unique solution $u_n \in \mathcal{X}_n$, and it satisfies*

$$\|u_n\| \le \frac{1}{c_e} \|\ell\| \tag{7.4.244}$$

The matrix A_n is Hermitian and positive definite, and

$$\eta^* A_n \eta \ge c_e \|v\|^2, \quad v = \sum_{j=1}^{d_n} \bar{\eta}_j \varphi_{n,j} \tag{7.4.245}$$

For the error in u_n,

$$\|u - u_n\| \le \frac{c_\mathcal{A}}{c_e} \inf_{v \in \mathcal{X}_n} \|u - v\| \tag{7.4.246}$$

Proof. The Lax-Milgram theorem can be applied to (7.4.239) with respect to the space $\mathcal{X}_n$. From it directly, we have that (7.4.239) has a unique solution satisfying (7.4.244). The bound (7.4.244) shows that the finite element method is stable.

The result that A_n is Hermitian follows directly from the definition $(A_n)_{i,j} = \mathcal{A}(\varphi_{n,i}, \varphi_{n,j})$ and the property **A1** of $\mathcal{A}$. To show A_n is positive definite, write $v = \sum \bar{\eta}_j \varphi_{n,j}$ and compute

$$\mathcal{A}(v, v) = \eta^* A_n \eta \qquad (7.4.247)$$

Use the strong ellipticity assumption **A3** to obtain (7.4.245), and in turn, this inequality shows clearly that A_n is positive definite.

Recall the original variational equation (7.4.218) and compare it with the finite dimensional variational equation (7.4.239). Then for the true solutions u and u_n of the two problems,

$$\mathcal{A}(v, u) = \ell(v), \quad v \in \mathcal{X}$$
$$\mathcal{A}(v, u_n) = \ell(v), \quad v \in \mathcal{X}_n$$

Subtracting,

$$\mathcal{A}(v, u - u_n) = 0, \quad v \in \mathcal{X}_n \qquad (7.4.248)$$

This implies that u_n is the orthogonal projection into $\mathcal{X}_n$ with respect to the inner product $(\cdot, \cdot)_\mathcal{A}$ of (7.4.220); thus

$$\|u\|_\mathcal{A}^2 = \|u_n\|_\mathcal{A}^2 + \|u - u_n\|_\mathcal{A}^2 \qquad (7.4.249)$$

For arbitrary $v \in \mathcal{X}_n$, write

$$\mathcal{A}(u - u_n, u - u_n) = \mathcal{A}(u - u_n, u - v) + \underbrace{\mathcal{A}(u - u_n, v - u_n)}_{=0 \text{ by } (7.4.248)}$$
$$= \mathcal{A}(u - u_n, u - v)$$

Using **A2** and **A3**,

$$c_e \|u - u_n\|^2 \leq c_\mathcal{A} \|u - u_n\| \|u - v\|$$

Cancel $\|u - u_n\|$ to obtain

$$\|u - u_n\| \leq \frac{c_\mathcal{A}}{c_e} \|u - v\| \qquad (7.4.250)$$

This leads directly to (7.4.246), and it completes the proof.   □

The finite element solution as a projection

Given $u \in \mathcal{X}$, define a linear functional by

$$\ell_u(v) = \mathcal{A}(v, u), \quad v \in \mathcal{X} \tag{7.4.251}$$

This linear functional is bounded, with

$$\|\ell_u\| \le c_A \|u\| \tag{7.4.252}$$

Then immediately, u is the solution of the variational equation (7.4.218) with linear functional $\ell = \ell_u$. Now consider the finite element approximation of this problem: find $u_n \in \mathcal{X}_n$ such that

$$\mathcal{A}(v, u_n) = \ell_u(v), \quad v \in \mathcal{X}_n \tag{7.4.253}$$

Write

$$u_n = \mathcal{G}_n(u) \tag{7.4.254}$$

The function $\mathcal{G}_n : \mathcal{X} \to \mathcal{X}_n$ can be shown to be linear, which we leave to the reader. Moreover,

$$u \in \mathcal{X}_n \Rightarrow \mathcal{G}_n u = u$$

To see this, consider the unique solution $u_n \in \mathcal{X}_n$ of (7.4.253):

$$\mathcal{A}(v, u_n) = \mathcal{A}(v, u), \quad v \in \mathcal{X}_n$$

Clearly it is $u_n = u$, and therefore $\mathcal{G}_n u = u_n = u$. This shows $\mathcal{G}_n$ is a projection from $\mathcal{X}$ to $\mathcal{X}_n$. As was stated earlier, preceding (7.4.249), $\mathcal{G}_n u$ is the orthogonal projection of u into $\mathcal{X}_n$ with respect to the inner product $(\cdot, \cdot)_A$ of (7.4.220).

To bound $\mathcal{G}_n$, use (7.4.244).

$$\|\mathcal{G}_n u\| = \|u_n\| \le \frac{1}{c_e} \|\ell_u\| \le \frac{c_A}{c_e} \|u\|$$

Thus

$$\|\mathcal{G}_n\| \le \frac{c_A}{c_e} \tag{7.4.255}$$

At this point, we can give applications to partial differential equations as well as to boundary integral equations, but we pursue only the latter.

7.4.3. Boundary element methods for boundary integral equations

Apply the preceding theory to a pseudodifferential operator equation $\mathcal{L}\varphi = \psi$ with

$$\mathcal{L} : H^\alpha(2\pi) \xrightarrow[\text{onto}]{1-1} H^{-\alpha}(2\pi)$$

such as was discussed in (7.4.199)–(7.4.202) and reformulated in (7.4.208) as a variational problem. We assume the properties (7.4.204)–(7.4.206) are satisfied. Using the schema of Theorem 7.4.2, we can obtain numerical methods with any desired speed of convergence. When finite element methods are applied to BIE, they are often called boundary element methods; although this applies more generally to any numerical method for solving BIE using a decomposition of the boundary into small elements.

If $\alpha \leq 0$, then let $\mathcal{X}_n$ be a space of piecewise polynomial functions of degree $\leq r$, with r a given integer, $r \geq 0$. To be more precise, let a partition of $[0, 2\pi]$ be given:

$$0 = t_{n,0} < t_{n,1} < \cdots < t_{n,n} = 2\pi$$

If $v \in \mathcal{X}_n$, then the restriction of v to each subinterval $[t_{n,i-1}, t_{n,i}]$ is to be a polynomial of degree $\leq r$, for $i = 1, \ldots, n$. All such functions $v \in L^2(0, 2\pi)$, and therefore $\mathcal{X}_n \subset H^\alpha(2\pi)$. Many authors also require that elements of $\mathcal{X}_n$ have additional smoothness, namely, that v be spline function of degree r, with continuous derivatives $v^{(j)}$ for $j = 1, \ldots, r - 1$. This leads to results on convergence of derivatives of the approximate solution, but we omit most such results here.

If $\alpha > 0$, then again we let $\mathcal{X}_n$ be a set of piecewise polynomial functions of degree $\leq r$. But now we must require all $v \in \mathcal{X}_n$ to have sufficient smoothness in order that $\mathcal{X}_n \subset H^\alpha(2\pi)$. For example, if $\alpha = \frac{1}{2}$, then it is sufficient to require that each $v \in \mathcal{X}_n$ be periodic, continuously differentiable, and piecewise polynomial of degree $\leq r$.

We will need to have $\mathcal{X}_n$ be so defined that all elements of $H^\alpha(2\pi)$ can be approximated by a sequence of elements from

$$\bigcup_{n\geq 1} \mathcal{X}_n$$

More precisely, we assume that for each $u \in H^\alpha(2\pi)$,

$$\lim_{n\to\infty} \inf_{v\in\mathcal{X}_n} \|u - v\|_\alpha = 0 \qquad (7.4.256)$$

Theorem 7.4.2 now asserts the existence of the numerical solution u_n for the problem (7.4.218), the unique solvability of the approximating system (7.4.241), the stability of the numerical scheme, in (7.4.244), and it gives the error bound

$$\|u - u_n\|_\alpha \le c \inf_{v \in \mathcal{X}_n} \|u - v\|_\alpha \qquad (7.4.257)$$

for some $c > 0$. (Throughout the remaining presentation of this section, the letter c denotes a generic constant, varying in its meaning in different formulas.) To say more about the convergence of u_n to u, we now consider separately the cases $\alpha < 0$, $\alpha = 0$, and $\alpha > 0$.

Case 1. Let $\alpha = 0$. Then we have

$$\mathcal{L} : L^2(0, 2\pi) \xrightarrow[onto]{1-1} L^2(0, 2\pi)$$

An example is the Cauchy transform $\mathcal{C}$ of (7.3.126). The approximation assumption (7.4.256) is exactly of the type studied earlier in §3.3 of Chapter 3. When u possesses additional smoothness, we can improve on the rate of convergence of (7.4.256)–(7.4.257). With only minimal assumptions on $\mathcal{X}_n$, we can show

$$\inf_{v \in \mathcal{X}_n} \|u - v\|_0 \le c \|u\|_{r+1} h^{r+1}, \quad u \in H^{r+1}(2\pi) \qquad (7.4.258)$$

where

$$h \equiv h_n = \max_{j=1,\dots,n} (t_{n,j} - t_{n,j-1})$$

and c is independent of u. This bound can be proven by using Taylor's theorem to approximate u on each subinterval $[t_{n,i-1}, t_{n,i}]$ with a polynomial of degree $\le r$; we omit the details.

The results (7.4.257) and (7.4.258) imply u_n converges to u with a rate of at least h^{r+1}. This is often referred to as an *optimal rate of convergence*, because it is the same order of convergence as the error in the orthogonal projection of u into $\mathcal{X}_n \subset L^2(0, 2\pi)$.

Case 2. Let $\alpha < 0$. An example is the single layer operator (7.3.107), with $\alpha = -\frac{1}{2}$. For this case, (7.4.256)–(7.4.257) implies convergence of $\|u - u_n\|_\alpha$ to zero. However, we usually want results for the convergence to zero of $\|u - u_n\|_0$, as this is a more standard norm for measuring errors.

We begin by discussing the convergence to zero of $\|u - u_n\|_\alpha$. Our approximating subspaces $\mathcal{X}_n$ usually satisfy

$$\inf_{v \in \mathcal{X}_n} \|u - v\|_0 \to 0, \quad u \in L^2(0, 2\pi) \tag{7.4.259}$$

as was discussed in the case for $\alpha = 0$. From the result that

$$H^\alpha(2\pi) \subset H^0(2\pi) \equiv L^2(0, 2\pi)$$

is a continuous embedding, we have

$$\|w\|_\alpha \le c(\alpha)\|w\|_0, \quad w \in H^\alpha(2\pi)$$

for a suitable $c(\alpha) > 0$. From this,

$$\|u - v\|_\alpha \le c(\alpha)\|u - v\|_0 \tag{7.4.260}$$

for all $u \in L^2(0, 2\pi)$, $v \in \mathcal{X}_n$. When combined with (7.4.257) and (7.4.259), this proves $\|u - u_n\|_\alpha \to 0$ for all such u. If $u \in H^\alpha(2\pi)$, but $u \notin L^2(0, 2\pi)$, then other means must be used to prove (7.4.256) and the convergence $\|u - u_n\|_\alpha \to 0$. To wit, we can use the denseness of $L^2(0, 2\pi)$ in $H^\alpha(2\pi)$ to prove the needed convergence; we omit the details.

We can also combine (7.4.260) and (7.4.258) to prove

$$\|u - u_n\|_\alpha \le c\|u\|_{r+1}h^{r+1}, \quad u \in H^{r+1}(2\pi) \tag{7.4.261}$$

for a suitable $c > 0$, independent of u and n. This error bound is not optimal, as the exponent $r + 1$ can be increased to $r + 1 - \alpha$. The proof of this last result requires more knowledge of properties of the approximation of u by elements of $\mathcal{X}_n$, and these are given below as properties **P1** and **P2**.

Assume the mesh $\{t_{n,j}\}$ satisfies

$$h_n \le c \min_{j=1,\dots,n} (t_{n,j} - t_{n,j-1}) \tag{7.4.262}$$

with some c independent of n. This says that the smallest and the largest of the subintervals $[t_{n,i-1}, t_{n,i}]$ converge to zero at the same rate, and therefore $\{t_{n,j}\}$ is not a graded mesh.

Assume there is an integer μ, $0 \le \mu \le r - 1$, such that

$$\mathcal{X}_n \subset C_p^\mu(2\pi) \tag{7.4.263}$$

Functions $v \in \mathcal{X}_n$ with $\mu = r - 1$ are spline functions of degree r. Using these assumptions, we can prove the following important properties.

P1 (Inverse property). Let $-(\mu+1) \le s \le t \le \mu+1$. Then

$$\|v\|_t \le c(s,t)\|v\|_s h^{s-t}, \quad v \in \mathcal{X}_n \tag{7.4.264}$$

with $c(s,t)$ independent of n.

P2 (Approximation property). Let $-(\mu+2) \le t \le s \le r+1$, $-(\mu+1) \le s$, and $t \le \mu+1$. Then for any $u \in H^s(2\pi)$ and any n, there is an element $w_n \in \mathcal{X}_n$ with

$$\|u - w_n\|_t \le c(s,t)\|u\|_s h^{s-t} \tag{7.4.265}$$

where $c(s,t)$ is independent of u, n, and w_n, and with w_n independent of s,t.

For proofs and discussions of these results, see Arnold and Wendland [25, pp. 352, 359], Babuška and Aziz [71], Helfrich [263], and Nitsche [401].

We can now give a convergence result for $\|u - u_n\|_0$. Recall from (7.4.254) that $u_n = \mathcal{G}_n u$. Then

$$\begin{aligned}
\|u - u_n\|_0 &= \|u - \mathcal{G}_n u\|_0 \\
&= \|(u-v) - \mathcal{G}_n(u-v)\|_0, \quad v \in \mathcal{X}_n \\
&\le \|u - v\|_0 + \|\mathcal{G}_n(u-v)\|_0 \\
&\le \|u - v\|_0 + c h^\alpha \|\mathcal{G}_n(u-v)\|_\alpha, \quad \text{using } \mathbf{P1} \\
&\le \|u - v\|_0 + c h^\alpha \|\mathcal{G}_n\|\|u-v\|_\alpha
\end{aligned}$$

Assume $u \in H^{r+1}(2\pi)$, and then let $v = w_n$ from **P2** to construct bounds for both $\|u - v\|_0$ and $\|u - v\|_\alpha$. Recall the bound (7.4.255) for $\|\mathcal{G}_n\|$. When these results are combined with the final inequality above, we have

$$\begin{aligned}
\|u - u_n\|_0 &\le c_1 h^{r+1}\|u\|_{r+1} + c_2 h^\alpha \|\mathcal{G}_n\|\|u\|_{r+1} h^{r+1-\alpha} \\
&\le \bar{c}\|u\|_{r+1} h^{r+1}, \quad u \in H^{r+1}(2\pi)
\end{aligned} \tag{7.4.266}$$

for appropriate constants c_1, c_2, and $\bar{c}$.

This form of argument can be generalized to give the result

$$\|u - u_n\|_t \le \bar{c}(s,t)\|u\|_s h^{s-t}, \quad u \in H^s(2\pi) \tag{7.4.267}$$

with $0 \le t \le s \le r+1$ and $t \le \mu+1$. The details are left to the reader. The bounds (7.4.266)–(7.4.267) are considered to be optimal bounds with respect to the rate of convergence shown. This shows that the finite element method converges for boundary integral equations of negative index $\alpha < 0$.

Case 3. Let $\alpha > 0$. An example is the hypersingular integral operator $\mathcal{H}$ of (7.3.137), with $\alpha = \frac{1}{2}$. For this case, the approximating subspaces must satisfy

$$\mathcal{X}_n \subset H^\alpha(2\pi) \tag{7.4.268}$$

and a sufficient condition for this is that (7.4.263) is satisfied with some $\mu \geq \alpha$. From (7.4.256)–(7.4.257), we have $\|u - u_n\|_\alpha \to 0$ for all $u \in H^\alpha(2\pi)$. With sufficient additional differentiability for u, the property **P2** implies

$$\|u - u_n\|_\alpha \leq c \, h^{r+1-\alpha} \|u\|_{r+1}, \quad u \in H^{r+1}(2\pi) \tag{7.4.269}$$

Again, we would like to obtain convergence results for $\|u - u_n\|_0$. This is a more complicated argument, needing additional results concerning the pseudodifferential operator $\mathcal{L}$, but it can still be shown that

$$\|u - u_n\|_0 \leq c \, h^{r+1} \|u\|_{r+1} \tag{7.4.270}$$

under suitable assumptions on u and $\mathcal{L}$. For a proof, see Wendland [565, pp. 252–253].

For many boundary integral equations, the finite element method is the only numerical method for which a complete and rigorous error analysis has been given. This approach to solving boundary integral equations has been developed primarily by W. Wendland and his colleagues, including M. Costabel, G. Hsiao, and E. Stephan; for example, see Refs. [129], [137], [141], [273], [277], [278], [536], [558], [560], [562], and [565]. For a survey and much more extensive discussion of finite element methods for solving boundary integral equations, see Sloan [509].

Additional remarks

To have the finite element method of Theorem 7.4.2 be a practical numerical method, we must also show that the associated linear system (7.4.241) is sufficiently well-conditioned. Moreover, we must give a numerical method for evaluating the Galerkin integrals in (7.4.241), and this quadrature method must be accurate enough to preserve the accuracy shown in statements such as (7.4.266), while not wasting time computing the integrals too accurately. For the conditioning of the linear system, see Wendland [565, p. 254]; and for some discussion of the quadrature errors, see Wendland [559, p. 299], [560]. A special modification of the finite element method, called the *qualocation method*, is discussed at some length in Sloan [509, §7]; this method incorporates specially designed quadrature methods to preserve accuracy and minimize computational cost.

Discussion of the literature

Over the past two decades, the most popular topic in the numerical analysis of integral equations has been the development and analysis of methods for solving boundary integral equation reformulations of boundary value problems for elliptic partial differential equations (PDE). Only a few such PDE are amenable to treatment using BIE, but these are the most important ones for applications, especially Laplace's equation $\Delta u = 0$, the Helmholtz equation $\Delta u - \lambda u = 0$, and the biharmonic equation $\Delta^2 u = 0$.

The use of BIE has been an important tool in understanding physical problems for over 150 years, going back to important works of Green and others in the nineteenth century. The means of solving PDE have varied greatly during this time, and for much of this period the most popular methods were based on the use of separation of variables and other techniques to obtain analytical series approximations. The advent of digital computers led first to the use of finite difference approximations and subsequently to finite element approximations, both approximating the PDE over its domain of definition. Boundary integral equation methods have always been used for special types of problems, especially the solution of potential flow calculations in fluid mechanics, but during the past thirty years the use of BIE as a general tool has become much more popular in the engineering literature. Most uses of BIE in engineering applications have involved the approximation of the solution by piecewise polynomial functions over a decomposition of the boundary into "elements;" and such methods have come to be called boundary element methods or BEM.

The BEM has a large literature, and we give only a few examples. For general introductions, see Banerjee and Watson [76]; Beskos [79]; Brebbia [89]–[91]; Chen and Zhou [115]; and Cruse, Pitko, and Armen [144]. For uses of BIE in elasticity, including crack problems, see Blue [81]; Blum [82]; Cruse [143]; Jaswon and Symm [286]; Krishnasamy, et al. [330]; Muskhelishvili [391]; Rizzo, Shippy, and Rezayat [468]; Rudolphi et al. [470]; and Watson [555]. For acoustic and electromagnetic wave propagation, see Canning [101]; Colton and Kress [127]; Harrington [258]; Kleinman and Roach [314]; and Stephan [529]. For fluid flow calculations, see Hackbusch [250]; Hess [268]; Hess and Smith [269]; Hsiao and Kress [275]; Hsin, Kerwin, and Newman [279]; Newman [400]; Power [431]; Pozrikidis [432]; and Tuck [545]. The work of Hess and Smith has been very important in the development of codes for potential flow calculations in the aerospace industry. The construction of a conformal mapping has often been approached by converting the problem to one of solving a BIE, and an introduction to this area is given by Gaier [207].

The numerical analysis of BIE for planar regions with a smooth boundary has been the most popular topic of study within the area of BIE over the past twenty years, and there is an enormous literature on it. Rather than cite all of the individual papers in this area within our bibliography, we instead will list the names of a number of the major participants in the area, leaving it to the reader to peruse the bibliography for the individual contributions. In alphabetical order, look at the contributions of Amini, Arnold, Brakhage, Chandler, Cheng, Colton, Costabel, Elschner, Graham, Hsiao, Kress, Kußmaul, McLean, von Petersdorf, Prößdorf, Rathsfeld, Ruotsalainen, Saranen, Sloan, Stephan, Vainikko, Wendland, and Werner. An especially nice introduction to this topic with an extensive bibliography is given by Sloan [509].

A major feature of the work of the past twenty years has been the use of Galerkin's method, with its analysis based on generalizing to BIE the finite element analysis developed for PDE. This began with the work of Stephan and Wendland [536], and subsequently it was extended by Wendland and his co-workers to most strongly elliptic pseudodifferential operators; for example, see Wendland [560], [565]. Before this time, BIE of the second kind were the favorite form of equation in applications, but this work gave methods of analysis for BIE of the first kind, as is discussed in this chapter. Extending this work to collocation methods for BIE of the first kind has been more difficult, but there have been a number of papers written on this topic, and one such method was discussed earlier in the chapter, in §7.3.3.

Recently, there has been an increase in interest in the numerical solution of the hypersingular integral equation $\mathcal{H}\rho = f$ of (7.3.137). Among the papers studying this equation are Amini and Maines [8], [9], Chien and Atkinson [122], Kress [327], and Rathsfeld, Kieser, and Kleeman [455].

Laplace's equation is sometimes solved with the nonlinear boundary condition

$$\frac{\partial u(P)}{\partial \mathbf{n}_P} = f(P, u(P))$$

with $f(P, v)$ a given function. This boundary value problem for $\Delta u = 0$ has been examined in recent years in a number of papers, including Atkinson and Chandler [57]; Eggermont and Saranen [174]; Ruotsalainen [471]; Ruotsalainen and Saranen [473]; and Ruotsalainen and Wendland [474].

The use of boundary integral equations is immediately applicable to only homogeneous partial differential equations. For problems such as solving the Dirichlet problem for the Poisson equation,

$$\Delta u(P) = f(P), \quad P \in D$$
$$u(P) = g(P), \quad P \in S$$

it is necessary to convert this to a problem involving a homogeneous equation. The usual procedure is to construct a particular solution of the Poisson equation, say $u_0(P)$, and to then use this to convert the original problem to a new Dirichlet problem for Laplace's equation. In this case, let $v = u - u_0$. Then v satisfies

$$\Delta v(P) = 0, \qquad\qquad P \in D$$
$$v(P) = g(P) - u_0(P), \quad P \in S$$

Methods by which u_0 can be constructed are discussed in Ref. [44], Golberg [221], and Golberg and Chen [223].

It is the personal opinion of the author that the most efficient numerical methods for solving BIE on smooth planar boundaries are those based on trigonometric polynomial approximations, and such methods are sometimes called *spectral methods*. When calculations using piecewise polynomial approximations are compared with those using trigonometric polynomial approximations, the latter are almost always the more efficient. When the data is smooth, a fixed order BEM method will converge like $O(n^{-p})$ for some fixed p, with n the order of the associated linear system. In contrast, spectral methods will converge faster than $O(n^{-p})$ for any $p > 0$, provided the unknown function and the boundary are both C^∞. One code implementing a Nyström method with the trapezoidal rule for solving BIE of the second kind, with automatic error control, is given in Jeon and Atkinson [291], and it has proven very efficient. The Nyström method with the trapezoidal rule on a closed smooth boundary is essentially a spectral method. Nonetheless, the study of boundary element methods for the smooth boundary case has been of great importance in understanding similar methods for boundaries that are only piecewise smooth, which is the case taken up in the following chapter.

8

Boundary integral equations on a piecewise smooth planar boundary

We continue with the study of the numerical solution of boundary integral equation reformulations of boundary value problems for Laplace's equation in the plane. The boundary value problems to be studied are the same as were given in §7.1 of Chapter 7, and much of the notation is also the same as that used earlier. In the present chapter we consider such problems on domains D with a boundary S that is only piecewise continuously differentiable. For such a piecewise smooth boundary S, the properties of the BIE of Chapter 7 are changed in significant ways. In this chapter we give ways of studying and numerically solving such BIE. The first major result in understanding such BIE goes back to the work of Radon in [440] in 1919, but most of the important numerical analysis papers on this problem date from 1984 onwards. Some of these papers are cited in the developments of this chapter, and a further discussion of the literature is given at the end of the chapter.

In §8.1 we discuss the properties of the BIE when S is only piecewise smooth. There are significant differences from the smooth boundary case, both in the behavior of the solutions and in the properties of the integral operators. We give some results on the qualitative behavior of the solutions of the BIE in the neighborhood of corner points of the bounday. Further, we discuss the function space properties of the integral operators, and an important theoretical connection to Wiener-Hopf integral operators is described. In §§8.2–8.4 numerical methods are developed for solving a BIE of the second kind for the Dirichlet problem for Laplace's equation. In §8.2 the Galerkin method is developed, with the approximating subspace consisting of piecewise polynomial functions defined on a graded mesh. In §8.3 we develop collocation methods, again using piecewise polynomial functions on a graded mesh. Nyström methods of a similar nature are developed in §8.4.

8.1. Theoretical behavior

We begin by discussing the behavior of solutions to the interior Dirichlet problem for Laplace's equation. The problem is to find $u \in C(\bar{D}) \cap C^2(D)$ that satisfies

$$\begin{aligned} \Delta u(P) &= 0, & P &\in D \\ u(P) &= f(P), & P &\in S \end{aligned}$$

(8.1.1)

with $f \in C(S)$ a given boundary function. Generally the data function is assumed to have several bounded continuous derivatives with respect to arc length on each smooth section of the boundary. It is then of interest to know the behavior of the derivatives of u as one approaches corner points of S from within D, since this differentiability affects the accuracy of our approximations of u. We begin by developing some intuition for the problem.

Consider the pie-shaped region D shown in Figure 8.1. The two sides meeting at the origin are both straight line segments of length 1, and the remaining portion of the boundary is to be a smooth joining of the two line segments. The central angle is ϕ, with the assumption $0 < \phi < 2\pi$, $\phi \neq \pi$. We would like to examine functions that are harmonic on D and have boundary values that are smoothly differentiable on S, except at the corner. To develop some intuition, we begin with an important example.

It is well known that if $F(z)$, with $z = x + iy$, is an analytic complex function on D (considered as a domain in the complex plane $\mathbf{C}$), then the real and imaginary components of $F(z)$ are harmonic functions on D. Consider the case $F(z) = z^\alpha$ for some $\alpha > 0$. We can always choose the branch cut for this

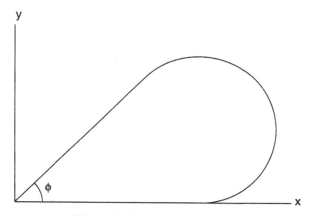

Figure 8.1. Pie-shaped region D.

analytic function to be outside of D, so that we can take this to be analytic on D. Using the polar coordinates representation of $z = re^{i\theta}$, we have

$$F(z) = r^\alpha \cos \alpha\theta + ir^\alpha \sin \alpha\theta$$

The function $r^\alpha \cos \alpha\theta$ is not smooth when restricted to the straight line boundary $\theta = 0$, for any noninteger power α. But the harmonic function

$$u(x, y) = r^\alpha \sin \alpha\theta, \quad r > 0, \ 0 < \theta < \phi \qquad (8.1.2)$$

will sometimes have smooth boundary data, provided α is chosen suitably.

Choose $\alpha > 0$ to be as small as possible, while having the restriction of u to S be smooth. We do this by forcing the boundary values to be zero on the straight line sections of S, and easily, u is smooth on the remaining portion of the boundary. Along the portion of the boundary where $\theta = 0$, we have $u \equiv 0$. We choose α to force

$$\sin \alpha\phi = 0.$$

This is trivially a smooth boundary function for u along $\theta = \phi$. The needed value of α will be as small as possible if we choose $\alpha\phi = \pi$, or

$$\alpha = \frac{\pi}{\phi} \qquad (8.1.3)$$

The resulting u is harmonic on D, and it has smooth boundary data. The boundedness of the derivatives of the $u(x, y)$ depends on the size of α. For the immediately following discussion, we need to assume α is not an integer, and later we consider the case where α is an integer by using a suitable modification of (8.1.2).

1. Let $0 < \phi < \pi$. If ϕ is small, then α is quite large and u has many continuous partial derivatives. As ϕ approaches π, then α approaches 1 from above. Thus the first partial derivatives of u with respect to x and y remain continuous as (x, y) approaches the origin from within D, but the second partial derivatives become unbounded if ϕ is sufficiently close to π.

2. Let $\pi < \phi < 2\pi$. Then α satisfies

$$\frac{1}{2} < \alpha < 1 \qquad (8.1.4)$$

and $\alpha \to \frac{1}{2}$ as $\phi \to 2\pi$. Thus the first partial derivatives of u with respect to x and y are not continuous or bounded as (x, y) approaches the origin from within D. Corners with such an angle ϕ are called *re-entrant corners*, and they present major problems to the use of finite difference and finite element approximations of Laplace's equation on such regions.

If the angle ϕ is an integral fraction of π, then the number α is a positive integer; for example, $\phi = \frac{\pi}{3}$ leads to $\alpha = 3$. For α a positive integer, the function z^α is a polynomial in z, and the function u of (8.1.2) is a polynomial in (x, y), which of course is infinitely differentiable for all (x, y). For this case, an example of ill-behavior around $r = 0$ can be based on using the imaginary part of the complex function $F(z) = z^\alpha \log z$ namely,

$$u(x, y) = (r^\alpha \log r) \sin(\alpha\theta) + \alpha\theta r^\alpha \cos(\alpha\theta) \qquad (8.1.5)$$

This is smooth on both $\theta = 0$ and $\theta = \phi$, and it is continuous at $r = 0$ when α is a positive integer. The derivatives of u of order α will be singular as $r \to 0$, thus again showing an unbounded behavior as (x, y) approaches the origin from within D.

The above behavior in (8.1.2) and (8.1.5) is typical of the general case of solutions u to (8.1.1) on regions with corners. Assume that the boundary S has a corner at P_0, much as shown in Figure 8.1, and for convenience, let P_0 be the origin. Let the boundary be divided into three portions, which we label S_1, S_2, and S_3, and let S_1 and S_2 connect at P_0, with S_3 connecting the remaining ends of S_1 and S_2. We do not require that S_1 and S_2 be linear as in Figure 8.1, but we still denote the central angle by ϕ. Assume that the parametrization $\mathbf{r}(s)$ of the boundary with respect to arc length s is analytic on both S_1 and S_2, and for the Dirichlet boundary data f from (8.1.1), assume that $f(\mathbf{r}(s))$ is analytic on both S_1 and S_2. Then in the vicinity of the corner point, the solution u of (8.1.1) satisfies

$$u(x, y) = \begin{cases} O(r^\alpha), & \alpha \neq m, \text{ a positive integer} \\ O(r^\alpha \log r), & \alpha = m, \text{ an integer} \end{cases} \qquad (8.1.6)$$

for $0 \leq r \leq \epsilon$, some $\epsilon > 0$, and points (x, y) inside of D. This is proven by Wasow [554]. If there are multiple corner points, then the behavior of u in the vicinity of each one can be obtained using (8.1.6). Other more general approaches to estimating the behavior of solutions to elliptic partial differential equations are given in Lehman [339] and Grisvard [240].

8.1.1. Boundary integral equations for the interior Dirichlet problem

If we return to the derivations of the Green's representation formulas of §7.1 in Chapter 7 (which were based on the Divergence theorem), we can generalize those formulas to the case where S is a piecewise smooth boundary for a simply connected bounded region D. We study the behavior of both the solutions of these BIE and the integral operators in them.

Begin by letting

$$\mathbf{r}(t) = (\xi(t), \eta(t)), \quad 0 \le t \le L$$

be a parametrization of S, with $\mathbf{r}(t)$ periodic with period exactly L (and thus $\mathbf{r}(0) = \mathbf{r}(L)$). We assume the interval $[0, L]$ can be subdivided as

$$0 = L_0 < L_1 < \cdots < L_J = L \tag{8.1.7}$$

with $\mathbf{r}(t)$ being several times continuously differentiable on each of the subintervals $[L_{i-1}, L_i]$, $i = 1, \ldots, J$. We assume the points $\mathbf{r}(L_i)$ are "corners" of the boundary. We use S_i to denote the section of S obtained by traveling along it from $\mathbf{r}(L_{i-1})$ to $\mathbf{r}(L_i)$, and $S_J \equiv S_0$, $S_{J+1} \equiv S_1$. At each corner $\mathbf{r}(L_i)$, form tangents to the boundary on both the section S_i and S_{i+1}, and let $(1 - \chi_i)\pi$ be the angle interior to D formed by these two tangent lines. We assume

$$-1 < \chi_i < 1, \quad \chi_i \ne 0, \quad i = 0, 1, \ldots, J \tag{8.1.8}$$

and of course, $\chi_0 \equiv \chi_J$. The choice $\chi_i = 0$ would correspond to a smooth boundary, and the values $\chi_i = \pm 1$ would correspond to "cusps," yielding boundary value problems and BIE that are much more difficult to treat, both theoretically and numerically.

Assuming $u \in C^2(D) \cap C^1(\bar{D})$, we have the Green's representation formula:

$$u(A) = \frac{1}{2\pi} \int_S \left[\frac{\partial u(Q)}{\partial \mathbf{n}_Q} \log|A - Q| - u(Q) \frac{\partial}{\partial \mathbf{n}_Q} [\log|A - Q|] \right] dS_Q,$$
$$A \in D \tag{8.1.9}$$

Letting $A \to P \in S$, it is straightforward to show that

$$\lim_{A \to P} \int_S \frac{\partial u(Q)}{\partial \mathbf{n}_Q} \log|A - Q| \, dS_Q = \int_S \frac{\partial u(Q)}{\partial \mathbf{n}_Q} \log|P - Q| \, dS_Q \tag{8.1.10}$$

However, the result of letting $A \to P \in S$ in the double layer potential is slightly different than that of (7.1.9) in Chapter 7. Namely,

$$\lim_{\substack{A \to P \\ A \in D}} \int_S u(Q) \frac{\partial}{\partial \mathbf{n}_Q} [\log|A - Q|] \, dS_Q$$

$$= (-2\pi + \Omega(P)) u(P) + \int_S u(Q) \frac{\partial}{\partial \mathbf{n}_Q} [\log|P - Q|] \, dS_Q \tag{8.1.11}$$

with $\Omega(P)$ the interior angle to S at P. At points P at which S is smooth, $\Omega(P) = \pi$, and at the corner points $\mathbf{r}(L_i)$ we have

$$\Omega(\mathbf{r}(L_i)) = (1 - \chi_i)\pi$$

Combining (8.1.10) and (8.1.11), the formula (8.1.9) becomes

$$\int_S \left[\frac{\partial u(Q)}{\partial \mathbf{n}_Q} \log |P - Q| - u(Q) \frac{\partial}{\partial \mathbf{n}_Q} [\log |P - Q|] \right] dS_Q = \Omega(P)u(P),$$

$$P \in S \qquad (8.1.12)$$

This is equivalent to (7.1.10) in Chapter 7 when S is a smooth curve.

Introduce the operator

$$\mathcal{K}u(P) = (-\pi + \Omega(P))\, u(P) + \int_S u(Q) \frac{\partial}{\partial \mathbf{n}_Q} [\log |P - Q|]\, dS_Q,$$

$$P \in S \qquad (8.1.13)$$

for $u \in C(S)$. This maps $C(S)$ into $C(S)$, as can be proven by using (8.1.12) and results such as (8.1.6) on the smoothness of harmonic functions. When S is smooth, $\mathcal{K}$ is the compact double layer operator of (7.1.31)–(7.1.35) in Chapter 7. When S is only piecewise smooth, $\mathcal{K}$ is no longer compact, and this is examined later in more detail. Note that

$$-\pi + \Omega(P) = \begin{cases} 0, & S \text{ smooth at } P \\ -\chi_i \pi, & P = \mathbf{r}(L_i) \end{cases} \qquad (8.1.14)$$

The discontinuity in this formula is balanced by a corresponding discontinuity in the double layer integral on the right side of (8.1.13).

The equation (8.1.12) can now be written in the equivalent symbolic form

$$(\pi + \mathcal{K})u = \mathcal{S}v, \quad v = \frac{\partial u}{\partial \mathbf{n}_Q} \qquad (8.1.15)$$

with $\mathcal{S}$ the single layer operator on the right side of (8.1.10). The integral equation (8.1.15) is a boundary integral equation, and depending on whether one is solving the Dirichlet or Neumann problem, it can be regarded as a BIE of the first or second kind, respectively.

8.1.2. An indirect method for the Dirichlet problem

In this chapter, rather than considering all possible BIE for piecewise smooth boundaries, we consider only a few selected equations, chosen to illustrate important points about such problems. Using the same kind of approach as was used in Chapter 7, (cf. (7.1.40)–(7.1.42)) we can solve the interior Dirichlet problem (8.1.1) by assuming the solution is represented as a double layer potential:

$$u(A) = \int_S \rho(Q) \frac{\partial}{\partial \mathbf{n}_Q} [\log |A - Q|] \, dS_Q, \quad A \in D \qquad (8.1.16)$$

Then using (8.1.11), (8.1.13), and the boundary condition in (8.1.1), we have that ρ satisfies the BIE

$$(-\pi + \mathcal{K})\rho = f \qquad (8.1.17)$$

As earlier in (7.1.40)–(7.1.42) of Chapter 7, the solution ρ is the difference of the boundary values of two harmonic functions, on the interior and exterior of S. Combine this with the regularity result (8.1.6) to look at the behavior of solution ρ in the vicinity of a corner point $\mathbf{r}(L_i)$. We find that both the interior angle $(1 - \chi_i)\pi$ and the exterior angle $(1 + \chi_i)\pi$ lead to components in the solution of the respective orders

$$\rho(t) = O(d^\alpha) + O(d^{\alpha_e}), \quad d = |\mathbf{r}(t) - \mathbf{r}(L_i)|$$

where

$$\alpha = \frac{1}{1 - \chi_i} \qquad \alpha_e = \frac{1}{1 + \chi_i}$$

Thus, we are virtually assured of a singularity

$$\rho(t) = O(d^\beta), \quad \beta = \frac{1}{1 + |\chi_i|} \qquad (8.1.18)$$

The exponent β satisfies $1/2 < \beta < 1$. There will almost always be a singularity in the first derivative of $\rho(t)$ in the vicinity of corner points. A direct approach to (8.1.18), based on studying only the boundary integral equation $(-\pi + \mathcal{K})\rho = f$, is given in Chandler [106].

8.1.3. A BIE on an open wedge

To better understand the mathematical analysis and the numerical analysis of the integral equation (8.1.17), we begin with a detailed look at a special case. We take the boundary S to be the simple wedge shown in Figure 8.2. Each arm is a straight line segment of length 1, and the central angle is $\phi = (1 - \chi)\pi$. We further assume, for simplicity, that $0 < \chi < 1$. (This is assumed only for the development of the present example with S a wedge boundary.) The integral

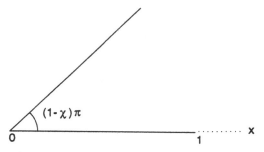

Figure 8.2. An open wedge with central angle $\phi = (1 - \chi)\pi$.

equation (8.1.17) no longer corresponds to any boundary value problem, but we can investigate exactly the solvability of the integral equation.

The integral equation (8.1.17) can now be written as a linear system of integral equations. These equations are obtained by considering separately the cases of

$$P = (x, 0), (r \cos \phi, r \sin \phi), (0, 0)$$

where P is the field point, as in (8.1.13). Note that when P and Q are on the same arm of the wedge, then the kernel function is identically zero, since then

$$\frac{\partial}{\partial \mathbf{n}_Q}[\log |P - Q|] = \frac{(Q - P) \cdot \mathbf{n}_Q}{|Q - P|^2} = 0 \qquad (8.1.19)$$

because $Q - P$ and $\mathbf{n}_Q$ are orthogonal. The integral equations are

$$-\pi\rho(x, 0) - \int_0^1 \frac{x \sin(\chi\pi)\rho(r \cos \phi, r \sin \phi)\, d\theta}{(r \cos \phi - x)^2 + r^2 \sin^2 \phi} = f(x, 0), \quad 0 < x \le 1$$

$$-\pi\rho(r \cos \phi, r \sin \phi) - \int_0^1 \frac{r \sin(\chi\pi)\rho(x, 0)\, d\theta}{(r \cos \phi - x)^2 + r^2 \sin^2 \phi}$$

$$= f(r \cos \phi, r \sin \phi), \quad 0 < r \le 1$$

$$-\pi\rho(0, 0) - \chi\pi\rho(0, 0) = f(0, 0) \qquad (8.1.20)$$

Introduce

$$\upsilon_1(x) = \rho(x, 0), \qquad \upsilon_2(r) = \rho(r \cos \phi, r \sin \phi), \quad 0 \le r, x \le 1$$

and define f_1, f_2 similarly. From (8.1.20), note that

$$\upsilon_1(0) = \upsilon_2(0) = \frac{-1}{(1 + \chi)\pi} f(0, 0) \qquad (8.1.21)$$

The above integral equations now become

$$-\pi v_1(x) - \int_0^1 \frac{x \sin(\chi\pi)v_2(y)\, dy}{x^2 + 2xy \cos(\chi\pi) + y^2} = f_1(x)$$

$$-\pi v_2(x) - \int_0^1 \frac{x \sin(\chi\pi)v_1(y)\, dy}{x^2 + 2xy \cos(\chi\pi) + y^2} = f_2(x) \qquad (8.1.22)$$

for $0 < x \le 1$.

We further simplify these by reducing them to two single integral equations. Introduce

$$\psi_1 = v_1 + v_2, \qquad \psi_2 = v_1 - v_2$$

By adding and subtracting the equations in (8.1.22), we obtain the two single equations

$$-\pi \psi_1(x) - \int_0^1 \frac{x \sin(\chi\pi)\psi_1(y)\, dy}{x^2 + 2xy \cos(\chi\pi) + y^2} = F_1(x), \quad 0 < x \le 1$$

$$-\pi \psi_2(x) + \int_0^1 \frac{x \sin(\chi\pi)\psi_2(y)\, dy}{x^2 + 2xy \cos(\chi\pi) + y^2} = F_2(x), \quad 0 < x \le 1 \quad (8.1.23)$$

The right-hand functions are the sum and difference, respectively, of f_1 and f_2. Let $\mathcal{L}$ denote the integral operator in these two equations. We write these two equations symbolically as

$$(-\pi \pm \mathcal{L})\psi_j = F_j, \quad j = 1, 2 \qquad (8.1.24)$$

It is straightforward to see that the solvability of $(-\pi + \mathcal{K})\rho = f$ in (8.1.17) is completely equivalent to (8.1.24). In particular,

$$\sigma(\mathcal{K}) = \sigma(\mathcal{L}) \cup \sigma(-\mathcal{L}) \qquad (8.1.25)$$

where $\sigma(\mathcal{K})$ denotes the spectrum of the operator $\mathcal{K}$ on whatever space of functions is being used to study these equations.

To further investigate $\mathcal{L}$, we make a further modification by introducing the change of variables

$$x = e^{-\xi}, \qquad y = e^{-\eta}, \quad 0 \le \xi, \eta < \infty$$

Introduce $\hat{\psi}(\eta) = \psi(e^{-\eta})$ and similarly for other functions in (8.1.23). Dropping the subscripts, the equation (8.1.24) can now be written

$$-\pi \hat{\psi}(\xi) \pm \int_0^\infty \frac{\sin(\chi\pi)\hat{\psi}(\eta)\, d\eta}{e^{\xi-\eta} + 2\cos(\chi\pi) + e^{\eta-\xi}} = \hat{F}(\xi), \quad 0 \le \xi < \infty \qquad (8.1.26)$$

This is a Wiener-Hopf integral equation, about which a great deal is known (cf. (1.1.14) in Chapter 1). Denote this integral operator by $\hat{\mathcal{L}}$.

The most appropriate function space on which to study (8.1.26) is $C_l[0, \infty)$, the space of functions that are continuous on $[0, \infty)$ and possess a limit at ∞. The norm used is simply the standard sup norm $\|\cdot\|_\infty$, and the resulting function space is isometric to $C[0, 1]$. From Krein [322], the spectrum of $\hat{\mathcal{L}}$ on this space is given by the closure of the range of

$$\int_{-\infty}^{\infty} e^{i\mu\eta} \frac{\sin(\chi\pi) \, d\eta}{e^{-\eta} + 2\cos(\chi\pi) + e^\eta} = \pi \frac{\sinh(\chi\pi\mu)}{\sinh(\pi\mu)}, \quad -\infty < \mu < \infty$$

From this, the spectrum of $\hat{\mathcal{L}}$ is $[0, \chi\pi]$, and from (8.1.25),

$$\sigma(\mathcal{K}) = [-\chi\pi, \chi\pi] \tag{8.1.27}$$

From the properties of compact operators developed in Chapter 1, neither the operator $\mathcal{K}$ nor any power of it can possibly be compact, as its spectrum is not discrete. For further information on the spectral properties of $\hat{\mathcal{L}}$, see Atkinson and deHoog [64, §2].

By the substitution $y = xs$,

$$\int_0^1 \frac{x\sin(\chi\pi) \, dy}{x^2 + 2xy\cos(\chi\pi) + y^2} = \int_0^{\frac{1}{x}} \frac{\sin(\chi\pi) \, ds}{1 + 2s\cos(\chi\pi) + s^2}$$

for $0 < x \le 1$. Thus this integral increases as x decreases to 0, and

$$\lim_{x \to 0} \int_0^1 \frac{x\sin(\chi\pi) \, dy}{x^2 + 2xy\cos(\chi\pi) + y^2} = \int_0^\infty \frac{\sin(\chi\pi) \, ds}{1 + 2s\cos(\chi\pi) + s^2}$$
$$= \chi\pi$$

From the positivity of the kernel in $\mathcal{K}$, [cf. (8.1.22)], it then follows that

$$\|\mathcal{K}\| = \chi\pi \tag{8.1.28}$$

for the wedge operator $\mathcal{K}$ when it is considered as an operator from $C(S)$ to $C(S)$ for the wedge S. From this and the geometric series theorem (cf. Theorem A.1 in the Appendix), we have that the integral equation $(-\pi + \mathcal{K})\rho = f$ in (8.1.17) is uniquely solvable on $C(S)$, with

$$\|(-\pi + \mathcal{K})^{-1}\| \le \frac{1}{\pi - \|\mathcal{K}\|} = \frac{1}{(1 - \chi)\pi} \tag{8.1.29}$$

This suggests that we should consider dealing with the case of a general boundary S by considering separately the boundary integral operator on a region around a corner and the boundary integral operator on the remaining part of the boundary.

8.1.4. A decomposition of the boundary integral equation

The integral operator $\mathcal{K}$ in (8.1.17) is defined in (8.1.13). To deal with $\mathcal{K}$ being noncompact, due to the presence of the corner points on S, we convert (8.1.17) into a system of equations in which the results of the preceding subsection on the wedge equation play a major role. We then decompose the system in such a way as to emphasize the role of the wedge equation (8.1.21)–(8.1.22). For the purpose only of simplifying the presentation, *we assume S is a polygonal boundary*, throughout this and the following section.

Begin by subdividing S into wedges and sections connecting the wedges. Define

$$\ell = \min_{j=1,\dots,J} \text{ length } S_j$$

and $\epsilon = \frac{1}{3}\ell$. For the parametrization interval $[0, L]$, introduce the "wedge subintervals"

$$[L_1 - \epsilon, L_1 + \epsilon], [L_2 - \epsilon, L_2 + \epsilon], \dots, [L_J - \epsilon, L_J + \epsilon] \qquad (8.1.30)$$

where the last subinterval $[L_J - \epsilon, L_J + \epsilon]$ is equivalent to $[L_J - \epsilon, L_J] \cup [0, \epsilon]$ in terms of the portion of S to which it refers. Denote the corresponding wedge sections of S by $W_1, \dots, W_J$, respectively. For the parametrization interval $[0, L]$, also introduce the "connecting sections"

$$[\epsilon, L_1 - \epsilon], [L_1 + \epsilon, L_2 - \epsilon], \dots, [L_{J-1} + \epsilon, L_J - \epsilon] \qquad (8.1.31)$$

and denote the corresponding sections of S by $C_1, \dots, C_J$, respectively. Collectively, denote the wedge and connecting sections by $\Gamma_1, \dots, \Gamma_{2J}$, beginning with W_1 and proceeding around S in a counterclockwise direction:

$$\Gamma_j = \begin{cases} C_k, & j \text{ even}, \quad k = j/2 \\ W_k, & j \text{ odd}, \quad k = (j+1)/2 \end{cases}$$

For a function $u \in C(S)$, we use the notation u^j to denote the restriction of u to the side Γ_j, $j = 1, \dots, 2J$. Introduce the space

$$\mathcal{X} = C(\Gamma_1) \oplus \cdots \oplus C(\Gamma_{2J}) \qquad (8.1.32)$$

which is complete with the norm

$$\|(u^1, \dots, u^{2J})\|_\infty \equiv \max_{i=1,\dots,2J} \|u^i\|_\infty$$

Let $\mathcal{E} : C(S) \to \mathcal{X}$ be the mapping $u \mapsto \bar{u} \equiv (u^1, \dots, u^{2J})$.

We can now reformulate the boundary integral equation $(-\pi + \mathcal{K})\rho = f$ of (8.1.17) as a system of integral equations on $\mathcal{X}$:

$$-\pi\rho^i(P) + \sum_{j=1}^{2J} \int_{\Gamma_j} \rho^j(Q)\frac{\partial}{\partial \mathbf{n}_Q}[\log|P - Q|]\,dS_Q = \tilde{f}^i(P), \quad P \in \Gamma_i$$
(8.1.33)

for $i = 1, \ldots, 2J$ and $\tilde{f} \equiv \mathcal{E}f$. This formula must be modified if P is a corner point, as in (8.1.13) and (8.1.20). We write this in operator form as

$$(-\pi + \tilde{\mathcal{K}})\tilde{\rho} = \tilde{f}$$

with $\tilde{\rho} = (\rho^1, \ldots, \rho^{2J})$, and $\tilde{\mathcal{K}}$ is considered to be a matrix of the integral operators in (8.1.33).

Introduce the operator $\tilde{\mathcal{W}}$ on $\mathcal{X}$: for i odd, define

$$(\tilde{\mathcal{W}}\tilde{\rho})^i(P) = \int_{\Gamma_i} \rho^i(Q)\frac{\partial}{\partial \mathbf{n}_Q}[\log|P - Q|]\,dS_Q, \quad P \in \Gamma_i$$
(8.1.34)

and for i even, define

$$(\tilde{\mathcal{W}}\tilde{\rho})^i(P) = 0, \quad P \in \Gamma_i$$

The operator $\tilde{\mathcal{W}}$ is a diagonal matrix of wedge integral operators of the type studied in the preceding subsection. Also, introduce the operator Γ,

$$\tilde{\mathcal{S}}\tilde{\rho} = \tilde{\mathcal{K}}\tilde{\rho} - \tilde{\mathcal{W}}\tilde{\rho}$$

Then

$$(\tilde{\mathcal{S}}\tilde{\rho})^i(P) = \sum_{\substack{j=1 \\ j \neq i}}^{2J} \int_{\Gamma_j} \rho^j(Q)\frac{\partial}{\partial \mathbf{n}_Q}[\log|P - Q|]\,dS_Q, \quad P \in \Gamma_i$$
(8.1.35)

Note that this formula is true for i even because

$$\frac{\partial}{\partial \mathbf{n}_Q}[\log|P - Q|] = 0, \quad P, Q \in \Gamma_i$$

The kernel of each of the integral operators in (8.1.35) is infinitely differentiable, and therefore $\tilde{\mathcal{S}}$ is a compact operator on $\mathcal{X}$ to $\mathcal{X}$. It is also compact on virtually any Banach space of interest in applications. For example, this includes the Hilbert space

$$L^2(\Gamma_1) \oplus \cdots \oplus L^2(\Gamma_{2J})$$

with a standard inner product norm derived from that of the component subspaces $L^2(\Gamma_i)$.

The boundary integral equation $(-\pi + \mathcal{K})\rho = f$ of (8.1.17) can now be
written as

$$(-\pi + \tilde{\mathcal{W}} + \tilde{\mathcal{S}})\tilde{\rho} = \tilde{f} \tag{8.1.36}$$

From the result (8.1.28), we can show

$$\|\tilde{\mathcal{W}}\| = \max_{1 \le j \le J} \pi |\chi_j| \tag{8.1.37}$$

By the assumption (8.1.8), we have

$$\|\tilde{\mathcal{W}}\| < \pi \tag{8.1.38}$$

and thus $(-\pi + \tilde{\mathcal{W}})^{-1}$ exists and is bounded by the geometric series theorem
(cf. Theorem A.1 in the Appendix) with

$$\|(-\pi + \tilde{\mathcal{W}})^{-1}\| \le \frac{1}{\pi - \|\tilde{\mathcal{W}}\|}$$

This allows us to reformulate (8.1.36) as

$$\tilde{\rho} + (-\pi + \tilde{\mathcal{W}})^{-1}\tilde{\mathcal{S}}\tilde{\rho} = (-\pi + \tilde{\mathcal{W}})^{-1}\tilde{f} \tag{8.1.39}$$

The operator $(-\pi + \tilde{\mathcal{W}})^{-1}\tilde{\mathcal{S}}$ is compact on $\mathcal{X}$ to $\mathcal{X}$, and thus the Fredholm
alternative theorem (cf. Theorem 1.3.1 from Chapter 1) applies to this equation:
the equation is uniquely solvable for all right sides in $\mathcal{X}$ if and only if the
homogeneous equation has only the zero solution. The homogeneous equation
is equivalent to $(-\pi + \mathcal{K})\rho = 0$, and by the same type of potential theory
argument as is used with a smooth boundary S, it can be shown that $\rho = 0$
(for example, see Kress [325, p. 70]). Therefore $(-\pi + \mathcal{K})\rho = f$ is uniquely
solvable for all $f \in \mathcal{X}$, and in turn, this gives an existence proof for the interior
Dirichlet problem (8.1.1).

The generalization of the above to a general piecewise smooth boundary
will require the approximation of each "wedge-like" corner boundary by a true
wedge boundary, and this is explored in detail in both Atkinson and Graham [61]
and Atkinson and deHoog [64]. This general approach to showing the existence
of solutions to Laplace's equation for regions with a piecewise smooth boundary
is due to Radon [440] in 1919, generalizing the earlier results of Ivar Fredholm
for regions with a smooth boundary. This framework has also been used as the
basis for the error analysis of numerical methods for solving $(-\pi + \mathcal{K})\rho = f$

for boundaries S that are only piecewise smooth, beginning with Bruhn and Wendland [93] in 1967 and Wendland [557] in 1968.

We sketch how the above framework is used in the analysis of projection methods for solving $(-\pi + \mathcal{K})\rho = f$. Let $\{\mathcal{X}_n\}$ denote a sequence of approximating subspaces for functions in $\mathcal{X}$, and let $\{\tilde{\mathcal{P}}_n\}$ be corresponding projections of $\mathcal{X}$ onto $\mathcal{X}_n$, $n \geq 1$. Then (8.1.36) is approximated by

$$(-\pi + \tilde{\mathcal{P}}_n\tilde{\mathcal{W}} + \tilde{\mathcal{P}}_n\tilde{\mathcal{S}})\tilde{\rho}_n = \tilde{\mathcal{P}}_n\tilde{f} \qquad (8.1.40)$$

The error analysis begins by showing the existence and uniform boundedness of $(-\pi + \tilde{\mathcal{P}}_n\tilde{\mathcal{W}})^{-1}$ for all sufficiently large n, with the details varying with whether the method is a collocation or Galerkin method. Then the equation (8.1.40) is converted to

$$(I + (-\pi + \tilde{\mathcal{P}}_n\tilde{\mathcal{W}})^{-1}\tilde{\mathcal{P}}_n\tilde{\mathcal{S}})\tilde{\rho}_n = (-\pi + \tilde{\mathcal{P}}_n\tilde{\mathcal{W}})^{-1}\tilde{\mathcal{P}}_n\tilde{f} \qquad (8.1.41)$$

This is analyzed as an approximation to (8.1.39), using the techniques of Chapter 3.

The key to this error analysis is the existence and uniform boundedness of $(-\pi + \tilde{\mathcal{P}}_n\tilde{\mathcal{W}})^{-1}$, and except in the simplest of cases, this has been a difficult problem. We discuss this and associated topics in greater detail in the next three sections of this chapter.

8.2. The Galerkin method

To simplify the presentation, as earlier, we assume the boundary S is polygonal, but much of the discussion applies equally well to a general piecewise smooth boundary. In the approximation (8.1.40) of the double layer equation $(-\pi + \mathcal{K})\rho = f$ of (8.1.17), we take the projection $\tilde{\mathcal{P}}_n$ to be an orthogonal projection operator from

$$\mathcal{X} = L^2(\Gamma_1) \oplus \cdots \oplus L^2(\Gamma_{2J})$$

to $\mathcal{X}_n \subset \mathcal{X}$. The approximating subspace $\mathcal{X}_n$ is to be a set of piecewise polynomial functions, which we define below. Around the corners a graded mesh is needed to compensate for the likely bad behavior of the solution ρ around the corner points, as described in (8.1.7). The reader may wish to review the material on graded meshes from subsection 4.2.5 of Chapter 4. The material of this section is due, in large part, to Chandler [106].

Select an integer $r \geq 0$, which is to be the maximum degree of the polynomials being used. On each of the parametrization subintervals of (8.1.31) for the connecting sections $C_j \subset S$, impose a uniform mesh with n subintervals. For each associated partial boundary $\Gamma_i = C_j$, $i = 2j$, let $\mathcal{X}_{n,i}$ denote the set of all functions $\rho^i \in L^\infty(\Gamma_i)$ that are polynomials of degree $\leq r$ in the arc length parametrization variable t on each of the n subintervals in the mesh for Γ_i. There is no requirement of continuity for such functions ρ^i at the mesh points on Γ_i, and thus the dimension of $\mathcal{X}_{n,i}$ is $n(r + 1)$.

For the wedge region W_j whose parametrization subinterval is $[L_j - \epsilon, L_j + \epsilon]$, we introduce a mesh graded towards the corner $\mathbf{r}(L_j)$. Consider first the subinterval $[L_j, L_j + \epsilon]$ and introduce the graded mesh

$$L_j, L_j + \left(\frac{1}{n}\right)^q \epsilon, L_j + \left(\frac{2}{n}\right)^q \epsilon, \ldots, L_j + \left(\frac{n}{n}\right)^q \epsilon \equiv L_j + \epsilon, \quad (8.2.42)$$

with q the grading parameter, $q \geq 1$. Using the arc length parametrization function $\mathbf{r}(s)$, map this graded mesh onto a mesh for the linear section of S connecting $\mathbf{r}(L_j)$ and $\mathbf{r}(L_j + \epsilon)$. Proceed similarly with the subinterval $[L_j - \epsilon, L_j]$, except grade the mesh towards the right end L_j. With this mesh for $W_j = \Gamma_i$, $i = 2j - 1$, let $\mathcal{X}_{n,i}$ denote the set of all functions $\rho^i \in L^\infty(\Gamma_i)$ that are polynomials of degree $\leq r$ in the arc length parametrization variable t on each of the $2n$ subintervals defined by the mesh for Γ_i. Again there is no requirement of continuity for such functions ρ^i at the mesh points on Γ_i, and now the dimension of $\mathcal{X}_{n,i}$ is $2n(r + 1)$.

Define

$$\mathcal{X}_n = \mathcal{X}_{n,1} \oplus \cdots \oplus \mathcal{X}_{n,2J} \quad (8.2.43)$$

Its dimension is $N = 3nJ(r + 1)$. Define $\tilde{\mathcal{P}}_n$ to be orthogonal projection of $\mathcal{X}$ onto $\mathcal{X}_n$. Also, let $\mathcal{P}_{n,i}$ denote the orthogonal projection operator of $L^2(\Gamma_i)$ onto $\mathcal{X}_{n,i}$. The operator $\tilde{\mathcal{P}}_n$ can be considered as a diagonal matrix of operators, with

$$\tilde{u} = (u^1, \ldots, u^{2J}) \mapsto \tilde{\mathcal{P}}_n \tilde{u} = (\mathcal{P}_{n,1} u^1, \ldots, \mathcal{P}_{n,2J} u^{2J})$$

Return to the approximation (8.1.40),

$$(-\pi + \tilde{\mathcal{P}}_n \tilde{\mathcal{W}} + \tilde{\mathcal{P}}_n \tilde{\mathcal{S}}) \tilde{\rho}_n = \tilde{\mathcal{P}}_n \tilde{f} \quad (8.2.44)$$

and consider the invertibility of $-\pi + \tilde{\mathcal{P}}_n \tilde{\mathcal{W}}$. Since $\tilde{\mathcal{W}}$ is a diagonal matrix operator, we can easily reduce the invertibility of $-\pi + \tilde{\mathcal{W}}$ on $\mathcal{X}$ to that of the unique solvability of

$$-\pi \rho^j(P) + \int_{W_j} \rho^j(Q) \frac{\partial}{\partial \mathbf{n}_Q}[\log |P - Q|] dS_Q = f^j(P), \quad P \in W_j$$
$$(8.2.45)$$

on the space $L^2(W_j)$ for each wedge W_j, $j = 1, 2, \ldots, n$. Similarly, the invertibility of $-\pi + \tilde{P}_n \tilde{W}$ can be reduced to the unique solvability of Galerkin's method applied to (8.2.45), considered separately on each wedge W_j.

From the material of subsection 8.1.3, with particular reference to (8.1.23)–(8.1.24), we can write (8.2.45) in the equivalent form

$$(-\pi \pm \mathcal{L})\psi_k = F_k, \quad k = 1, 2 \tag{8.2.46}$$

with

$$\mathcal{L}\psi(x) = \int_0^1 \frac{x \sin(\chi\pi)\,\psi(y)\,dy}{x^2 + 2xy\cos(\psi\pi) + y^2}, \quad 0 < x \le 1$$
$$\mathcal{L}\psi(0) = \chi\pi\psi(0) \tag{8.2.47}$$

When considered relative to a corner point on S, this uses a change from the arc length parametrization variable t to the variable x, with $t = \epsilon x, 0 \le x \le 1$, but this does not lead to any essential change in the mapping properties of the integral operator. In (8.2.46)–(8.2.47), we are letting $\mathcal{L}$ and χ denote the operator and angle parameter associated with a general wedge W_i.

In Chandler [106, Lemma 1], it is shown that when $\mathcal{L}$ is considered as an operator on $L^2(0, 1)$ to $L^2(0, 1)$,

$$\|\mathcal{L}\| \le \pi \left| \sin\left(\frac{\chi\pi}{2}\right) \right| < \pi \tag{8.2.48}$$

From this, the equations (8.2.46) are uniquely solvable on $L^2(0, 1)$ with

$$\|(-\pi \pm \mathcal{L})^{-1}\| \le \frac{1}{\pi - \|\mathcal{L}\|} \tag{8.2.49}$$

The proof of (8.2.48) begins by showing an isometric equivalence of $\mathcal{L}$ to a Wiener-Hopf integral operator on $L^2(0, \infty)$. Next, the norm of this Wiener-Hopf integral operator is bounded using results of Krein [322], wherein the problem is reduced to that of finding the maximum size of the Fourier transform of the kernel function generating the integral operator. We refer the reader to Ref. [106] for the details.

The Galerkin method applied to (8.2.45) reduces similarly to

$$(-\pi \pm \mathcal{P}_n\mathcal{L})\psi_{k,n} = \mathcal{P}_n F_k, \quad k = 1, 2 \tag{8.2.50}$$

In this the operator $\mathcal{P}_n$ is the orthogonal projection operator from $L^2(0, 1)$ onto the approximating subspace $\mathcal{S}_n$ associated with the graded mesh (8.2.42) relative to $[0, 1]$. More precisely, introduce the mesh

$$x_{n,i} = \left(\frac{i}{n}\right)^q, \quad i = 0, 1, \dots, n \qquad (8.2.51)$$

Then $\varphi \in \mathcal{S}_n$ means that over each subinterval $[x_{n,i-1}, x_{n,i}]$, $\varphi(x)$ is a polynomial of degree $\leq r$. The dimension of $\mathcal{S}_n$ is $n(r+1)$.

Since $\mathcal{P}_n$ is orthogonal, it follows that $\|\mathcal{P}_n\| = 1$. Using (8.2.48), we have

$$\|\mathcal{P}_n \mathcal{L}\| \leq \|\mathcal{L}\| < \pi$$

Then $(-\pi \pm \mathcal{P}_n \mathcal{L})^{-1}$ exists and is uniformly bounded,

$$\|(-\pi \pm \mathcal{P}_n \mathcal{L})^{-1}\| \leq \frac{1}{\pi - \|\mathcal{L}\|}, \quad n \geq 1 \qquad (8.2.52)$$

By standard arguments (cf. (3.1.31) in subsection 3.1.3 of Chapter 3), we have the following error bound for the solution of (8.2.50) when compared to (8.2.46):

$$\|\psi_k - \psi_{k,n}\| \leq \frac{\pi}{\pi - \|\mathcal{L}\|}\|(I - \mathcal{P}_n)\psi_k\|, \quad n \geq 1, \quad k = 1, 2 \qquad (8.2.53)$$

The norm is that of $L^2(0, 1)$.

Returning to the system (8.2.44), the above argument implies the invertibility of $-\pi + \tilde{\mathcal{P}}_n \tilde{\mathcal{W}}$, with

$$\|(-\pi + \tilde{\mathcal{P}}_n \tilde{\mathcal{W}})^{-1}\| \leq \frac{1}{\pi \left(1 - \max_{1 \leq j \leq J} \left| \sin\left(\frac{\chi_j \pi}{2}\right)\right|\right)}, \quad n \geq 1 \qquad (8.2.54)$$

This justifies converting (8.2.44) to the equivalent system

$$\left(I + (-\pi + \tilde{\mathcal{P}}_n \tilde{\mathcal{W}})^{-1}\tilde{\mathcal{P}}_n \tilde{\mathcal{S}}\right)\tilde{\rho}_n = (-\pi + \tilde{\mathcal{P}}_n \tilde{\mathcal{W}})^{-1}\tilde{\mathcal{P}}_n \tilde{f} \qquad (8.2.55)$$

This will be analyzed as a numerical approximation to the modified original equation (8.1.39):

$$\tilde{\rho} + (-\pi + \tilde{\mathcal{W}})^{-1}\tilde{\mathcal{S}}\tilde{\rho} = (-\pi + \tilde{\mathcal{W}})^{-1}\tilde{f} \qquad (8.2.56)$$

Theorem 8.2.1. *Assume the interior angles* $(1 - \chi_j)\pi$ *for S satisfy*

$$|\chi_j| < 1, \quad j = 1, \dots, J$$

Then for all sufficiently large n, say $n \geq n_0$, the matrix operator $-\pi + \tilde{\mathcal{P}}_n \tilde{\mathcal{K}}$ has an inverse, and moreover, it is uniformly bounded in n,

$$\|(-\pi + \tilde{\mathcal{P}}_n \tilde{\mathcal{K}})^{-1}\| \leq c_1 < \infty, \quad n \geq n_0 \tag{8.2.57}$$

In addition,

$$\|\tilde{\rho} - \tilde{\rho}_n\| \leq c_2 \|(I - \tilde{\mathcal{P}}_n)\rho\|, \quad n \geq n_0 \tag{8.2.58}$$

for suitable constants $c_1, c_2 > 0$.

Proof. We assert that

$$E_n \equiv \|(-\pi + \tilde{\mathcal{W}})^{-1}\tilde{\mathcal{S}} - (-\pi + \tilde{\mathcal{P}}_n\tilde{\mathcal{W}})^{-1}\tilde{\mathcal{P}}_n\tilde{\mathcal{S}}\| \to 0 \quad \text{as } n \to \infty \tag{8.2.59}$$

To show this, write

$$\begin{aligned} E_n &\leq \|(-\pi + \tilde{\mathcal{W}})^{-1}\tilde{\mathcal{S}} - (-\pi + \tilde{\mathcal{P}}_n\tilde{\mathcal{W}})^{-1}\tilde{\mathcal{S}}\| \\ &\quad + \|(-\pi + \tilde{\mathcal{P}}_n\tilde{\mathcal{W}})^{-1}\tilde{\mathcal{S}} - (-\pi + \tilde{\mathcal{P}}_n\tilde{\mathcal{W}})^{-1}\tilde{\mathcal{P}}_n\tilde{\mathcal{S}}\| \\ &\leq \|[(-\pi + \tilde{\mathcal{W}})^{-1} - (-\pi + \tilde{\mathcal{P}}_n\tilde{\mathcal{W}})^{-1}]\tilde{\mathcal{S}}\| \\ &\quad + \|(-\pi + \tilde{\mathcal{P}}_n\tilde{\mathcal{W}})^{-1}\|\|(I - \tilde{\mathcal{P}}_n)\tilde{\mathcal{S}}\| \\ &\equiv E_n^{(1)} + E_n^{(2)} \end{aligned} \tag{8.2.60}$$

The term $\|(I - \tilde{\mathcal{P}}_n)\tilde{\mathcal{S}}\| \to 0$ because $\tilde{\mathcal{S}}$ is a compact operator and $\tilde{\mathcal{P}}_n$ is pointwise convergent to I. Combined with the uniform boundedness result (8.2.54), this proves

$$E_n^{(2)} \to 0 \quad \text{as } n \to \infty$$

For the remaining term $E_n^{(1)}$ on the right side of (8.2.60), use a straightforward algebraic manipulation to write

$$E_n^{(1)} = -(-\pi + \tilde{\mathcal{P}}_n\tilde{\mathcal{W}})^{-1}(I - \tilde{\mathcal{P}}_n)\tilde{\mathcal{W}}(-\pi + \tilde{\mathcal{W}})^{-1}\tilde{\mathcal{S}}$$

$$\|E_n^{(1)}\| \leq \|(-\pi + \tilde{\mathcal{P}}_n\tilde{\mathcal{W}})^{-1}\| \, \|(I - \tilde{\mathcal{P}}_n)\tilde{\mathcal{W}}(-\pi + \tilde{\mathcal{W}})^{-1}\tilde{\mathcal{S}}\|$$

The operator $\tilde{\mathcal{W}}(-\pi + \tilde{\mathcal{W}})^{-1}$ is bounded on $\mathcal{X}$, and therefore the operator $\tilde{\mathcal{W}}(-\pi + \tilde{\mathcal{W}})^{-1}\tilde{\mathcal{S}}$ is compact. As before, this implies

$$\|(I - \tilde{\mathcal{P}}_n)\tilde{\mathcal{W}}(-\pi + \tilde{\mathcal{W}})^{-1}\tilde{\mathcal{S}}\| \to 0 \quad \text{as } n \to \infty$$

With the uniform boundedness result (8.2.54), this proves

$$E_n^{(1)} \to 0 \quad \text{as } n \to \infty$$

This also completes the proof of (8.2.59).

As discussed earlier following (8.1.39), the operator $-\pi + \mathcal{K}$ is invertible on $L^2(S)$, and in turn, this implies that $I + (-\pi + \tilde{\mathcal{W}})^{-1}\tilde{\mathcal{S}}$ is invertible on $\mathcal{X}$. A standard perturbation argument now implies the invertibility of

$$I + (-\pi + \tilde{\mathcal{P}}_n\tilde{\mathcal{W}})^{-1}\tilde{\mathcal{P}}_n\tilde{\mathcal{S}}$$

together with its uniform boundedness,

$$\|(I + (-\pi + \tilde{\mathcal{P}}_n\tilde{\mathcal{W}})^{-1}\tilde{\mathcal{P}}_n\tilde{\mathcal{S}})^{-1}\| \le c < \infty$$

for all sufficiently large n, say $n \ge n_0$, with a suitable constant $c > 0$. When combined with (8.2.54), this in turn implies the existence and uniform boundedness of $(-\pi + \tilde{\mathcal{P}}_n\tilde{\mathcal{K}})^{-1}$, as asserted in (8.2.57).

The bound (8.2.58) comes from the standard identity

$$\tilde{\rho} - \tilde{\rho}_n = -\pi(-\pi + \tilde{\mathcal{P}}_n\tilde{\mathcal{K}})^{-1}(I - \tilde{\mathcal{P}}_n)\rho$$

for second kind equations.   □

To obtain error bounds, we must examine the rate of convergence to zero of $(I - \tilde{\mathcal{P}}_n)\rho$. To examine this rate, we must know the differentiability properties of the solution ρ. We assume the Dirichlet data f is a smooth function on each linear section of S, say $f \in C^\infty$ on each such section, for simplicity. Recalling the discussion associated with (8.1.18), we have the following:

Case 4. On each connecting section $C_j = \Gamma_i$, $i = 2j$, $\rho \in C^\infty(\Gamma_i)$.

Case 5. On the wedge section $W_j = \Gamma_i$, $i = 2j - 1$,

$$\rho(t) = O((t - L_j)^{\beta_j}), \quad \beta_j = \frac{1}{1 + |\chi_j|} \tag{8.2.61}$$

for $j = 1, \ldots, J$. This can be extended to giving information about the growth of the derivatives of $\rho(t)$ with respect to the arc length parametrization t. In particular, these derivatives can be characterized in exactly the manner assumed in Lemma 4.2.3 of subsection 4.2.5 in Chapter 4. This can be proven by an extension of the ideas used in obtaining (8.1.18), and a separate method of proof, using only integral equations, is given in Chandler [106].

For the above Case 1, we have by standard approximation methods that

$$\|(I - \mathcal{P}_{n,i})\rho^i\| \leq O\left(n^{-(r+1)}\right) \tag{8.2.62}$$

For Case 2, we must use a graded mesh appropriate to the behavior given in (8.2.61). In the vicinity of the corner of wedge W_j, the grading parameter q_j should be chosen to satisfy

$$q_j > \frac{r+1}{\beta_j + \frac{1}{2}} \tag{8.2.63}$$

With this we will again have (8.2.62) satisfied on this portion of the boundary. The proof is a direct application of Lemma 4.2.3.

Corollary 8.2.1. *Assume the Dirichlet data $f \in C^\infty$ on each linear section of S. Moreover, assume the mesh around each corner is graded as described in (8.2.42), with the grading parameter for the corner of W_j satisfying (8.2.63). Then the Galerkin solution $\tilde{\rho}_n$ satisfies*

$$\|\tilde{\rho} - \tilde{\rho}_n\| \leq O\left(n^{-(r+1)}\right) \tag{8.2.64}$$

Proof. Apply (8.2.62), in the manner discussed preceding this theorem. □

From (8.2.61), it follows that $\beta_j > 1/2$ in all cases. Therefore, if we choose the grading parameters q_j to satisfy the uniform bound $q_j \geq r + 1$, then the order of convergence in (8.2.64) is assured. This has the advantage that the same type of grading can be used in all cases, regardless of the size of the angle.

8.2.1. *Superconvergence results*

The iterated Galerkin solution was introduced in subsection 3.4.1 of Chapter 3. In the present context the iterated Galerkin solution is defined by

$$\tilde{\rho}_n^* = \frac{1}{\pi}(\tilde{\mathcal{K}}\tilde{\rho}_n - f) \tag{8.2.65}$$

It satisfies $\tilde{\mathcal{P}}_n\tilde{\rho}_n^* = \tilde{\rho}_n$ and

$$(-\pi + \tilde{\mathcal{K}}\tilde{\mathcal{P}}_n)\tilde{\rho}_n^* = f$$

The invertibility of this approximation is assured by combining (8.2.57) with (3.4.84) from Chapter 3,

$$(-\pi + \tilde{\mathcal{K}}\tilde{\mathcal{P}}_n)^{-1} = \frac{1}{\pi}[-I + \tilde{\mathcal{K}}(-\pi + \tilde{\mathcal{P}}_n\tilde{\mathcal{K}})^{-1}\tilde{\mathcal{P}}_n]$$

It also follows that $(-\pi + \tilde{\mathcal{K}}\tilde{\mathcal{P}}_n)^{-1}$ is uniformly bounded for all sufficiently large n.

In formula (3.4.100) of subsection 3.4.2 it was noted that the iterated Galerkin solution converged with a rate of $O(n^{-2(r+1)})$, which is twice the rate of the original Galerkin solution. This faster rate of convergence can be reproduced in the present context if the mesh grading parameter is increased. More precisely, replace condition (8.2.63) with

$$q_j > \frac{2(r+1)}{\beta_j + \frac{1}{2}} \tag{8.2.66}$$

Then

$$\left\| \tilde{\rho} - \tilde{\rho}_n^* \right\| \leq O\left(n^{-2(r+1)}\right) \tag{8.2.67}$$

This is proven in Chandler [106, Theorem 6]. The argument uses the error bound in the first line of (3.4.93) from subsection 3.4.1 of Chapter 3:

$$\left\| \tilde{\rho} - \tilde{\rho}_n^* \right\| \leq \|(-\pi + \tilde{\mathcal{K}}\tilde{\mathcal{P}}_n)^{-1}\| \, \|\tilde{\mathcal{K}}(I - \tilde{\mathcal{P}}_n)\tilde{\rho}\| \tag{8.2.68}$$

The more severe grading parameter of (8.2.66) is needed to prove

$$\|\tilde{\mathcal{K}}(I - \tilde{\mathcal{P}}_n)\tilde{\rho}\| \leq O\left(n^{-2(r+1)}\right)$$

which is then used in (8.2.68) to prove (8.2.67).

An extension of (8.2.67) using the uniform norm is given in Chandler [107]. This latter reference also gives a less costly means of obtaining a superconvergent solution directly from values of $\tilde{\rho}_n$.

8.3. The collocation method

In this section we consider some graded mesh collocation methods for the numerical solution of the boundary integral equation $(-\pi + \mathcal{K})\rho = f$ of (8.1.17). Some early analyses of particular collocation methods were given in Benveniste [77], Bruhn and Wendland [93], Wendland [557], and Atkinson and deHoog [64]. The first satisfactory general treatment of graded mesh collocation methods appears to be that of Chandler and Graham [109] in 1988, and we follow their analysis in this section. A general extension of this work, together with a survey of the area, can be found in the important paper of Elschner [183].

From the work of the preceding sections it is clear that the key to understanding the numerical solution of the boundary integral equation $(-\pi + \mathcal{K})\rho = f$

is to understand the numerical solution of the reduced wedge equation

$$-\pi\psi(x) \pm \int_0^1 \frac{x\sin(\chi\pi)\psi(y)\,dy}{x^2 + 2xy\cos(\chi\pi) + y^2} = F(x), \quad 0 < x \le 1 \qquad (8.3.69)$$

of (8.1.23). At $x = 0$, this becomes

$$-\pi\psi(0) \pm \chi\pi\psi(0) = F(0) \qquad (8.3.70)$$

as in (8.1.20). As before, we write this symbolically as $(-\pi \pm \mathcal{L})\psi = F$. The two cases (based on the term $\pm\mathcal{L}$) are essentially the same for the purposes of our discussion (because $\|\pm\mathcal{L}\| = \pi|\chi| < \pi$), and therefore we deal henceforth with only the case

$$(-\pi + \mathcal{L})\psi = F \qquad (8.3.71)$$

in this and the following section. It should be noted, however, that when $\chi \to 1$, the norm of $(-\pi + \mathcal{L})^{-1}$ becomes much larger than that of $(-\pi - \mathcal{L})^{-1}$, and an analogous statement is true when $\chi \to -1$, with the roles of $(-\pi + \mathcal{L})^{-1}$ and $(-\pi - \mathcal{L})^{-1}$ being reversed.

Early on, it was recognized by Chandler and Graham [109] and by Costabel and Stephan [134] that it is useful to rewrite (8.3.69) as an equivalent *Mellin convolution equation*:

$$-\pi\psi(x) + \int_0^1 L\left(\frac{x}{y}\right)\psi(y)\frac{dy}{y} = F(x), \quad 0 < x \le 1 \qquad (8.3.72)$$

with

$$L(z) = \frac{z\sin(\chi\pi)}{z^2 + 2z\cos(\chi\pi) + 1} \qquad (8.3.73)$$

A satisfactory numerical analysis can be developed for (8.3.72) by assuming only that the function L possesses certain properties, without particular reference to the particular case (8.3.73). Other problems can also be written in the form (8.3.72), as is illustrated in Ref. [109] with a problem from plane strain elasticity theory. A solvability theory for (8.3.72) can be based on both the use of Mellin transforms and the theory of Wiener-Hopf integral equations. Since we are interested primarily in the case (8.3.73), for which we already understand the solvability of (8.3.72), we do not consider the general case here and we refer the reader to Elschner [183].

The Mellin convolution equation (8.3.72) can be transformed to an equivalent Wiener-Hopf integral equation, using the same change of variables as was used in (8.1.26) of §8.1 for the wedge equation. Introduce

$$\hat{\psi}(\eta) = \psi(e^{-\eta}), \quad 0 \leq \eta < \infty$$

Then (8.3.72) becomes

$$-\pi\hat{\psi}(\eta) + \int_0^\infty L\left(e^{-(\eta-\xi)}\right)\hat{\psi}(\xi)\,d\xi = \hat{F}(\eta), \quad 0 \leq \eta < \infty \qquad (8.3.74)$$

Such equations are well understood from the work of Krein [322]. In addition, it has been an important problem for Wiener-Hopf equations to understand the "finite section" approximation

$$-\pi\hat{\psi}_\gamma(\eta) + \int_0^\gamma L\left(e^{-(\eta-\xi)}\right)\hat{\psi}_\gamma(\xi)\,d\xi = \hat{F}(\eta), \quad 0 \leq \eta \leq \gamma \qquad (8.3.75)$$

for $\gamma > 0$. In this approximation one is interested in whether the equation is solvable when γ is sufficiently large, and if it is solvable, what is the relation of $\hat{\psi}_\gamma(\eta)$ to $\hat{\psi}(\eta)$ as $\gamma \to \infty$? A complete error analysis for this was given by Anselone and Sloan in [20], and this theory plays an important role in the error analysis of some Nyström methods for solving the Mellin convolution equation (8.3.72) and of generalizations of the present work on collocation methods.

8.3.1. *Preliminary definitions and assumptions*

The solutions of (8.3.72) are likely to be poorly behaved near the origin, as is illustrated in (8.2.61). To help deal with such functions, we introduce some suitable function spaces. Let $k \geq 0$ be an integer, and let $\beta > 0$. Introduce the seminorm

$$|f|_{k,\beta} = \begin{cases} \sup\limits_{0 < x \leq 1} |x^{k-\beta} f(x)|, & k > \beta \\ \left\| f^{(k)} \right\|_\infty, & k \leq \beta \end{cases} \qquad (8.3.76)$$

and the function space

$$C_\beta^k = \{ f \in C^k(0, 1] \mid |f|_{k,\beta} < \infty \}$$

With the norm

$$\| f \|_{k,\beta} = \max_{0 \leq j \leq k} |f|_{j,\beta}$$

C_β^k is a Banach space. The principal case of interest to us is the case $0 < \beta < 1$, and t^α belongs to C_β^k for $\alpha \geq \beta$, but not for $\alpha < \beta$.

These function spaces are very similar to the spaces $C^{(\ell,\alpha)}[0, 1]$ introduced in §4.2.5 of Chapter 4 (cf. (4.2.85) in Chapter 4), which were used in analyzing graded mesh approximations. If $0 < \beta < 1, k > 0$, and if $t^k f^{(k)}(t) \in C^{(0,\beta)}[0, 1]$, then $f \in C_\beta^k$. We omit the proof. The hypotheses of this result are exactly those used in Theorem 4.2.2 on graded mesh approximations. Conversely, if $f \in C_\beta^k$, then $t^j f^{(j)}(t) \in C^{(0,\beta)}[0, 1]$ for $j = 0, 1, \ldots, k - 1$, and $t^k f^{(k)}(t)$ satisfies a Hölder condition with exponent β about $t = 0$,

$$\left| t^k f^{(k)}(t) \right| \leq ct^\beta, \quad c = |f|_{k,\beta} \tag{8.3.77}$$

The latter result for $t^k f^{(k)}(t)$ is slightly weaker than is assumed in the hypotheses of Lemma 4.2.3 on graded mesh approximations, but a careful examination of the proof of that lemma shows that (8.3.77) is sufficient for obtaining the results given there.

For the function L of (8.3.72), we make the following assumptions.

A1. For all integers $k \geq 0$, $L(\xi)$ is infinitely differentiable and

$$B_k \equiv \int_0^\infty \xi^k |L^{(k)}(\xi)| \frac{d\xi}{\xi} < \infty \tag{8.3.78}$$

A2.

$$B_0 = \int_0^\infty |L(\xi)| \frac{d\xi}{\xi} < \pi \tag{8.3.79}$$

A3. For the solution $\psi(y)$ of (8.3.72), there exists $\beta^* > 0$ such that for all integers $k \geq 0$, $\psi \in C_{\beta^*}^k$.

Let the integral operator of (8.3.72) be denoted by $\mathcal{L}$. Then $\mathcal{L}: C[0, 1] \to C[0, 1]$, and

$$\|\mathcal{L}\| = \sup_{0 \leq y \leq 1} \int_0^1 \left| L\left(\frac{x}{y}\right) \right| \frac{dy}{y} = \int_0^\infty |L(\xi)| \frac{d\xi}{\xi} \tag{8.3.80}$$

using the change of variables $\xi = x/y$. Thus **A2** implies $-\pi + \mathcal{L}$ is invertible by the geometric series theorem (cf. Theorem A.1 in the Appendix), with

$$\|(-\pi + \mathcal{L})^{-1}\| \leq \frac{1}{\pi - B_0} \tag{8.3.81}$$

Generalizations that remove the restriction **A2** are given in Elschner [183]. The assumption **A3** corresponds to the result (8.2.61) on the differentiability of the double layer density in the vicinity of corner points of the boundary. The assumption **A1** can be proven in a straightforward manner for rational function examples such as (8.3.73), and we leave this as an exercise for the reader. We also need the following consequences of the above assumption **A1**. Although nontrivial, we leave the proof to the reader.

Lemma 8.3.1. Let

$$L_x(y) := L\left(\frac{x}{y}\right)\frac{1}{y}, \quad 0 < y \le 1$$

(a) For all $k \ge 0$, there is a constant C_k independent of (x, y) for which

$$\left| y^{k+1} L_x^{(k)}(y) \right| \le C_k, \quad 0 < y \le 1 \tag{8.3.82}$$

(b) For all $v \in C[0, 1]$, $\mathcal{L}v \in C^\infty(0, 1]$. Moreover, for all $k \ge 0$,

$$|x^k (D^k \mathcal{L}v)(x)| \le B_k \|v\|_\infty, \quad 0 < x \le 1 \tag{8.3.83}$$

with D denoting differentiation.

Graded meshes

Introduce the mesh $\{x_i^{(n)}\}$,

$$0 = x_0^{(n)} < x_1^{(n)} < \cdots < x_n^{(n)} = 1 \tag{8.3.84}$$

with the superscript n usually omitted. Let $I_i = (x_{i-1}, x_i)$, and $h_i = x_i - x_{i-1}$. To compensate in our approximation scheme for functions with unbounded derivatives around $x = 0$, we need to introduce graded meshes. In doing so, we follow the notation of Chandler and Graham [109] and of several other authors.

Let $k \ge 1$ be an integer, and let $\beta \in (0, k]$. A mesh $\{x_i\}$ is said to be (k, β)-*graded* if there is a constant γ independent of n for which

$$h_i < \frac{\gamma}{n} x_{i-}^{1-\beta/k}, \quad i = 1, \ldots, n \tag{8.3.85}$$

In this, $i- := \max\{i - 1, 1\}$. The example of such meshes of most interest is

$$x_i = \left(\frac{i}{n}\right)^q \tag{8.3.86}$$

This mesh is (k, β)-graded if $q \ge k/\beta$. This mesh is also the one used in (8.2.42) and in §4.2.5 of Chapter 4.

For a function v defined on $[0, 1]$, let $(v)_i$ denote the restriction of v to I_i. For a given integer $r \geq 0$, introduce the approximating space

$$\mathcal{S}_n = \{\phi \mid (\phi)_i \text{ is a polynomial of degree } \leq r \text{ on } I_i, 1 \leq i \leq n\} \quad (8.3.87)$$

The elements $\phi \in \mathcal{S}_n$ need not be continuous, and the dimension of $\mathcal{S}_n$ is $n(r + 1)$.

To define the collocation points, let a set of points $\{\xi_i\}$,

$$0 \leq \xi_0 < \xi_1 < \cdots < \xi_r \leq 1 \quad (8.3.88)$$

be given. Define

$$x_{i,j} = x_{i-1} + h_i \xi_j, \quad j = 0, \ldots, r, \quad i = 1, \ldots, n$$

For $v \in C[0, 1]$, let $\mathcal{P}_n v$ denote the unique function in $\mathcal{S}_n$ that interpolates v at the points $\{x_{i,j}\}$. The operator $\mathcal{P}_n$ is simply the standard interpolatory projection operator mapping $C[0, 1]$ onto $\mathcal{S}_n$, and examples of it have been studied earlier in Chapter 3.

Let $v \in C_\beta^{r+1}$, and assume the mesh satisfies (8.3.85) with $k = r + 1$. Then

$$\|v - \mathcal{P}_n v\|_\infty = O\left(n^{-(r+1)}\right), \quad v \in C_\beta^{r+1} \quad (8.3.89)$$

This follows from Rice [461] or from a generalization of the proof used in Lemma 4.2.3 of §4.2.5 in Chapter 4. For the mesh (8.3.86) the equivalent assumption on the grading parameter is that $q \geq (r + 1)/\beta$, just as in (4.2.94) of Lemma 4.2.3. From **A3** above, our solution $\psi \in C_{\beta*}^k$, and we need to approximate ψ with a $(r+1, \beta^*)$-graded mesh to preserve the full order of convergence associated with approximations from $\mathcal{S}_n$.

To obtain the best possible order of convergence in the error bounds derived below, we make the assumption that the mesh in (8.3.88) satisfies

$$\int_0^1 \left(\prod_{i=0}^r (\xi - \xi_i)\right) g(\xi) \, d\xi = 0 \quad (8.3.90)$$

for all polynomials $g(\xi)$ of degree $\leq r'$, with $r' \geq 0$ [use $r' = -1$ if (8.3.90) is not true for $g(\xi)$ a constant]. If the points $\{\xi_0, \ldots, \xi_r\}$ are the zeros of the Legendre polynomial of degree $r + 1$ (normalized to the interval $[0, 1]$), then (8.3.90) is true with $r' = r$. Let $\mathcal{Q}v(\xi)$ denote the polynomial of degree $\leq r$ that interpolates $v(\xi)$ at the points in $\{\xi_0, \ldots, \xi_r\}$. Then (8.3.90) is equivalent to requiring

$$\int_0^1 g(\xi)\,d\xi = \int_0^1 \mathcal{Q}g(\xi)\,d\xi$$

for all polynomial $g(\xi)$ of degree $\leq r + r' + 1$. And in turn, this is equivalent
to saying that the numerical integration formula

$$\int_0^1 g(\xi)\,d\xi \approx \int_0^1 \mathcal{Q}g(\xi)\,d\xi, \quad g \in C[0,1]$$

has degree of precision $r + r' + 1$.

8.3.2. The collocation method

For solving the integral equation (8.3.71), our initial collocation method is identical to that considered in earlier chapters, especially Chapter 3. Symbolically,
the collocation method amounts to solving

$$(-\pi + \mathcal{P}_n \mathcal{L})\psi_n = \mathcal{P}_n F \tag{8.3.91}$$

The principal difficulty in doing an error analysis of this method is in showing
its stability, that is

$$\|(-\pi + \mathcal{P}_n \mathcal{L})^{-1}\| \leq C_S, \quad n \geq N \tag{8.3.92}$$

with some $C_S < \infty$ and some $N > 0$. If this stability is known, then error bounds
can be found based on Theorem 3.1.1 of §3.1 of Chapter 3:

$$\|\psi - \psi_n\|_\infty \leq \pi C_S \|\psi - \mathcal{P}_n \psi\|_\infty \tag{8.3.93}$$

Recall the interated collocation solution:

$$\hat{\psi}_n = \frac{1}{\pi}(-F + \mathcal{L}\psi_n) \tag{8.3.94}$$

This was introduced in §3.4 of Chapter 3. It satisfies

$$(-\pi + \mathcal{L}\mathcal{P}_n)\hat{\psi}_n = F \tag{8.3.95}$$

$$\mathcal{P}_n \hat{\psi}_n = \psi_n \tag{8.3.96}$$

From the latter,

$$\hat{\psi}_n(x_{i,j}) = \psi_n(x_{i,j}), \quad j = 0, 1, \ldots, r, \ i = 1, \ldots, n \tag{8.3.97}$$

For the error in $\hat{\psi}_n$,

$$\|\psi - \hat{\psi}_n\|_\infty \le \|(-\pi + \mathcal{L}\mathcal{P}_n)^{-1}\| \, \|\mathcal{L}(I - \mathcal{P}_n)\psi\|_\infty \qquad (8.3.98)$$

The uniform boundedness of $\|(-\pi + \mathcal{L}\mathcal{P}_n)^{-1}\|$ will follow from (8.3.93) by using (3.4.84) of §3.4. Error bounds for $\{\psi_n(x_{i,j})\}$ can be found by using (8.3.98), and this often leads to improved rates of convergence when compared with the rates found based on (8.3.93).

From **A3** the true solution $\psi \in C_{\beta^*}^{r+1}$. Let the mesh (8.3.84) be $(r + 1, \beta^*)$-graded. Then (8.3.89) and (8.3.93) imply

$$\|\psi - \psi_n\|_\infty \le O\left(n^{-(r+1)}\right) \qquad (8.3.99)$$

In McLean [368] it is shown that

$$\|\mathcal{L}(I - \mathcal{P}_n)\psi\|_\infty \le O\left(n^{-(r+r'+2)}\right)$$

provided the mesh $\{x_i\}$ of (8.3.84) is $(r + r' + 2, \beta^*)$-graded. Thus, (8.3.98) implies

$$\|\psi - \hat{\psi}_n\|_\infty \le O\left(n^{-(r+r'+2)}\right) \qquad (8.3.100)$$

If $r' \ge 0$, then this and (8.3.97) imply that $\psi_n \to \psi$ is superconvergent at the collocation node points.

We return to showing the stability result in (8.3.92). The simplest approach is to use

$$\|\mathcal{P}_n\mathcal{L}\| \le \|\mathcal{P}_n\| \, \|\mathcal{L}\| \qquad (8.3.101)$$

If the right side is bounded by some number $\delta < \pi$ independent of n, then (8.3.92) follows trivially from the geometric series theorem (cf. Theorem A.1 in the Appendix), with

$$\|(-\pi + \mathcal{P}_n\mathcal{L})^{-1}\| \le \frac{1}{\pi - \delta} \equiv C_S \qquad (8.3.102)$$

Since $\|\mathcal{P}_n\| \ge 1$, the use of (8.3.101) is not a good general strategy unless $\|\mathcal{P}_n\| = 1$, as for example, when using piecewise constant interpolation to define $\mathcal{P}_n v$ for $v \in C[0, 1]$. A more general discussion of (8.3.101) is given in Ref. [64].

There are other ways of showing stability in particular cases. But there is no general way to show stability under the above framework and assumptions, and for good reason. It is shown by Chandler and Graham [109, §4] that there

are choices of $\{\xi_i\}$ for which the stability result (8.3.92) is not true. To obtain
a general theory of convergent schemes, they introduce a minor but significant
modification of the above collocation scheme. In the following, we follow quite
closely the presentation given in Ref. [109, §4]

A modified collocation method

A *modification* of our collocation scheme is determined by a seqence $\{i(n)|n \geq 1\}$ with $0 \leq i(n) < n$. The modification is called (k, β)-*acceptable* if

$$\left(x_{i(n)}^{(n)}\right)^{\beta/k} \leq \frac{\bar{\gamma}}{n} \tag{8.3.103}$$

for some constant $\bar{\gamma}$ independent of n. The approximation space S_n is modified
to

$$\tilde{S}_n = \{\phi \in S_n \mid i \leq i(n) \Rightarrow (\phi)_i \text{ a constant}\} \tag{8.3.104}$$

The function $\phi \in \tilde{S}_n$ is piecewise constant on $[0, x_{i(n)}^{(n)}]$. It can be shown that if
the coarser mesh $\{0, x_{i(n)}, \ldots, 1\}$ is (k, β)-graded, then the mesh $\{0, x_1, \ldots, 1\}$
is (k, β)-acceptable. The proof is straightforward. Also, it is straightforward
to show that the mesh (8.3.86) is (k, β)-acceptable if and only if

$$i(n) \leq \bar{\gamma}' n^{1-k/\beta q} \tag{8.3.105}$$

for some constant $\bar{\gamma}'$. In this case this inequality also implies that k, β, q must
satisfy

$$q \geq \frac{k}{\beta}$$

The error analysis of the modified collocation method introduced below in
(8.3.113) will require the following technical result.

Lemma 8.3.2. Suppose the mesh $\{x_i^{(n)}\}$ is (k, β)-graded. Then for all $\epsilon > 0$ and
for all sufficiently large n (depending on ϵ), there exists $i(\epsilon, n) \leq n$ for which

$$\left(x_{i(\epsilon,n)}^{(n)}\right)^{\frac{\beta}{k}} \leq \left(1 + \frac{1}{\epsilon}\right)\frac{\gamma}{n} \tag{8.3.106}$$

and

$$i > i(\epsilon, n) \Rightarrow \frac{h_i}{x_{i-}} \leq \epsilon \tag{8.3.107}$$

Proof. Define

$$i(\epsilon, n) = \min\left\{\ell \geq 0 \mid i > \ell \Rightarrow \frac{h_i}{x_{i-}} \leq \epsilon\right\} \qquad (8.3.108)$$

For a given ϵ, the set on the right side is nonempty for all sufficiently large n. In particular, it contains n, since

$$\frac{h_n}{x_{n-}} = \frac{h_n}{1 - h_n} = O\left(\frac{1}{n}\right)$$

because $h_n = O(n^{-1})$ from (8.3.85). The definition ensures that (8.3.107) is satisfied, and we check only (8.3.106).

With $i(\epsilon, n) = 0$, (8.3.106) is trivial. For $i(\epsilon, n) = 1$, recall (8.3.85) with $i = 1$:

$$h_1 < \frac{\gamma}{n} x_{1-}^{1-\beta/k} = \frac{\gamma}{n} h_1^{1-\beta/k}$$

which implies

$$\left(x_1^{(n)}\right)^{\frac{\beta}{k}} = h_1^{\beta/k} < \frac{\gamma}{n}$$

and proves (8.3.106).

With $i(\epsilon, n) > 1$, the definition (8.3.108) implies

$$\frac{h_{i(\epsilon,n)}}{x_{i(\epsilon,n)-}} > \epsilon$$

$$x_{i(\epsilon,n)} = x_{i(\epsilon,n)-} + h_{i(\epsilon,n)}$$

$$\leq \left(1 + \frac{1}{\epsilon}\right) h_{i(\epsilon,n)}$$

Returning to the mesh grading assumption (8.3.85), apply it with $i = i(\epsilon, n)$:

$$h_{i(\epsilon,n)} < \frac{\gamma}{n} x_{i(\epsilon,n)-}^{1-\beta/k}$$

$$x_{i(\epsilon,n)} < \frac{\gamma}{n} x_{i(\epsilon,n)-}^{1-\beta/k}\left(1 + \frac{1}{\epsilon}\right)$$

$$\leq \left(1 + \frac{1}{\epsilon}\right)\frac{\gamma}{n}$$

because $1 - \beta/k \geq 0$ from the definition of $\{x_i^{(n)}\}$ being (k, β)-graded and $x_{i(\epsilon,n)-} \leq 1$. $\qquad\square$

Introduce a modified projection operator for interpolation on $\bar{S}_n$. For $v \in C[0, 1]$, define

$$(\bar{P}_n v)_i = \begin{cases} (\mathcal{P}_n v)_i, & i > i(n) \\ v(x_{i-\frac{1}{2}}), & i \leq i(n) \end{cases} \tag{8.3.109}$$

with $x_{i-\frac{1}{2}} = \frac{1}{2}(x_{i-1} + x_i)$. The interpolation $\bar{P}_n v$ is simply piecewise constant interpolation on $[0, x_{i(n)}]$. The error of such interpolation is given by the following.

Lemma 8.3.3. Let $\bar{S}_n$ be an acceptable modification of S_n.

(a) If $v \in C^{r+1}[x_{i-1}, x_i]$ with $i > i(n)$, then

$$\| (v - \bar{P}_n v)_i \|_\infty \leq C h_i^{r+1} \| (D^{r+1} v)_i \|_\infty \tag{8.3.110}$$

(b) If $v \in C_\beta^1$ for some $\beta \leq 1$, then for all $i \leq i(n)$,

$$\| (v - \bar{P}_n v)_i \|_\infty \leq C h_i^\beta \| v \|_{1,\beta} \tag{8.3.111}$$

(c) If the mesh underlying S_n is $(r + 1, \beta)$-graded and $\bar{S}_n$ is an $(r + 1, \beta)$-acceptable modification for $\beta \leq 1$, then for all $v \in C_\beta^{r+1}$,

$$\| v - \bar{P}_n v \|_\infty \leq \frac{C}{n^{r+1}} \| v \|_{r+1,\beta} \tag{8.3.112}$$

Proof. A proof may be given along the lines of that given earlier in Lemma 4.2.3 of §4.2.5 in Chapter 4. We omit it here. $\qquad\square$

With this modified interpolatory projection operator, we can repeat the definitions and results in (8.3.91) and (8.3.94)–(8.3.96) for the collocation method. The modified collocation solution is obtained from solving

$$(-\pi + \bar{P}_n \mathcal{L})\bar{\psi}_n = \bar{P}_n F \tag{8.3.113}$$

The iterated modified collocation is defined by

$$\hat{\psi}_n = \frac{1}{\pi}(-F + \mathcal{L}\bar{\psi}_n)$$

and it satisfies

$$(-\pi + \mathcal{L}\bar{P}_n)\hat{\psi}_n = F$$

$$\mathcal{P}_n \hat{\psi}_n = \bar{\psi}_n$$

The modified collocation method (8.3.113) is as easy to implement as the original collocation method (8.3.91). The error analysis for (8.3.113) is given in the following sequence of results, taken from Ref. [109].

Lemma 8.3.4. Assume the mesh (8.3.84) is $(r + 1, \beta)$-graded, and assume

$$\|\mathcal{L}\| < \delta < \pi \qquad (8.3.114)$$

Then there exists a $(r + 1, \beta)$-acceptable modification $\{i_\delta(n)\}$ for which

$$\|\bar{\mathcal{P}}_n \mathcal{L}v\|_\infty \leq \delta \|v\|_\infty, \quad v \in C[0, 1] \qquad (8.3.115)$$

for all sufficiently large n. Moreover, (8.3.115) is satisfied for any other modification $\{i(n)\}$ for which $i(n) \geq i_\delta(n)$ for all n.

Proof. We prove (8.3.115) by showing

$$\|(\bar{\mathcal{P}}_n \mathcal{L}v)_i\|_\infty \leq \delta \|v\|_\infty, \quad i = 1, \ldots, n, \quad v \in C[0, 1] \qquad (8.3.116)$$

Let $\{i(n)\}$ be any modification. For $i = 1, \ldots, i(n)$,

$$\|(\bar{\mathcal{P}}_n \mathcal{L}v)_i\|_\infty = \left|(\mathcal{L}v)\left(x_{i-\frac{1}{2}}\right)\right| \leq \|\mathcal{L}v\|_\infty \leq \|\mathcal{L}\|\|v\|_\infty$$

and the result (8.3.116) follows from (8.3.114).

Now let $i > i(n)$, and write

$$\|(\bar{\mathcal{P}}_n \mathcal{L}v)_i\|_\infty \leq \|(\mathcal{L}v)_i\|_\infty + \|((I - \bar{\mathcal{P}}_n)\mathcal{L}v)_i\|_\infty$$

From (8.3.83) of Lemma 8.3.1,

$$\|(D^{r+1}\mathcal{L}v)_i\|_\infty \leq \frac{B_{r+1}}{(x_{i-1})^{r+1}}\|v\|_\infty$$

Combining this with (8.3.110) of Lemma 8.3.3,

$$\|(\bar{\mathcal{P}}_n \mathcal{L}v)_i\|_\infty \leq \|\mathcal{L}\|\|v\|_\infty + C h_i^{r+1}\|(D^{r+1}\mathcal{L}v)_i\|_\infty$$

$$\leq \|\mathcal{L}\|\|v\|_\infty + C_1 \left(\frac{h_i}{x_{i-}}\right)^{r+1}\|v\|_\infty \qquad (8.3.117)$$

with $C_1 := CB_{r+1}$ independent of n, i, and v. To apply Lemma 8.3.2, let

$$\epsilon = \left(\frac{\delta - \|\mathcal{L}\|}{C_1}\right)^{\frac{1}{r+1}}$$

For the resulting $(r + 1, \beta)$-graded sequence $\{i(\epsilon, n)\}$, define

$$i_\delta(n) := \max\{1, i(\epsilon, n)\}$$

From (8.3.106), it is $(r + 1, \beta)$-acceptable. In addition, by (8.3.107),

$$\left(\frac{h_i}{x_{i-}}\right)^{r+1} \leq \epsilon^{r+1} = \frac{\delta - \|\mathcal{L}\|}{C_1}$$

Combining this with (8.3.117), we have (8.3.116), as desired.

If we replace $\{i_\delta(n)\}$ with $\{i(n)\}$, with $i(n) \geq i_\delta(n)$ for all n, then the above proof is still valid, without any changes. □

Lemma 8.3.5. Assume $v \in C_\beta^{r+r'+2}$ for some $\beta \leq 1$. Assume the mesh $\{x_i^{(n)}\}$ is $(r + r' + 2, \eta)$-graded for some $\eta < \beta$. Then for any $(r + r' + 2, \beta)$-acceptable modification,

$$\|\mathcal{L}(I - \bar{\mathcal{P}}_n)v\| \leq \frac{C}{n^{r+r'+2}}\|v\|_{r+r'+2,2} \tag{8.3.118}$$

Proof. This is a fairly lengthy technical proof, and we just reference [109, p. 133]. □

Theorem 8.3.1. *Suppose the mesh $\{x_i^{(n)}\}$ is $(r + r' + 2, \beta)$-graded for some $\beta \leq 1$ with $\beta < \beta^*$ [cf. **A3** following (8.3.79)]. Then there exists a modification $\{i(n)\}$ such that for all n sufficiently large, $\bar{\psi}_n$ and $\hat{\psi}_n$ are uniquely defined and*

$$\|\psi - \hat{\psi}_n\|_\infty \leq \frac{C}{n^{r+r'+2}} \tag{8.3.119}$$

Proof. This follows from the two preceding lemmas. Since $\{x_i^{(n)}\}$ is $(r + r' + 2, \beta)$-graded, it is also $(r + r' + 2, \bar{\beta}^*)$-graded with $\bar{\beta}^* := \min\{\beta^*, 1\}$. Use Lemma 8.3.4 to show there is an $(r + r' + 2, \bar{\beta}^*)$-acceptable modification with $\|\bar{\mathcal{P}}_n\mathcal{L}\| \leq \delta$ for some $\|\mathcal{L}\| < \delta < \pi$. Then use the geometric series theorem [cf. (A.1) in the Appendix] to obtain the existence and uniform boundedness of $(-\pi + \bar{\mathcal{P}}_n\mathcal{L})^{-1}$,

$$\|(-\pi + \bar{\mathcal{P}}_n\mathcal{L})^{-1}\| < \frac{1}{\pi - \delta} \equiv C_S \tag{8.3.120}$$

as in (8.3.102). Use Lemma 8.3.5 to complete the proof. □

Corollary 8.3.1. *Suppose the mesh $\{x_i^{(n)}\}$ is generated with the mesh (8.3.86).
Assume q is so chosen that*

$$q > \frac{r+r'+2}{\beta*}, \quad q \geq r+r'+2 \tag{8.3.121}$$

*Then there is an integer constant $i^* \geq 0$ such that the modification $i(n) \equiv i^*$ is
sufficient in obtaining Theorem (8.3.1) and its consequences.*

Proof. With the assumption (8.3.121), the mesh $\{x_i^{(n)}\}$ of (8.3.86) is $(r+r'+2,$
$(r+r'+2)/q)$-graded. [Note this amounts to using $k = r+r'+2$ and $\beta = (r+r'+2)/q$ in the sentence following (8.3.86), and then $q = k/\beta$.] Using Lemma
8.3.2, there is an $(r+r'+2, \beta*)$-acceptable mesh, and from (8.3.105), we can
choose it so that

$$i(n) \leq \bar{\gamma}' \tag{8.3.122}$$

for some constant $\bar{\gamma}'$, since we have $1 - k/\beta q = 0$ in the bound of (8.3.105).
This inequality implies it is sufficient to choose $i(n)$ to be the largest integer
$i^* \leq \bar{\gamma}'$, for all values n. ☐

The above theorem and corollary gives a complete convergence and stability
analysis, although it still leaves unspecified the choice of modification $\{i(n)\}$.
The theorem says that stability and optimal convergence can be guaranteed by
using a slight change in the approximating functions $\phi \in S_n$ in a very small
neighborhood of 0. With the choice of (8.3.86) for the mesh, we need use
piecewise constant interpolation on only a fixed number i^* of subintervals
around 0, but again i^* is not specified. In practice, with the usual choices of $\{\xi_i\}$,
this modification appears unnecessary. Therefore, we advise using the original
collocation scheme of (8.3.91). In the seemingly rare case that your method
seems to not be converging, or if the condition number of the linear system
associated with (8.3.91) seems to be much too large, then use the modification
of (8.3.113) as described above.

In converting (8.3.113) to a linear system, we must calculate the collocation
integrals of the coefficient matrix. In most cases this requires a further numerical
integration, and this should be done so as to (*i*) minimize the computation
time, and (*ii*) preserve the order of convergence associated with the collocation
method. This can be done, but rather than doing so here, we note that it is
also possible to define a completely discretized Nyström method with the same
order of convergence and a similar number of unknowns in the solution process.
This has been done for polygonal regions in the paper of Graham and Chandler

[230], and it is extended to general piecewise smooth boundaries in Atkinson and Graham [61]. An alternative variant of this work is given in the following section.

8.4. The Nyström method

In this section, we give a completely discrete numerical method due to Rathsfeld [450] for solving the integral equation $(-\pi + \mathcal{K})\rho = f$ of (8.1.17). In some ways, the completely discrete method of Graham and Chandler [230] is more general and applies to a greater variety of integral equations. But the method of Rathsfeld applies to a number of important equations, and it shows an alternative approach to dealing with the singular behavior at corner points of the boundary and singularities in a kernel function.

In the Green's formula (8.1.12) of §8.1, choose the harmonic function $u \equiv 1$, thus obtaining

$$\Omega(P) = -\int_S \frac{\partial}{\partial \mathbf{n}_Q}[\log|P - Q|] \, dS_Q, \quad P \in S \qquad (8.4.123)$$

Then the double layer integral operator $\mathcal{K}$ of (8.1.13) can be written in the equivalent form

$$\mathcal{K}\rho(P) = -\pi\rho(P) + \int_S [\rho(Q) - \rho(P)]\frac{\partial}{\partial \mathbf{n}_Q}[\log|P - Q|] \, dS_Q,$$
$$P \in S \qquad (8.4.124)$$

for $\rho \in C(S)$. When the interior Dirichlet problem (8.1.1) is solved by writing it as a double layer potential, as in (8.1.16),

$$u(A) = \int_S \rho(Q)\frac{\partial}{\partial \mathbf{n}_Q}[\log|A - Q|] \, dS_Q, \quad A \in D$$

the density ρ is obtained by solving the second kind equation of (8.1.17),

$$(-\pi + \mathcal{K})\rho = f$$

Using (8.4.124), this can be written as

$$-2\pi\rho(P) + \int_S [\rho(Q) - \rho(P)]K(P, Q) \, dS_Q = f(P), \quad P \in S$$
$$(8.4.125)$$

with

$$K(P, Q) = \frac{\partial}{\partial \mathbf{n}_Q}[\log|P - Q|] \qquad (8.4.126)$$

We apply numerical integration to this equation to obtain a Nyström method for its solution. For the presentation and analysis of the method, we follow closely the framework of Rathsfeld [450].

Assume the boundary S has a single corner, say at the origin $\mathbf{0}$. Let S be parameterized by $\mathbf{r}(t)$, for $-1 \leq t \leq 1$, with $\mathbf{r}(0) = \mathbf{0}$. Extend $\mathbf{r}(t)$ to $-\infty < t < \infty$ by periodicity, and for simplicity assume $\mathbf{r} \in C^{\infty}(0, 2)$. For the derivative $\mathbf{r}'(t)$ assume that

$$|\mathbf{r}'(t)| = \frac{L}{2}, \quad -1 \leq t < 0 \quad \text{and} \quad 0 < t \leq 1$$

so that the arclength from $\mathbf{0}$ to $\mathbf{r}(t)$ is $\frac{1}{2}L|t|$, for $-2 \leq t \leq 2$. Introduce the parametrization mesh points

$$s_j^{(n)} = \left(\frac{j}{n}\right)^q, \quad s_{-j} = -s_j, \quad j = 0, 1, \dots, n \tag{8.4.127}$$

and define corresponding points on S,

$$\mathbf{r}_j^{(n)} = \mathbf{r}\left(s_j^{(n)}\right)$$

The exponent $q \geq 1$ is a constant that is determined later. Extend these grid points periodically, with $s_{j+2n}^{(n)} = s_j^{(n)}$, $\mathbf{r}_{j+2n}^{(n)} = \mathbf{r}_j^{(n)}$, for all integers j. From here on, we dispense with the superscript n, although understanding implicitly its presence.

Choose an interpolatory quadrature formula on $C[0, 1]$,

$$\int_0^1 \phi(\eta)\, d\eta \approx \sum_{j=1}^m \omega_j \phi(\eta_j) \tag{8.4.128}$$

with $0 \leq \eta_1 < \eta_2 < \cdots < \eta_m \leq 1$, and assume the quadrature weights ω_j are all positive. Further assume this formula has degree of precision R, with $m - 1 \leq R \leq 2m - 1$. We apply this formula to integration over the sections of S corresponding to parametrization subintervals $[s_{i-1}, s_i]$:

$$\int_{\mathbf{r}_{j-1}}^{\mathbf{r}_j} \phi(\xi)\, dS_\xi \approx \sum_{j=1}^m \omega_{i,j} \phi(\mathbf{r}_{i,j}) \tag{8.4.129}$$

with

$$\omega_{i,j} = \frac{1}{2}L\omega_j(s_i - s_{i-1}), \quad s_{i,j} = (1 - \eta_j)s_{i-1} + \eta_j s_i, \quad \mathbf{r}_{i,j} = \mathbf{r}(s_{i,j})$$

for $j = 1, \dots, m$.

Let i^* be a nonnegative integer, fixed independently of n. In the integral of (8.4.125) delete the portion of the integral over that part of the boundary corresponding to the subinterval $[-s_{i^*}, s_{i^*}]$. Then approximate the remaining portion of the integral of (8.4.125) with (8.4.129), yielding the approximation

$$-2\pi\rho_n(P) + \sum_{i=i^*+1}^{2n-i^*} \sum_{\substack{j=1 \\ P\neq \mathbf{r}_{i,j}}} \omega_{i,j}[\rho_n(\mathbf{r}_{i,j}) - \rho_n(P)]K(P, \mathbf{r}_{i,j}) = f(P),$$

$$P \in S \qquad (8.4.130)$$

The integrand is taken as zero in the case $P = \mathbf{r}_{i,j}$, thus avoiding the apparent singularity in the kernel function K. This is a Nyström method, with the natural interpolation formula obtained by solving for $\rho_n(P)$ in (8.4.125). Re-arranging the above, we have

$$\left(-2\pi - \sum_{i=i^*+1}^{2n-i^*} \sum_{\substack{j=1 \\ P\neq \mathbf{r}_{i,j}}}^{m} \omega_{i,j}K(P, \mathbf{r}_{i,j})\right)\rho_n(P)$$

$$+ \sum_{i=i^*+1}^{2n-i^*} \sum_{\substack{j=1 \\ P\neq \mathbf{r}_{i,j}}}^{m} \omega_{i,j}\rho_n(\mathbf{r}_{i,j})K(P, \mathbf{r}_{i,j}) = f(P), \quad P \in S \qquad (8.4.131)$$

It can be shown that if i^* is chosen sufficiently large, then the quantity

$$-2\pi - \sum_{i=i^*+1}^{2n-i^*} \sum_{\substack{j=1 \\ P\neq \mathbf{r}_{i,j}}} \omega_{i,j}K(P, \mathbf{r}_{i,j})$$

is uniformly bounded away from zero as $n \to \infty$; thus, one can solve for $\rho_n(P)$ without fear of dividing by zero. Collocating at the node points $\mathbf{r}_{k,l}$, one obtains the equivalent linear system

$$\left(-2\pi - \sum_{i=i^*+1}^{2n-i^*} \sum_{\substack{j=1 \\ P\neq \mathbf{r}_{i,j}}}^{m} \omega_{i,j}K(\mathbf{r}_{k,l}, \mathbf{r}_{i,j})\right)\rho_n(\mathbf{r}_{k,l})$$

$$+ \sum_{i=i^*+1}^{2n-i^*} \sum_{\substack{j=1 \\ P\neq \mathbf{r}_{i,j}}}^{m} \omega_{i,j}\rho_n(\mathbf{r}_{i,j})K(\mathbf{r}_{k,l}, \mathbf{r}_{i,j}) = f(\mathbf{r}_{k,l}) \qquad (8.4.132)$$

for $k = 1, \ldots, 2n$, $l = 1, \ldots, m$.

8.4.1. Error analysis

The error analysis is divided into two main steps. First, approximate the integral equation (8.4.125) by

$$-2\pi \rho_n^*(P) + \int_{S^{(n)}} \left[\rho_n^*(Q) - \rho_n^*(P) \right] K(P, Q) \, dS_Q = f(P), \quad P \in S$$
(8.4.133)

In this, $S^{(n)} = S - \{ \mathbf{r}(t) \mid -s_{i^*} < t < s_{i^*} \}$, which amounts to removing from S a small portion of the boundary about the corner $\mathbf{0}$, and for this we assume $i^* > 0$. As $n \to \infty$, the portion of S removed to form $S^{(n)}$ becomes smaller, approaching zero in length, since

$$s_{i^*} = s_{i^*}^{(n)} = \left(\frac{i^*}{n} \right)^q$$

Second, use numerical integration to approximate the equation (8.4.133) as in (8.4.130). We begin by considering the approximation (8.4.133).

Let c_n denote the characteristic function of $S^{(n)}$,

$$c_n(P) = \begin{cases} 1, & P \in S^{(n)} \\ 0, & P \notin S^{(n)} \end{cases}$$

Then the equations (8.4.125) and (8.4.133) can be written, respectively, as

$$\mathcal{A}\rho \equiv (-2\pi - \mathcal{K}1)\,\rho + \mathcal{K}\rho = f$$
(8.4.134)

$$\mathcal{A}^{(n)}\rho_n^* \equiv (-2\pi - \mathcal{K}c_n)\,\rho_n^* + \mathcal{K}(c_n \rho_n^*) = f$$
(8.4.135)

In this,

$$(\mathcal{K}c_n)\,(P) = \int_S K(P, Q) c_n(Q)\, dS_Q = \int_{S^{(n)}} K(P, Q)\, dS_Q, \quad P \in S$$

and $\mathcal{K}1$ is defined analogously. By $\mathcal{K}(c_n \rho_n^*)$, we intend

$$\mathcal{K}(c_n \rho_n^*)(P) = \int_S [c_n(Q)\rho_n^*(Q)] K(P, Q)\, dS_Q$$
$$+ [-\pi + \Omega(P)]\, c_n(P)\rho_n^*(P)$$
$$= \int_{S^{(n)}} \rho_n^*(Q) K(P, Q)\, dS_Q$$

which is equivalent to replacing ρ by $c_n \rho_n^*$ in (8.4.124).

Lemma 8.4.1. There is an integer $n_0 > 0$ such that for $n \geq n_0$, $\mathcal{A}^{(n)}$ is invertible on $C(S)$ with

$$\left\| \left(\mathcal{A}^{(n)} \right)^{-1} \right\| \leq c < \infty, \quad n \geq n_0 \qquad (8.4.136)$$

Proof. The proof begins with a decomposition of the operator $\mathcal{K}$ of (8.4.124), to relate it to the wedge operator that characterizes the behavior of the operator in the vicinity of the corner at $\mathbf{0}$. Let $\phi = (1 - \chi)\pi$ denote the angle at $\mathbf{0}$, with $-1 < \chi < 1$. Let W denote a wedge of the type studied in §8.1 and pictured in Figure 8.2. Let its arms each have length $\frac{1}{2}L$, and let these arms be tangent to the two sides of S at $\mathbf{0}$. Introduce a mapping

$$\mathbf{r}_0 : [-1, 1] \to W$$

which is to be linear on both $[-1, 0]$ and $[0, 1]$. Moreover, let $\mathbf{r}_0(0) = \mathbf{0}$, $\mathbf{r}_0([-1, 0))$ correspond to $\mathbf{r}([-1, 0))$, $\mathbf{r}_0((0, 1])$ correspond to $\mathbf{r}((0, 1])$, and $|\mathbf{r}_0'(t)| = \frac{1}{2}L, t \neq 0$. In essence, $\mathbf{r}_0$ is the mapping introduced following (8.1.19), used to reduce the double layer potential operator to the equations given in (8.1.22).

Using the parametrization $\mathbf{r}(t)$ of S, the operator $\mathcal{K}$ of (8.1.2) takes the form

$$\mathcal{K}\rho(\mathbf{r}(t)) = -\pi\rho(\mathbf{r}(t)) + \int_{-1}^{1} [\rho(\mathbf{r}(s)) - \rho(\mathbf{r}(t))] \frac{\mathbf{n}(s) \cdot (\mathbf{r}(s) - \mathbf{r}(t))}{|\mathbf{r}(s) - \mathbf{r}(t)|^2} \frac{L}{2} \, ds$$

for $-1 \leq t \leq 1$, with $\mathbf{n}(s)$ the inner normal to S at $\mathbf{r}(s)$. Using the parametrization $\mathbf{r}_0(t)$ of W, decompose $\mathcal{K}$ as

$$\mathcal{K} = \mathcal{W} + \mathcal{S} \qquad (8.4.137)$$

with

$$\mathcal{W}\rho(\mathbf{r}(t))$$
$$= -\pi\rho(\mathbf{r}(t)) + \int_{-1}^{1} [\rho(\mathbf{r}(s)) - \rho(\mathbf{r}(t))] \frac{\mathbf{n}_0(s) \cdot (\mathbf{r}_0(s) - \mathbf{r}_0(t))}{|\mathbf{r}_0(s) - \mathbf{r}_0(t)|^2} \frac{L}{2} \, ds$$
$$\qquad (8.4.138)$$

for $-1 \leq t \leq 1$, with $\mathbf{n}_0(s)$ the inner normal to W at $\mathbf{r}_0(s)$. In essence, $\mathcal{W}$ is the wedge operator studied in §8.1.3, and as an operator on $C[-1, 1]$ to $C[-1, 1]$, we have

$$\|\mathcal{W}\| = |\chi|\pi \qquad (8.4.139)$$

The operator S can be shown to be a compact operator on $C[-1, 1]$ to $C[-1, 1]$, by the argument given in [64, §7]. Further smoothing properties of S are proved in Ref. [61].

From (8.4.124), $(\mathcal{K}1)\,(P) \equiv -\pi$. Using this, rewrite $\mathcal{A}^{(n)}\rho_n^* = f$ as

$$-\pi\rho_n^*(P) + [\mathcal{K}(1 - c_n)(P)]\,\rho_n^*(P) + \mathcal{K}\left(c_n\rho_n^*\right)(P) = f(P), \quad P \in S$$

Introduce the decomposition (8.4.137) into this, writing it as

$$\mathcal{A}^{(n)}\rho_n^* \equiv (-\pi + \tilde{\mathcal{W}}_n + \tilde{\mathcal{S}}_n)\rho_n^* = f \tag{8.4.140}$$

with

$$\tilde{\mathcal{W}}_n\rho_n^*(P) = [\mathcal{W}(1 - c_n)(P)]\rho_n^*(P) + \mathcal{W}\left(c_n\rho_n^*\right)(P)$$

and similarly for $\tilde{\mathcal{W}}_n\rho_n^*$. For general $\rho \in C(S)$, we have

$$\begin{aligned}
\left|\tilde{\mathcal{W}}_n\rho_n^*(P)\right| &= |\mathcal{W}[(1 - c_n)\rho(P) + c_n\rho](P)| \\
&\le \|\mathcal{W}\|\,\|(1 - c_n)\rho(P) + c_n\rho\|_\infty \\
&\le |\chi|\pi\|\rho\|_\infty
\end{aligned}$$

The last step follows from (8.4.139) and

$$[(1 - c_n)\rho(P) + c_n\rho](Q) = \begin{cases} \rho(Q), & Q \in S^{(n)} \\ \rho(P), & Q \notin S^{(n)} \end{cases}$$

Thus

$$\|\tilde{\mathcal{W}}_n\| \le |\chi|\pi$$

and $-\pi + \tilde{\mathcal{W}}_n$ is invertible, with

$$\|(-\pi + \tilde{\mathcal{W}}_n)^{-1}\| \le \frac{1}{\pi(1 - |\chi|)} \tag{8.4.141}$$

for all n.

To show the solvability of (8.4.140), consider the identity

$$\begin{aligned}
\mathcal{A}^{(n)}&\{(-\pi + \tilde{\mathcal{W}}_n)^{-1} - \mathcal{A}^{-1}\tilde{\mathcal{S}}_n(-\pi + \tilde{\mathcal{W}}_n)^{-1}\} \\
&= I + \tilde{\mathcal{S}}_n(-\pi + \tilde{\mathcal{W}}_n)^{-1} - \mathcal{A}^{(n)}\mathcal{A}^{-1}\tilde{\mathcal{S}}_n(-\pi + \tilde{\mathcal{W}}_n)^{-1} \\
&= I + \left(\mathcal{A} - \mathcal{A}^{(n)}\right)\mathcal{A}^{-1}\tilde{\mathcal{S}}_n(-\pi + \tilde{\mathcal{W}}_n)^{-1} \tag{8.4.142}
\end{aligned}$$

Write

$$(\mathcal{A} - \mathcal{A}^{(n)})\mathcal{A}^{-1}\tilde{\mathcal{S}}_n$$
$$= (\mathcal{A} - \mathcal{A}^{(n)})\mathcal{A}^{-1}\mathcal{S} + (\mathcal{A} - \mathcal{A}^{(n)})\mathcal{A}^{-1}(\tilde{\mathcal{S}}_n - \mathcal{S}) \qquad (8.4.143)$$

It is straightforward to show that

$$(\mathcal{A} - \mathcal{A}^{(n)})\rho \to 0 \quad \text{as } n \to \infty, \quad \text{all } \rho \in C(S) \qquad (8.4.144)$$

Moreover, the compactness and other properties of $\mathcal{S}$ imply

$$\|\tilde{\mathcal{S}}_n - \mathcal{S}\| \to 0 \quad \text{as } n \to \infty \qquad (8.4.145)$$

as follows from the results in Ref. [64, §7]. Combining (8.4.143)–(8.4.145) with (8.4.141) implies

$$\|(\mathcal{A} - \mathcal{A}^{(n)})\mathcal{A}^{-1}\tilde{\mathcal{S}}_n(-\pi + \tilde{\mathcal{W}}_n)^{-1}\| \to 0 \quad \text{as } n \to \infty$$

This implies that the right side of the final equation in (8.4.142) is invertible. Returning to (8.4.142), the above proves $\mathcal{A}^{(n)}$ is one-to-one. From (8.4.140),

$$\mathcal{A}^{(n)} = (-\pi + \tilde{\mathcal{W}}_n)[I + (-\pi + \tilde{\mathcal{W}}_n)^{-1}\tilde{\mathcal{S}}_n]$$

Since $\tilde{\mathcal{S}}_n$ is compact, $I + (-\pi + \tilde{\mathcal{W}}_n)^{-1}\tilde{\mathcal{S}}_n$ will satisfy the standard Fredholm alternative theorem (see Theorem 1.3.1 in Chapter 1). Since $\mathcal{A}^{(n)}$ is one-to-one from $C(S)$ to $C(S)$, this implies it is also *onto*, and therefore $(\mathcal{A}^{(n)})^{-1}$ exists as a bounded operator on $C(S)$ to $C(S)$. Using (8.4.142), we can solve for $(\mathcal{A}^{(n)})^{-1}$ and show that it is uniformly bounded in n, as asserted in (8.4.136). $\square$

The remaining part of the error analysis for the approximate equation (8.4.130) is summarized in the following.

Theorem 8.4.1.

(a) *Suppose the integer i^* of (8.4.130) is chosen sufficiently large. Then the Nyström method (8.4.130) is stable.*

(b) *Assume the solution ρ is C^∞ on both $\{\mathbf{r}(s) \mid -1 - \epsilon \leq s < 0\}$ and $\{\mathbf{r}(s) \mid 0 < s \leq 1 + \epsilon\}$ for some $\epsilon > 0$. Moreover, assume ρ satisfies*

$$\left|\left(\frac{d}{ds}\right)^k \rho(\mathbf{r}(s))\right| \leq C|s|^{\beta-k}, \quad k = 1, 2, \ldots$$

with $\beta = (1 + |\chi|)^{-1}$. *Let* ρ_n *denote the solution of (8.4.130). Then*

$$\|\rho - \rho_n\|_\infty \le C \begin{cases} n^{-(R+1)}, & q > (R+1)/\beta \\ n^{-(R+1)} \log n, & q = (R+1)/\beta \\ n^{-q\beta}, & q < (R+1)/\beta \end{cases} \qquad (8.4.146)$$

In this, q *is the grading exponent used in (8.4.127) in defining the mesh* $\{s_j^{(n)}\}$, *and* R *is the degree of precision of the quadrature formula (8.4.128).*

Proof. This is an algebraically complicated proof, and it builds on the results of the preceding lemma. We refer the reader to [450, Theorem 3.1, §5] for the details. □

As is noted in Rathsfeld [450], as well as in Graham and Chandler [230] and Ref. [61], the choice of $i^* = 0$ seems sufficient empirically to have stability and convergence of (8.4.130). However, iterative variants of the Nyström method are given in both Refs. [450] and [61], and both of them show empirically that $i^* > 0$ is necessary to have convergence of the iterative method.

Since the orders of convergence of (8.4.146) are essentially identical to those obtained with the Galerkin and collocation methods, when using a suitably graded mesh, the Nyström method appears to be the most practical choice of a numerical method for solving boundary integral equations on a piecewise smooth boundary.

Discussion of the literature

For a discussion of the use of BIE in engineering applications, including a brief bibliography of such applications, we refer the reader to the introductory paragraphs in the Discussion of the Literature at the end of Chapter 7. We note particularly the survey paper of Blum [82] on the solution of corner and crack problems in elasticity. For the theoretical foundations of the study of boundary integral equations on piecewise smooth planar curves, see the important book of Grisvard [240].

The numerical analysis of boundary integral equations defined on piecewise smooth planar curves is of recent vintage, with the first satisfactory error analyses being given only in the mid-1980s. A significant aspect of this research has been in accounting for the behavior of the solution, thus leading to the need for an appropriately graded mesh. Earlier research is given in Atkinson and deHoog [63], [64]; Benveniste [77]; Bruhn and Wendland [93]; Cryer [145]; and Miller [383]. Most of this work did not not consider the use of graded meshes.

Almost all of the research on BIE problems on curves with corners has been for integral equations of the second kind. The first analyses were those of Chandler [106], [107] and Costabel and Stephan [134], all for Galerkin methods with piecewise polynomial functions on polygonal boundaries, much as described in §8.2. The analysis of the collocation method was more difficult, and it depended crucially on the error analysis of Anselone and Sloan [20] on the use of *finite section approximations* for solving Wiener-Hopf integral equations on the half-line. Another approach to this latter theory is given in de Hoog and Sloan [163]. The first analysis of collocation methods for solving BIE on polygonal boundaries is given in Chandler and Graham [109]; which also showed the somewhat surprising result that not all consistent collocation methods are convergent. A convergent Nyström method for such BIE problems on polygonal boundaries is given in Graham and Chandler [230], and it is extended to general piecewise smooth boundaries in Atkinson and Graham [61]. Other results on the numerical analysis of such BIE problems, again using suitably graded meshes and piecewise polynomial approximations, are given in Amini and Sloan [10]; Bourland et al. [84]; Costabel and Stephan [134]; Elschner [182], [183]; Elschner et al. [191]; and Rathsfeld [452].

Another approach to such problems is due to Kress [326], who introduces a suitable parametrization of the boundary. He can then use a standard uniform mesh and still obtain an optimal rate of convergence. This approach is extended in Jeon [289], in which a parametrization is given that is independent of the degree of the piecewise polynomial approximants being used.

For BIE of the second kind defined on planar curves with piecewise smooth boundaries, iteration methods for solving various discretizations are given in Atkinson and Graham [61]; Rathsfeld [450], [454]; and Schippers [489], [490], [491].

BIE of the first kind defined on planar curves with piecewise smooth boundaries have been among the most difficult problems to solve and analyze. Some early papers on this problem are Yan [578] and Yan and Sloan [581]. Recently, Elschner and Graham [188], [189] have given new collocation methods for solving BIE of the first kind on piecewise smooth curves, with no restriction on the of piecewise polynomials being used. This is also expected to lead to other new numerical schemes for such BIE problems, as in the paper of Elschner and Stephan [190] in which the discrete collocation method of §7.3.3 is generalized to piecewise smooth curves.

9
Boundary integral equations in three dimensions

The study of boundary integral equation reformulations of Laplace's equation in three dimensions is quite an old one, with the names of many well-known physicists, engineers, and mathematicians associated with it. The development of practical numerical methods for the solution of such boundary integral equations lagged behind and is of more recent vintage, with most of it dating from the mid-1960s. In the 1980s there was an increased interest in the numerical analysis of such equations, and it has been quite an active area of research in the 1990s.

These boundary integral equations are defined on surfaces in space, and there is a far greater variety to such surfaces than is true of boundaries for planar problems. The surfaces may be either smooth or piecewise smooth, and when only piecewise smooth, there is a large variation as to the structure of edges and vertices present on the surface. In addition, most numerical methods require approximations of a piecewise polynomial nature over triangulations of the surface, in the manner developed in Chapter 5. These numerical methods lead to the need to solve very large linear systems, and until recently most computers had great difficulty in handling such problems. The practical aspects of setting up and solving such linear systems are more onerous for boundary integral equations in three dimensions, and this means that the numerical analysis problems of concern are often of a different nature than for planar boundary integral equations. In addition, there are some quite difficult and deep problems in understanding the behavior of boundary integral equations on piecewise smooth surfaces, and it is important to have this understanding both to prove stability of numerical schemes and to know how to properly grade the mesh in the vicinity of edges and corners.

In the limited space of this chapter we can only begin to lay a foundation for the study of boundary integral equations on surfaces in three dimensions. In

§9.1 we give Green's representation formula, and then we apply it to define the most common direct and indirect methods for solving boundary integral equations. We describe the mapping properties of the boundary integral operators, and we summarize some regularity results for the solutions of the boundary integral equations. In §9.2 we consider collocation methods for solving boundary integral equations of the second kind over smooth surfaces, using approximations that are piecewise polynomial. It is usually possible to give complete error analyses of these numerical methods for solving such BIE defined on smooth surfaces, whereas collocation methods are often more difficult to understand when the surface is only piecewise smooth. In addition, it is usually possible to give a complete error analysis for iterative methods for solving the associated linear systems when the surface is smooth. Finally, we discuss the very important and difficult problem of calculating the collocation matrix, as this is usually the most time-consuming part of the solution process. We extend collocation methods to piecewise smooth surfaces in §9.3, although some of the numerical analysis for these problems is incomplete. Galerkin finite element approaches are discussed and analyzed in §9.4, for BIE of both the first and second kind. The chapter concludes with §9.5, in which we introduce methods based on using spherical polynomial approximations to solve problems on smooth surfaces, for integral equations of both the first and second kind.

9.1. Boundary integral representations

Let D denote a bounded open simply connected region in $\mathbf{R}^3$, and let S denote its boundary. Assume $\bar{D} = D \cup S$ is a region to which the divergence theorem can be applied. Let $D' = \mathbf{R}^3 \backslash \bar{D}$ denote the region complementary to D. Sometimes, we will use the notations D_i and D_e in place of D and D', respectively, to refer to the regions interior and exterior to S. At a point $P \in S$, let $\mathbf{n}_P$ denote the unit normal directed into D, provided such a normal exists. In addition, we will assume that S is a *piecewise smooth surface*. This means that S can be written as

$$S = S_1 \cup S_2 \cup \cdots \cup S_J \qquad (9.1.1)$$

with each S_j being the image of a smooth parametrization over a planar polygon. More precisely, for each j, there is a function

$$F_j : R_j \xrightarrow[onto]{1-1} S_j \qquad (9.1.2)$$

with R_j a closed planar polygon and $F_j \in C^m(R_j)$ for some $m \geq 2$. For simplicity, we often take $m = \infty$. Moreover, we assume that the only possible intersection of S_i and S_j, $i \neq j$, is along some portion of their boundaries (which correspond to the boundaries of R_i and R_j, respectively). When S is a smooth surface, we may still choose to write it in the form (9.1.1), to simplify constructing triangulations of S and to introduce a surface parametrization into the equation being studied. This general framework for dealing with piecewise smooth surfaces is the same as that used earlier in §5.3 of Chapter 5. The assumption that D is simply connected is more restrictive than necessary, but it simplifies the presentation of the solvability theory of many of the boundary integral equations we derive.

The boundary value problems in which we are interested are as follows, just as in Chapters 7 and 8.

The Dirichlet problem. Find $u \in C(\bar{D}) \cap C^2(D)$ that satisfies

$$
\begin{aligned}
\Delta u(P) &= 0, & P &\in D \\
u(P) &= f(P), & P &\in S
\end{aligned}
\tag{9.1.3}
$$

with $f \in C(S)$ a given boundary function.

The Neumann problem. Find $u \in C^1(\bar{D}) \cap C^2(D)$ that satisfies

$$
\begin{aligned}
\Delta u(P) &= 0, & P &\in D \\
\frac{\partial u(P)}{\partial \mathbf{n}_P} &= f(P), & P &\in S
\end{aligned}
\tag{9.1.4}
$$

with $f \in C(S)$ a given boundary function.

There is a corresponding set of problems for the exterior region D', and we discuss those later in the section. For the solvability of the above problems, we have the following.

Theorem 9.1.1. *Let S be a piecewise smooth surface, with $\bar{D}$ a region to which the divergence theorem can be applied. Assume the function $f \in C(S)$. Then:*

(a) *The Dirichlet problem (9.1.3) has a unique solution, and*
(b) *The Neumann problem (9.1.4) has a unique solution, up to the addition of an arbitrary constant, provided*

$$
\int_S f(Q)\, dS = 0
\tag{9.1.5}
$$

Proof. These results are contained in many textbooks, often in a stronger form than is given here. For example, see Kress [325, Chap. 6] or Mikhlin [380,

Chaps. 12, 16]. The proof is often based on a BIE reformulation of these problems. Also note that it can be shown that $u \in C^\infty(D)$ for both problems (9.1.3) and (9.1.4).   □

9.1.1. Green's representation formula

As in §7.1 of Chapter 7, the divergence theorem can be used to obtain a representation formula for functions that are harmonic inside the region D. If we assume $u \in C^2(D) \cap C^1(\bar{D})$ and that $\Delta u \equiv 0$ on D, then

$$
\int_S u(Q) \frac{\partial}{\partial \mathbf{n}_Q} \left[\frac{1}{|P - Q|} \right] dS(Q) - \int_S \frac{\partial u(Q)}{\partial \mathbf{n}_Q} \frac{dS(Q)}{|P - Q|}
$$
$$
= \begin{cases} 4\pi u(P), & P \in D \\ \Omega(P)u(P), & P \in S \\ 0, & P \in D' \end{cases} \tag{9.1.6}
$$

In this, $\Omega(P)$ denotes the interior solid angle at $P \in S$. To measure it, begin by defining

$$
\hat{D}(P) = \{Q \in D \mid L[P, Q] \subset D\}
$$

where $L[P, Q]$ denotes the open line segment joining P and Q:

$$
L[P, Q] = \{\lambda P + (1 - \lambda)Q \mid 0 < \lambda < 1\}
$$

The set $\hat{D}(P)$ consists of all points of D that can be seen from P on a direct line that does not intersect the boundary S of D. Define the *interior tangent cone* at $P \in S$ by

$$
\Gamma_i(P) = \{P + \lambda(Q - P) \mid \lambda > 0, \ Q \in \hat{D}(P)\} \tag{9.1.7}
$$

and define the *exterior tangent cone* at $P \in S$ as the complement of $\Gamma_i(P)$:

$$
\Gamma_e(P) = \mathbf{R}^3 \backslash \overline{\Gamma_i(P)} \tag{9.1.8}
$$

The boundary of $\Gamma_i(P)$ is denoted by $\Gamma(P)$:

$$
\Gamma(P) = \text{boundary} \left[\Gamma_i(P) \right] \tag{9.1.9}
$$

Now intersect $\Gamma_i(P)$ with the unit sphere centered at P. The surface area of this intersection is called the *interior solid angle* at $P \in S$, and it is denoted by $\Omega(P)$. If S is smooth at P, then $\Omega(P) = 2\pi$. For a cube the corners have interior solid angles of $\frac{1}{2}\pi$, and the edges have interior solid angles of π. We always assume

$$
0 < \Omega(P) < 4\pi \tag{9.1.10}
$$

which eliminates the possibility of a cuspidal type of behavior on S.

The first and third relationships in (9.1.6) follow from the Green's identities, which in turn follow from the divergence theorem. The second relationship follows from the following limit:

$$\lim_{\substack{A \to P \\ A \in D}} \int_S u(Q) \frac{\partial}{\partial \mathbf{n}_Q} \left[\frac{1}{|A - Q|} \right] dS(Q)$$

$$= [4\pi - \Omega(P)]u(P) + \int_S u(Q) \frac{\partial}{\partial \mathbf{n}_Q} \left[\frac{1}{|P - Q|} \right] dS(Q) \qquad (9.1.11)$$

A proof of this relation can be found in most textbooks on partial differential equations. For example, see Kress [325, p. 68] or Mikhlin [380, p. 360] for the case with S a smooth surface; the proof of the general case is a straightforward generalization.

Referring to (9.1.6), integrals of the form

$$v(A) = \int_S \psi(Q) \frac{dS(Q)}{|A - Q|}, \qquad\qquad A \in \mathbf{R}^3 \qquad (9.1.12)$$

$$w(A) = \int_S \rho(Q) \frac{\partial}{\partial \mathbf{n}_Q} \left[\frac{1}{|A - Q|} \right] dS(Q), \quad A \in \mathbf{R}^3 \qquad (9.1.13)$$

are called single layer and double layer potentials, respectively; and the functions ψ and ρ are called single layer density and double layer density functions, respectively. The functions $v(A)$ and $w(A)$ are harmonic at all $A \notin S$. For a complete introduction to the properties of such layer potentials, see Günter [246], Kellogg [307], Kress [325, Chap. 6], or Mikhlin [380, Chap. 18]. The Green's representation formula (9.1.6) says that every function u harmonic in D can be represented as a particular combination of single and double layer potentials. It is also usually possible to represent a function u harmonic in D as just one such potential, as we relate below.

The existence of the single and double layer potentials

For well-behaved density functions and for $A \notin S$, the integrands in (9.1.12) and (9.1.13) are nonsingular, and thus there is no difficulty in showing the integrals exist. However, for the case $A = P \in S$, the integrands become singular. With the single layer potential v in (9.1.12), it is straightforward to show the integral exists; moreover, assuming ψ is bounded on S,

$$\sup_{A \in \mathbf{R}^3} |v(A)| \le c \|\psi\|_\infty$$

The essential argument relies on the simple fact for planar integrals that

$$\int_{x^2 + y^2 \le 1} \frac{dx \, dy}{(\sqrt{x^2 + y^2})^\alpha} < \infty \qquad (9.1.14)$$

for all $\alpha < 2$. We can also show the asymptotic result that

$$v(A) = \frac{c(\psi)}{|A|} + O(|A|^{-2}) \quad \text{as } |A| \to \infty \qquad (9.1.15)$$

$$c(\psi) = \int_S \psi(Q) \, dS(Q)$$

For the double layer potential (9.1.13), the argument of the existence of $w(A)$, $A \in S$, is more delicate. For $A = P \in S$, the kernel function in (9.1.13) is given by

$$\frac{\partial}{\partial \mathbf{n}_Q} \left[\frac{1}{|P - Q|} \right] = \frac{\mathbf{n}_Q \cdot (P - Q)}{|P - Q|^3} = \frac{\cos(\theta_{P,Q})}{|P - Q|^2} \qquad (9.1.16)$$

In this, $\theta_{P,Q}$ is the angle between $\mathbf{n}_Q$ and $P - Q$, with

$$\cos(\theta_{P,Q}) = \mathbf{n}_Q \cdot \frac{P - Q}{|P - Q|}$$

If P is a point at which S is smooth, then it can be shown that

$$|\cos(\theta_{P,Q})| \le c|P - Q| \qquad (9.1.17)$$

for all Q in a neighborhood of P. When combined with (9.1.16), this implies

$$\left| \frac{\partial}{\partial \mathbf{n}_Q} \left[\frac{1}{|P - Q|} \right] \right| \le \frac{c}{|P - Q|}, \quad P, Q \in S$$

for some constant c. The existence of the double layer potential $w(A)$ of (9.1.13) at $A = P$ then follows from (9.1.14), just as for the single layer $v(P)$. To show (9.1.17), see Günter [246, p. 6] or Mikhlin [380, pp. 352–355]. When P is a point at which $\Omega(P) \ne 2\pi$, an even more careful argument is needed to show the existence of $W(P)$, and we omit it here.

In analogy with (9.1.15), it is straightforward to show that

$$w(A) = \frac{1}{|A|^3} \int_S \rho(Q) \, \mathbf{n}_Q \cdot A \, dS(Q) + O(|A|^{-3})$$

$$= O(|A|^{-2}) \quad \text{as } |A| \to \infty \qquad (9.1.18)$$

Exterior problems and the Kelvin transform

Define a transformation $\mathcal{T} : \mathbf{R}^3 \backslash \{0\} \to \mathbf{R}^3 \backslash \{0\}$ by

$$\mathcal{T}(x, y, z) = (\xi, \eta, \zeta) \equiv \frac{1}{r^2}(x, y, z), \quad r = \sqrt{x^2 + y^2 + z^2} \qquad (9.1.19)$$

In spherical coordinates,

$$T(r \cos \phi \sin \theta, r \sin \phi \sin \theta, r \cos \theta) = \frac{1}{r}(\cos \phi \sin \theta, \sin \phi \sin \theta, \cos \theta)$$

for $0 < r < \infty, 0 \le \phi \le 2\pi, 0 \le \theta \le \pi$. The inverse transformation is given by

$$T^{-1}(\xi, \eta, \zeta) = (x, y, z) \equiv \frac{1}{\rho^2}(\xi, \eta, \zeta), \qquad \rho = \sqrt{\xi^2 + \eta^2 + \zeta^2}$$

and $T^{-1} = T$. The point (ξ, η, ζ) is called the reflection of (x, y, z) through the unit sphere, and $r\rho = 1$.

Assume the bounded open region $D_i \equiv D$ contains the origin **0**. For a function $u \in C(\bar{D}_e), D_e \equiv D'$, define

$$\hat{u}(\xi, \eta, \zeta) = \frac{1}{\rho}u(x, y, z) \quad \text{with} \quad (x, y, z) = T(\xi, \eta, \zeta) \qquad (9.1.20)$$

and $\rho = \sqrt{\xi^2 + \eta^2 + \zeta^2}$. The function $\hat{u}$ is called the *Kelvin transformation* of u.

Introduce the region

$$\hat{D} = \{T(x, y, z) \mid (x, y, z) \in D_e\} \qquad (9.1.21)$$

and let $\hat{S}$ denote its boundary. By our assumptions on D, the region $\hat{D}$ is an open, bounded region containing the origin **0**. Then it can be shown that

$$\Delta \hat{u} = r^5 \Delta u, \qquad r = \sqrt{x^2 + y^2 + z^2}$$

In this,

$$\Delta u(x, y, z) = \frac{\partial^2 u}{\partial x^2} + \frac{\partial^2 u}{\partial y^2} + \frac{\partial^2 u}{\partial z^2}, \qquad \Delta \hat{u}(\xi, \eta, \zeta) = \frac{\partial^2 \hat{u}}{\partial \xi^2} + \frac{\partial^2 \hat{u}}{\partial \eta^2} + \frac{\partial^2 \hat{u}}{\partial \zeta^2}$$

Thus

$$\Delta u \equiv 0 \quad \text{on} \quad D_e \quad \Longleftrightarrow \quad \Delta \hat{u} \equiv 0 \quad \text{on} \quad \hat{D} \qquad (9.1.22)$$

This can be used to better understand the Dirichlet problem for Laplace's equation on the exterior region D_e.

The exterior Dirichlet problem. Find $u \in C(\bar{D}_e) \cap C^2(D_e)$ that satisfies

$$\Delta u(P) = 0, \qquad P \in D_e$$
$$u(P) = f(P), \qquad P \in S \qquad (9.1.23)$$
$$|u(P)| = O(|P|^{-1}), \qquad |P| \to \infty$$

with $f \in C(S)$ a given boundary function.

Using the above discussion on the Kelvin transform, the exterior Dirichlet problem is equivalent to the following interior Dirichlet problem:

$$\Delta \hat{u}(\xi, \eta, \zeta) = 0, \qquad\qquad\qquad (\xi, \eta, \zeta) \in \hat{D}$$

$$\hat{u}(\xi, \eta, \zeta) = \frac{1}{\sqrt{\xi^2 + \eta^2 + \zeta^2}} f(\mathcal{T}(\xi, \eta, \zeta)), \quad (\xi, \eta, \zeta) \in \hat{S} \qquad (9.1.24)$$

The condition on $u(P)$ as $|P| \to \infty$ can be used to show that $\hat{u}(\xi, \eta, \zeta)$ has a removable singularity at the origin, and Theorem 9.1.1 guarantees the unique solvability of this interior Dirichlet problem. Therefore, the exterior Dirichlet problem (9.1.23) also has a unique solution; moreover,

$$u(P) = \frac{\hat{u}(0)}{|P|} + (|P|^{-2}) \quad \text{as} \quad |P| \to \infty \qquad (9.1.25)$$

will be the value of $u(P)$ as $|P| \to \infty$. It would seem advantageous to use the Kelvin transformation to reformulate all exterior Dirichlet problems as interior problems on an equivalent domain, but this seems to be a seldom used approach to the numerical solution of such problems.

The exterior Neumann problem does not convert to an equivalent interior Neumann problem, unlike the planar case developed in §7.1.2.

The exterior Neumann problem. Find $u \in C^1(\bar{D}_e) \cap C^2(D_e)$ that satisfies

$$\Delta u(P) = 0, \qquad P \in D_e$$

$$\frac{\partial u(P)}{\partial \mathbf{n}_P} = f(P), \quad P \in S \qquad (9.1.26)$$

$$u(P) = O(|P|^{-1}), \qquad \frac{\partial u(P)}{\partial r} = O(|P|^{-2}) \qquad (9.1.27)$$

as $r = |P| \to \infty$, uniformly in $P/|P|$.

Note that there is no auxiliary condition on $f(P)$ needed for solvability, in contrast to the condition (9.1.5) needed for the interior problem. The above problem has a unique solution u for each $f \in C(S)$. For a proof, see Kress [325, p. 73] or Mikhlin [380, pp. 374–376].

Green's representation formula for exterior regions

Assume $u \in C^2(D_e) \cap C^1(\bar{D}_e)$, and assume $\Delta u \equiv 0$ on D_e. Further assume u satisfies the growth conditions of (9.1.27). Then

$$\int_S u(Q) \frac{\partial}{\partial \mathbf{n}_Q} \left[\frac{1}{|P - Q|} \right] dS(Q) - \int_S \frac{\partial u(Q)}{\partial \mathbf{n}_Q} \frac{dS(Q)}{|P - Q|}$$

$$= \begin{cases} 0, & P \in D_i \\ -[4\pi - \Omega(P)]u(P), & P \in S \\ -4\pi u(P), & P \in D_e \end{cases} \qquad (9.1.28)$$

This can be proven by using the earlier interior formula (9.1.6), applying it to the region between S and a large sphere of radius R. Then let $R \to \infty$ and use the growth conditions (9.1.27) to obtain (9.1.28).

We can combine the two representation formulas (9.1.6) and (9.1.28) to obtain another quite useful representation formula. Let $u^i \in C^2(D_i) \cap C^1(\bar{D}_i)$ and $u^e \in C^2(D_e) \cap C^1(\bar{D}_e)$, and further assume u^e satisfies the growth conditions of (9.1.27). Define

$$u^i(P) = \lim_{\substack{A \to P \\ A \in D_i}} u^i(A), \qquad u^e(P) = \lim_{\substack{A \to P \\ A \in D_e}} u^e(A), \quad P \in S$$

$$[u(P)] = u^i(P) - u^e(P), \quad P \in S$$

Similarly, define

$$\frac{\partial u^i(P)}{\partial \mathbf{n}_P} = \lim_{\substack{A \to P \\ A \in D_i}} \mathbf{n}_P \cdot \nabla u^i(A), \qquad \frac{\partial u^e(P)}{\partial \mathbf{n}_P} = \lim_{\substack{A \to P \\ A \in D_e}} \mathbf{n}_P \cdot \nabla u^e(A), \quad P \in S$$

$$\left[\frac{\partial u(P)}{\partial \mathbf{n}_P} \right] = \frac{\partial u^i(P)}{\partial \mathbf{n}_P} - \frac{\partial u^e(P)}{\partial \mathbf{n}_P}, \quad P \in S$$

Subtract (9.1.28) from (9.1.6), obtaining

$$\int_S [u(Q)] \frac{\partial}{\partial \mathbf{n}_Q} \left[\frac{1}{|P-Q|} \right] dS(Q) - \int_S \left[\frac{\partial u(Q)}{\partial \mathbf{n}_Q} \right] \frac{dS(Q)}{|P-Q|}$$

$$= \begin{cases} 4\pi u^i(P), & P \in D_i \\ \Omega(P)u^i(P) + [4\pi - \Omega(P)]u^e(P), & P \in S \\ 4\pi u^e(P), & P \in D_e \end{cases} \qquad (9.1.29)$$

This is used in obtaining *indirect boundary integral equation* methods.

9.1.2. *Direct boundary integral equations*

Using (9.1.6), the interior Dirichlet problem (9.1.3) can be solved by first solving the integral equation

$$\mathcal{S}\psi(P) \equiv \int_S \psi(Q) \frac{dS(Q)}{|P-Q|} = g(P), \quad P \in S \qquad (9.1.30)$$

in which

$$\psi(Q) = \frac{\partial u(Q)}{\partial \mathbf{n}_Q}$$

$$g(P) = \int_S f(Q) \frac{\partial}{\partial \mathbf{n}_Q} \left[\frac{1}{|P-Q|} \right] dS(Q) - \Omega(P)f(P)$$

The first kind integral equation (9.1.30) with single layer operator S is discussed in greater detail later in the section. For now, suffice it to say that when S is a smooth surface, this integral equation has a unique solution $\psi(Q)$ for any continuous function $f(P)$. After obtaining $\rho(Q)$, the solution $u(P)$ can be generated from (9.1.6):

$$u(P) = \frac{1}{4\pi} \int_S f(Q) \frac{\partial}{\partial \mathbf{n}_Q} \left[\frac{1}{|P - Q|} \right] dS(Q) - \frac{1}{4\pi} \int_S \psi(Q) \frac{dS(Q)}{|P - Q|},$$
$$P \in D$$

If we wish to solve the interior Neumann problem (9.1.4), then (9.1.6) leads to the need to solve

$$-\Omega(P)u(P) + \int_S u(Q) \frac{\partial}{\partial \mathbf{n}_Q} \left[\frac{1}{|P - Q|} \right] dS(Q)$$
$$= \int_S f(Q) \frac{dS(Q)}{|P - Q|}, \quad P \in S \tag{9.1.31}$$

The difficulty with using this equation is that it is not uniquely solvable. In particular, $u \equiv 1$ is a solution of the homogeneous form of the equation, and the null space for the homogeneous equation consists entirely of the constant functions. There are ways of modifying this equation to make it uniquely solvable, but it is easier to use the indirect method (9.1.36) introduced below.

For the exterior Dirichlet and Neumann problems the equation (9.1.28) also leads to integral equations of the first and second kind, respectively. For the Dirichlet problem this leads again to the first kind equation (9.1.30). For the Neumann problem we obtain the integral equation (9.1.31)

$$[4\pi - \Omega(P)]u(P) + \int_S u(Q) \frac{\partial}{\partial \mathbf{n}_Q} \left[\frac{1}{|P - Q|} \right] dS(Q)$$
$$= -\int_S \frac{\partial u(Q)}{\partial \mathbf{n}_Q} \frac{dS(Q)}{|P - Q|}, \quad P \in S \tag{9.1.32}$$

We usually write it in the form

$$2\pi u + \mathcal{K}u = g \tag{9.1.33}$$
$$g(P) = -\int_S \frac{\partial u(Q)}{\partial \mathbf{n}_Q} \frac{dS(Q)}{|P - Q|}$$
$$\mathcal{K}u(P) = \int_S u(Q) \frac{\partial}{\partial \mathbf{n}_Q} \left[\frac{1}{|P - Q|} \right] dS(Q)$$
$$+ [2\pi - \Omega(P)]u(P), \quad P \in S \tag{9.1.34}$$

The double layer operator $\mathcal{K}$ is important, and later in the section, it is discussed further. The equation (9.1.33) is uniquely solvable in most function spaces of interest, and it is a practical approach to solving the exterior Neumann problem for Laplace's equation.

9.1.3. Indirect boundary integral equations

Combining (9.1.29) with the type of arguments used in §7.1.4 of Chapter 7, we can justify representing harmonic functions as single or double layer potentials. For the interior Dirichlet problem, assume the solution u is a double layer potential,

$$u(A) = \int_S \rho(Q) \frac{\partial}{\partial \mathbf{n}_Q} \left[\frac{1}{|A - Q|} \right] dS(Q), \quad A \in D \tag{9.1.35}$$

Then using (9.1.11), the unknown function ρ must satisfy

$$2\pi\rho + \mathcal{K}\rho = f \tag{9.1.36}$$

with f the given Dirichlet boundary data and $\mathcal{K}$ the integral operator of (9.1.34). This is the same uniquely solvable integral equation as in (9.1.33), with a different right-hand side. This is among the most-studied of boundary integral equations in the literature of the past 150 years, and almost any book on partial differential equations contains a discussion of it.

For the interior Dirichlet problem we can also assume u can be represented as a single layer potential,

$$u(A) = \int_S \frac{\psi(Q)}{|A - Q|} dS(Q), \quad A \in D \tag{9.1.37}$$

In this case, the density function ψ must be chosen to satisfy the first kind integral equation

$$\mathcal{S}\psi = f \tag{9.1.38}$$

which is basically the same equation as in (9.1.30). The equation (9.1.36) is the classical BIE approach to solving the interior Dirichlet problem, but recently there has been a great deal of interest in (9.1.38). The operator $\mathcal{S}$ is symmetric, and that makes it easier to consider the solution of (9.1.38) by a Galerkin method. An important example of such is the paper of Nedelec [396], which we discuss in §9.4.

For the interior Neumann problem, consider representing its solution as a single layer potential, as in (9.1.37). Then imposing the boundary condition from (9.1.4), we have

$$\lim_{\substack{A \to P \\ A \in D}} \mathbf{n}_P \cdot \nabla \left[\int_S \frac{\psi(Q)}{|A - Q|} \, dS(Q) \right] = f(P)$$

for all $P \in S$ at which the normal $\mathbf{n}_P$ exists (which implies $\Omega(P) = 2\pi$). Using a limiting argument much the same as that used in obtaining (9.1.11), we obtain the second kind integral equation

$$-2\pi \psi(P) + \int_S \psi(Q) \frac{\partial}{\partial \mathbf{n}_P} \left[\frac{1}{|P - Q|} \right] dS(Q) = f(P), \quad P \in S^*$$
$$(9.1.39)$$

The set S^* is to contain all points $P \in S$ for which $\Omega(P) = 2\pi$. For S a smooth surface, $S^* = S$; otherwise, $S \backslash S^*$ is a set of measure zero.

As with the earlier equation (9.1.31) for the Neumann problem, this equation has the difficulty that it is not uniquely solvable, and moreover, it is solvable if and only if the Neumann data f satisfy the integral condition (9.1.5). The solutions can be shown to differ by a constant. For this particular equation, there is a simple fix to the lack of unique solvability. Consider the integral equation

$$-2\pi C(P) + \int_S C(Q) \frac{\partial}{\partial \mathbf{n}_P} \left[\frac{1}{|P - Q|} \right] dS(Q) + C(P^*) = f(P), \quad P \in S^*$$
$$(9.1.40)$$

with P^* some fixed point from S^*. It is shown in Ref. [29] that this equation is uniquely solvable, and in the case f satisfies (9.1.5), the solution C is one of the solutions of (9.1.38). The integral operator in both (9.1.39) and (9.1.40) is the adjoint $\mathcal{K}^*$ of that in (9.1.36). When S is smooth, $\mathcal{K}$ and $\mathcal{K}^*$ possess much the same mapping properties when considered as operators on common function spaces such as $C(S)$ and $L^2(S)$.

We can also develop similar integral equations for the exterior Dirichlet and Neumann problems, but we omit the details as they are quite similar to what has already been done. It is worth noting, however, that the solution of the exterior Dirichlet problem cannot be represented as a double layer potential in the manner of (9.1.36). From the asymptotic behavior (9.1.16) of such potentials, they cannot represent all possible solutions. For example, the harmonic function

$$u = \frac{1}{r}$$

cannot be represented as a double layer potential. The best way to solve the exterior Dirichlet problem using BIE is to either represent the solution as a single layer potential or to use the Kelvin transformation to convert the problem to an equivalent interior Dirichlet problem, as in (9.1.24).

9.1.4. Properties of the integral operators

The integral operators $\mathcal{K}$ and $\mathcal{S}$ have been much studied in the literature, especially for the case that S is a smooth surface. With S sufficiently smooth, both $\mathcal{K}$ and $\mathcal{S}$ are compact operators from $C(S)$ to $C(S)$, and from $L^2(S)$ to $L^2(S)$. For proofs, see Kress [325, Theorem 2.22] or Mikhlin [380, Chap. 7]. Consequently, the Fredholm alternative theorem (Theorem 1.3.1 of §1.3 in Chapter 1) can be used in developing a solvability theory for the second kind equation $2\pi\rho + \mathcal{K}\rho = f$ of (9.1.36). In fact, the entire theory of Fredholm integral equations was motivated by the need to understand the solvability of boundary integral equations of the second kind. This, in turn, was a means to give existence results for solving Laplace's equation and other elliptic partial differential equations. For such existence theorems, see Refs. [325, §6.4] or [380, Chap. 18]; for a history of the early work on such integral equations in the first decades of this century, see Bernkopf [78].

The function spaces $L^2(S)$ and $C(S)$ are the most widely used function spaces, but we need to also introduce the Sobolev spaces $H^r(S)$. Assuming S is smooth and that D is a bounded simply connected region in space, there are several equivalent ways to define $H^r(S)$. For r a positive integer, consider the space $C^r(S)$ of all r-times continuously differentiable functions on S. Introduce the inner product norm

$$\|g\|_r = \left[\sum_{k=0}^{r} \sum_{|j|=k} \|D^j g\|_{L^2}^2 \right]^{\frac{1}{2}}, \quad g \in C^r(S) \tag{9.1.41}$$

where $\|\cdot\|_{L^2}$ denotes the standard norm for $L^2(S)$. In this definition the index j is to be a multiindex, so that $D^j g$ ranges over a maximal independent set of possible derivatives of g of order k. The space $H^r(S)$ is defined to be the completion of $C^r(S)$ with respect to this norm. (Strictly speaking, we should divide S into "patches," say $S = S_1 \cup \cdots \cup S_J$, each of which is the image of a parametrization domain R_j in the plane. Apply the above definition (9.1.41) over each such domain R_j, and then combine these norms over the R_js into a new norm, for example, by using their maximum.) For r not an integer, the definitions are a bit more complicated, and we refer the reader to Grisvard [240, p. 17].

Another approach, used more seldom but possibly more intuitive, is based on the Laplace expansion of a function g defined on the unit sphere U. Recalling (5.5.207) of §5.5 in Chapter 5, the Laplace expansion of g is

$$g(Q) = \sum_{n=0}^{\infty} \sum_{k=1}^{2n+1} (g, S_n^k) S_n^k(Q), \quad g \in L^2(U) \tag{9.1.42}$$

The function $S_n^k(Q)$ is a spherical harmonic of degree k, as defined in (5.5.203)–(5.5.205) of §5.5. The function $g \in L^2(U)$ if and only if

$$\sum_{n=0}^{\infty} \sum_{k=1}^{2n+1} |(g, S_n^k)|^2 < \infty \tag{9.1.43}$$

We define the Sobolev space $H^r(U)$ to be the set of functions whose Laplace series (9.1.42) satisfies

$$\|g\|_{*,r} \equiv \left[\sum_{n=0}^{\infty} (2n+1)^{2r} \sum_{k=1}^{2n+1} |(g, S_n^k)|^2 \right]^{\frac{1}{2}} < \infty \tag{9.1.44}$$

and this definition is satisfactory for any real number $r \geq 0$. For r a positive integer, the norm $\|g\|_{*,r}$ can be shown to be equivalent to the norm $\|g\|_r$ of (9.1.41).

Assume there is a mapping between our more general surface S and the unit sphere U:

$$M : U \xrightarrow[\text{onto}]{1-1} S \tag{9.1.45}$$

with M having the smoothness of S. Recall the mapping $\mathcal{M} : L^2(S) \xrightarrow[\text{onto}]{1-1} L^2(U)$ of (5.5.195)–(5.5.197) in §5.5, namely $g \mapsto \hat{g} \equiv \mathcal{M}g$, with

$$\hat{g}(P) = g(M(P)), \quad P \in U \tag{9.1.46}$$

Define the Sobolev space $H^r(S)$ to be the set of functions $g \in L^2(S)$ for which $\mathcal{M}g \in H^r(U)$. The Sobolev space $H^r(S)$ with $r < 0$ is defined to be the dual space of $H^{-r}(S)$. It can be shown that the spaces $H^r(U)$ with $r < 0$ can also be identified with the set of all Laplace expansions (9.1.42) for which (9.1.44) is satisfied. All such expansions are sometimes called *generalized functions* or *distributions*. A further discussion of these results is given in Mikhlin and Prößdorf [381, §10.6]. This way of defining $H^r(S)$ generalizes the definition of Sobolev spaces for planar curves as given in §7.4 of Chapter 7.

To motivate the properties of the single and double layer operators S and $\mathcal{K}$ on a general smooth surface S, we first look at their properties when $S = U$. By direct evaluation in this case,

$$S = 2\mathcal{K} \tag{9.1.47}$$

Also, it can be shown from results in MacRobert [360] that

$$\mathcal{K} S_n^k = \frac{2\pi}{2n+1} S_n^k, \quad k = 1, \ldots, 2n+1, \quad n = 0, 1, \ldots \tag{9.1.48}$$

From this, the mapping properties of $\mathcal{K}$ and S can be obtained very precisely. In particular, let $\rho \in H^r(U)$, and write it in terms of its Laplace series,

$$\rho = \sum_{n=0}^{\infty} \sum_{k=1}^{2n+1} \left(\rho, S_n^k\right) S_n^k \tag{9.1.49}$$

Then

$$S\rho = 4\pi \sum_{n=0}^{\infty} \frac{1}{2n+1} \sum_{k=1}^{2n+1} \left(\rho, S_n^k\right) S_n^k \tag{9.1.50}$$

From the definition of $H^r(U)$ involving (9.1.44), this implies immediately that

$$S : H^r(U) \xrightarrow[onto]{1-1} H^{r+1}(U), \quad r \geq 0 \tag{9.1.51}$$

In addition, the formula (9.1.50) can be used to extend the definition of S to negative order Sobolev spaces, whereas the integral formula (9.1.30) no longer makes sense with any standard definition of the integral.

These results also allow us to solve exactly the single layer equation $S\psi = f$ of (9.1.38) and the double layer equation $2\pi\rho + \mathcal{K}\rho = f$ of (9.1.36) when $S = U$. Assume the solution in the form of a Laplace series (9.1.42), expand f in a Laplace series, substitute these series into the equation of interest, apply (9.1.48), match coefficients of corresponding spherical harmonics S_n^k, and solve for the Laplace series coefficients of ρ or ψ. For the single layer equation $S\psi = f$, we obtain

$$\psi = \frac{1}{2\pi} \sum_{n=0}^{\infty} (2n+1) \sum_{k=1}^{2n+1} \left(f, S_n^k\right) S_n^k, \quad f \in H^r(U) \tag{9.1.52}$$

This can be used with any real number r, but for $r < 1$, the solution must be interpreted as a distribution or generalized function.

From (9.1.47)–(9.1.51), both the operators $\mathcal{K}$ and $\mathcal{S}$ are smoothing operators, with (9.1.51) giving the precise amount of smoothing when $S = U$. This result generalizes to general surfaces S that are the boundary of a bounded simply connected region. As before, it is possible to extend uniquely the definition of $\mathcal{K}$ and $\mathcal{S}$ to the larger spaces $H^r(S)$ with $r < 0$, even though the integral formulas no longer apply. Finally, it can be proven that

$$\mathcal{K}, \; \mathcal{S} : H^r(S) \xrightarrow[\text{onto}]{1-1} H^{r+1}(S), \quad -\infty < r < \infty \qquad (9.1.53)$$

just as for $S = U$. See Dautray and Lions [159, pp. 124, 128] and Nedelec and Planchard [397].

Consider $\mathcal{S}\psi = f$ for a boundary S as specified in the preceding paragraph. Then for the equation to be solvable with $\psi \in L^2(S)$, it is necessary and sufficient that $f \in H^1(S)$. Moreover, if $f \in H^r(S)$, $r \geq 1$, then $\psi \in H^{r-1}(S)$. Consider now the second kind equation $2\pi\rho + \mathcal{K}\rho = f$. With (9.1.53), we can show that for a solution $\rho \in L^2(S)$ to exist, it is necessary and sufficient that $f \in L^2(S)$. Moreover, if $f \in H^r(S)$, $r \geq 0$, then $\rho \in H^r(S)$. There are additional regularity results for this second kind equation, based on looking at spaces of Hölder continuous functions on S. For these and other results on single and double layer integral operators, see Günter [246] and Hackbusch [249, Chap. 8].

9.1.5. *Properties of $\mathcal{K}$ and $\mathcal{S}$ when S is only piecewise smooth*

The topic of studying the mapping properties of $\mathcal{K}$ and $\mathcal{S}$ when S is only piecewise smooth is currently an active area of research, and there is much that is still not understood. The first significant work in this area was given by Wendland [557] in 1968. In this paper he assumed the surface satisfied a number of properties. The major one pertained to the tangent cones formed at all points of the surface S. Recall the definitions (9.1.7) and (9.1.8) of the interior tangent cone $\Gamma_i(P)$ and the exterior tangent cone $\Gamma_e(P)$, respectively.

Assumption Ξ. For each $P \in S$, either $\Gamma_i(P)$ or $\Gamma_e(P)$ is convex.

With this and other assumptions on S, Wendland examined the mapping properties of $\mathcal{K}$, and he also proposed and analyzed a piecewise constant collocation method for solving $2\pi\rho + \mathcal{K}\rho = f$. In particular, he showed

$$\mathcal{K} : C(S) \rightarrow C(S) \qquad (9.1.54)$$

is a bounded operator, and this generalizes to most other piecewise smooth surfaces S. The principal benefit of Ξ is that the kernel function $K(P, Q)$ will be of one sign with respect to Q varying over S, for each $P \in S$. This allows the easy calculation of the operator norm of $\mathcal{K}$ on the space $C(S)$, and it leads to invertibility results for $2\pi + \mathcal{K}$ when the surface is restricted to a neighborhood of an edge or corner of S. More details are given later. As an example of a region D that does not satisfy Ξ, consider

$$D = ((0, 1) \times (0, 2) \times (0, 2)) \backslash ([0, 1] \times [1, 2] \times [1, 2])$$

At points $P = (0, 1, 1)$ and $(1, 1, 1)$ on the boundary, the assumption Ξ is not satisfied. The 1968 paper of Wendland was generalized in Kral and Wendland [317] to handle this case, although there are many remaining polyhedral surfaces that do not satisfy Ξ or its generalization. This type of assumption remains a restrictive requirement when analyzing numerical methods for solving BIE on S.

In Chapter 8 we used a local wedge operator to isolate the essential part of the operator $\mathcal{K}$, with the remaining part a compact perturbation of the wedge operator. This can also be done here, with the wedge operator replaced by an operator over the tangent cone associated with all points at which $\Omega(P) \neq 2\pi$. However, the full theory of the double layer operator on such conical surfaces is not yet fully understood. Some of the major results can be found in the papers of Elschner [183] and Rathsfeld [447]–[449], [451]. Rather than give a complete development of their results, we will cite them as needed.

To discuss the differentiability of the solutions to the single layer equation $\mathcal{S}\psi = f$ of (9.1.38) and the double layer equation $2\pi\rho + \mathcal{K}\rho = f$ of (9.1.36), there are two current approaches. First, the solutions of these equations are related to the solutions of certain boundary value problems for Laplace's equation. Results on the regularity of the solutions of such boundary value problems then yield corresponding results when solving the boundary integral equation. Second, the BIE is studied directly, and regularity results are obtained from the mapping properties of the operators, as was done for smooth surfaces in (9.1.53) and the paragraph following it.

Considering the first approach in greater detail, direct BIE methods involve explicitly the solution of a boundary value problem. With indirect methods, in contrast, the solution is generally the difference of the boundary values of the solutions of exterior and interior boundary value problems. As an example, the double layer representation (9.1.35) of the interior Dirichlet problem can be obtained from the formula (9.1.29) as follows. Let $u^i \in C^2(S) \cap C^1(\bar{S})$ be the solution of the interior Dirichlet problem (9.1.3). Then solve the exterior Neumann problem with the Neumann data generated from the normal derivative of u^i. Referring to (9.1.29), this leads to

$$u(A) = \frac{1}{4\pi} \int_S [u(Q)] \frac{\partial}{\partial \mathbf{n}_Q} \left[\frac{1}{|P - Q|} \right] dS(Q) \qquad (9.1.55)$$

with $[u(Q)] = u^i(Q) - u^e(Q)$. Identifying with (9.1.35), we have that the solution ρ of $2\pi\rho + \mathcal{K}\rho = f$ is also equal to $[u(Q)]$. With results on the solutions of the interior and exterior problems, one then can obtain results on the behavior of $\rho(P)$ for P near an edge or corner of S. Such results for Laplace's equation on polyhedral regions are given by Petersdorff and Stephan in [418] and Grisvard [240, §8.2], and applications to the regularity of solutions of the boundary integral equation are given in Petersdorff and Stephan [419] and Stephan [531]. We do not give these regularity results, as they require a more sophisticated use of Sobolev spaces than we wish to introduce here, but we note the importance of such regularity results in determining how to grade the mesh in the approximation of whatever boundary integral equation is under consideration.

The second approach to giving regularity results for the solutions of boundary integral equations is to look directly to the mapping properties of the boundary integral operators. This approach has been used with success by Elschner [184] for the case of the double layer operator $\mathcal{K}$ when looking at the solutions of $2\pi\rho + \mathcal{K}\rho = f$ on polyhedra in space. As was done for planar problems, the operator $\mathcal{K}$ is first considered for the special case of the boundary $\Gamma \equiv \Gamma(P)$ of the interior tangent cones $\Gamma_i(P)$ constructed at edge and corner points $P \in S$. Such a cone Γ can either be finite in extent, in analogy with our finite wedge in Chapter 8, or it can be infinite in extent. In the former case, we can regard the associated double layer operator $\mathcal{W}$ as being defined on the space $C(\Gamma)$, and in the latter, $\mathcal{W}$ can be regarded as being defined on $C_0(\Gamma)$, the continuous functions with a uniform limit of 0 at ∞. Elschner takes Γ to be the infinite cone, and he introduces a variety of weighted Sobolev spaces on which he analyzes the invertibility of $2\pi + \mathcal{W}$.

The central arguments of Elschner [184] depend on interpreting $\mathcal{W}$ as a Mellin transform (cf. [8.3.72] in §8.3) on appropriate function spaces. Assume Γ has its vertex at the origin, and let the intersection of Γ with the unit sphere U be noted by γ. The curve γ is composed of great circle arcs. Parameterize γ by $\omega(s), 0 \leq s \leq L$, and let

$$0 = s_0 < s_1 < \cdots < s_J = L$$

correspond to the corner points of γ, namely $\omega(s_1), \ldots, \omega(s_J) = \omega(s_0)$. Let the open planar faces of Γ be denoted by $F_1, \ldots, F_J$, with $\bar{F}_j \cap U = \gamma_j$, the arc connecting $\omega(s_{j-1})$ and $\omega(s_j)$. Points P on Γ are parameterized by using

$$P = r\omega(s), \quad r = |P|, \quad \omega(s) = \frac{1}{r}P$$

For $g \in C_0(\Gamma)$, the integral formula for $\mathcal{W}g$ can be rewritten as follows:

$$\mathcal{W}g(r\omega(s)) = \int_0^\infty \int_\gamma g(\tau\omega(\sigma))\frac{r\tau\mathbf{n}_{\omega(\sigma)} \cdot \omega(s)}{|r\omega(s) - \tau\omega(\sigma)|^3}\, d\sigma\, d\tau,$$

$$0 \le r < \infty, \quad 0 \le s \le L$$

in which the original integration variable Q has been replaced by $\tau\omega(\sigma)$. The quantity $\mathbf{n}_{\omega(\sigma)}$ denotes a unit normal to γ at $\omega(\sigma)$, taken in the direction interior to γ and tangent to the sphere U. This integral can be rewritten as

$$\mathcal{W}g(r\omega(s)) = \int_0^\infty \mathcal{B}\left(\frac{r}{\tau}\right)g(\tau\omega(s))\frac{d\tau}{\tau}. \tag{9.1.56}$$

with $\mathcal{B}(t)$ an operator on $C(\gamma)$:

$$\mathcal{B}(t)h(\omega(s)) = \int_\gamma h(\omega(\sigma))\frac{t\mathbf{n}_{\omega(\sigma)} \cdot \omega(s)}{|t\omega(s) - \omega(\sigma)|^3}\, d\sigma, \quad 0 \le t < \infty \tag{9.1.57}$$

for $h \in C(\gamma)$. The operator $\mathcal{W}$ in (9.1.56) can now be viewed as a Mellin convolution operator, and its mapping properties can be explored using tools from the general theory for such operators, which is the direction taken in Ref. [184].

We now describe one of the regularity results of Ref. [184]. Choose a function $\varphi \in C_p[0, L]$ that satisfies the following:

1. When restricted to each subinterval $[s_{j-1}, s_j]$, $\varphi \in C^\infty$;
2. φ vanishes only at the points $s_0, s_1, \ldots, s_J$, corresponding to the corner points of γ;
3. $\varphi(s)$ coincides with $|s - s_j|$ in some small neighborhood of s_j, for each point $j = 0, 1, \ldots, J$.

It is straightforward to construct such a function, but we merely assert its existence. Introduce $C_{pw}^\infty(\Gamma)$, the set of all continuous functions that have compact support on Γ and are C^∞ on each closed face $\bar{F}_j$. For any real number β, $0 < \beta < \frac{1}{2}$, and any integer $k \ge 0$, we introduce the weighted Sobolev space $Y_\beta^k(\Gamma)$. For all $u \in C_{pw}^\infty(\Gamma)$, introduce the norm

$$\|u\|_{k,\beta} = \sum_{0 \le i+l \le k} \sum_{1 \le j \le J} \left\| r^{l-\beta}\left(\frac{\partial}{\partial r}\right)^l \varphi^{i-\beta}\left(\frac{\partial}{\partial s}\right)^i u \right\|_{L^2(F_j)} \tag{9.1.58}$$

The space $Y_\beta^k(\Gamma)$ is the completion of $C_{pw}^\infty(\Gamma)$ with respect to this norm. Functions in $Y_\beta^k(\Gamma)$ permit a certain type of singular behavior in the neighborhood of both $\mathbf{0}$ and of each edge of the cone, of a character typified by r^β and $|s - s_j|^\beta$.

The regularity of solutions of the double layer equation $(2\pi + \mathcal{W})\rho = f$ is given by the following theorem.

Theorem 9.1.2. *Assume the framework of the preceding few paragraphs. Then*

$$2\pi + \mathcal{W} : Y_\beta^k(\Gamma) \xrightarrow[onto]{1-1} Y_\beta^k(\Gamma) \qquad (9.1.59)$$

Proof. See Ref. [184, Theorem 3.1]. More general regularity results are also proven in that paper.   □

The parameter β is related to the size of angles at the corners of γ, and the choice $0 < \beta < \frac{1}{2}$ works for all such angles. With planar problems, the analogous type of result is embodied in (8.1.18) of §8.1.2. In that case, the choice $\beta = \frac{1}{2}$ would give a result uniform over all possible angles. For an angle-dependent result in the present case, see Ref. [184, §2]. The result (9.1.59) says, in particular, that if the Dirichlet data $f \in C_{pw}^\infty(\Gamma)$, then the double layer density function $\rho \in Y_\beta^k(\Gamma)$, for all $k \geq 0$ and all $0 < \beta < \frac{1}{2}$. This theorem can be generalized to handle the boundary S of a closed polyhedron, and it is a relatively straightforward application of the above theorem. It amounts to extending the above result to each of the corners of S. See Ref. [184, §4].

9.2. Boundary element collocation methods on smooth surfaces

In this section we consider the numerical solution of the second kind double layer equation $2\pi\rho + \mathcal{K}\rho = f$, from (9.1.33) or (9.1.36). The methods of discretization can be used in approximating a wide variety of BIE, but the convergence theory presented here will generalize to only second kind BIE. We also restrict the methods of discretization to piecewise polynomial forms of approximating ρ, deferring to §9.5 the use of spherical polynomials. Methods of discretization for BIE that break the surface into small elements (as in this section) are often called boundary element methods or BEM, but we do not make much use of this notation in this work.

The smoothness of S implies that $\mathcal{K}$ is a compact operator both from $C(S)$ to $C(S)$ and from $L^2(S)$ to $L^2(S)$, and this was discussed above in §9.1.4. The compactness of $\mathcal{K}$ implies that the full body of theory developed in Chapters 3 and 5 for projection methods can be brought to bear on the numerical solution of $2\pi\rho + \mathcal{K}\rho = f$. Applying a collocation method leads to a linear system in which all elements are surface integrals over S, and a Galerkin method leads to a linear system in which all elements are double surface integrals over S. Because the Galerkin systems are far more expensive to set up, most BIE on surfaces are solved using collocation methods, and that is the point of view in both this section and the following one. A discussion of Galerkin methods is given in §§9.4 and 9.5, where they are especially useful in solving boundary integral equations of the first kind. With BIE of the first kind for three-dimensional problems, we know essentially nothing about the stability and convergence of collocation methods, whereas there is a well-developed theory for Galerkin methods for such problems.

We use the notation of Chapter 5, especially that of §§5.4.2 and 5.4.3. The collocation and Galerkin methods for solving $2\pi\rho + \mathcal{K}\rho = f$ are written abstractly as

$$(2\pi + \mathcal{P}_n\mathcal{K})\rho_n = \mathcal{P}_n f \tag{9.2.60}$$

with $\mathcal{P}_n$ an interpolatory projection or orthogonal projection, respectively. Since $\mathcal{K}$ is compact, complete error analyses for collocation and Galerkin methods are given in Theorems 5.4.3 and 5.4.5, respectively. For bounds on the rate of convergence for these methods, see (5.4.167) and (5.4.187), respectively.

In the remainder of this section we consider the particular case of collocation with $\mathcal{P}_n g$ defined using piecewise quadratic interpolation. This is done in part to make the presentation a bit more intuitive, and in part because this is an important practical case that is generally much superior to the use of piecewise linear interpolation. The results generalize to other degrees of polynomial interpolation, but we omit these generalizations.

The piecewise quadratic interpolation operator $\mathcal{P}_n g$ is based on the interpolation formula (5.1.18) of §5.1 over the unit simplex σ:

$$G(s, t) \approx \sum_{j=1}^{6} G(q_j)\ell_j(s, t), \quad (s, t) \in \sigma \tag{9.2.61}$$

with the interpolation nodes $\{q_j\}$ as shown in Figure 5.1 and the basis functions $\{\ell_j\}$ given in (5.1.17). Let $\mathcal{T}_n = \{\Delta_k\}$ be a triangulation of S, and assume $\mathcal{T}_n$ satisfies **T1–T3** given near the beginning of §5.3, making it a conforming triangulation. The formula (9.2.61) is converted to a definition over a triangulation of S by means of (5.3.102):

$$(\mathcal{P}_n g)(m_k(s, t)) = \sum_{j=1}^{6} g(m_k(q_j))\ell_j(s, t), \quad (s, t) \in \sigma, \quad k = 1, \ldots, n \tag{9.2.62}$$

The parametrization mapping $m_k : \sigma \to \Delta_k$ is defined in (5.3.99). With this definition, $\mathcal{P}_n$ maps $C(S)$ to $C(S)$, and the approximating subspace $\mathcal{X}_n$ is defined to be the range of $\mathcal{P}_n$ when applied to $C(S)$. Later in the chapter we consider definitions of $\mathcal{P}_n$ in which $\mathcal{P}_n g$ is only piecewise continuous. Also, recall that the collocation nodes of the triangulation are denoted by

$$\mathcal{V}_n = \{v_i \mid i = 1, \ldots, n_v\}$$
$$= \{v_{k,j} \mid j = 1, \ldots, 6, \ k = 1, \ldots, n\}$$

with $v_{k,j} = m_k(q_j)$, $j = 1, \ldots, 6$, the six collocation nodes in Δ_k, corresponding to the six nodes $\{q_j\}$ pictured in Figure 5.1.

We apply Theorem 5.4.3 from §5.4.2 in the present context for $\mathcal{P}_n$ defined using (9.2.62).

Theorem 9.2.1. *Let S be a smooth boundary, $S \in C^4$, and assume S is parameterized as in (5.3.96)–(5.3.97). Let $\mathcal{K}$ be the double layer integral operator of (9.1.34), which is compact on $C(S)$ to $C(S)$. Let $\mathcal{T}_n = \{\Delta_k\}$ be a conforming triangulation of S, as described above, and assume the triangulations $\mathcal{T}_n$ are such that $\hat{\delta}_n \to 0$ as $n \to \infty$ (cf. Theorem 5.3.1 for the definition of $\hat{\delta}_n$). Define the quadratic interpolatory projection operator $\mathcal{P}_n$ using (9.2.62), and consider the approximate solution of $2\pi\rho + \mathcal{K}\rho = f$ by means of the collocation approximation (9.2.60). Then:*

(a) *The inverse operators $(2\pi + \mathcal{P}_n\mathcal{K})^{-1}$ exist and are uniformly bounded for all sufficiently large n, say $n \geq N$.*
(b) *The approximation ρ_n has the error*

$$\rho - \rho_n = 2\pi(2\pi + \mathcal{P}_n\mathcal{K})^{-1}(I - \mathcal{P}_n)\rho \qquad (9.2.63)$$

and thus $\rho_n \to \rho$ as $n \to \infty$.
(c) *Assume $\rho \in C^3(S)$. Then*

$$\|\rho - \rho_n\|_\infty \leq c\hat{\delta}_n^3, \quad n \geq N \qquad (9.2.64)$$

Proof. This is a straightforward application of Theorem 5.4.3, together with the known unique solvability of $2\pi\rho + \mathcal{K}\rho = f$ and the compactness of $\mathcal{K}$. ☐

A generalization of this theorem to more general BIE of the second kind is given in Wendland [561, §9.4].

Theorem 5.4.3 in §5.4.2 also contained a result on superconvergence of ρ_n at the collocation node points $\{v_i\}$ of the triangulation $\mathcal{T}_n$. The proof given there depended crucially on the differentiability of the kernel function, whereas the kernel of the double layer operator $\mathcal{K}$ is singular. Nonetheless, it is possible to prove a superconvergence result.

Theorem 9.2.2. *Assume the hypotheses of the preceding theorem, with $S \in C^5$ and $\rho \in C^4(S)$. Further assume the refinement process for the triangulations T_n is such that they are symmetric triangulations. Then for sufficiently large n, say $n \geq N$,*

$$\max_{1 \leq i \leq n_v} |\rho(v_i) - \rho_n(v_i)| \leq c\hat{\delta}_n^4 \log \hat{\delta}_n^{-1}, \quad n \geq N \qquad (9.2.65)$$

Proof. A complete derivation is given in Atkinson and Chien [59, Theorem 3.1] (although there are some minor but correctable errors in that derivation). Below, we sketch that proof. For the definition of symmetric triangulation, see the discussion preceding and following (5.1.52) in §5.1.2 of Chapter 5.

(a) Recall the iterated collocation solution $\hat{\rho}_n$ of (5.4.164) in §5.4.2. It satisfies

$$\hat{\rho}_n = \frac{1}{2\pi}(f - \mathcal{K}\rho_n)$$
$$(2\pi + \mathcal{K}\mathcal{P}_n)\hat{\rho}_n = f \qquad (9.2.66)$$
$$\rho_n = \mathcal{P}_n\hat{\rho}_n$$

or equivalently,

$$\rho_n(v_i) = \hat{\rho}_n(v_i), \quad i = 1, \ldots, n_v$$

By simple manipulation, the error $\hat{\rho}_n$ satisfies

$$(2\pi + \mathcal{K}\mathcal{P}_n)(\rho - \hat{\rho}_n) = -\mathcal{K}(I - \mathcal{P}_n)\rho \qquad (9.2.67)$$

The linear system associated with this is

$$(2\pi + K_n)e_n = -\epsilon_n \qquad (9.2.68)$$

with

$$e_{n,i} = \rho(v_i) - \hat{\rho}_n(v_i) = \rho(v_i) - \rho_n(v_i)$$
$$\epsilon_{n,i} = \mathcal{K}(I - \mathcal{P}_n)\rho(v_i), \quad i = 1, \ldots, n_v$$

From §3.4.5 in Chapter 3, note that the matrix of coefficients for (9.2.67) is the same as that for the linear system associated with the collocation equation (9.2.60).

From (3.4.84) in §3.4 of Chapter 3, we have

$$(2\pi + \mathcal{K}\mathcal{P}_n)^{-1} = \frac{1}{2\pi}[I - \mathcal{K}(2\pi + \mathcal{P}_n\mathcal{K})^{-1}\mathcal{P}_n] \qquad (9.2.69)$$

The uniform boundedness of $(2\pi + \mathcal{P}_n\mathcal{K})^{-1}$, for all sufficiently large n, then implies the same of $(2\pi + \mathcal{K}\mathcal{P}_n)^{-1}$. The iterated collocation equation (9.2.66) can be interpreted as a Nyström method defined using product integration. From the generalization to product integration of (4.1.53) in §4.1 of Chapter 4, we have

$$\|(2\pi + K_n)^{-1}\| \le \|(2\pi + \mathcal{K}\mathcal{P}_n)^{-1}\|$$

Combining these results, we have

$$c_B \equiv \sup_{n \ge N} \|(2\pi + K_n)^{-1}\| < \infty \qquad (9.2.70)$$

Returning to (9.2.68), this proves

$$\|e_n\|_\infty \le c_B \|\epsilon_n\|_\infty, \quad n \ge N \qquad (9.2.71)$$

Thus to bound the maximum of the errors $\rho(v_i) - \rho_n(v_i)$, we can instead look at the maximum of the errors $\mathcal{K}(I - \mathcal{P}_n)\rho(v_i)$.

(b) Using the parametrization $m_k : \sigma \to \Delta_k$, the unit normal to S at $Q = m_k(s, t)$ is given by

$$\mathbf{n}_Q \equiv \mathbf{n}(s, t) = \pm \frac{D_s m_k \times D_t m_k}{|D_s m_k \times D_t m_k|}$$

with the sign so chosen that $\mathbf{n}_Q$ is the inner normal at Q. For simplicity we take this sign to be negative. The compact operator $\mathcal{K}$ can now be written as

$$\mathcal{K}\rho(P) = \sum_{k=1}^{n} \int_\sigma \rho(m_k(s, t)) \frac{(D_s m_k \times D_t m_k) \cdot [m_k(s, t) - P]}{|m_k(s, t) - P|^3} \, d\sigma$$

For the errors ϵ_i, we examine

$$\mathcal{K}(I - \mathcal{P}_n)\rho(P) = \sum_{k=1}^{n} \int_\sigma [\rho(m_k(s, t)) - (\mathcal{P}_n\rho)(m_k(s, t)]$$

$$\times \frac{(D_s m_k \times D_t m_k) \cdot [m_k(s, t) - P]}{|m_k(s, t) - P|^3} \, d\sigma \qquad (9.2.72)$$

with P either a vertex or a midpoint of some triangular element Δ_ℓ.

For the moment, assume P belongs to some Δ_ℓ, and for simplicity assume $P = m_\ell(0, 0)$. Then for the integral in (9.2.72) corresponding to integration over Δ_ℓ, the singularity in the integrand occurs at $(s, t) = (0, 0)$. It is fairly straightforward to show that such an integral is of size $O(\hat{\delta}_n^4)$, with $\hat{\delta}_n$ the mesh size of the triangulation in the parametrization domain. For a given vertex P, there is a finite number of such triangular elements Δ_ℓ, and therefore their sum is also $O(\hat{\delta}_n^4)$.

As in the last paragraph, let $P = m_\ell(0, 0)$ be a vertex of one or more elements Δ_ℓ, and consider the remaining integrals of (9.2.72),

$$\int_\sigma [\rho(m_k(s, t)) - (\mathcal{P}_n\rho)(m_k(s, t)] \frac{(D_s m_k \times D_t m_k) \cdot [m_k(s, t) - P]}{|m_k(s, t) - P|^3} d\sigma$$
(9.2.73)

for which the integrand is nonsingular (meaning that P does not belong to the associated element Δ_k). We can show that the sum of these integrals is $O(\hat{\delta}_n^4 \log \hat{\delta}_n^{-1})$, assuming the method of refinement of the triangulations $\mathcal{T}_n$ produces symmetric triangulations. The proof uses expansions of various parts of the integrand about $(s, t) = (0, 0)$. This is done in much the same manner as in §5.2.1 to show superconvergence of the iterated collocation method.

For the integral in (9.2.72) that corresponds to integration over Δ_k, introduce

$$G_k(s, t) = \rho(m_k(s, t))$$

and expand it in a Taylor formula about $(s, t) = (0, 0)$:

$$G_k(s, t) = H_k(s, t) + J_k(s, t) + L_k(s, t)$$

with $H_k(s, t)$ of degree 2 in (s, t),

$$J_k(s, t) = \left(s\frac{\partial}{\partial \xi} + t\frac{\partial}{\partial \eta} \right)^3 G_k(\xi, \eta) \Big|_{\substack{\xi=0 \\ \eta=0}}$$

$$L_k(s, t) = \frac{1}{6} \int_0^1 (1 - v)^3 \frac{d^4 G_k(vs, vt)}{dv^4} dv$$

Then for the polynomial interpolation error portion of the integrand of (9.2.73), we have

$$\rho(m_k(s, t)) - (\mathcal{P}_n\rho)(m_k(s, t) = E_k(s, t; J_k) + E_k(s, t; L_k)$$
(9.2.74)

$$E_k(s, t; J_k) = J_k(s, t) - \sum_{j=1}^6 J_k(s_j, t_j)\ell_j(s, t)$$

$$E_k(s, t; L_k) = L_k(s, t) - \sum_{j=1}^6 L_k(s_j, t_j)\ell_j(s, t)$$

with $q_j \equiv (s_j, t_j)$.

Recall that $S = S_1 \cup \cdots \cup S_J$, as in (9.1.1). Assume the element $\Delta_k \subset S_l$ for some l, and associated with S_l there is a C^5 function $F_l : R_l \rightarrow S_l$ for some polygonal planar region R_l. The parametrization function m_k is defined by

$$m_k(s, t) = F_l(\hat{v}_{k,1} + t(\hat{v}_{k,2} - \hat{v}_{k,1}) + s(\hat{v}_{k,3} - \hat{v}_{k,1})) \tag{9.2.75}$$

with $\{\hat{v}_{k,1}, \hat{v}_{k,2}, \hat{v}_{k,3}\}$ the vertices of $\hat{\Delta}_k$, and $\hat{\Delta}_k \subset R_l$ the preimage of Δ_k under F_l. In analogy with the proof of Theorem 5.3.1 of §5.3.1,

$$\max_k \|E_k(\cdot; J_k)\|_\infty = O(\hat{\delta}_n^3) \tag{9.2.76}$$

$$\max_k \|E_k(\cdot; L_k)\|_\infty = O(\hat{\delta}_n^4) \tag{9.2.77}$$

The error function involving J_k has an additional special property when the triangulations are symmetric, as summarized in (5.2.74). This is used later in the proof. From (9.2.75), note also that

$$\|D_s m_k\|_\infty, \|D_t m_k\|_\infty = O(\hat{\delta}_n) \tag{9.2.78}$$

Denote the remaining part of the integrand in (9.2.73) by

$$\kappa_k(s, t) \equiv \frac{(D_s m_k \times D_t m_k) \cdot [m_k(s, t) - P]}{|m_k(s, t) - P|^3} \tag{9.2.79}$$

Also introduce

$$d_k = \text{distance}(P, \Delta_k).$$

By the uniformity of the refinement scheme for symmetric triangulations, we can show there are constants $\alpha > 0$ for which

$$\alpha \hat{\delta}_n \leq d_k \tag{9.2.80}$$

with this uniform in n, k, and for any collocation point P.

For the numerator of (9.2.79),

$$(D_s m_k \times D_t m_k) \cdot [m_k(s, t) - P] = |D_s m_k \times D_t m_k| \, |m_k(s, t) - P| \cos \theta$$

with θ the angle between $m_k(s, t) - P$ and the unit normal $\nu_k(s, t)$. It is well known that

$$|\cos \theta| \leq c |m_k(s, t) - P|$$

with c independent of P and $Q = m_k(s, t)$ on S. Consequently,

$$|\kappa_k(s, t)| = O\left(\hat{\delta}_n^2 d_k^{-1}\right) \tag{9.2.81}$$

uniformly in (s, t), k, and P.

By our assumption that $P \notin \Delta_k$, we can differentiate $\kappa_k(s, t)$ over σ. In particular, we can show

$$D_s \kappa_k(s, t)$$

$$= \frac{[D_s(D_s m_k \times D_t m_k)] \cdot [m_k(s, t) - P]}{|m_k(s, t) - P|^3}$$

$$+ 3 \frac{[(D_s m_k \times D_t m_k) \cdot (m_k(s, t) - P)][D_s m_k \cdot (m_k(s, t) - P)]}{|m_k(s, t) - P|^5}$$

Then

$$D_s \kappa_k(s, t) = O\left(\hat{\delta}_n^3 d_k^{-2}\right)$$

uniformly in (s, t), k, and P. Using Taylor's theorem, we can write

$$\kappa_k(s, t) = \kappa_k(0, 0) + O\left(\hat{\delta}_n^3 d_k^{-2}\right) \tag{9.2.82}$$

Substitute (9.2.74) $\mathcal{T}_n'$ and (9.2.82) into (9.2.73), and following this, apply the results to (9.2.72). Let $\mathcal{T}_n'$ denote the elements in $\mathcal{T}_n$ that do not contain P. Recalling the discussion of the singular integrals in $\mathcal{K}(I - \mathcal{P}_n)\rho(P)$, in the paragraph following (9.2.72), we have

$$\mathcal{K}(I - \mathcal{P}_n)\rho(P) = O\left(\hat{\delta}_n^4\right) + \sum_{\Delta_k \in \mathcal{T}_n'} \int_\sigma [E_k(s, t; J_k) + E_k(s, t; L_k)]$$

$$\times \left[\kappa_k(0, 0) + O\left(\hat{\delta}_n^3 d_k^{-2}\right)\right] d\sigma \tag{9.2.83}$$

Multiply the various terms in the integrand and consider each one separately. We begin with

$$\mathcal{E}_1 \equiv \sum_{\Delta_k \in \mathcal{T}_n'} \kappa_k(0, 0) \int_\sigma E_k(s, t; J_k) \, d\sigma$$

Partition the set $\mathcal{T}_n'$ into two disjoint sets $\mathcal{T}_{n,1}'$ and $\mathcal{T}_{n,2}'$. The set $\mathcal{T}_{n,1}'$ is based on partitioning $\mathcal{T}_n'$ into symmetric pairs, as illustrated earlier in Figure 5.2 and discussed preceding (5.1.53). $\mathcal{T}_{n,2}'$ consists of the remaining elements of $\mathcal{T}_n'$ that are not in $\mathcal{T}_{n,1}'$. The number of elements in $\mathcal{T}_{n,1}'$ is $O(n)$, and that in $\mathcal{T}_{n,2}'$ is

$O(\sqrt{n})$. It should be added that the symmetric pairs can be so chosen that the remaining elements are at a bounded distance from P, independent of n. For a symmetric pair of triangles in $T'_{n,1}$, say Δ_k and Δ_l, it can be shown that

$$\kappa_k(0,0)E_k(s,t;J_k) + \kappa_l(0,0)E_l(s,t;J_l) = 0$$

for much the same reasons as in (5.2.74).

Thus

$$\begin{aligned}
\mathcal{E}_1 &= \sum_{\Delta_k \in T'_{n,2}} \kappa_k(0,0) \int_\sigma E_k(s,t;J_k)\,d\sigma \\
&= \sum_{\Delta_k \in T'_{n,2}} O\left(\hat{\delta}_n^2 d_k^{-1}\right)O\left(\hat{\delta}_n^3\right)
\end{aligned}$$

In this instance the term $O(d_k^{-1})$ can be bounded by a constant, since all elements of $T'_{n,2}$ are at a distance from P that is bounded away from 0,

$$d_k \ge c > 0, \qquad \Delta_k \in T'_{n,2}$$

independent of n and P. Thus

$$\mathcal{E}_1 = O(\sqrt{n})O\left(\hat{\delta}_n^2\right)O\left(\hat{\delta}_n^3\right) = O\left(\hat{\delta}_n^4\right) \tag{9.2.84}$$

since $\sqrt{n} = O(\hat{\delta}_n^{-1})$.

Referring to (9.2.83), consider next

$$\mathcal{E}_2 \equiv \sum_{\Delta_k \in T'_n} \kappa_k(0,0) \int_\sigma E_k(s,t;L_k)\,d\sigma$$

Combining (9.2.81) and (9.2.76),

$$\mathcal{E}_2 = \sum_{\Delta_k \in T'_n} O\left(\hat{\delta}_n^3 d_k^{-2}\right)O\left(\hat{\delta}_n^3\right) = \sum_{\Delta_k \in T'_n} O\left(\hat{\delta}_n^6 d_k^{-2}\right).$$

To add this, we first simplify the derivation by assuming that the values of d_k are

$$d_k = \alpha\hat{\delta}_n, (1+\alpha)\hat{\delta}_n, (2+\alpha)\hat{\delta}_n, \ldots, (N+\alpha)\hat{\delta}_n$$

with some $\alpha > 0$ and with $N\hat{\delta}_n$ proportional to the diameter of S. Moreover, we simplify by assuming that at a distance of $d_k = (l+\alpha)\hat{\delta}_n$, the number of such elements is $O(l)$, say $\nu(l+1)$ for some $\nu > 0$ that is not dependent on n. These assumptions are true in somewhat of an asymptotic sense, and rigorous bounds can be constructed without these assumptions. Using these assumptions, we can rewrite $\mathcal{E}_2$ as

$$|\mathcal{E}_2| \le c \sum_{l=1}^{N} \sum_{j=1}^{6l} \frac{\hat{\delta}_n^6}{[(l+\alpha)\hat{\delta}_n]^2} = vc\hat{\delta}_n^4 \sum_{l=1}^{N} \frac{l+1}{(l+\alpha)^2}$$

The last sum is $O(\log N) = O(\log \hat{\delta}_n^{-1})$. Thus

$$\mathcal{E}_2 = O\left(\hat{\delta}_n^4 \log \hat{\delta}_n^{-1}\right) \tag{9.2.85}$$

Returning to (9.2.83) and using analogous arguments, we obtain

$$\sum_{\Delta_k \in T_n'} \int_\sigma E_k(s, t; J_k) O\left(\hat{\delta}_n^3 d_k^{-2}\right) d\sigma = O\left(\hat{\delta}_n^4 \log \hat{\delta}_n^{-1}\right)$$

$$\sum_{\Delta_k \in T_n'} \int_\sigma E_k(s, t; L_k) O\left(\hat{\delta}_n^3 d_k^{-2}\right) d\sigma = O\left(\hat{\delta}_n^5 \log \hat{\delta}_n^{-1}\right)$$

Combine these results with (9.2.84) and (9.2.85) to complete the proof that

$$\mathcal{K}(I - \mathcal{P}_n) \rho(P) = O\left(\hat{\delta}_n^4 \log \hat{\delta}_n^{-1}\right)$$

for each vertex P among the collocation points.

For the case that $P = v_i$ is a midpoint of a side of some element Δ_ℓ, we can split the element into two smaller triangles with P as a vertex of each, use an affine transformation of the parametrization domain onto the unit simplex, and then return to the case just treated. When combined with (9.2.71), this completes the proof of (9.2.65).   □

9.2.1. The linear system

The linear system associated with solving (9.2.60) is

$$2\pi \rho_n(v_i) + \sum_{k=1}^{n} \sum_{j=1}^{6} \rho_n(v_{k,j})$$

$$\times \int_\sigma \ell_j(s, t) \frac{(D_s m_k \times D_t m_k) \cdot [m_k(s, t) - v_i]}{|m_k(s, t) - v_i|^3} d\sigma = f(v_i) \tag{9.2.86}$$

for $i = 1, \ldots, n_v$, and we denote it abstractly by

$$(2\pi + K_n)\underline{\rho}_n = \mathbf{f}_n \tag{9.2.87}$$

The numerical integration of the coefficients of this system is the computation-ally most expensive part of solving for ρ_n. In addition, we often make a further approximation by replacing the parametrization functions m_k by interpolatory approximations $\tilde{m}_k$, just as was discussed in §5.3.3:

$$\tilde{m}_k(s, t) = \sum_{j=1}^{6} m_k(q_j)\ell_j(s, t) = \sum_{j=1}^{6} v_{k,j}\ell_j(s, t) \tag{9.2.88}$$

In this case we actually solve the linear system

$$2\pi \tilde{\rho}_n(v_i) + \sum_{k=1}^{n} \sum_{j=1}^{6} \tilde{\rho}_n(v_{k,j})$$

$$\times \int_{\sigma} \ell_j(s, t) \frac{(D_s\tilde{m}_k \times D_t\tilde{m}_k) \cdot [\tilde{m}_k(s, t) - v_i]}{|\tilde{m}_k(s, t) - v_i|^3}\, d\sigma = f(v_i) \tag{9.2.89}$$

for $i = 1, \ldots, n_v$. We denote it abstractly by

$$(2\pi + \tilde{K}_n)\underline{\tilde{\rho}}_n = \mathbf{f}_n \tag{9.2.90}$$

In Ref. [53, p. 1102] an argument is sketched to show

$$\|K_n - \tilde{K}_n\| = O(\hat{\delta}_n) \tag{9.2.91}$$

It is based on arguments similar to those used in the proof of the preceding Theorem 9.2.2.

Using (9.2.91), we can use a standard perturbation argument to prove the existence and uniform boundedness of $(2\pi + \tilde{K}_n)^{-1}$,

$$\sup_{n \geq N} \|(2\pi + \tilde{K}_n)^{-1}\| < \infty \tag{9.2.92}$$

provided N is chosen sufficiently large. To prove this, begin by writing

$$2\pi + \tilde{K}_n = (2\pi + K_n)[I + (2\pi + K_n)^{-1}(K_n - \tilde{K}_n)]$$

Then combine (9.2.70), (9.2.91), and the geometric series theorem (cf. Theorem A.1 in the Appendix) to obtain (9.2.92).

To prove convergence of $\underline{\tilde{\rho}}_n$ to ρ, use the identity

$$\rho - \underline{\tilde{\rho}}_n = (\rho - \underline{\rho}_n) + (\underline{\rho}_n - \underline{\tilde{\rho}}_n) \tag{9.2.93}$$

$$\underline{\rho}_n - \underline{\tilde{\rho}}_n = -(2\pi + \tilde{K}_n)^{-1}(K_n - \tilde{K}_n)\underline{\rho}_n \tag{9.2.94}$$

with ρ the vector of values of ρ at the collocation node points. From (9.2.64), we have

$$\rho - \rho_n = O\left(\hat{\delta}_n^3\right) \tag{9.2.95}$$

for general triangulations $\mathcal{T}_n$, and from (9.2.65), we have

$$\rho - \rho_n = O\left(\hat{\delta}_n^4 \log \hat{\delta}_n^{-1}\right) \tag{9.2.96}$$

if $\mathcal{T}_n$ is symmetric. In both cases this assumes ρ is sufficiently smooth on S.

To obtain a good rate of convergence for $(K_n - \tilde{K}_n)\rho_n$ in (9.2.94), use

$$(K_n - \tilde{K}_n)\rho_n = (K_n - \tilde{K}_n)\rho - (K_n - \tilde{K}_n)(\rho - \rho_n)$$

$$= (K\rho - \tilde{K}_n\rho) - (K\rho - K_n\rho) - (K_n - \tilde{K}_n)(\rho - \rho_n)$$

By $K\rho$, we mean the vector with components $K\rho(v_i)$, $i = 1, \ldots, n_v$. From (9.2.91) and Theorem 9.2.1, the term $(K_n - \tilde{K}_n)(\rho - \rho_n)$ is $O(\hat{\delta}_n^5 \log \hat{\delta}_n^{-1})$. Using the proof of Theorem 9.2.1, $(K - K_n)\rho$ is $O(\hat{\delta}_n^4 \log \hat{\delta}_n^{-1})$. The term $K\rho - \tilde{K}_n\rho$ requires a separate proof, along the lines of that of Theorem 9.2.1. We have been able to show only that

$$(K - \tilde{K}_n)\rho = O\left(\hat{\delta}_n^2\right) \tag{9.2.97}$$

although empirically it appears to be $O(\hat{\delta}_n^3)$ (cf. Atkinson and Chien [59], Chien [119]). We conjecture that (9.2.97) is correct when using general triangulations, but that it should be $O(\hat{\delta}_n^3)$ when using symmetric triangulations. Combining all of the above results into (9.2.93) and (9.2.94), we have

$$\rho - \tilde{\rho}_n = O\left(\hat{\delta}_n^2\right) \tag{9.2.98}$$

Our empirical results are much better than this, as illustrated later in Table 9.1.

Numerical integration of singular integrals

When $v_i \in \Delta_k$, the integral

$$\int_\sigma \ell_j(s, t) \frac{(D_s m_k \times D_t m_k) \cdot [m_k(s, t) - v_i]}{|m_k(s, t) - v_i|^3} \, d\sigma \tag{9.2.99}$$

[and its analog from (9.2.89) using the approximate surface $\tilde{m}_k$] has a singular integrand. To handle such singular integrals, we use a change of variable that removes the singularity.

Write the general integral as

$$I(g) = \int_\sigma g(s, t) \, d\sigma$$

and assume it has a singularity at $(s, t) = (0, 0)$. Introduce the change of variables

$$s = (1 - y)x, \qquad t = yx, \quad 0 \leq x, y \leq 1 \qquad (9.2.100)$$

With this,

$$I(g) = \int_0^1 \int_0^1 x g((1 - y)x, yx) \, dx \, dy \qquad (9.2.101)$$

As a simple example of its effectiveness,

$$\int_\sigma \frac{d\sigma}{\sqrt{s^2 + t^2}} = \int_0^1 \int_0^1 \frac{dx \, dy}{\sqrt{(1 - y)^2 + y^2}}$$

and the latter is nonsingular, together with all of its derivatives.

This type of improvement in differentiability can also be shown to be true for double integrals of the form (9.2.99) when $v_i = m_k(0, 0)$. It also applies to single layer integrals

$$\int_\sigma \psi(m_k(s, t)) \frac{|D_s m_k \times D_t m_k|}{|m_k(s, t) - v_i|} \, d\sigma \qquad (9.2.102)$$

with smooth functions ψ. We show the resulting smoothness for this single layer integral, and a similar argument can be given for the double layer integral (9.2.99).

For (9.2.102), the transformed integral is

$$\int_0^1 \int_0^1 x \psi(m_k((1 - y)x, yx)) \frac{|D_s m_k \times D_t m_k|}{|m_k((1 - y)x, yx) - v_i|} \, d\sigma \qquad (9.2.103)$$

Assume $v_i = m_k(0, 0)$. Then

$$m_k(s, t) - m_k(0, 0) = \int_0^1 \frac{d}{dr} m(rs, rt) \, dr$$

Make the substitution (9.2.100), obtaining

$$|m_k(r(1 - y)x, ryx) - v_i|$$
$$= \left| \int_0^1 [(1 - y)x D_1 m_k(rs, rt) + yx D_2 m_k(rs, rt)] \, dr \right|$$

with $D_1 m_k(s, t) \equiv (\partial/\partial s) \, m_k(s, t)$ and analogously for $D_2 m_k(s, t)$. The remaining values of s and t in this formula are also to be replaced using (9.2.100).

With respect to (9.2.103),

$$\overline{\frac{x}{|m_k(r(1-y)x, ryx) - v_i|}}$$

$$= \Big| (1-y) \int_0^1 D_1 m_k(r(1-y)x, ryx)\, dr$$

$$+ y \int_0^1 D_2 m_k(r(1-y)x, ryx)\, dr \Big|^{-1} \qquad (9.2.104)$$

The term

$$\mathbf{v}_1(x, y) \equiv \int_0^1 D_1 m_k(r(1-y)x, ryx)\, dr$$

is an average of the vectors $D_1 m_k(p, q)$ as (p, q) varies along the straight line joining $(0, 0)$ and $(s, t) = ((1 - y)x, yx)$, and the same can be said of

$$\mathbf{v}_2(x, y) \equiv \int_0^1 D_2 m_k(r(1-y)x, ryx)\, dr$$

The vectors $\mathbf{v}_1$ and $\mathbf{v}_2$ can be shown to be independent, based on the nature of the mapping $m_k(s, t)$ from σ to Δ_k, but we omit the proof. The statement (9.2.104) says that

$$\overline{\frac{x}{|m_k(r(1-y)x, ryx) - v_i|}} = |(1 - y)\mathbf{v}_1(x, y) + y\mathbf{v}_2(x, y)|^{-1}$$

The argument $(1 - y)\mathbf{v}_1 + y\mathbf{v}_2$ is a linear combination of $\mathbf{v}_1$ and $\mathbf{v}_2$, and since the latter are linearly independent, the combination cannot be zero. Differentiability of the fraction can be shown by similar means, since $\mathbf{v}_1$ and $\mathbf{v}_2$ are smooth functions of x and y. Returning to (9.2.103), the integrand is well-behaved.

To approximate (9.2.101), we apply a standard product Gauss–Legendre quadrature formula:

$$I(g) \approx \sum_{i=1}^{m} \sum_{j=1}^{m} w_{m,i} w_{m,j} x_{m,i} g((1 - x_{m,j})x_{m,i}, x_{m,j} x_{m,i}) \qquad (9.2.105)$$

Usually we need only a relatively small value of m, say $5 \le m \le 10$. In using this to evaluate those integrals (9.2.99) that are singular, let the integration parameter m be denoted by M_s. The use of (9.2.100) to obtain (9.2.101) is called the *Duffy transformation*. For an extensive analysis of its use in the area of boundary integral equations, see Schwab and Wendland [497].

Note that the number of singular integrals is only $O(n)$, whereas the total number of integrals to be evaluated is $O(n^2)$. With an effective quadrature method such as the above, the main cost of setting up the collocation matrix K_n is in the evaluation of the integrals (9.2.99) that are nonsingular.

Numerical integration of nonsingular integrals

When the field point $v_i \notin \Delta_k$, the integrand in (9.2.99) is nonsingular, but it becomes increasingly ill-behaved as the distance $d_k(v_i)$ between v_i and Δ_k decreases towards zero. In contrast, when the distance $d_k(v_i)$ is well away from zero, the integrand in (9.2.99) is quite well-behaved, and a simple quadrature method often gives excellent accuracy. For reasons of efficiency, it is a poor idea to use the same integration formula for all such nonsingular integrals. Instead, one should make the numerical integration formula increasingly sophisticated as $d_k(v_i) \to 0$, so as to compensate for the increasing ill-behavior in the integrand. The question of how to evaluate these nonsingular integrals is still an active area of investigation in the boundary element literature, and we merely sketch what we have found to be a reasonably efficient and accurate method for the numerical approximation of these integrals.

We introduce a composite numerical integration formula for integrals over σ, and it is based on the 7-point formula (5.1.46) of §5.1.2 of Chapter 5,

$$I(f) \equiv \int_\sigma f(s, t) \, d\sigma \approx \sum_{j=1}^{7} w_j f(s_j, t_j) \qquad (9.2.106)$$

which has degree of precision 5. For a given integer $L \geq 0$, subdivide σ into $L_* \equiv 4^L$ congruent smaller triangles, calling them $\sigma_1^{(L)}, \ldots, \sigma_{L_*}^{(L)}$. For example, see Figure 9.1 for the case $L = 2$. To approximate $I(g)$, write

$$I(g) = \sum_{j=1}^{L_*} \int_{\sigma_j} g(s, t) \, d\sigma$$

Apply a properly scaled and translated form of (9.2.106) to each of these integrals. Denote the resulting numerical approximation by $I_L(g)$. It is straightforward to show

$$|I(g) - I_L(g)| \leq ch^6, \quad h = 2^{-L} \qquad (9.2.107)$$

for all $g \in C^6(\sigma)$. The quantity h is the length of the shorter sides of the triangles $\sigma_j^{(L)}$, with the mesh size $\sqrt{2}h$. The constant c depends on the sixth order derivatives of g.

Return now to the approximation of the integrals in (9.2.99). Let

$$\delta_n = \max_{\Delta_k \in T_n} \text{diameter}(\Delta_k)$$

denote the mesh size of T_n. Let M_{ns} be a given nonnegative integer. Again let

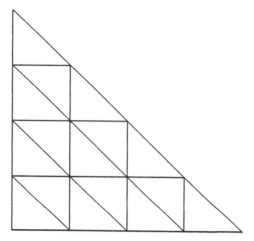

Figure 9.1. Subdivision of σ into 4^2 congruent subdomains.

the field point $v_i \in \Delta_l$, and for other elements Δ_k with $v_i \notin \Delta_k$, define

$$d_k \equiv d_k(v_i) = \text{distance}(v_i, \Delta_k) \qquad (9.2.108)$$

1. If $0 < d_k \le \delta_n$ then approximate (9.2.99) by applying the above composite quadrature formula $I_L(g)$ with $L = M_{ns}$.
2. If $\delta_n < d_k \le 2\delta_n$ then approximate (9.2.99) by applying the above composite quadrature formula $I_L(g)$ with $L = \max\{M_{ns} - 1, 0\}$.
3. If $2\delta_n < d_k \le 3\delta_n$ then approximate (9.2.99) by applying the above composite quadrature formula $I_L(g)$ with $L = \max\{M_{ns} - 2, 0\}$.
4. Continue moving outward from Δ_l until reaching a point at which $L = 0$ in applying $I_L(g)$. Then use $I_0(g)$ in approximating all remaining integrals (9.2.99).

Experimentally this has proven superior to using the same integration formula for all nonsingular integrals (9.2.99), in the sense that equivalent accuracy is attained with a far smaller number of arithmetic operations. We give some indication of this below in some numerical examples. We have used integration formulas with a lower degree of precision, and the above choice appears to be more efficient. One can also vary somewhat the definition of the distance d_k, while maintaining the general strategy of using simpler formulas as the distance increases between v_i and Δ_k.

There is still the question of how to choose M_{ns} as the parameter n increases. We note that our method of refinement of $\mathcal{T}_n$ leads to the increase $n \rightarrow 4n$ in the number of triangular elements with each successive refinement. Theoretical calculations done by Graeme Chandler suggest that M_{ns} should be increased by 1 with every two successive refinements of the triangulation. We usually begin with $M_{ns} = 0$ or 1, and empirically, the Chandler strategy appears to be a sound one.

All of the ideas of this section are incorporated into the boundary element package BIEPACK described in Ref. [52]. We use it in the numerical examples given later in the section. For more specific details on the use of numerical integration in BIEPACK, see the user's guide in Ref. [52].

9.2.2. Solving the linear system

The approximating equations $(2\pi + K_n)\underline{\rho}_n = \mathbf{f}_n$ of (9.2.87) and $(2\pi + \tilde{K}_n)\underline{\tilde{\rho}}_n = \mathbf{f}_n$ of (9.2.90) are of the second kind, and we can apply the two-grid and multi-grid methods of Chapter 6. The use of two-grid methods is discussed at length in Chapter 6 and in Ref. [53], and we summarize those results here. Multigrid methods are discussed and illustrated in Hackbusch [249, p. 348]. As in Chapter 6, let $m < n$ and assume that $(2\pi + K_m)^{-1}$ is known (or equivalently, that we can solve directly the linear systems associated with the coarse mesh coefficient matrix $2\pi + K_m$). We solve the fine mesh equation $(2\pi + K_n)\underline{\rho}_n = \mathbf{f}_n$ by iteration.

Recall the discussion of two-grid methods for collocation methods given in §6.3, including the definitions of prolongation operators and restriction operators. Apply the two-grid method of §6.3.2 to iteratively solve $(2\pi + K_n)\underline{\rho}_n = \mathbf{f}_n$, and in particular, apply formulas (6.3.81)–(6.3.111) to define the iterates $\{\underline{\rho}_n^{(k)}\}$. The convergence of this process, provided m is chosen sufficiently large, is assured by Theorem 6.3.1:

$$\left[\underline{\rho}_n - \underline{\rho}_n^{(k+1)}\right] = M_{m,n}\left[\underline{\rho}_n - \underline{\rho}_n^{(k)}\right], \quad k = 0, 1, \ldots$$

$$\|M_{m,n}\| \leq c\|(\mathcal{P}_n - \mathcal{P}_m)\mathcal{K}\|$$

Choose m such that $\|M_{m,n}\| < 1$ for all $n \geq m$. This assures convergence of $\underline{\rho}_n^{(k)}$ to $\underline{\rho}_n$, uniformly for $n > m$. A complete formula for $M_{m,n}$ can be obtained by combining (6.3.115) and (6.3.123) from §6.3.81.

The convergence of the process can be extended to the iterative solution of $(2\pi + \tilde{K}_n)\underline{\tilde{\rho}}_n = \mathbf{f}_n$ by means of (9.2.91), that $\|K_n - \tilde{K}_n\| = O(\hat{\delta}_n)$.

Write

$$\left[\underline{\tilde{\rho}}_n - \underline{\tilde{\rho}}_n^{(k+1)}\right] = \tilde{M}_{m,n}\left[\underline{\tilde{\rho}}_n - \underline{\tilde{\rho}}_n^{(k)}\right]$$

and extend the definition (6.3.124) to give an explicit formula for $\tilde{M}_{m,n}$. It then follows that

$$\|M_{m,n} - \tilde{M}_{m,n}\| = O(\hat{\delta}_m), \quad n \geq m$$

By choosing m sufficiently large, we obtain $\|\tilde{M}_{m,n}\| < 1$ for all $n \geq m$, and this assures the convergence of $\underline{\tilde{\rho}}_n^{(k)}$ to $\underline{\tilde{\rho}}_n$, uniformly for $n > m$.

A numerical example of the iterative solution of $(2\pi + \tilde{K}_n)\underline{\tilde{\rho}}_n = \mathbf{f}_n$ is given following (6.3.125), for the equation (9.1.33) for the exterior Neumann problem on an ellipsoidal domain. Empirical rates of convergence are given in Table 6.6 for varying values of m and n. In the numerical examples given below, we compare empirically the cost of the iteration method and the cost of the setup of the collocation matrix $2\pi + \tilde{K}_n$.

Example. We solve the boundary integral equation

$$2\pi u + \mathcal{K}u = g \tag{9.2.109}$$

$$g(P) = -\int_S \frac{f(Q)\,dS(Q)}{|P - Q|} \tag{9.2.110}$$

associated with the exterior Neumann problem for Laplace's equation [cf. (9.1.26)–(9.1.27)]. The surface S is the ellipsoid

$$\left(\frac{x}{a}\right)^2 + \left(\frac{y}{b}\right)^2 + \left(\frac{z}{c}\right)^2 = 1$$

and we choose $(a, b, c) = (1, 1.5, 2)$. As a test case, we choose the harmonic function

$$u(x, y, z) = \frac{1}{\sqrt{x^2 + y^2 + z^2}} \tag{9.2.111}$$

to be the solution of our Neumann problem, and the Neumann data f is determined from this function.

We use the collocation method described above to solve (9.2.109), including the approximation of the boundary using the quadratic interpolation of (9.2.88). We denote this collocation approximation by

$$(2\pi + \tilde{K}_n)\underline{\tilde{u}}_n = \mathbf{g}_n \tag{9.2.112}$$

Table 9.1. *Errors E_n in the collocation solution of (9.2.109)*

n	$M_{ns} = 0$	$M_{ns} = 1$	$M_{ns} = 2$	$M_{ns} = 3$	Order
8	3.63E−2	3.30E−2	3.31E−2	3.31E−2	
32	3.78E−3	3.79E−3	3.78E−3	3.78E−3	3.13
128	5.49E−4	2.92E−4	2.91E−4	2.91E−4	3.70
512	2.62E−4	1.88E−5	1.87E−5	1.87E−5	3.96

In addition, we further modify the coefficient matrix $\tilde{K}_n$. The integral operator $\mathcal{K}$ satisfies $\mathcal{K}1 \equiv 2\pi$. To maintain this identity, we force each row of $\tilde{K}_n$ to sum to 2π. We do this by adding a correction term to the diagonal entry in each row. Symbolically,

$$\tilde{K}_n 1 = 2\pi 1 \tag{9.2.113}$$

where **1** denotes the column vector with components all equal to 1. Below in Table 9.5 we illustrate how this improves the order of convergence. This modification arises naturally when dealing with piecewise smooth surfaces, because we need to approximate the solid angle at each point of the surface. This is discussed in §9.3.

With this modification to the collocation matrix, we give some numerical results in Table 9.1 for $M_s = 10$, with increasing n and M_{ns}. The errors given in the table are

$$E_n = \max_{1 \le i \le n_v} |u(v_i) - \tilde{u}_{n,i}|$$

The final column, labeled *Order*, is the empirical order of convergence of the collocation method, based on the errors in the column labeled $M_{ns} = 3$. According to Theorem 9.2.2, this should equal 4 if the exact surface S were being used. It is interesting that it still appears to be approaching 4, although we can prove no greater than $O(\hat{\delta}_n^2)$, which was given in (9.2.98).

We conjecture that the rate of $O(\hat{\delta}_n^2)$ is what one obtains with a general triangulation, and that the use of symmetric triangulations leads to $O(\hat{\delta}_n^3)$. Furthermore, we conjecture the modification embodied in (9.2.113) leads to a further improvement to $O(\hat{\delta}_n^4 \log \hat{\delta}_n^{-1})$.

To give some notion of the machine resources required in solving (9.2.112), consider the number of entries in the coefficient matrix. For $n = 512$, the number of collocation node points is $n_v = 1026$. Then the number of elements in the matrix is $n_v^2 = 1,052,676$, and this is the number of surface integrals (9.2.99) that must be evaluated numerically. When storing this in double precision on a standard workstation, this requires $8n_v^2 = 8,421,408$ bytes of storage. The

Table 9.2. *Timings for setup of collocation matrix, with $M_s = 10$*

n	$M_{ns} = 0$	$M_{ns} = 1$	$M_{ns} = 2$	$M_{ns} = 3$	Iteration
8	0.86	0.86	1.6	4.0	
32	3.9	6.1	20.8	59.5	<0.01
128	31.2	42.7	123	439	0.13
512	388	424	754	2240	2.05

Table 9.3. *Setup timings with a uniform quadrature mesh for nonsingular integrals*

n	$M_{ns} = 0$	$M_{ns} = 1$	$M_{ns} = 2$	$M_{ns} = 3$
8	0.86	0.90	1.6	4.0
32	3.9	7.2	20.8	76.0
128	31.2	89.3	317	1230
512	388	1350	5030	19800

next case would be $n = 2048$, and this would require approximately 16 times the storage, or about 134 megabytes. This far surpasses the memory available on most current computers, especially workstations, and it is the main reason we stopped with $n = 512$ faces in the triangulation of S.

We illustrate the cost associated with various choices of M_{ns} and M_s by giving some timings using a Hewlett-Packard workstation. In Table 9.2 we give timings in seconds for the setup of the collocation matrix, and the final column is the cost of a single iteration with the coarse mesh parameter $m = 8$. The costs are quite large in some cases, although we can reduce this by using various optimization features of the compiler. The important point is the relative costs of the various choices for M_{ns}, and these will remain the same after invoking the compiler optimization options. The setup costs increase quite rapidly with increasing M_{ns}, and clearly, the iteration costs are far less significant than the costs of the numerical integrations.

To see the increased efficiency associated with our numerical method for evaluating nonsingular integrals (9.2.99), consider using the quadrature method $I_L(g)$ with $L = M_{ns}$ for all triangles Δ_k, rather than as described following (9.2.108). The errors are exactly the same as those given in Table 9.1, and the timings are given in Table 9.3. Note the very significant increase in cost, especially for larger values of n and M_{ns}, and as there is no change in the accuracy of the solution $\tilde{u}_n$, this extra computation time is wasted.

We illustrate the costs of the evaluation of the singular integrals by comparing the use of $M_s = 10$ and $M_s = 15$, both with $M_{ns} = 3$. Note that the cost in

Table 9.4. *Cost comparison*
for singular integrations

n	$M_s = 10$	$M_s = 15$
8	4.0	4.85
32	59.5	62.2
128	439	451
512	2240	2290

Table 9.5. *Errors E_n for unmodified and modified*
linear system

n	Unmodified	*Order*	Modified	*Order*
8	1.94E$-$1		3.31E$-$2	
32	2.33E$-$2	3.06	3.78E$-$3	3.13
128	5.66E$-$3	2.04	2.91E$-$4	3.70
512	7.87E$-$4	2.85	1.87E$-$5	3.96

operations of the Gaussian quadrature scheme (9.2.105) is $O(M_s^2)$. Therefore, the increase in running time associated with the numerical integration of the singular integrals (9.2.99), comparing $M_s = 10$ and $M_s = 15$, should be by a factor of approximately $(15/10)^2 = 2.25$. The timings are given in Table 9.4. In passing from $M_s = 10$ to $M_s = 15$, there was no change in the accuracy of the solution $\tilde{u}_n$ as given in Table 9.1. Also, the increase in total setup cost is quite small; therefore, it is the cost of evaluating the nonsingular integrals that is most significant in setting up the collocation matrix. The choice of M_s is unlikely to seriously affect the costs of the overall computation, and M_s should be chosen large enough to ensure the accuracy of the singular integrations using (9.2.105).

Earlier, we introduced a modification of the collocation matrix, to ensure the identity (9.2.113). In Table 9.5 we give a comparison of the error, both without and with this modification. Again, the original, unmodified collocation system is given in (9.2.89), and the modified system has correction terms added to the diagonal elements so as to ensure (9.2.113). The corresponding errors are given in columns 2 and 4, respectively, and the columns titled *Order* give the empirical order of convergence. In calculating the table, the integration parameters were $M_s = 10$ and $M_{ns} = 2$. The results seem to indicate that without the modification, the empirical order of convergence is only $O(\hat{\delta}_n^3)$, whereas the modified procedure has a rate consistent with $O(\hat{\delta}_n^4 \log \hat{\delta}_n^{-1})$.

Table 9.6. *Solving (9.2.114) using piecewise
quadratic collocation*

n	n_v	E_n	Ratio	Order	Cond
8	18	2.545E−1			15.4
32	66	5.447E−2	4.67	2.22	27.7
128	258	4.340E−3	12.6	3.65	66.2
512	1026	4.059E−4	10.7	3.42	146

9.2.3. Experiments for a first kind equation

Consider solving the first kind single layer equation

$$\int_S \frac{\psi(Q)}{|P - Q|} dS(Q) = f(P), \quad P \in S \tag{9.2.114}$$

The right-hand function f is to have the form

$$f(P) \equiv 2\pi u(P) + \int_S u(Q) \frac{\partial}{\partial \mathbf{n}_Q} \left[\frac{1}{|P - Q|} \right] dS(Q)$$

with u a function harmonic on the exterior region D_e and S the boundary of D_e. From (9.1.29), the solution is

$$\psi(Q) = \frac{\partial u(Q)}{\partial \mathbf{n}_Q}$$

We give some empirical results for the numerical solution of (9.2.114) with two collocation methods. At this time, there is no error analysis for the use of any collocation method for solving the first kind equation (9.2.114).

Begin by using piecewise quadratic polynomial approximations of ψ, with collocation at the interpolation node points to determine an approximating solution ψ_n. This is just the scheme used above in (9.2.86) for a BIE of the second kind. We also compute f numerically, but with sufficient accuracy that it does not significantly affect the accuracy in ψ_n. In Table 9.6 we give the errors when S is the unit sphere and

$$u(x, y, z) = \frac{x}{r^3}, \quad r = \sqrt{x^2 + y^2 + z^2}$$

The column labeled n denotes the number of triangular elements, as before; n_v denotes the number of collocation node points. The column labeled E_n is defined as

$$E_n \equiv \max_{P \in \mathcal{V}_n} |\psi(P) - \psi_n(P)|$$

Table 9.7. *Solving (9.2.114) using collocation*
at centroids

n	E_n	Ratio	Order	Cond
8	1.379E$-$1			4.6
32	5.363E$-$2	2.57	1.36	14.0
128	3.049E$-$2	1.76	.82	28.6
512	1.788E$-$2	1.71	.77	57.2

The column *Ratio* gives the ratio of the successive values of E_n, and *Order* gives the empirical power for which

$$E_n = O(h^{Order})$$

The column labeled *Cond* denotes the condition number for the linear system associated with the collocation method for (9.2.114), based on using the matrix row norm. We note that the package from Atkinson [52] was used to set up and solve the collocation method in this case; consequently, the surface S was approximated with piecewise quadratic interpolation, as with (9.2.89) earlier in this section.

The empirical rate of convergence appears to be either $O(h^3)$ or $O(h^4)$, and the condition number appears to be of size $O(h^{-1})$. We conjecture the rate of convergence is $O(h^3)$. These numerical results are consistent with results obtained when solving over ellipsoidal surfaces.

For our second collocation method we use the centroid rule. Thus we approximate ψ as a piecewise constant function, and we collocate at the centroids of the triangular elements. Thus n denotes both the number of triangular elements and the number of collocation points. The numerical results are given in Table 9.7. In this case, E_n denotes the maximum of the errors in ψ_n at the centroids of the elements of the triangulation. As before, the package from Ref. [52] was used to set up and solve the collocation method, and again the surface was approximated by piecewise quadratic interpolation. In the quadratic case the collocation nodes are still located on S, but in the centroid case the collocation nodes are on the approximate surface and are not, in general, on S.

The results in Table 9.7 indicate a rate of convergence of $O(h)$, at best. For the condition number, again it appears to be of size $O(h^{-1})$. We were concerned that the use of the interpolation of the surface S made a difference in the accuracy, and for that reason we recomputed the problem with the exact surface being used. This led to essentially the same results, with virtually identical results for $n = 128$ and 512.

These are interesting experimental results, and they indicate there is some kind of nontrivial behavior in the error. Let $A_n \underline{\psi}_n = \mathbf{f}_n$ denote the linear system associated with the collocation method for solving (9.2.114). If we use a standard error analysis for $\underline{\psi}_n$ at the node points, we have

$$E_n \leq \left\| A_n^{-1} \right\| \max_i |\mathcal{S}\psi(P_i) - (A_n \underline{\psi})_i| \qquad (9.2.115)$$

with $\underline{\psi}$ the vector of values of ψ at the collocation node points $\{P_i\}$. From the empirical result for the condition numbers, we obtain the empirical result

$$\left\| A_n^{-1} \right\| = O(h^{-1})$$

Thus if the discretization error

$$D_n \equiv \max_i |\mathcal{S}\psi(P_i) - (A_n \underline{\psi})_i| = O(h^p) \qquad (9.2.116)$$

for some p, we would have

$$E_n \leq O(h^{p-1})$$

From other work, it seems reasonable that the centroid rule leads to $D_n = O(h^2)$, and this would then lead to $E_n = O(h)$. The results seen in Table 9.7 appear somewhat consistent with this. With piecewise quadratic interpolation, we have observed with other numerical experiments that $D_n = O(h^4)$, provided a symmetric triangulation scheme is used. This would then lead to the result $E_n = O(h^3)$, which is consistent with Table 9.6. We have, however, been able to prove only that $D_n = O(h^3)$.

First kind equations such as (9.2.114) are being solved by collocation on a regular basis. Based on our experiments, it seems important that users of such methods be quite cautious as regards the accuracy in the solutions they obtain.

9.3. Boundary element collocation methods on piecewise smooth surfaces

Solving boundary integral equations on piecewise smooth surfaces has more potential difficulties associated with it than does the smooth surface case. These difficulties involve both the approximation of the BIE and the smoothness of the solution function. This is currently an active area of research, as was noted earlier in §9.1.5, and we only summarize some of the main research in this area. The first important research paper for understanding such BIEs of the second kind and the collocation method for their solution is that of Wendland [557].

We discuss some of the ideas of this paper, and we extend them to higher order collocation methods. Later in the section, we consider the ideas of Rathsfeld [447]–[453], which yield another analysis of collocation methods for BIE of the second kind on polyhedral surfaces.

As in the preceding section, we consider the numerical solution of the second kind double layer equation $2\pi\rho + \mathcal{K}\rho = f$, from (9.1.33) or (9.1.36). For piecewise smooth surfaces S the operator $\mathcal{K}$ takes the form

$$\mathcal{K}u(P) = \int_S u(Q)\frac{\partial}{\partial \mathbf{n}_Q}\left[\frac{1}{|P - Q|}\right]dS(Q) + [2\pi - \Omega(P)]u(P),$$

$$P \in S \qquad (9.3.117)$$

which was introduced earlier in (9.1.34). Wendland [557] proves that

$$\mathcal{K} : C(S) \to C(S)$$

is a well-defined and bounded operator. To show this, certain assumptions about the surface S are needed, and we discuss briefly those assumptions in the following paragraphs.

As in (9.1.1)–(9.1.2), assume

$$S = S_1 \cup S_2 \cup \cdots \cup S_J$$

with each S_j being the image of a smooth parametrization over a planar polygon. More precisely, for each j, assume there is a function

$$F_j : R_j \xrightarrow[onto]{1-1} S_j \qquad (9.3.118)$$

with R_j a closed planar polygon and $F_j \in C^m(R_j)$ for some $m \geq 2$. The boundary of each S_j is piecewise smooth, and it is assumed that each corner point of this boundary is not a cusp.

We modify slightly the assumption (9.1.10) on the solid angle $\Omega(P)$, assuming now that

$$\omega \equiv \sup_{P \in S} |2\pi - \Omega(P)| < 2\pi \qquad (9.3.119)$$

We also assume that the interior tangent cone $\Gamma_i(P)$ of (9.1.7) and the exterior tangent cone $\Gamma_e(P)$ of (9.1.8) satisfy

Assumption Ξ. For each $P \in S$, either $\Gamma_i(P)$ or $\Gamma_e(P)$ is convex.

This last assumption was discussed earlier in §9.1.5 where we noted that Ξ was weakened somewhat in Kral and Wendland [317] for "rectangular" corners that are obtained by intersecting planes parallel to the coordinate planes.

However, for most surfaces Ξ remains a restrictive assumption when analyzing the collocation method for solving $2\pi\rho + \mathcal{K}\rho = f$, as there are many interesting surfaces that do not satisfy Ξ.

To deal with the double layer operator $\mathcal{K}$, we decompose it into simpler parts that are easier to handle. For $P \in S$ and $\delta > 0$, introduce

$$F_\delta(P) = \{Q \in S \mid |P - Q| \leq \delta\}$$

Introduce

$$\mathcal{H}_\delta u(P) = \int_{F_\delta(P)} u(Q) \frac{\partial}{\partial \mathbf{n}_Q} \left[\frac{1}{|P - Q|} \right] dS(Q)$$

$$+ [2\pi - \Omega(P)]u(P), \quad P \in S \qquad (9.3.120)$$

for $u \in C(S)$. It is shown in Ref. [557, Theorem 2] that for any given $\epsilon > 0$, it is possible to choose a sufficiently small δ such that

$$\|\mathcal{H}_\delta\| \leq \omega + \epsilon \qquad (9.3.121)$$

The argument of Wendland in [557, Theorem 2] is excellent, but fairly technical and difficult, and we omit it here. Noting (9.3.119), we usually choose $\epsilon < 2\pi - \omega$, and then

$$\|\mathcal{H}_\delta\| < 2\pi \qquad (9.3.122)$$

Introduce $\mathcal{F}_\delta = \mathcal{K} - \mathcal{H}_\delta$. Then

$$\mathcal{F}_\delta u(P) = \int_{S \setminus F_\delta(P)} u(Q) \frac{\partial}{\partial \mathbf{n}_Q} \left[\frac{1}{|P - Q|} \right] dS(Q), \quad P \in S, \ u \in C(S)$$

It is straightforward that this defines a compact operator on $C(S)$ to $C(S)$.

We have

$$\mathcal{K} = \mathcal{H}_\delta + \mathcal{F}_\delta \qquad (9.3.123)$$

The equation $2\pi\rho + \mathcal{K}\rho = f$ is replaced by

$$2\pi\rho + \mathcal{H}_\delta\rho + \mathcal{F}_\delta\rho = f \qquad (9.3.124)$$

Using (9.3.122), it follows from the geometric series theorem (cf. Theorem A.1 in the Appendix) that $(2\pi + \mathcal{H}_\delta)^{-1}$ is well-defined and

$$\|(2\pi + \mathcal{H}_\delta)^{-1}\| \leq \frac{1}{2\pi - \|\mathcal{H}_\delta\|}$$

The equation (9.3.124) is equivalent to

$$[I + (2\pi + \mathcal{H}_\delta)^{-1}\mathcal{F}_\delta]\rho = (2\pi + \mathcal{H}_\delta)^{-1}f \qquad (9.3.125)$$

The operator $(2\pi + \mathcal{H}_\delta)^{-1}\mathcal{F}_\delta$ is compact, since it is the product of the bounded operator $(2\pi + \mathcal{H}_\delta)^{-1}$ and the compact operator $\mathcal{F}_\delta$. The equation (9.3.125) has the structure of a standard integral equation of the second kind as studied in Chapters 1 through 6.

9.3.1. The collocation method

Let $\{\mathcal{P}_n\}$ be a sequence of pointwise convergent interpolatory projection operators on $C(S)$ to $C(S)$. As in §9.2, we approximate $2\pi\rho + \mathcal{K}\rho = f$ by using the collocation method:

$$2\pi\rho_n + \mathcal{P}_n\mathcal{K}\rho_n = \mathcal{P}_n f \qquad (9.3.126)$$

The decomposition (9.3.123) leads to

$$2\pi\rho_n + \mathcal{P}_n\mathcal{H}_\delta\rho_n + \mathcal{P}_n\mathcal{F}_\delta\rho_n = \mathcal{P}_n f \qquad (9.3.127)$$

To analyze the convergence of ρ_n, we make the assumption that

$$\beta_\delta \equiv \left[\sup_n \|\mathcal{P}_n\|\right] \|\mathcal{H}_\delta\| < 2\pi \qquad (9.3.128)$$

Alternatively, we can assume

$$\omega \cdot \sup_n \|\mathcal{P}_n\| < 2\pi \qquad (9.3.129)$$

This implies (9.3.128) by choosing a sufficiently small δ in (9.3.121).

With (9.3.128), it follows that $(2\pi + \mathcal{P}_n\mathcal{H}_\delta)^{-1}$ exists and is uniformly bounded in n,

$$\|(2\pi + \mathcal{P}_n\mathcal{H}_\delta)^{-1}\| \le \frac{1}{2\pi - \beta_\delta} \qquad (9.3.130)$$

The equation (9.3.127) is equivalent to

$$[I + (2\pi + \mathcal{P}_n\mathcal{H}_\delta)^{-1}\mathcal{P}_n\mathcal{F}_\delta]\rho_n = (2\pi + \mathcal{P}_n\mathcal{H}_\delta)^{-1}\mathcal{P}_n f \qquad (9.3.131)$$

This must be compared to (9.3.125) to prove convergence of ρ_n to ρ.

We begin by showing

$$\|(2\pi + \mathcal{H}_\delta)^{-1}\mathcal{F}_\delta - (2\pi + \mathcal{P}_n\mathcal{H}_\delta)^{-1}\mathcal{P}_n\mathcal{F}_\delta\| \to 0 \qquad (9.3.132)$$

as $n \to \infty$. Write

$$(2\pi + \mathcal{H}_\delta)^{-1}\mathcal{F}_\delta - (2\pi + \mathcal{P}_n\mathcal{H}_\delta)^{-1}\mathcal{P}_n\mathcal{F}_\delta$$

$$= [(2\pi + \mathcal{H}_\delta)^{-1}\mathcal{F}_\delta - (2\pi + \mathcal{P}_n\mathcal{H}_\delta)^{-1}\mathcal{F}_\delta]$$
$$+ (2\pi + \mathcal{P}_n\mathcal{H}_\delta)^{-1}(I - \mathcal{P}_n)\mathcal{F}_\delta$$

$$= (2\pi + \mathcal{P}_n\mathcal{H}_\delta)^{-1}(\mathcal{P}_n - I)\mathcal{H}_\delta(2\pi + \mathcal{H}_\delta)^{-1}\mathcal{F}_\delta$$
$$+ (2\pi + \mathcal{P}_n\mathcal{H}_\delta)^{-1}(I - \mathcal{P}_n)\mathcal{F}_\delta$$

$$\equiv \mathcal{E}_1 + \mathcal{E}_2$$

For the term $\mathcal{E}_2$,

$$\|\mathcal{E}_2\| = \|(2\pi + \mathcal{P}_n\mathcal{H}_\delta)^{-1}(I - \mathcal{P}_n)\mathcal{F}_\delta\|$$
$$\leq \|(2\pi + \mathcal{P}_n\mathcal{H}_\delta)^{-1}\|\|(I - \mathcal{P}_n)\mathcal{F}_\delta\|$$
$$\leq \frac{\|(I - \mathcal{P}_n)\mathcal{F}_\delta\|}{2\pi - \beta_\delta}$$

The term $\|(I - \mathcal{P}_n)\mathcal{F}_\delta\|$ converges to zero because $\{\mathcal{P}_n\}$ is a pointwise convergent sequence on $C(S)$ and $\mathcal{F}_\delta$ is compact (cf. Lemma 3.1.2 in §3.1.3). For the term $\mathcal{E}_1$, first note that $\mathcal{H}_\delta(2\pi + \mathcal{H}_\delta)^{-1}\mathcal{F}_\delta$ is compact, since it is the product of the bounded operator $\mathcal{H}_\delta(2\pi + \mathcal{H}_\delta)^{-1}$ and the compact operator $\mathcal{F}_\delta$. Then again by Lemma 3.1.2 in §3.1.3,

$$\|(\mathcal{P}_n - I)\mathcal{H}_\delta(2\pi + \mathcal{H}_\delta)^{-1}\mathcal{F}_\delta\| \to 0$$

Combine this with (9.3.130) to prove $\|\mathcal{E}_1\| \to 0$; this also completes the proof of (9.3.132).

Combine (9.3.132) with the geometric series theorem (cf. Theorem A.1 in the Appendix) to obtain the existence of the inverse of $I + (2\pi + \mathcal{P}_n\mathcal{H}_\delta)^{-1}\mathcal{P}_n\mathcal{F}_\delta$ along with its uniform boundedness for sufficiently large n. We omit the details. This also leads to a proof of the existence and uniform boundedness of $(2\pi + \mathcal{P}_n\mathcal{K})^{-1}$ for all sufficiently large n. Simply use the identity

$$(2\pi + \mathcal{P}_n\mathcal{K})^{-1} = [I + (2\pi + \mathcal{P}_n\mathcal{H}_\delta)^{-1}\mathcal{P}_n\mathcal{F}_\delta]^{-1}(2\pi + \mathcal{P}_n\mathcal{H}_\delta)^{-1}$$

and take bounds, obtaining

$$\|(2\pi + \mathcal{P}_n\mathcal{K})^{-1}\| \leq c < \infty, \quad n \geq N \tag{9.3.133}$$

for N chosen sufficiently large.

For convergence, simply use the standard proof from Theorem 3.1.1 of §3.1.3 to obtain

$$\rho - \rho_n = 2\pi \, (2\pi + \mathcal{P}_n \mathcal{K})^{-1} \, (\rho - \mathcal{P}_n \rho) \qquad (9.3.134)$$

Thus, the speed of convergence of ρ_n to ρ is the same as that of $\mathcal{P}_n \rho$ to ρ, and this agrees with the result in (9.2.63) for the smooth surface case in which the integral operator is compact.

Applications to various interpolatory projections

If the interpolatory projection $\mathcal{P}_n$ is based on piecewise linear interpolation over triangulations of S, then $\|\mathcal{P}_n\| = 1$ and (9.3.129) follows immediately from (9.3.119). The collocation method in Ref. [557] used piecewise constant interpolation (which requires working in $L^\infty(S)$ rather than $C(S)$), and with this, we again have $\|\mathcal{P}_n\| = 1$, and (9.3.129) is also immediate.

If we use the piecewise quadratic interpolation of the preceding section [cf. (9.2.62)], then

$$\|\mathcal{P}_n\| = \frac{5}{3} \qquad (9.3.135)$$

For (9.3.129) to be true, the angles on S must be so restricted that

$$\frac{5}{3} |2\pi - \Omega(P)| < 2\pi$$

or equivalently,

$$\frac{4\pi}{5} < \Omega(P) < \frac{16\pi}{5} \qquad (9.3.136)$$

This is quite restrictive. For example, it does not allow the case that D is a cubical surface, as the lower inequality is not satisfied at the corner points, at which $\Omega(P) = \frac{1}{2}\pi$. Nonetheless, the collocation method (9.3.126) is still empirically convergent in this case, and the collocation methods of Rathsfeld have been shown by him to be convergent in this case.

Numerical integration and surface approximation

The remaining discussion assumes $\mathcal{P}_n$ is the piecewise quadratic interpolatory projection (9.2.62) used in the preceding section. In line with this, we assume the parametrization functions F_j of (9.3.118) all belong to C^4. The linear system to be solved is the same as in (9.2.86), with the exception that the term $[2\pi - \Omega(v_i)] \, u(v_i)$ must be included, consistent with (9.3.117). The collocation integrals are evaluated in exactly the same manner as that described following (9.2.99). In addition, if the surface needs to be approximated, then we use the scheme described in and following (9.2.88). The only difficulty is the estimation of $\Omega(v_i)$, which we now describe.

Returning to the Green's formula (9.1.6), substitute the harmonic function $u \equiv 1$. Then

$$\Omega(P) = \int_S \frac{\partial}{\partial \mathbf{n}_Q} \left[\frac{1}{|P - Q|} \right] dS(Q), \quad P \in S \tag{9.3.137}$$

We approximate $\Omega(v_i)$ using

$$\Omega_n(v_i) = \sum_{k=1}^{n} \sum_{j=1}^{6} \int_\sigma \ell_j(s, t) \frac{(D_s \tilde{m}_k \times D_t \tilde{m}_k) \cdot [\tilde{m}_k(s, t) - v_i]}{|\tilde{m}_k(s, t) - v_i|^3} \, d\sigma \tag{9.3.138}$$

which includes the use of the approximate surface. Later we give some numerical examples of this approximation.

The use of $\Omega(v_i) \approx \Omega_n(v_i)$, along with the evaluation of the remaining integrals in (9.2.86), leads to a linear system

$$(2\pi + \bar{K}_n)\tilde{\rho}_n = \mathbf{f}_n \tag{9.3.139}$$

in analogy with (9.2.90) in §9.2.1. In addition, using $\Omega(v_i) \approx \Omega_n(v_i)$ amounts to forcing

$$\bar{K}_n \mathbf{1} = 2\pi \mathbf{1} \tag{9.3.140}$$

just as in (9.2.113). The proof is straightforward, and we omit it.

With the definition of $\Omega_n(v_i)$ using (9.3.138), one can prove

$$\max_i |\Omega(v_i) - \Omega_n(v_i)| = O\left(\hat{\delta}_n^2\right) \tag{9.3.141}$$

(cf. Atkinson and Chien [59, Theorem 5.2], Chien [119]). We usually observe a faster rate of convergence, as indicated in numerical examples given below, but this is most likely an artifact of using symmetric triangulations in constructing our examples. In theory, the approximation of $\Omega(v_i)$ is a problem only when the surface S is being approximated. Otherwise, the numerical integration of the integrals in (9.3.138) is being done with such accuracy as to have $\Omega_n(v_i) = \Omega(v_i)$ for all practical purposes. For this reason we consider only numerical examples for surfaces S that are not polyhedral.

Table 9.8. *Errors in solid angle approximation $\Omega_n(v_i)$ on an ellipsoid*

i	E_8	E_{32}	E_{128}	E_8/E_{32}	E_{32}/E_{128}	E_{128}/E_{512}
1	1.52E−1	2.01E−2	2.54E−3	6.58	7.58	7.89
2	8.26E−2	1.09E−2	1.38E−3	6.64	7.58	7.89
3	1.17E−1	1.55E−2	1.96E−3	6.62	7.58	7.89
7	1.73E−1	2.35E−2	3.01E−3	8.13	7.34	7.82
8	1.47E−1	2.00E−2	2.56E−3	7.55	7.36	7.82
9	2.34E−1	3.19E−2	4.09E−3	8.29	7.32	7.81

Example 1. The surface S is the ellipsoid

$$\left(\frac{x}{a}\right)^2 + \left(\frac{y}{b}\right)^2 + \left(\frac{z}{c}\right)^2 = 1$$

and we choose $(a, b, c) = (2, 2.5, 3)$. For smooth surfaces, $\Omega(P) \equiv 2\pi$ on S. We give values of $\Omega(v_i) - \Omega_n(v_i)$ at the node points

$$v_1 = (0, 0, 3), \qquad v_2 = (2, 0, 0), \qquad v_3 = (0, 2.5, 0)$$
$$v_7 = (\sqrt{2}, 0, \sqrt{4.5}), \quad v_8 = (\sqrt{2}, \sqrt{3.125}, 0), \quad v_9 = (0, \sqrt{3.125}, \sqrt{4.5})$$

The indices are those from the BIEPACK package of Ref. [52], which was used in computing the examples of this and the preceding section.

Table 9.8 gives the errors $E_n \equiv \Omega(v_i) - \Omega_n(v_i)$ and their ratios for increasing n. These ratios are consistent with a speed of convergence of $O(\hat{\delta}_n^3)$, which is better than (9.3.141). Similar empirical results are true for other smooth surfaces.

Example 2. The surface S is an elliptical paraboloid with a "cap":

$$\left(\frac{x}{a}\right)^2 + \left(\frac{y}{b}\right)^2 = z, \quad 0 \le z \le c$$
$$\left(\frac{x}{a}\right)^2 + \left(\frac{y}{b}\right)^2 \le c, \quad z = c \tag{9.3.142}$$

and the surface parameters are $(a, b, c) = (2, 2, 1)$. We give values of $\Omega(v_i) - \Omega_n(v_i)$ at the node points

$$v_1 = (0, 0, 0), \qquad v_2 = (2, 0, 1), \qquad v_6 = (0, 0, 1)$$
$$v_7 = (1, 0, 0.25), \quad v_8 = (\sqrt{2}, \sqrt{2}, 1), \quad v_{15} = (1, 0, 1) \tag{9.3.143}$$

The points v_6 and v_{15} are on the interior of the cap at $z = 1$, v_1 and v_7 are on the lateral surface, and v_2 and v_8 are edge points at $z = 1$. The angles $\Omega(v_i)$

Table 9.9. *Errors in solid angle approximation* $\Omega_n(v_i)$ *on a paraboloid*

i	$\Omega(v_i)$	E_{32}	E_{128}	E_{512}	E_8/E_{32}	E_{32}/E_{128}	E_{128}/E_{512}
1	2π	$-2.82\text{E}{-9}$	$-2.82\text{E}{-9}$	$-2.82\text{E}{-9}$	1.52	1.00	1.00
2	$\pi/2$	$1.98\text{E}{-2}$	$2.62\text{E}{-3}$	$3.33\text{E}{-4}$	6.47	7.54	7.88
6	2π	$-2.82\text{E}{-9}$	$-2.82\text{E}{-9}$	$-2.82\text{E}{-9}$	3.29	1.00	1.00
7	2π	$5.67\text{E}{-2}$	$-3.97\text{E}{-4}$	$-2.59\text{E}{-3}$	12.9	-143	0.15
8	$\pi/2$	$1.59\text{E}{-1}$	$4.08\text{E}{-2}$	$1.00\text{E}{-2}$	1.60	3.88	4.07
15	2π	$9.72\text{E}{-8}$	$-2.82\text{E}{-9}$	$7.18\text{E}{-9}$	-15.3	-34.5	-0.39

are given in Table 9.9, along with the errors $E_n \equiv \Omega(v_i) - \Omega_n(v_i)$. Referring to notation introduced in §9.2.1, the integration parameters used in obtaining Table 9.9 were $M_{ns} = 3$ and $M_s = 10$, for all values of n.

In general, there appears to be no uniform pattern to the rate at which the error E_n decreases. The accuracy is excellent at v_1, v_6, and v_{15}. This appears to be due to effects of symmetry in the surface, and the accuracy seems limited only by the accuracy of the numerical integration method. The much slower convergence at the remaining smooth node point v_7 is less easily understood, and it may well be more typical of what happens at most smooth points of the surface. The errors at the edge points v_2 and v_8 are only slowly converging to zero. The rate at v_2 is consistent with $O(\hat{\delta}_n^3)$, whereas that for v_8 is only $O(\hat{\delta}_n^2)$. For such empirical measurements of rates of convergence, larger values of n are needed to get a more complete idea of what is happening. We conjecture the rate of convergence at v_7 is $O(\hat{\delta}_n^3)$.

Example 3. We solve the exterior Neumann problem (9.1.26)–(9.1.27) by means of its direct BIE reformulation in (9.1.33):

$$2\pi u + \mathcal{K}u = g \tag{9.3.144}$$

$$g(P) = -\int_S \frac{f(Q)}{|P - Q|}\, dS(Q), \quad P \in S$$

with

$$f(Q) \equiv \frac{\partial u(Q)}{\partial \mathbf{n}_Q}$$

the given normal derivative. The function g is evaluated by the same scheme as is used in the discretization, namely

$$g(v_i) \approx g_n(v_i) \equiv \sum_{k=1}^{n} \sum_{j=1}^{6} f(v_{k,j}) \int_\sigma \frac{\ell_j(s,t)|D_s \tilde{m}_k \times D_t \tilde{m}_k|}{|\tilde{m}_k(s,t) - v_i|}\, d\sigma \tag{9.3.145}$$

Table 9.10. *Errors in solving the exterior Neumann problem*

n	$E_n^{(1)}$	Ratio	$E_n^{(2)}$	Ratio
8	5.73E−2		1.87E−1	
32	1.94E−2	2.95	5.39E−2	3.48
128	4.49E−3	4.33	8.68E−3	6.21
512	2.06E−3	2.18	9.00E−3	0.96

for $i = 1, \ldots, n_v$. It is shown in Ref. [59, Theorem 3.4] that

$$\max_i |g(v_i) - g_n(v_i)| = O(\hat{\delta}_n^3) \tag{9.3.146}$$

with a proof based on a combination of techniques from the proofs of Theorem 9.2.2 from §9.2, and Theorem 5.3.4 from §5.3.3. The integrals in (9.3.145) are evaluated with the same scheme as that described in §9.2.1.

We solve the Neumann problem on the elliptical paraboloid of (9.3.142), and the integration parameters are $M_{ns} = 3$ and $M_s = 10$. We solved the BIE (9.3.144) with the normal derivative f generated from the harmonic functions

$$u^{(1)}(x, y, z) = \frac{1}{r}, \qquad r = \sqrt{x^2 + y^2 + \left(z - \frac{1}{2}c\right)^2} \tag{9.3.147}$$

$$u^{(2)}(x, y, z) = \frac{1}{r} \exp\left(\frac{x}{r^2}\right) \cos\left(\frac{1}{r^2}\left(z - \frac{1}{2}c\right)\right)$$

The maximum errors

$$E_n^{(k)} \equiv \max_i \left|u^{(k)}(v_i) - u_n^{(k)}(v_i)\right|, \quad k = 1, 2$$

are given in Table 9.10.

From combining (9.3.134) with the type of analysis leading to (9.2.98), we can prove

$$E_n^{(k)} = O\left(\hat{\delta}_n^2\right)$$

From Table 9.10, the errors $E_n^{(k)}$ appear to be converging at a rate slower than $O(\hat{\delta}_n^2)$, but with larger values of n, we expect to see the ratios approach 4 or greater. To give some support to this, we give in Table 9.11 the pointwise errors

$$e_{n,i} \equiv u^{(1)}(v_i) - u_n^{(1)}(v_i)$$

Table 9.11. *Pointwise errors at v_i in solving the exterior Neumann problem*

i	$e_{8,i}$	$e_{32,i}$	$e_{128,i}$	$e_{512,i}$	$e_{8,i}/e_{32,i}$	$e_{32,i}/e_{128,i}$	$e_{128,i}/e_{512,i}$
1	3.42E−2	1.30E−2	3.39E−3	1.04E−3	2.63	3.84	3.26
2	3.16E−3	2.14E−3	6.06E−4	1.02E−4	1.48	3.52	5.93
6	5.73E−2	1.94E−2	4.41E−3	6.91E−4	2.95	4.40	6.39
7	2.04E−2	5.90E−3	9.82E−4	2.01E−4	3.46	6.01	4.88
8	5.35E−3	3.34E−3	7.92E−4	1.26E−4	1.60	4.21	6.27
15	3.61E−2	1.04E−2	2.39E−3	3.35E−4	3.48	4.36	7.13

for the node points of (9.3.143). These appear to indicate a rate of $O(\hat{\delta}_n^2)$ or faster.

Note that our paraboloid surface S does not satisfy the restriction (9.3.136) on the solid angle function $\Omega(P)$, as $\Omega(P) = \frac{1}{2}\pi$ along the edge of S at $z = 1$. Nonetheless, the theory still appears to be valid empirically, and our method of analyzing the convergence seems flawed by the need to make the assumption (9.3.129).

From these and other numerical results, we recommend that when solving a BIE on a piecewise smooth surface S, the interpolation of the surface should be of degree 4, provided piecewise quadratic interpolation is used to define the collocation method. This will ensure that the rate of convergence of the collocation method will be at least $O(\hat{\delta}_n^3)$. This is in contrast to the smooth surface case, in which piecewise quadratic interpolation of the surface appears to be adequate. The difficulty in using degree 4 interpolation is its much greater complexity, as it uses 15 basis functions in writing the quartic interpolation formula analogous to the quadratic formula (9.2.62). As an alternative to using quartic interpolation of the surface, use the exact surface S if possible. This too should ensure a rate of convergence of $O(\hat{\delta}_n^3)$ for solution functions ρ that are C^3 on each smooth section of S.

9.3.2. *Iterative solution of the linear system*

Two-grid iteration for solving the linear system $(2\pi + \bar{K}_n)\underline{\tilde{\rho}}_n = \mathbf{g}_n$ associated with the collocation solution of $2\pi u + \mathcal{K}u = g$ was discussed in §6.3.2 of Chapter 6. For the actual two-grid algorithm, see (6.3.109)–(6.3.111). Recall that the fine mesh system $(2\pi + \bar{K}_n)\underline{\tilde{\rho}}_n = \mathbf{g}_n$ is solved iteratively by using a sequence of corrections that are obtained by solving the coarse mesh system $(2\pi + \bar{K}_m)\underline{\tilde{\rho}}_m = \mathbf{g}_m$, some $m < n$, for a sequence of right-hand sides $\mathbf{g}_m$. A numerical example of two-grid iteration was given for (9.3.144) with S an

Table 9.12. *Two-grid iteration for solving (9.3.144)*
on a circular paraboloid

(a, b, c)	m	m_v	n	n_v	Ratio
$(.5, .5, 1)$	8	18	32	66	.615
	8	18	128	258	.592
	8	18	512	1026	.586
	32	66	128	258	.594
	32	66	512	1026	.586
$(1, 1, 1)$	8	18	32	66	.668
	8	18	128	258	.658
	8	18	512	1026	.652
$(2, 2, 1)$	8	18	32	66	.761
	8	18	128	258	.758
	8	18	512	1026	.754
	32	66	128	258	.757
	32	66	512	1026	.754
$(2.5, 2.5, 1)$	8	18	32	66	89.326
	32	66	128	258	.790[a]
	32	66	512	1026	.862

[a]A geometric average of several of the fractions
(9.3.148).

ellipsoid, and the numerical results were given in Table 6.6. We begin this
subsection by giving a numerical example using this two-grid iteration method
to solve (9.3.144) when S is only piecewise smooth. The example shows that the
resulting lack of compactness in the integral operator $\mathcal{K}$ affects the convergence
behavior of the two-grid iteration.

We solve (9.3.144) with S the elliptical paraboloid of (9.3.142), with several
choices for (a, b, c). The Neumann data is taken from the function $u^{(1)}$ of
(9.3.147). Table 9.12 contains information on the two-grid iteration, showing
the behavior as m and n are varied. The quantities m_v and n_v denote the number
of collocation nodes associated with the m and n faces, respectively, in the
triangulation of the surface. The column *Ratio* is an empirical estimate of the
rate at which the two-grid iteration is converging:

$$Ratio \doteq \lim_{k \to \infty} \frac{\left\| \tilde{\rho}_n^{(k+1)} - \tilde{\rho}_n^{(k)} \right\|_\infty}{\left\| \tilde{\rho}_n^{(k)} - \tilde{\rho}_n^{(k-1)} \right\|_\infty} \qquad (9.3.148)$$

with $\tilde{\rho}_n^{(k)}$ denoting iterate k. The value in the table is the above fraction using
the final computed iterates (except for the case marked[a] which is obtained as a
geometric average of several of the above fractions (9.3.148)).

Unlike the behavior seen in Table 6.6 of Chapter 6, the rates of convergence in Table 9.12 do not improve as m is increased [except for the case $(a, b, c) =$ $(2.5, 2.5, 1)$, in which the iteration diverges for $m = 8$]. To give an empirical formula for the convergence rates given in the Table 9.12, first introduce the solid angle Ω_a at the edge for the surface S with $(a, b, c) = (a, a, c)$:

$$\Omega_a = 2 \arctan\left(\frac{2\sqrt{c}}{a}\right)$$

A rough approximation to the empirical rate of linear convergence r_a is given by

$$r_a \approx 1 - .157\Omega_a \qquad (9.3.149)$$

There is some sense in this, since $r_a \approx 0$ implies $\Omega_a \approx 2\pi$, and the latter would imply the surface is smooth, thus making the two-grid iteration behave in the manner seen in Table 6.6.

The main feature of (9.3.149) is that the rate of convergence does not improve as m is increased, unlike what is seen when S is smooth. The source of difficulty is the lack of compactness in $\mathcal{K}$. In Theorem 6.3.1 of §6.3.2 the formula (6.3.115) for the rate of convergence contains the factor $\|(\mathcal{P}_n - \mathcal{P}_m)\mathcal{K}\|$, and this factor converges to zero because of the compactness of $\mathcal{K}$. This is no longer true when S possesses edges or corners, and the analysis of Theorem 6.3.1 no longer applies.

To obtain a convergent iteration method, we look at a way to return to a second kind equation with an integral operator that is compact, in the manner of (9.3.125). Given $\eta > 0$, introduce

$$S_\eta = \{P \in S \mid \text{the distance from } P \text{ to an edge or corner of } S \text{ is } \leq \eta\}$$
$$(9.3.150)$$

We modify the definition (9.3.120) of $\mathcal{H}_\delta$ as follows. Introduce

$$\mathcal{K}_\eta^{1,1} u(P) = \int_{S_\eta} u(Q) \frac{\partial}{\partial \mathbf{n}_Q}\left[\frac{1}{|P - Q|}\right] dS(Q)$$
$$+ [2\pi - \Omega(P)]u(P), \quad P \in S_\eta \qquad (9.3.151)$$

for all $u \in C(S_\eta)$. This is an operator from $C(S_\eta)$ to $C(S_\eta)$, and it captures the essential behavior of $\mathcal{K}$ in the vicinity of edges and corners of S. In analogy with (9.3.121) for $\mathcal{H}_\delta$,

$$\left\|\mathcal{K}_\eta^{1,1}\right\| \approx \omega \qquad (9.3.152)$$

and thus $2\pi I + \mathcal{K}_\eta^{1,1}$ is invertible on $C(S_\eta)$.

For general $u \in C(S)$, consider it as an element of $C(S_\eta) \oplus C(S_\eta^*)$, with $S_\eta^* \equiv S \backslash S_\eta$, and with the association

$$u \simeq \begin{bmatrix} u^1 \\ u^2 \end{bmatrix}, \qquad u^1 = u \,|\, S_\eta, \qquad u^2 = u \,|\, S_\eta^*$$

The equation $(2\pi + \mathcal{K})\rho = g$ is now written as

$$\begin{bmatrix} 2\pi I + \mathcal{K}_\eta^{1,1} & \mathcal{K}_\eta^{1,2} \\ \mathcal{K}_\eta^{2,1} & 2\pi I + \mathcal{K}_\eta^{2,2} \end{bmatrix} \begin{bmatrix} \rho^1 \\ \rho^2 \end{bmatrix} = \begin{bmatrix} g^1 \\ g^2 \end{bmatrix} \qquad (9.3.153)$$

The operators $\mathcal{K}_\eta^{1,2}$, $\mathcal{K}_\eta^{2,1}$, and $\mathcal{K}_\eta^{2,2}$ are defined in a straightforward manner. For example,

$$\mathcal{K}_\eta^{2,1} \rho^1(P) = \int_{S_\eta} \rho(Q) \frac{\partial}{\partial \mathbf{n}_Q} \left[\frac{1}{|P - Q|} \right] dS(Q), \qquad P \in S_\eta^*$$

is an operator from $C(S_\eta)$ to $C(S_\eta^*)$. The operators $\mathcal{K}_\eta^{1,2}$ and $\mathcal{K}_\eta^{2,1}$ have continuous kernel functions, and thus they are compact integral operators. The operator $\mathcal{K}_\eta^{2,2}$ can be analyzed in much the same manner as in §9.2 for the smooth surface case. We consider (9.3.153) as an equation over $C(S_\eta) \oplus C(S_\eta^*)$. Multiply row 1 of the matrix in (9.3.153) by $2\pi(2\pi I + \mathcal{K}_\eta^{1,1})^{-1}$, obtaining the equivalent equation

$$\begin{bmatrix} 2\pi I & 2\pi \left(2\pi I + \mathcal{K}_\eta^{1,1}\right)^{-1} \mathcal{K}_\eta^{1,2} \\ \mathcal{K}_\eta^{2,1} & 2\pi I + \mathcal{K}_\eta^{2,2} \end{bmatrix} \begin{bmatrix} u^1 \\ u^2 \end{bmatrix} = \begin{bmatrix} 2\pi \left(2\pi I + \mathcal{K}_\eta^{1,1}\right)^{-1} g^1 \\ g^2 \end{bmatrix}$$

$$(9.3.154)$$

This is of the general form

$$(2\pi + \mathcal{L})\rho = h \qquad (9.3.155)$$

with $\mathcal{L}$ a compact operator. We now partition our approximating linear system in an analogous manner.

Begin by partitioning the node points $\{v_i\}$ into two sets, to reflect closeness to an edge or corner of S. For a given $\eta \geq 0$, define

$$\mathcal{V}_n^1 = \{v_i \mid \text{the distance from } v_i \text{ to an edge or corner of } S \text{ is } \leq \eta\}$$

$$(9.3.156)$$

$$\mathcal{V}_n^2 = \mathcal{V}_n \backslash \mathcal{V}_n^1$$

With this, decompose any vector $\mathbf{f}_n \in \mathbf{R}^{v_v}$ by

$$\mathbf{f}_n^1 = \mathbf{f}_n \,|\, \mathcal{V}_n^1, \qquad \mathbf{f}_n^2 = \mathbf{f}_n \,|\, \mathcal{V}_n^2 \qquad (9.3.157)$$

and for convenience, write

$$\mathbf{f}_n = \begin{bmatrix} \mathbf{f}_n^1 \\ \mathbf{f}_n^2 \end{bmatrix}$$

Recalling the collocation linear system $(2\pi + K_n)\underline{\rho}_n = \mathbf{g}_n$ (or the system $(2\pi + \check{K}_n)\tilde{\underline{\rho}}_n = \mathbf{g}_n$), we use the above decomposition to rewrite the linear system as

$$2\pi \begin{bmatrix} \rho_n^1 \\ \rho_n^2 \end{bmatrix} + \begin{bmatrix} K_n^{1,1} & K_n^{1,2} \\ K_n^{2,1} & K_n^{2,2} \end{bmatrix} \begin{bmatrix} \rho_n^1 \\ \rho_n^2 \end{bmatrix} = \begin{bmatrix} \mathbf{g}_n^1 \\ \mathbf{g}_n^2 \end{bmatrix} \tag{9.3.158}$$

The matrix $2\pi I + K_n^{1,1}$ is the discretization associated with the integral equation defined over the portion of the boundary S within η of an edge or corner of the surface. We assume that the partition has been so chosen that $2\pi I + K_n^{1,1}$ is nonsingular. In fact, we would expect to have

$$\left\| K_n^{1,1} \right\| \approx \left\| \mathcal{P}_n \right\| \omega \tag{9.3.159}$$

in analogy with (9.3.129), and we assume

$$\left[\sup_n \left\| \mathcal{P}_n \right\| \right] \sup_n \left\| K_n^{1,1} \right\| < 2\pi \tag{9.3.160}$$

This implies that $2\pi I + K_n^{1,1}$ is nonsingular, uniformly in n.

Multiply the first component of (9.3.158) by $2\pi (2\pi I + K_n^{1,1})^{-1}$, to obtain

$$\begin{bmatrix} 2\pi I & 2\pi \left(2\pi I + K_n^{1,1}\right)^{-1} K_n^{1,2} \\ K_n^{2,1} & 2\pi I + K_n^{2,2} \end{bmatrix} \begin{bmatrix} \rho_n^1 \\ \rho_n^2 \end{bmatrix}$$

$$= \begin{bmatrix} 2\pi \left(2\pi I + K_n^{1,1}\right)^{-1} \mathbf{g}_n^1 \\ \mathbf{g}_n^2 \end{bmatrix} \tag{9.3.161}$$

Write this equation abstractly as

$$(2\pi + L_n)\underline{\rho}_n = \mathbf{h}_n \tag{9.3.162}$$

This can be regarded as an approximation of (9.3.155). Since $\mathcal{L}$ is a compact operator, we can develop convergent two-grid iteration methods for (9.3.162) along the lines described in §6.3.2 of Chapter 6.

Let m denote the coarse mesh parameter, with $m < n$. There are analogs of (9.3.162) with m replacing n, and we apply directly the two-grid algorithm of

(6.3.109)–(6.3.111) in §6.3.2. In practice, the matrix inverse $(2\pi I + K_n^{1,1})^{-1}$ is not calculated explicitly; rather, we calculate

$$\mathbf{v}^1 \equiv 2\pi \left(2\pi I + K_n^{1,1}\right)^{-1} K_n^{1,2} \mathbf{g}_n^1$$

by solving the linear system

$$\left(2\pi I + K_n^{1,1}\right) \mathbf{v}^1 = 2\pi K_n^{1,2} \mathbf{g}_n^1 \qquad (9.3.163)$$

This involves fewer calculations per iterate, once the LU-factorization of $2\pi I + K_n^{1,1}$ has been calculated and stored. The two-grid algorithm for (9.3.162) is complicated to implement but still possible, and an example of such is given in BIEPACK in Ref. [52].

Let n_1 and n_2 denote the number of points in $\mathcal{V}_n^1$ and $\mathcal{V}_n^2$, respectively, and $n_v = n_1 + n_2$. The cost of calculating the LU-factorization of $2\pi I + K_n^{1,1}$ is approximately $\frac{2}{3} n_1^3$ operations, and approximately $2n_1^2$ operations are needed for subsequent solutions of (9.3.163). The cost of the remaining operations in the two-grid iteration for solving (9.3.162) is $O(n_v^2)$ operations per iterate. To keep the total costs per iterate at $O(n_v^2)$ operations, it is necessary to so choose n_1 that

$$O\left(n_1^3\right) = O\left(n_v^2\right)$$

Thus n_1 should be chosen to satisfy

$$n_1 = O\left(n_v^{2/3}\right) \qquad (9.3.164)$$

Our examples have used values for n_1 that increased like $\sqrt{n_v}$; our iteration method may have performed better had we allowed (9.3.164).

Example 1. Consider again the example (9.3.144) with S a paraboloid with cap, and recall the two-grid iteration results given in Table 9.12. We now use the two-grid iteration method applied to the fine grid equation (9.3.162), and we refer to it as the *modified two-grid iteration*. The corresponding coarse grid equation is

$$(2\pi + L_m) \underline{\rho}_m = \mathbf{h}_m$$

with a definition of L_m corresponding to that of L_n in (9.3.161). In defining $\mathcal{V}_n^1$ and $\mathcal{V}_n^1$ in (9.3.156), we chose $\eta = 0$, and this resulted in $n_1 = O(\sqrt{n_v})$. From a theoretical perspective, it might have been better to have chosen $\eta > 0$, so as to have the discrete modification (9.3.162) correspond to the original continuous case (9.3.155).

Table 9.13. *Modified two-grid iteration for solving (9.3.144)*
on a circular paraboloid

(a, b, c)	m	m_v	n	n_v	n_1	R	NI
(2, 2, 1)	8	18	32	66	16	.25	14
	8	18	128	258	32	.41	22
	8	18	512	1026	64	.50	28
(2.5, 2.5, 1)	8	18	32	66	16	.31	16
	8	18	128	258	32	.46	25
	8	18	512	1026	64	.55	32
	32	66	128	258	32	.32	16
	32	66	512	1026	64	.47	25
(3, 3, 1)	8	18	32	66	16	.43	18
	8	18	128	258	32	.53	28
	8	18	512	1026	64	.60	36
	32	66	128	258	32	.42	18
	32	66	512	1026	64	.56	28
(4, 4, 1)	8	18	32	66	16	3.45	
	32	66	128	258	32	.74	41
	32	66	512	1026	64	.90	102

The numerical results are given in Table 9.13, and we have three new column headings: n_1, R, and NI. We take R to be the geometric mean of several of the final values of the ratio in (9.3.148). For this particular example, these ratios oscillated, and the oscillations did not become smaller as the number of iterates was increased. The number of iterates computed is denoted by NI, where we iterated until

$$\left\| \tilde{\rho}_n^{(k)} - \tilde{\rho}_n^{(k-1)} \right\|_\infty \leq 10^{-8} \qquad (9.3.165)$$

was satisfied, beginning with $\tilde{\rho}_n^{(0)} = \tilde{\rho}_m$.

A graph of k vs. $-\log \left\| \tilde{\rho}_n^{(k)} - \tilde{\rho}_n^{(k-1)} \right\|_\infty$ is given in Figure 9.2, and we iterated to a smaller error tolerance than that of (9.3.165). The problem parameters were $m = 8$ and $(a, b, c) = (2, 2, 1)$. This graph clearly indicates the modified two-grid iteration has a linear rate of convergence, and it becomes worse with increasing n. Looking at the results in Table 9.13 and at other examples, it appears that with this modified two-grid iteration (and with $\eta = 0$), the rate of convergence is a function of m/n, rather than being a function of m and n separately. Also, since the iteration rate is becoming worse with increasing n, then either we should increase η or we should increase m. Certainly, we could increase η, while still maintaining the restriction (9.3.164), and this is probably the preferable course of action.

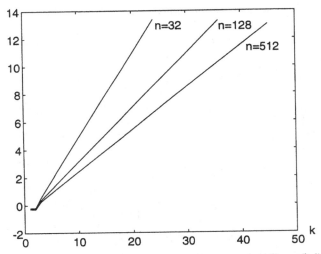

Figure 9.2. For iteration with S a paraboloid, k vs. $-\log \| \tilde{\underline{\rho}}_n^{(k)} - \tilde{\underline{\rho}}_n^{(k-1)} \|_\infty$.

Example 2. Solve equation (9.3.144) with D the solid simplex with vertices $(0, 0, 0)$, $(a, 0, 0)$, $(0, b, 0)$, $(0, 0, c)$. The true solution was chosen to be

$$u(x, y, z) = \frac{1}{\left| (x, y, z) - \frac{1}{4}(a, b, c) \right|}$$

and the normal derivative data was obtained from this function. Again we used $\eta = 0$, and the numerical results are given in Table 9.14. In this case many of the ratios in (9.3.148) approached a limit as k increased.

An extensive comparison of this modified two-grid iteration and other iteration methods for solving the equation $2\pi u + \mathcal{K}u = g$ of (9.3.144) is given in Rathsfeld [453].

9.3.3. Collocation methods for polyhedral regions

The above theory for collocation methods for solving $2\pi\rho + \mathcal{K}\rho = f$ is based on the rather crude estimate

$$\| \mathcal{P}_n \mathcal{H}_\delta \| \le \| \mathcal{P}_n \| \, \| \mathcal{H}_\delta \| \le \beta_\delta \tag{9.3.166}$$

and the assumption (9.3.128) that $\beta_\delta < 2\pi$, together with the decomposition $\mathcal{K} = \mathcal{H}_\delta + \mathcal{F}_\delta$ of (9.3.123). To improve on this, Rathsfeld has introduced a deeper analysis along the lines introduced by Chandler and Graham for the

Table 9.14. *Modified two-grid iteration for solving*
(9.3.144) on a simplex

(a, b, c)	m	m_v	n	n_v	n_1	R	NI
$(3, 3, 3)$	4	10	16	34	22	.29	18
	4	10	64	130	46	.34	20
	4	10	256	514	94	.55	31
$(1, 1, 1)$	4	10	16	34	22	.29	19
	4	10	64	130	46	.35	21
	4	10	256	514	94	.55	33
	16	34	64	130	46	.28	17
	16	34	256	514	94	.43	25
$(1, 1, 2)$	4	10	16	34	22	.34	20
	4	10	64	130	46	.55	33
	4	10	256	514	94	.69	51
	16	34	64	130	46	.35	20
	16	34	256	514	94	.54	31
$(1, 1, 3)$	4	10	16	34	22	.42	24
	4	10	64	130	46	.62	42
	4	10	256	514	94	.74	65
	16	34	64	130	46	.42	24
	16	34	256	514	94	.62	35

planar problem on regions with corners (cf. §8.3). In the papers [447]–[453], Rathsfeld analyses the approximation of $2\pi\rho + \mathcal{K}\rho = f$ using special collocation and Nyström methods for a variety of polyhedral surfaces S. In addition, he introduces graded meshes to improve on the uniform meshes we have used earlier in this chapter.

Recall the formulas of §9.1.5 in which the double layer equation was studied for a conical surface Γ, and in this instance, the double layer integral operator was denoted by $\mathcal{W}$. In particular, recall (9.1.56)–(9.1.57) in which

$$\mathcal{W}g(r\omega(s)) = \int_0^\infty \mathcal{B}\left(\frac{r}{\tau}\right) g(\tau\omega(s))\frac{d\tau}{\tau}, \quad r \geq 0, \ 0 \leq s \leq L$$

with $\mathcal{B}(t)$ an operator on $C(\gamma)$:

$$\mathcal{B}(t)h(\omega(s)) = \int_\gamma h(\omega(\sigma))\frac{t\mathbf{n}_{\omega(\sigma)} \cdot \omega(s)}{|t\omega(s) - \omega(\sigma)|^3}\, d\sigma,$$

$$h \in C(\gamma), \ 0 \leq t < \infty$$

For any $\epsilon > 0$, introduce

$$\mathcal{W}_\epsilon g(r\omega(s)) = \int_\epsilon^\infty \mathcal{B}\left(\frac{r}{\tau}\right) g(\tau\omega(\sigma))\frac{d\tau}{\tau}, \quad r \geq 0, \ 0 \leq s \leq L \quad (9.3.167)$$

Basic to the work of Rathsfeld [453, cf. (3.1)] is the assumption that $(2\pi + \mathcal{W}_\epsilon)^{-1}$ is invertible for all sufficiently small ϵ, say $0 < \epsilon \leq \epsilon_0$, and in addition, that

$$\sup_{0 < \epsilon \leq \epsilon_0} \|(2\pi + \mathcal{W}_\epsilon)^{-1}\| < \infty \qquad (9.3.168)$$

In showing this assumption, we are to consider the tangent cones Γ formed at each of the corners of the polyhedral surface S. The assumption (9.3.168) is straightforward for surfaces S satisfying the assumption Ξ of Wendland, as given earlier in §9.3 following (9.3.119). It is also known to be true for some other surfaces, and this is discussed at greater length in Ref. [453, p. 929]. If we return to the analysis of collocation methods for the comparable problem in the plane, in §8.3 of Chapter 8, then the assumption (9.3.168) has an analog that can be proven using the results in Anselone and Sloan [20] for *finite section Wiener–Hopf equations*. Unfortunately, analogous results are not known for problems on conical surfaces.

In his papers [447]–[453] Rathsfeld defines collocation and Nyström methods for solving $2\pi\rho + \mathcal{K}\rho = f$, and these use meshes that are graded as one approaches an edge or corner of S. The grading is needed to obtain stability of the numerical method, in addition to the assumption (9.3.168). The grading of the mesh is also needed to compensate for likely ill-behavior in the solution function about edges and corners of S (cf. Theorem 9.1.2 in §9.1.5) and thus to obtain optimal rates of convergence. The portion of Rathsfeld's work that is based on using grading to improve the approximation of the solution can also be used with the methods given earlier in this section, provided one has (9.3.166) with $\beta_\delta < 2\pi$. But to give truly general results that do not need the assumption $\beta_\delta < 2\pi$, Rathsfeld's approach to showing stability appears superior. The types of graded meshes being used are described in the following §9.4 in connection with the approximation of $(2\pi + \mathcal{K})\rho = f$ by Galerkin's method.

It is important to note, however, that Rathsfeld's approach requires modifying the collocation method as given earlier in this section to change the interpolation being used in the vicinity of edges and corners of S. Such modifications were also necessary with the collocation methods of Chandler and Graham for planar problems with piecewise smooth boundaries, as presented earlier in §8.3. A detailed presentation of Rathsfeld's methods and error analysis results is omitted here, primarily for reasons of space, as his results are quite technical in both their presentation and analysis; instead we refer the reader to his original papers [447]–[453].

9.4. Boundary element Galerkin methods

In this text, Galerkin methods have been studied from two perspectives: that of a projection method (as in Chapter 3) and that of the minimization of an "energy function" (as in §7.4 of Chapter 7). Both perspectives are used in this section to analyze Galerkin methods for boundary integral equations of both the first and second kind. We begin with the second kind equation

$$2\pi\rho + \mathcal{K}\rho = f \qquad (9.4.169)$$

of (9.1.33) and (9.1.36), for which collocation methods were analyzed in the preceding two sections. Following that, we consider the numerical solution of the first kind single layer equation

$$\mathcal{S}\psi = f \qquad (9.4.170)$$

of (9.1.38).

The analysis of the Galerkin method for solving (9.4.169) with S a smooth surface is relatively straightforward, with an error analysis similar to that of Theorem 9.2.1 for collocation methods on smooth surfaces. We consider the more difficult case, that S is only piecewise smooth, and in particular, that S is a Lipschitz polyhedral surface. In addition to being composed of polygonal faces, this also implies that at each edge or corner point $P \in S$, the surface can be represented locally about P as a function of two variables. Elschner [185] presents and analyzes a Galerkin method that generalizes and greatly extends ideas used by Chandler and Graham in their analysis of collocation methods for planar problems with polygonal boundaries (cf. §8.3), and we describe below this work of Elschner.

As was discussed near the end of §9.1, in Theorem 9.1.2, the solution ρ of (9.4.169) is usually not well-behaved in the neighborhood of edges and corners of S. As with the planar case, one can compensate for this by grading the mesh for S, using a relatively smaller mesh around edges and corners of S. To describe the mesh, consider a single face S_j of S. Let d be the minimum of the lengths of the edges of S, and let ϵ be chosen to satisfy

$$0 < \epsilon \le \frac{d}{3}$$

We describe in several steps the triangulation of S_j, with some of the steps illustrated in Figure 9.3.

1. Draw lines parallel to the edges of S_j at a distance of ϵ from them. This is illustrated in Figure 9.3(a). Small quadrilateral-shaped regions will be

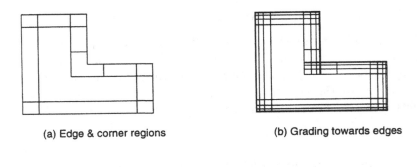

(a) Edge & corner regions   (b) Grading towards edges

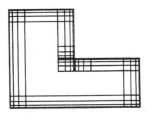   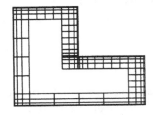

(c) Grading at concave corner   (d) Subdividing remaining edge regions

Figure 9.3. Steps in the triangulation of S_j.

created around each convex corner. For concave corners of S_j we require
a slight modification to create a neighboring L-shaped region, as illustrated
in Figure 9.3(a).

2. Let $m \geq 2$ and $q \geq 1$ be given constants. Inside S_j draw lines parallel to
   the edges of S_j at distances of

$$\left(\frac{i}{m}\right)^q \epsilon, \quad i = 0, 1, \dots, m \qquad (9.4.171)$$

as illustrated in Figure 9.3(b) with $m = 4$ and $q = 2$. We refer to this as
"grading towards the edges" of S_j. This produces a doubly graded mesh of
quadrilaterals in neighborhoods of convex corners of S_j.

3. To complete the grading needed around concave corners, we must draw in
   additional lines as illustrated in Figure 9.3(c), again based on (9.4.171).

4. Consider the edge regions between the corner regions. Divide these uniformly with m lines drawn perpendicular to the edge, as illustrated in
   Figure 9.3(d).

5. This leaves only an inner region of S_j to be subdivided. For it, create a uniform subdivision into convex quadrilaterals, with the number being $O(m^2)$.

6. Finally, subdivide all quadrilaterals into triangles, to obtain a triangulation of S_j with $O(m^2)$ triangular elements.

If these are collected together for each of the faces of S, then we obtain a triangulation $\mathcal{T}_n$, with $n = O(m^2)$.

To define the approximating subspace for the Galerkin method, choose some integer $r \geq 0$ and let $\mathcal{X}_n$ denote all functions ψ that are piecewise polynomial of degree $\leq r$ over $\mathcal{T}_n$, with no continuity restrictions. We do, however, require that $\psi(P) \equiv 0$ over the ϵ_0-neighborhood of the edges of each face S_j,

$$\epsilon_0 = \left(\frac{i_0}{m} \right)^q \epsilon$$

with the neighborhood as illustrated in Figure 9.3(a) with ϵ replaced by ϵ_0. The constant i_0 is chosen independently of m (or n), and its choice is discussed below in Theorem 9.4.1. Let $\mathcal{P}_n$ be the orthogonal projection of $L^2(S)$ onto $\mathcal{X}_n$.

The Galerkin method is given abstractly by

$$(2\pi + \mathcal{P}_n\mathcal{K})\rho_n = \mathcal{P}_n f$$

Elschner [185] transforms the problem of inverting $2\pi + \mathcal{P}_n\mathcal{K}$ over $L^2(S)$ to that of inverting it over $L^2(\Gamma(P))$ for each of the corner vertices P of S, with $\Gamma(P)$ the infinite tangent cone generated by the surface S at P. To carry out an error analysis for this process, an assumption must be made of the kind assumed by Rathsfeld in (9.3.168).

Let Γ be an infinite polyhedral cone, and let Γ_δ be formed by removing from Γ all points within a distance δ of an edge of Γ. Let the resulting double layer integral operator over $L^2(\Gamma_\delta)$ be defined by $\mathcal{W}_\delta$. Then the analysis in Ref. [185] requires that for all tangent cones $\Gamma \equiv \Gamma(P)$ with P a vertex of S, and for some $\delta > 0$, $2\pi + \mathcal{W}_\delta$ is invertible over $L^2(\Gamma_\delta)$. If this is true for some $\delta > 0$, then it is true for all $\delta > 0$, as can be proven by using the integral formula in (9.3.167). We refer to this as the *finite section assumption*. As with the earlier assumption of Rathsfeld in (9.3.168), the invertibility of $2\pi + \mathcal{W}_\delta$ is immediate if the surface S satisfies the assumption Ξ of §9.3, and then also the inverses are uniformly bounded with respect to δ.

Theorem 9.4.1. *Let the triangulations $\mathcal{T}_n$ and the approximating families $\mathcal{X}_n$ be constructed as above, with $n = O(m^2)$. Assume the finite section assumption is true. Then:*

(a) *if i_0 is chosen sufficiently large, the approximating operators $2\pi + \mathcal{P}_n \mathcal{K}$ are invertible on $L^2(S)$ for all sufficiently large values of n, say $n \geq n_0$; moreover,*

$$\| (2\pi + \mathcal{P}_n \mathcal{K})^{-1} \| \leq c, \quad n \geq n_0$$

for some constant $c < \infty$.

(b) *if the right side $f \in L^2(S)$ in (9.4.169) belongs to the classical Sobolev space of order $r + 2$ on each of the faces of S, and if the mesh grading parameter q satisfies $q \geq 2(r + 1)$ then the approximate solutions $\rho_n = (2\pi + \mathcal{P}_n \mathcal{K})^{-1} \mathcal{P}_n f$ satisfy*

$$\| \rho - \rho_n \| \leq \frac{c}{m^{r+1}} \tag{9.4.172}$$

Proof. The proof is given in Ref. [185]. Since it is quite lengthy, and since it utilizes a number of more sophisticated ideas that would also need to be developed, we omit the proof. □

The rate of convergence in (9.4.172) is comparable to that given earlier with uniform meshes. For example, with degree r approximations in §9.2, we had a rate of convergence of $O(\hat{\delta}_n^{r+1})$ in Theorem 9.2.1. That is comparable to the above, since then

$$n = O\big(\delta_n^{-2}\big) \quad \text{and} \quad n = O(m^2)$$

At this time, it is not clear whether or not a value of i_0 greater than zero is needed in practice. In the theorem, note also that stability is proven for all $q \geq 1$, provided only that i_0 is chosen sufficiently large. The grading is needed to compensate for possible bad behavior in the solution ρ in the neighborhood of edges and corners of S.

9.4.1. A finite element method for an equation of the first kind

Consider solving the first kind single layer integral equation

$$\int_S \frac{\psi(Q)}{|P - Q|} \, dS(Q) = f(P), \quad P \in S \tag{9.4.173}$$

and write it symbolically as

$$\mathcal{S}\psi = f \tag{9.4.174}$$

We assume S is the boundary of a simply connected bounded region $D \subset \mathbf{R}^3$, and we also assume S is a smooth surface. Examples of this equation were given earlier in (9.1.30) and (9.1.38) of §9.1. Mapping properties of S were discussed following (9.1.47), with (9.1.53) being particularly important for the following work.

Nedelec [396] introduced a finite element method to solve (9.4.173) using the type of framework given earlier in §7.4 of Chapter 7. We now describe his results. Using (9.1.53), we have

$$S : H^{-\frac{1}{2}}(S) \xrightarrow[onto]{1-1} H^{\frac{1}{2}}(S) \qquad (9.4.175)$$

making S a pseudodifferential operator of order -1. This means that (9.4.174) has a solution $\psi \in H^{-\frac{1}{2}}(S)$ for each given $f \in H^{\frac{1}{2}}(S)$, and ψ is the unique such solution in $H^{-\frac{1}{2}}(S)$.

Proceeding in analogy with what was done in §7.4, we begin by introducing the bilinear functional

$$\mathcal{A} : H^{-\frac{1}{2}}(S) \times H^{-\frac{1}{2}}(S) \to \mathbf{R} \qquad (9.4.176)$$

with

$$\mathcal{A}(\eta, \varphi) = (\eta, S\varphi) = \int_S \int_S \frac{\eta(P)\,\varphi(Q)}{|P - Q|}\, dS(Q)\, dS(P), \quad \eta, \varphi \in H^{-\frac{1}{2}}(S) \qquad (9.4.177)$$

Of course, the use of integration no longer makes sense if either η or ψ does not belong to $L^2(S)$, and we are instead using the integration notation to denote the bounded extension of S to $H^{-\frac{1}{2}}(S)$. The bilinear functional denoted by $(\eta, S\varphi)$ is the unique extension to $H^{-\frac{1}{2}}(S) \times H^{\frac{1}{2}}(S)$ of the standard inner product on $L^2(S)$, and it was denoted by $\langle \eta, S\varphi \rangle$ in §7.4.1 following (7.4.196). These ideas were discussed previously in §7.4 for planar problems, and we refer the reader to that earlier material. The form of (9.4.177) makes it clear that $\mathcal{A}$ is a symmetric operator.

In Nedelec and Planchard [397], it is also shown that $\mathcal{A}$ is strongly elliptic:

$$|\mathcal{A}(\varphi, \varphi)| \geq c_e \|\varphi\|^2_{-\frac{1}{2}}, \quad \varphi \in H^{-\frac{1}{2}}(S) \qquad (9.4.178)$$

and that $\mathcal{A}$ is bounded with

$$|\mathcal{A}(\eta, \varphi)| \leq c_{\mathcal{A}} \|\eta\|_{-\frac{1}{2}} \|\varphi\|_{-\frac{1}{2}} \qquad (9.4.179)$$

For $f \in H^{\frac{1}{2}}(S)$, introduce the linear functional $\ell : H^{-\frac{1}{2}}(S) \to \mathbf{R}$:

$$\ell(\eta) = (\eta, f), \quad \eta \in H^{-\frac{1}{2}}(S) \qquad (9.4.180)$$

This again uses the extension to $H^{-\frac{1}{2}}(S) \times H^{\frac{1}{2}}(S)$ of the standard inner product defined on $L^2(S)$, as discussed in the preceding paragraph. With the above definitions and results, it follows that the abstract hypotheses **A1–A4** of §7.4, following (7.4.215), are satisfied. Consequently, we can apply the Lax–Milgram theorem (Theorem 7.4.1) and Cea's lemma (Theorem 7.4.2) to the study and solution of $S\psi = f$.

Let $\mathcal{T}_n = \{\Delta_k\}$ denote a sequence of triangulations of S, for some sequence of integers $n \to \infty$, in the general manner discussed in §5.3 of Chapter 5, and let h denote the mesh size. Choose an integer $r \geq 0$ and let $\mathcal{X}_n$ denote the set of all functions φ that are piecewise polynomial of degree $\leq r$ in the parametrization variables, as discussed in §5.3. For example, see the interpolatory form of such functions given in (5.3.102) of §5.3.1. There are no continuity restrictions of the functions in $\mathcal{X}_n$. The dimension of $\mathcal{X}_n$ is nf_r with n the number of elements in $\mathcal{T}_n$ and with

$$f_r \equiv \frac{1}{2}(r+1)(r+2)$$

the number of independent polynomials of degree $\leq r$. Since there are no continuity restrictions, $\mathcal{X}_n \subset L^2(S) \equiv H^0(S)$ is the most we can say.

The finite element solution of $S\psi = f$ is obtained by finding the function $\psi_n \in \mathcal{X}_n$ for which

$$\mathcal{A}(\eta, \psi_n) = \ell(\eta), \quad \text{all } \eta \in \mathcal{X}_n \tag{9.4.181}$$

From Theorem 7.4.2, we have that there is a unique such ψ_n, and from (7.4.246) of that theorem,

$$\|\psi - \psi_n\|_{-\frac{1}{2}} \leq \frac{c_\mathcal{A}}{c_e} \inf_{\eta \in \mathcal{X}_n} \|\psi - \eta\|_{-\frac{1}{2}} \tag{9.4.182}$$

Easily, this implies the convergence of ψ_n to ψ as $n \to \infty$. This result can also be converted to a bound for $\|\psi - \psi_n\|_0$, using the standard norm of $L^2(S)$, following ideas used in the Case **2** of §7.4.3.

In the language of Babŭska and Aziz [71, Chap. 4], the approximating space $\mathcal{X}_n$ is an $(r+1, 0)$-regular system. For $\mathcal{X}_n$, it can be proven from the results of Ref. [71, Theorems 4.1.3, 4.1.5] that

P1 (Inverse property). Let $s \leq t \leq 0$. Then

$$\|\eta\|_t \leq c(s, t)\|\eta\|_s h^{s-t}, \quad \eta \in \mathcal{X}_n \tag{9.4.183}$$

with $c(s, t)$ independent of n.

P2 (Approximation property). Let $t \leq s \leq r + 1$, $t \leq 0$. Then for any $\varphi \in H^s(S)$ and any n, there is an element $\varphi_n \in \mathcal{X}_n$ with

$$\|\varphi - \varphi_n\|_t \leq c(s, t)\|\varphi\|_s h^{s-t} \qquad (9.4.184)$$

where $c(s, t)$ is independent of φ, φ_n, and n.

These are the analogs of (7.4.264) and (7.4.265) in §7.4.3 of Chapter 7.

Then the form of proof given in §7.4.3, following (7.4.265), can again be used to show

$$\|\psi - \psi_n\|_0 \leq c h^{r+1} \|\psi\|_{r+1}, \quad \psi \in H^{r+1}(S) \qquad (9.4.185)$$

This gives a complete convergence analysis for the Galerkin method of (9.4.181).

Nedelec [396] also looks at the effects of approximating the surface S and of approximating the right-hand function f in the equation $\mathcal{S}\psi = f$. In particular, we approximate S using interpolation of it over the triangulation $\mathcal{T}_n$, in the manner described in §5.3. Let the approximation of S be based on polynomial interpolation of degree k and denote the approximate surface by $\tilde{S}_n$. The Galerkin method (9.4.181) is approximated by the problem of finding $\tilde{\psi}_n \in \tilde{\mathcal{X}}_n$ such that

$$\int_{\tilde{S}_n} \int_{\tilde{S}_n} \frac{\eta(P)\tilde{\psi}_n(Q)}{|P - Q|} \, dS(Q) \, dS(P) = \int_{\tilde{S}_n} \eta(P) f_n(P) \, dS(P), \quad \text{all } \eta \in \tilde{\mathcal{X}}_n \qquad (9.4.186)$$

In this, $f_n \approx f$, and f_n is defined over $\tilde{S}_n$. The approximating subspace $\tilde{\mathcal{X}}_n$ is the analog of $\mathcal{X}_n$, but it is defined over the triangulation of $\tilde{S}_n$ resulting from the interpolation of the triangular elements in the triangulation $\mathcal{T}_n$ for S.

For the error bounds on $\tilde{\psi}_n$ we must compare it with ψ, which is defined on S. Introduce a mapping $\Psi_n : \tilde{S}_n \to S$ as follows. Let $\epsilon > 0$ be a sufficiently small parameter (to be chosen later) and consider those surfaces $\tilde{S}_n$ that are located entirely within an ϵ-neighborhood of S. At a point $P \in S$, construct a normal to S, and where this normal intersects $\tilde{S}_n$ we will call $\Psi_n^{-1}(P)$. If ϵ is sufficiently small, then each $P \in S$ will correspond to a distinct point on $\tilde{S}_n$, and Ψ_n will be well-defined. To simplify our error formulas, we assume that f has been extended to $f_n \in \tilde{S}_n$ by using

$$f_n(\tilde{P}) = f(\Psi_n(\tilde{P})), \quad \tilde{P} \in \tilde{S}_n \qquad (9.4.187)$$

One of the main convergence results of Nedelec [396, Theorem 2.1] is that

$$\left\| \psi - \tilde{\psi}_n \circ \Psi_n^{-1} \right\|_0 \leq c[h^{r+1}\|\psi\|_{r+1} + h^k\|\psi\|_0], \quad \psi \in H^{r+1}(S) \qquad (9.4.188)$$

The proof is quite complicated, and we refer the reader to the original paper [396]. That paper also considers the use of approximations other than (9.4.187).

One of the major consequences of (9.4.188) is that k should be chosen to equal $r + 1$, and thus the surface interpolation should be based on using polynomials of degree 1 greater than the approximation of the solution. This is in contrast to what we observed empirically in §9.2 for BIE of the second kind, in which interpolation of the surface with degree 2 polynomials and approximation of the unknown by quadratic polynomials led to an error of size $O(h^3)$ or larger in order.

Generalizations to other boundary integral equations

Wendland, his students and colleagues (including Costabel, Hsiao, Stephan, and others) developed a general theory of boundary element Galerkin methods for solving Laplace's equation, the Helmholtz equation, and many other classic partial differential equations. They obtained boundary integral equations that could be used to define bilinear functionals analogous to $\mathcal{A}$, and these were strongly elliptic. A convergence analysis similar to that given above by Nedelec was then given for a much larger class of partial differential equation boundary value problems. For a presentation of this general theory, see Wendland [558]–[560], [564], and [565].

9.5. Numerical methods using spherical polynomial approximations

We return to the global approximation methods of §5.5 of Chapter 5, applying them to the approximation of boundary integral equations over smooth surfaces. As in §5.5, we assume S is a smooth surface for which there is a smooth mapping

$$M : U \xrightarrow[\text{onto}]{1-1} S, \quad U = \text{unit sphere} \qquad (9.5.189)$$

with a smooth inverse M^{-1}. We convert boundary integral equations over S, for example,

$$\lambda \rho(P) - \int_S K(P, Q)\rho(Q)\,dS(Q) = f(P), \quad P \in S \qquad (9.5.190)$$

to equivalent integral equations over U, and then we discretize the latter integral equation by using spherical polynomials to approximate functions over U.

We begin by considering the second kind equation $(2\pi + \mathcal{K})\rho = f$ with $\mathcal{K}$ the double layer potential operator, and then we consider the first kind equation $\mathcal{S}\psi = f$ with $\mathcal{S}$ a single layer potential operator. As was noted earlier in §5.5, there does not exist an interpolation theory involving spherical polynomials, and therefore we cannot present a collocation theory for the boundary integral equations just mentioned. Also, the kernel functions for $\mathcal{K}$ and $\mathcal{S}$ are singular, and therefore the Nyström method of Theorem 5.5.1 in §5.5.3 is not suitable for integral equations containing $\mathcal{K}$ and $\mathcal{S}$. This forces us to consider only Galerkin methods and discrete variants of them, such as were discussed in and following Theorem 5.5.2 in §5.5.3.

The reader should review §5.5, as we will use the notation and results presented there. In particular, for functions $\rho \in C(S)$, let

$$\hat{\rho}(P) = \rho(M(P)), \quad P \in U$$

We write this transformation as

$$\mathcal{M}\rho = \hat{\rho}$$

defining the linear operator $\mathcal{M} : C(S) \overset{1-1}{\underset{onto}{\to}} C(U)$. Easily $\mathcal{M}$ is an isometric mapping,

$$\|\mathcal{M}\rho\|_\infty = \|\rho\|_\infty, \quad \rho \in C(S)$$

The integral equation (9.5.190) is transformed to

$$\lambda\hat{\rho}(P) - \int_U \hat{K}(P, Q)\hat{\rho}(Q)\, dS(Q) = \hat{f}(P), \quad P \in U \qquad (9.5.191)$$

with $\mathcal{M}f = \hat{f}, \mathcal{M}\rho = \hat{\rho}$,

$$\hat{K}(P, Q) = K(M(P), M(Q))J_M(Q) \qquad (9.5.192)$$

and $J_M(Q)$ the Jacobian of the transformation (9.5.189). From the assumptions on M, the Jacobian $J_M(Q)$ is a smooth nonzero function. Examples for S an ellipsoidal surface are given in the introductory paragraphs of §5.5.

Convert the double layer equation $(2\pi + \mathcal{K})\rho = f$ in the manner of (9.5.191) and denote it by

$$(2\pi + \hat{\mathcal{K}})\hat{\rho} = \hat{f} \qquad (9.5.193)$$

with $\hat{\mathcal{K}}$ the converted double layer operator. Using the compactness of $\mathcal{K}$ from $L^2(S)$ to $L^2(S)$, it is essentially immediate that $\hat{\mathcal{K}}$ is a compact mapping from $L^2(U)$ to $L^2(U)$. Similarly, $\hat{\mathcal{K}}$ is a compact mapping from $C(U)$ to $C(U)$.

Let $\mathcal{S}_n$ denote the set of spherical polynomials of degree $\leq n$. For information concerning them, see §5.5.1. The Galerkin method for solving (9.5.193) is given by

$$(2\pi + \mathcal{P}_n\hat{\mathcal{K}})\hat{\rho}_n = \mathcal{P}_n\hat{f} \qquad (9.5.194)$$

with $\mathcal{P}_n$ the orthogonal projection of $L^2(U)$ onto $\mathcal{S}_n$. It is straightforward to give an error analysis of (9.5.194) within the framework of $L^2(U)$, and it amounts essentially to Theorem 5.5.2 in §5.5.3. Summarizing, $(2\pi + \mathcal{P}_n\hat{\mathcal{K}})^{-1}$ exists for all sufficiently large n, say $n \geq n_0$, and it is uniformly bounded for all such n. In addition,

$$\|\rho - \rho_n\|_{L^2} \leq 2\pi \|(2\pi + \mathcal{P}_n\hat{\mathcal{K}})^{-1}\|\,\|\rho - \mathcal{P}_n\rho\|_{L^2} \qquad n \geq n_0 \qquad (9.5.195)$$

using the norm $\|\cdot\|_{L^2}$ of $L^2(U)$. We are more interested in carrying out an error analysis that leads to uniform error bounds, and this requires a more careful consideration of the mapping properties of the integral operator $\mathcal{K}$ with respect to the space $C(S)$.

A *Lyapunov* surface S is defined as follows. For each $P \in S$, let there exist a continuous mapping F_P from some open region $\Omega \subset \mathbf{R}^2$ to a neighborhood of P on S. In addition, assume that F_P is continuously differentiable, and that its first derivatives DF_P are Hölder continuous with exponent λ for some $0 < \lambda \leq 1$:

$$|DF_P(\mathbf{x}) - DF_P(\mathbf{y})| \leq c|\mathbf{x} - \mathbf{y}|^\lambda, \quad \mathbf{x}, \mathbf{y} \in \Omega \qquad (9.5.196)$$

The constants c and λ are assumed to be independent of P. The set of all such surfaces is denoted by $L_{1,\lambda}$, and $S \in L_{1,\lambda}$ is shorthand for saying S is a Lyapunov surface. For the parametrization function F_P, let $D^j F_P$ denote a generic derivative of order j. With some $k \geq 2$, assume the surface S is such that $D^j F_P$ exists and is continuous for all $j \leq k$, and further assume

$$|D^k F_P(\mathbf{x}) - D^k F_P(\mathbf{y})| \leq c|\mathbf{x} - \mathbf{y}|^\lambda, \quad \mathbf{x}, \mathbf{y} \in \Omega \qquad (9.5.197)$$

for some $0 < \lambda \leq 1$, with c and λ independent of P. Then we write $S \in L_{k,\lambda}$. The definition of such functions F_P can be obtained from the function M of (9.5.189), and the mapping properties of F_P and M are tied to each other. We can always limit our interest to only a finite number of such points P and associated functions F_P, say $P_1, \ldots, P_k$, chosen so that

$$\bigcup_1^k F_{P_i}(\Omega) = S$$

Let $S \in L_{k,\lambda}$ for some $k \geq 1$. Given a function $f \in C(S)$, we say f is *Hölder continuous* with exponent μ on S if for each $P \in S$, f satisfies

$$|f(F_P(\mathbf{x})) - f(F_P(\mathbf{y}))| \leq c(f)|\mathbf{x} - \mathbf{y}|^\lambda, \quad \mathbf{x}, \mathbf{y} \in \Omega \qquad (9.5.198)$$

We denote this by writing $f \in C^{0,\lambda}(S)$. If in addition, $D^j f(F_P(\mathbf{x}))$ exists and is continuous for all $j \leq k$, with

$$|D^k f(F_P(\mathbf{x})) - D^k f(F_P(\mathbf{y}))| \leq c(f)|\mathbf{x} - \mathbf{y}|^\lambda, \quad \mathbf{x}, \mathbf{y} \in \Omega$$

then we write $f \in C^{k,\lambda}(S)$.

Günter [246] gives the mapping properties of single and double layer integral operators in terms of the spaces $C^{k,\lambda}(S)$. In particular, we have the following results. From Ref. [246, p. 49],

$$\rho \in C(S), \ S \in L_{1,\lambda} \Rightarrow \mathcal{K}\rho \in \begin{cases} C^{0,\lambda}(S), & 0 < \lambda < 1 \\ C^{0,\lambda'}(S), & \lambda = 1 \end{cases} \qquad (9.5.199)$$

with $0 < \lambda' < 1$ arbitrary. For simplicity in notation, and with respect to only (9.5.199) and its consequences, we will let $\lambda' = \lambda$ whenever $0 < \lambda < 1$. Letting ρ be the solution of $(2\pi + \mathcal{K})\rho = f$, write

$$\rho = \frac{1}{2\pi}(f - \mathcal{K}\rho)$$

Then the result (9.5.199) implies

$$f \in C^{0,\lambda}(S), \ S \in L_{1,\lambda} \Rightarrow \rho \in C^{0,\lambda'}(S) \qquad (9.5.200)$$

If f has greater smoothness on S, then this regularity result can be improved. Let $k \geq 0$. From Ref. [246, p. 312],

$$\rho \in C^{k,\lambda}(S), \ S \in L_{k+2,\lambda} \Rightarrow \mathcal{K}\rho \in C^{k+1,\lambda'}(S) \qquad (9.5.201)$$

with $0 < \lambda' < \lambda$ arbitrary; and in general, $\lambda' \neq \lambda$. Using simple induction, we obtain the regularity result

$$f \in C^{k,\lambda}(S), \ S \in L_{k+1,\lambda} \Rightarrow \rho \in C^{k,\lambda'}(S), \quad k \geq 1 \qquad (9.5.202)$$

Günter [246] also contains results on the mapping properties of the single layer operator $\mathcal{S}$ and other boundary integral operators for Laplace's equation. Generalizations of these results to the double layer operator associated with the Helmholtz equation are given in Lin [343].

Returning to the Galerkin method (9.5.194), consider it and the original equation $(2\pi + \hat{\mathcal{K}})\hat{\rho} = \hat{f}$ in the framework of $C(U)$.

Theorem 9.5.1. *Assume $S \in L_{1,\lambda}$, and assume $\frac{1}{2} < \lambda \leq 1$. Then*

(i)

$$\|\hat{\mathcal{K}} - \mathcal{P}_n \hat{\mathcal{K}}\| \to 0 \quad \text{as } n \to \infty \qquad (9.5.203)$$

Consequently, $(2\pi + \mathcal{P}_n \hat{\mathcal{K}})^{-1}$ exists and is uniformly bounded on $C(U)$ for all sufficiently large n, say $n \geq n_0$.

(ii) If $f \in C^{k,\lambda}(S)$ and $S \in L_{k+1,\lambda}$, with $k + \lambda > \frac{1}{2}$, then

$$\|\hat{\rho} - \hat{\rho}_n\|_\infty \leq \frac{c}{n^{k+\lambda'-1/2}}, \quad n \geq n_0 \qquad (9.5.204)$$

with $\lambda' < \lambda$ arbitrary. The constant c depends on f, k, and λ'.

Proof. (i) To show (9.5.203), we must return to (9.5.199) and give a more precise statement of it. For $\rho \in C(S)$,

$$|\mathcal{K}\rho(F_P(\mathbf{x})) - \mathcal{K}\rho(F_P(\mathbf{y}))| \leq c\|\rho\|_\infty |\mathbf{x} - \mathbf{y}|^{\lambda'}, \quad \mathbf{x}, \mathbf{y} \in \Omega \qquad (9.5.205)$$

with c independent of ρ. This is proven in Günter [246, p. 49]. Using the assumption $S \in L_{1,\lambda}$, (9.5.205) implies an analogous result for $\hat{\mathcal{K}}\rho$:

$$|\hat{\mathcal{K}}\hat{\rho}(P_1) - \hat{\mathcal{K}}\hat{\rho}(P_2)| \leq c\|\hat{\rho}\|_\infty |P_1 - P_2|^{\lambda'}, \quad P_1, P_2 \in U \qquad (9.5.206)$$

We can choose $\lambda' > \frac{1}{2}$, using the assumption $\frac{1}{2} < \lambda \leq 1$ of the theorem.

Recall Lemma 5.5.1 from §5.5 of Chapter 5, on the approximation of Hölder continuous functions by spherical polynomials. Now

$$\|\hat{\mathcal{K}} - \mathcal{P}_n \hat{\mathcal{K}}\| = \sup_{\|\varphi\|_\infty \leq 1} \|(\hat{\mathcal{K}} - \mathcal{P}_n \hat{\mathcal{K}})\varphi\|$$

$$= \sup_{\psi \in \mathcal{F}} \|(I - \mathcal{P}_n)\psi\|_\infty \qquad (9.5.207)$$

with

$$\mathcal{F} = \{\hat{\mathcal{K}}\varphi \mid \|\varphi\|_\infty \leq 1, \ \varphi \in C(U)\}$$

Combining Lemma 5.5.1, (9.5.207), and the assumption $\lambda' > \frac{1}{2}$, we obtain

$$\|\hat{\mathcal{K}} - \mathcal{P}_n \hat{\mathcal{K}}\| \leq \frac{c}{n^{\lambda'}}(1 + \|\mathcal{P}_n\|)$$

Then use (5.5.210), that $\|\mathcal{P}_n\| = O(\sqrt{n})$, to obtain

$$\|\hat{\mathcal{K}} - \mathcal{P}_n \hat{\mathcal{K}}\| \leq \frac{c}{n^{\lambda'-1/2}}$$

This proves (9.5.203).

Using (9.5.203), it is straightforward to prove the existence and uniform boundedness of $(2\pi + \mathcal{P}_n\hat{\mathcal{K}})^{-1}$, using the form of proof used in earlier chapters, say Theorem 3.1.1 from §3.1. Simply write

$$2\pi + \mathcal{P}_n\hat{\mathcal{K}} = (2\pi + \hat{\mathcal{K}}) - (\hat{\mathcal{K}} - \mathcal{P}_n\hat{\mathcal{K}})$$
$$= (2\pi + \hat{\mathcal{K}})[I - (2\pi + \mathcal{P}_n\hat{\mathcal{K}})^{-1}(\hat{\mathcal{K}} - \mathcal{P}_n\hat{\mathcal{K}})]$$

Use (9.5.203) and the geometric series theorem (cf. Theorem A.1 in the Appendix) to obtain the existence of

$$[I - (2\pi + \mathcal{P}_n\hat{\mathcal{K}})^{-1}(\hat{\mathcal{K}} - \mathcal{P}_n\hat{\mathcal{K}})]^{-1}$$

for all sufficiently large n, say $n \geq n_0$; then write

$$(2\pi + \mathcal{P}_n\hat{\mathcal{K}})^{-1} = [I - (2\pi + \mathcal{P}_n\hat{\mathcal{K}})^{-1}(\hat{\mathcal{K}} - \mathcal{P}_n\hat{\mathcal{K}})]^{-1}(2\pi + \hat{\mathcal{K}})^{-1},$$

$$n \geq n_0$$

The uniform boundedness is immediate.

(ii) For convergence of $\hat{\rho}_n$ to $\hat{\rho}$, write the identity

$$\hat{\rho} - \hat{\rho}_n = 2\pi(2\pi + \mathcal{P}_n\hat{\mathcal{K}})^{-1}(\hat{\rho} - \mathcal{P}_n\hat{\rho}), \quad n \geq n_0$$

Using (9.5.202) and the assumptions on f and S, we have $\rho \in C^{k,\lambda'}(S)$, and equivalently, $\hat{\rho} \in C^{k,\lambda'}(U)$. Combining this with Lemma 5.5.1, the uniform boundedness of $(2\pi + \mathcal{P}_n\hat{\mathcal{K}})^{-1}$, and the assumption $k + \lambda > \frac{1}{2}$, we obtain (9.5.204). $\qquad\qquad\square$

For $f \in C^{0,\lambda}(S)$ with $0 < \lambda \leq \frac{1}{2}$, we cannot ensure the uniform convergence of ρ_n to ρ. Nonetheless, we still have convergence in $L^2(S)$, based on (9.5.195).

From (9.5.204), we expect very rapid convergence if $f \in C^\infty(S)$, as happens in many cases if S is a smooth surface. This approach to the numerical solution of the BIE $(2\pi + \mathcal{K})\rho = f$ is likely to be much more efficient computationally than are the piecewise polynomial methods of the preceding sections for such smooth surfaces S. An illustration of this is given in the numerical example given below at the end of the following subsection.

9.5.1. *The linear system for* $(2\pi + \mathcal{P}_n\hat{\mathcal{K}})\hat{\rho}_n = \mathcal{P}\hat{f}$

Recall the formula

$$\mathcal{P}_n g(Q) = \sum_{m=0}^{n} \sum_{k=1}^{2m+1} (g, S_m^k) S_m^k(Q) \tag{9.5.208}$$

from (5.5.209) of §5.5. This uses the basis $\{S_n^k\}$ of orthonormal spherical harmonics, given in (5.5.203). Using this, we write the solution of $(2\pi + \mathcal{P}_n \hat{\mathcal{K}})\hat{\rho}_n = \mathcal{P}_n \hat{f}$ in the form

$$\hat{\rho}_n(Q) = \sum_{m=0}^{n} \sum_{k=1}^{2m+1} \alpha_{m,k}\, S_m^k(Q)$$

The coefficients $\{\alpha_{m,k} \mid 1 \leq k \leq 2m+1,\ 0 \leq m \leq n\}$ are determined by solving

$$2\pi\alpha_{m,k} + \sum_{\mu=0}^{n} \sum_{\kappa=1}^{2\mu+1} \alpha_{\mu,\kappa}\left(\hat{\mathcal{K}}S_\mu^\kappa, S_m^k\right) = \left(f, S_m^k\right) \qquad (9.5.209)$$

with $1 \leq k \leq 2m+1,\ 0 \leq m \leq n$. The Galerkin coefficients $(\hat{\mathcal{K}}S_\mu^\kappa, S_m^k)$ can be written

$$\left(\hat{\mathcal{K}}S_\mu^\kappa, S_m^k\right) = \int_U S_m^k(P) \int_U \hat{K}(P, Q) S_\mu^\kappa(Q)\, dS(Q)\, dS(P) \qquad (9.5.210)$$

with $\hat{K}(P, Q)$ given by (9.5.192) and $K(P, Q)$ the double layer potential kernel. The Galerkin coefficient $(\hat{\mathcal{K}}S_\mu^\kappa, S_m^k)$ is a double surface integral, which is equivalent to a four-fold integral involving single integrals. Since $K(P, Q)$ is singular, the "inner integral" $\hat{\mathcal{K}}S_\mu^\kappa$ is a singular integral and must be evaluated with care. Since $\hat{\mathcal{K}}S_\mu^\kappa$ is itself a very smooth function, the inner product, which we call the "outer integral," is an integration involving a smooth integrand. As a consequence, a method such as the Gaussian quadrature method of §5.5.2 should be very suitable, with only a small number of node points being needed.

Returning to the singular integral $\hat{\mathcal{K}}S_\mu^\kappa(P)$, with a singularity in the integrand at $Q = P$, we must use some method that recognizes the singularity. Such a method is described and implemented in Ref. [45]. It begins by choosing a coordinate system for the integration variable Q such that $Q = P$ corresponds to either the north pole $(0, 0, 1)$ or the south pole $(0, 0, -1)$. Then spherical coordinates are used, with the trapezoidal rule used for the equatorial direction and a special integration formula used for the polar direction. We refer the reader to Ref. [45] for the details.

An algorithm implementing this method is given in Ref. [45]. One part of the program produces approximations of the Galerkin coefficients, and a second program then sets up and solves the linear system (9.5.209). These programs are also available from the author's anonymous ftp site.

Example. Let S be the ellipsoidal surface

$$x^2 + \left(\frac{y}{1.5}\right)^2 + \left(\frac{z}{2}\right)^2 = 1 \qquad (9.5.211)$$

and let D denote its interior. We solve the interior Dirichlet problem

$$\begin{aligned} \Delta u(P) &= 0, & P \in D \\ u(P) &= f(P), & P \in S \end{aligned} \qquad (9.5.212)$$

For demonstration purposes, use the harmonic function

$$u(x, y, z) = e^x \cos y + e^z \sin x \qquad (9.5.213)$$

to generate the Dirichlet data f.

Represent the solution as a double layer potential,

$$u(A) = \int_S \rho(Q) \frac{\partial}{\partial \mathbf{n}_Q} \left[\frac{1}{|A - Q|} \right] dS(Q), \qquad A \in D$$

as in (9.1.35), and solve the integral equation $(2\pi + \mathcal{K}) \rho = f$. After obtaining an approximate solution ρ_n, define an approximating potential solution by

$$u_n(A) = \int_S \rho_n(Q) \frac{\partial}{\partial \mathbf{n}_Q} \left[\frac{1}{|A - Q|} \right] dS(Q), \qquad A \in D$$

This is integrated numerically by using the product Gaussian quadrature method of §5.5.2 in Chapter 5. For the error in u_n, it is straightforward that

$$\max_{A \in \bar{D}} |u(A) - u_n(A)| \leq (2\pi + \|\mathcal{K}\|) \|\rho - \rho_n\|_\infty \qquad (9.5.214)$$

and for D convex, it follows that $\|\mathcal{K}\| = 2\pi$. Thus the speed of uniform convergence of u_n to u is at least as fast as that of ρ_n to ρ.

In Table 9.15 the errors $u(A) - u_n(A)$ are given at selected points $A = (x, y, z) \in D$, for several different degrees n. The column labeled α is a number for which (9.5.213)

$$\frac{1}{\alpha}(x, y, z) \in S$$

thus giving some idea of the proximity of (x, y, z) to the boundary. The numerical integrations used in computing the Galerkin coefficients and the other integrals were chosen to be of high accuracy, so that the errors $u - u_n$ in the table are not contaminated by numerical integration errors.

Table 9.15. *Solution of Dirichlet problem with u given by (9.5.213)*

(x, y, z)	α	$u - u_4$	$u - u_5$	$u - u_6$	$u - u_7$	$u - u_8$
$(0, 0, 0)$		5.66E−5	5.66E−5	3.20E−7	3.20E−7	1.12E−9
$(.1, .1, .1)$	.13	2.35E−3	1.37E−4	−4.42E−5	−1.93E−6	6.04E−7
$(.25, .25, .25)$	.33	5.07E−3	7.14E−4	−8.83E−5	−1.71E−5	6.27E−7
$(.5, .5, .5)$	.65	1.31E−3	2.53E−3	9.22E−5	−6.89E−5	−8.47E−6
$(0, .25, .25)$	.21	5.62E−6	5.62E−6	−1.11E−7	−1.11E−7	−8.14E−10
$(−.1, −.2, .2)$	.19	−2.01E−3	−1.73E−4	3.63E−5	4.72E−6	−3.59E−7
$(0, 0, 1)$	.50	5.66E−5	5.66E−5	3.20E−7	3.20E−7	1.12E−9
$(.5, 0, .5)$	.56	7.80E−3	3.64E−3	2.40E−4	−5.32E−5	−7.14E−6

The errors in Table 9.15 indicate a rapid rate of convergence of u_n to u, and by inference, a rapid rate of convergence of ρ to ρ_n. For this example the function $\rho \in C^\infty(S)$, and thus the projections $\mathcal{P}_n \hat{\rho}$ should converge to $\hat{\rho}$ with a rate faster than $O(n^{-p})$ for any $p > 0$. This then implies a similar rate of convergence for ρ to ρ_n. Note that the linear system (9.5.209) to be solved is order $(n + 1)^2$. Thus for the highest order case in Table 9.15, the linear system has order 81, which is quite small for such a problem. Moreover, the main cost is in calculating the Galerkin coefficients, and this need be done only once, regardless of the given boundary data f. Examples of this numerical method for other boundary values and other surfaces are given in Refs. [40], [41], and [45].

The methods of this section can also be applied to BIE reformulations of other boundary value problems for Laplace's equation. Details of such are given in the papers [40], [41] for BIE of the second kind, and below we extend this development to a BIE of the first kind. Extensions have also been given to BIE reformulations of the Helmholtz equation (cf. Lin [342]) and to nonlinear BIE (cf. Ganesh et al. [209]).

9.5.2. Solution of an integral equation of the first kind

As in §9.4.1, we consider the numerical solution of the first kind BIE

$$\mathcal{S}\psi(P) \equiv \int_S \frac{\psi(Q)}{|P - Q|}\, dS(Q) = f(P), \quad P \in S$$

In that earlier subsection, a finite element method was presented and analyzed. Here we also use the variational framework of §7.4, but we use an approximating subspace $\mathcal{X}_n$ constructed from spherical polynomials of degree $\leq n$. The surface is assumed to be of the same type as that assumed earlier in this section, and to further simplify the presentation, we assume S is a C^∞ surface. Much of the work of this subsection is based on the work in Chen [116], and we will refer to it for some of the needed proofs.

The problem of solving $S\psi = f$ is replaced by that of finding $\psi \in H^{-\frac{1}{2}}(S)$ such that

$$\mathcal{A}(\eta, \psi) = (\eta, f), \quad \text{all } \eta \in H^{-\frac{1}{2}}(S)$$

with $\mathcal{A}$ the bilinear functional

$$\mathcal{A}(\eta, \varphi) = (\eta, S\varphi) = \int_S \int_S \frac{\eta(P)\,\varphi(Q)}{|P - Q|}\, dS(Q)\, dS(P), \quad \eta, \varphi \in H^{-\frac{1}{2}}(S)$$

For the properties of S and $\mathcal{A}$, recall the discussion in §9.4.1.

As the approximating subspace, define $\mathcal{X}_n = \mathcal{M}^{-1}(S_n)$. For a basis, we use

$$B_m^k = \mathcal{M}^{-1}(S_m^k) \equiv S_m^k \circ M^{-1}, \quad k = 1, \ldots, 2m + 1, \quad m = 0, 1, \ldots, n$$
$$(9.5.215)$$

The numerical method is to find $\psi_n \in \mathcal{X}_n$ such that

$$\mathcal{A}(\eta, \psi_n) = (\eta, f), \quad \text{all } \eta \in \mathcal{X}_n \qquad (9.5.216)$$

To give a stability and convergence analysis, we begin with the type of analysis used earlier in §9.4.1. A number of preliminary results are needed first, however.

With the earlier simplifying assumption that S is a C^∞ surface, we can prove that

$$\mathcal{M} : H^t(S) \xrightarrow[\text{onto}]{1-1} H^t(U), \quad -\infty < t < \infty \qquad (9.5.217)$$

showing that the Sobolev spaces $H^t(S)$ and $H^t(U)$ are isomorphic. For a proof, see Ref. [116, Theorem 2.2.2]. We recall that the standard norm $\|g\|_t$, $g \in H^t(U)$, is equivalent to

$$\|g\|_{*,t} \equiv \left[\sum_{m=0}^{\infty} (2m + 1)^{2t} \sum_{k=1}^{2m+1} \left| (g, S_m^k) \right|^2 \right]^{\frac{1}{2}} < \infty, \quad g \in H^t(U)$$

based on the Laplace expansion

$$g(Q) = \sum_{m=0}^{\infty} \sum_{k=1}^{2m+1} (g, S_m^k) S_m^k(Q), \quad g \in H^t(U) \qquad (9.5.218)$$

This was discussed earlier in §9.1.4.

Using the characterization of $H^t(U)$ as the set of all Laplace expansions (9.5.218) for which $\|g\|_{*,t} < \infty$, we can prove the following approximation properties for S_n.

Lemma 9.5.1 (Approximation property). Let $-\infty < t \le s < \infty$. Then for any $u \in H^s(U)$, there is an element $u_n \in \mathcal{X}_n$ for which

$$\|u - u_n\|_{*,t} \le n^{t-s} \|u\|_{*,s}, \quad n \ge 1 \tag{9.5.219}$$

Namely, define

$$u_n = \sum_{m=0}^{n} \sum_{k=1}^{2m+1} (u, S_m^k) S_m^k \tag{9.5.220}$$

Proof. We use a manipulation of the formula for $\|u - u_n\|_{*,t}^2$.

$$
\begin{aligned}
\|u - u_n\|_{*,t}^2 &= \sum_{m=n+1}^{\infty} (2m+1)^{2t} \sum_{k=1}^{2m+1} \left|(u, S_m^k)\right|^2 \\
&= \sum_{m=n+1}^{\infty} (2m+1)^{2(t-s)} (2m+1)^{2s} \sum_{k=1}^{2m+1} \left|(u, S_m^k)\right|^2 \\
&\le (2n+3)^{2(t-s)} \sum_{m=n+1}^{\infty} (2m+1)^{2s} \sum_{k=1}^{2m+1} \left|(u, S_m^k)\right|^2 \\
&\le [c_n(s-t)]^2 \, n^{2(t-s)} \|u\|_{*,s}^2
\end{aligned}
$$

with

$$c_n(r) = \left(2 + \frac{3}{n}\right)^{-r}, \quad n \ge 1, \ r \ge 0$$

and $c_0(r) = 3^{-r}$. Easily,

$$\max_{r \ge 0} c_n(r) = 1$$

and we have (9.5.219). $\qquad\square$

Lemma 9.5.2 (Inverse property). Let $-\infty < t \le s < \infty$. Then for any $u \in \mathcal{X}_n$,

$$\|u\|_{*,s} \le 3^{s-t} n^{s-t} \|u\|_{*,t}, \quad n \ge 1 \tag{9.5.221}$$

Proof. For any $u \in \mathcal{X}_n$,

$$u = \sum_{m=0}^{n} \sum_{k=1}^{2m+1} (u, S_m^k) S_m^k$$

Then

$$\|u\|_{*,s}^2 = \sum_{m=0}^{n}(2m+1)^{2s}\sum_{k=1}^{2m+1}\left|(u, S_m^k)\right|^2$$

$$= \sum_{m=0}^{n}(2m+1)^{2(s-t)}(2m+1)^{2t}\sum_{k=1}^{n}\left|(u, S_m^k)\right|^2$$

$$\leq (2n+1)^{2(s-t)}\sum_{m=0}^{n}(2m+1)^{2t}\sum_{k=1}^{2m+1}\left|(u, S_m^k)\right|^2$$

$$\leq [c(s-t)]^2 \, n^{2(s-t)}\, \|u\|_{*,t}^2$$

with

$$c(r) = \sup_{n\geq 1}\left(2 + \frac{1}{n}\right)^r \leq 3^r \qquad\qquad \square$$

For the convergence analysis, we have the following, in analogy to that given in §9.4.1.

Theorem 9.5.2. *Consider the numerical solution of* $S\psi = f$, *with* $f \in H^{\frac{1}{2}}(S)$, *using the variational method (9.5.216). Then this has a unique solution for all* $n \geq 0$. *Moreover,*

$$\|\psi - \psi_n\|_{-\frac{1}{2}} \leq \frac{c_{\mathcal{A}}}{c_e}\inf_{\eta\in\mathcal{X}_n}\|\psi - \eta\|_{-\frac{1}{2}} \qquad (9.5.222)$$

with the constants $c_{\mathcal{A}}$ *and* c_e *taken from (7.4.216) and (7.4.217), respectively. In addition, if* $\psi \in H^s(S)$ *for some* $s \geq 0$, *then*

$$\|\psi - \psi_n\|_{L^2(S)} \leq \frac{c}{n^s}\|\psi\|_{*,s} \qquad (9.5.223)$$

Proof. From the use of Cea's Lemma (Theorem 7.4.2), and from the same type of argument that led to (9.4.182) in §9.4.1, we have the unique solvability of (9.5.216) and the inequality (9.5.222). We work with this to obtain error bounds in $L^2(S)$ and to show convergence of ψ_n to ψ, for all $\psi \in L^2(S)$. As another consequence of Cea's Lemma, recall from the discussion following (7.4.251) that

$$\psi_n = \mathcal{G}_n\psi \qquad (9.5.224)$$

with $\mathcal{G}_n$ a bounded projection of $H^{-\frac{1}{2}}(S)$ onto $\mathcal{X}_n$ (cf. (7.4.254)),

$$\|\mathcal{G}_n\| \leq \frac{c_{\mathcal{A}}}{c_e} \qquad (9.5.225)$$

Let $\hat{\psi} = \mathcal{M}\psi$, and let $\hat{\varphi}_n$ be the corresponding element specified by (9.5.220) and satisfying (9.5.219),

$$\|\hat{\psi} - \hat{\varphi}_n\|_{*,t} \le n^{t-s}\|\hat{\psi}\|_{*,s}, \quad n \ge 1, \ t \le s$$

Let $\varphi_n = \mathcal{M}^{-1}\hat{\varphi}_n \in \mathcal{X}_n$. It is straightforward from (9.5.217) that

$$\|\psi - \varphi_n\|_t \le cn^{t-s}\|\psi\|_s, \quad n \ge 1, \ t \le s \tag{9.5.226}$$

for a suitable constant c depending on t and s.

For any r, write

$$\|\psi - \psi_n\|_r \le \|\psi - \varphi_n\|_r + \|\varphi_n - \psi_n\|_r \tag{9.5.227}$$

Let $s \ge r \ge -\frac{1}{2}$. For the quantity $\|\varphi_n - \psi_n\|_r$,

$$\|\varphi_n - \psi_n\|_r \le c_1\|\hat{\varphi}_n - \hat{\psi}_n\|_r, \tag{9.5.217}$$
$$\le c_2 n^{r+\frac{1}{2}}\|\hat{\varphi}_n - \hat{\psi}_n\|_{*,-\frac{1}{2}}, \tag{9.5.221}$$
$$\le c_3 n^{r+\frac{1}{2}}\|\varphi_n - \psi_n\|_{-\frac{1}{2}}, \tag{9.5.217}$$
$$= c_3 n^{r+\frac{1}{2}}\|\mathcal{G}_n(\varphi_n - \psi)\|_{-\frac{1}{2}}, \tag{9.5.224}$$
$$\le c_4 n^{r+\frac{1}{2}}\|\varphi_n - \psi\|_{-\frac{1}{2}}, \tag{9.5.225}$$
$$\le c_5 n^{r-s}\|\psi\|_s, \tag{9.5.226}$$

with suitable constants $c_1, \ldots, c_5$. Combine this result with (9.5.226) and (9.5.227) to obtain

$$\|\psi - \psi_n\|_r \le cn^{r-s}\|\psi\|_s, \quad -\frac{1}{2} \le r \le s, \ \psi \in H^s(S) \tag{9.5.228}$$

with the constant c depending on t and s. The result (9.5.223) is a special case of this more general error bound. □

Represent the solution to the interior Dirichlet problem (9.1.3) as a simple layer potential:

$$u(A) = \int_S \frac{\psi(Q)}{|A - Q|}\, dS(Q), \quad A \in D$$

Using the boundary condition $u \mid S = f$, find ψ by solving $\mathcal{S}\psi = f$ and denote the solution by ψ_n, as above. Then introduce the approximating harmonic function

$$u_n(A) = \int_S \frac{\psi_n(Q)}{|A - Q|}\, dS(Q), \quad A \in D \tag{9.5.229}$$

Using the maximum principle, it is straightforward to show that

$$\max_{A \in \hat{D}} |u(A) - u_n(A)| = \|\mathcal{S}(\psi - \psi_n)\|_\infty$$

with the latter norm being the uniform norm over S. Since $C(S)$ is compactly embedded in $H^{1+\epsilon}(S)$, for any $\epsilon > 0$, we have

$$\|\mathcal{S}(\psi - \psi_n)\|_\infty \le c\|\mathcal{S}(\psi - \psi_n)\|_{1+\epsilon} \le c\|\mathcal{S}\|\|\psi - \psi_n\|_\epsilon$$

In this, $\|\mathcal{S}\|$ is the operator norm for the bounded mapping of $\mathcal{S}$ from $H^0(S)$ to $H^1(S)$, and c is a suitable finite constant that depends on ϵ. Combining these results with (9.5.228), we have

$$\max_{A \in \hat{D}} |u(A) - u_n(A)| \le cn^{\epsilon - s}\|\psi\|_s, \quad \psi \in H^s(S) \tag{9.5.230}$$

Implementation of the Galerkin method

To solve for ψ_n in (9.5.216), write

$$\psi_n = \sum_{m=0}^{n} \sum_{k=1}^{2m+1} \alpha_{m,k} B_m^k \tag{9.5.231}$$

Substitute into (9.5.216) and let $\eta = B_\mu^\kappa$, for $\kappa = 1, \ldots, 2\mu+1$, $\mu = 0, 1, \ldots, n$. This yields the linear system

$$\sum_{m=0}^{n} \sum_{k=1}^{2m+1} \alpha_{m,k} \mathcal{A}(B_\mu^\kappa, B_m^k) = (B_\mu^\kappa, f),$$

$$\kappa = 1, \ldots, 2\mu + 1, \quad \mu = 0, 1, \ldots, n \tag{9.5.232}$$

The Galerkin coefficients are given by

$$\mathcal{A}(B_\mu^\kappa, B_m^k) = \int_S \int_S \frac{B_\mu^\kappa(P) B_m^k(Q)}{|P - Q|} \, dS(Q) \, dS(P)$$

$$= \int_U \int_U \frac{S_\mu^\kappa(\hat{P}) S_m^k(\hat{Q}) J_M(\hat{Q}) J_M(\hat{P})}{|M(\hat{P}) - M(\hat{Q})|} \, dS_{\hat{Q}} \, dS_{\hat{P}} \tag{9.5.233}$$

These last integrals must be computed numerically. The statements given in connection with the numerical evaluation of the Galerkin coefficients of (9.5.210) are still valid for the above integrals in (9.5.233).

Write the linear system (9.5.232) in matrix form as

$$A_n \alpha = \mathbf{f}_n \tag{9.5.234}$$

Then the following is proven by Chen [116, §2.2.3].

Theorem 9.5.3. *The condition number of* A_n *satisfies*

$$\text{cond}(A_n) = O(n) \tag{9.5.235}$$

In this case the condition number is defined using the matrix norm induced by the Euclidean vector norm on $\mathbf{R}^{d_n}$, *with* $d_n \equiv (n+1)^2$ *the order of* A_n.

Proof. We only sketch the main ideas of the proof. The matrix A_n is symmetric and positive definite, and consequently, its eigenvalues λ are real and positive. For these eigenvalues, write

$$0 < \lambda_{\min} \leq \lambda \leq \lambda_{\max}$$

Then it is well known that

$$\text{cond}(A_n) = \frac{\lambda_{\max}}{\lambda_{\min}}$$

In Chen [116, §2.2.3], it is proven that

$$\lambda_{\max} \leq O(1), \qquad \lambda_{\min} \geq O\left(\frac{1}{n}\right) \tag{9.5.236}$$

This then proves (9.5.235).   □

This result shows that the condition number of the matrix A_n grows at a reasonably slow rate as a function of n. Moreover, it suggests that A_n is not ill-conditioned until n becomes fairly large.

Example. Let S be the ellipsoid of (9.5.211) and again solve the Dirichlet problem (9.5.212) with the boundary data generated from the harmonic function u of (9.5.213). The errors at selected points in D are given in Table 9.16.

The errors can be compared with those given in Table 9.15 at the same points of D. Doing so, we see that the errors in Table 9.15 are converging to zero at a rate faster than are those in Table 9.16. For example, with $n = 4$, the errors in the Table 9.15 are around 2 to 8 times smaller than those in the Table 9.16. When we increase to $n = 8$, the errors in Table 9.15 become 50 to 300 times those in the Table 9.16. We have no explanation for the slower rate of convergence in solving the first kind equation $\mathcal{S}\psi = f$. The convergence in Table 9.16 is still reasonably good, and it looks somewhat poor only because of the comparison with Table 9.15.

Table 9.16. *Solution of Dirichlet problem with u given by (9.5.213)*

(x, y, z)	α	$u - u_4$	$u - u_5$	$u - u_6$	$u - u_7$	$u - u_8$
$(0, 0, 0)$		9.51E$-$5	9.51E$-$5	-4.68E$-$6	-4.68E$-$6	3.40E$-$7
$(.1, .1, .1)$	.13	5.16E$-$3	1.97E$-$4	-4.23E$-$4	-2.52E$-$5	3.73E$-$5
$(.25, .25, .25)$	.33	1.23E$-$2	2.36E$-$3	-8.89E$-$4	-2.83E$-$4	5.97E$-$5
$(.5, .5, .5)$	.65	1.13E$-$2	1.14E$-$2	9.00E$-$4	-1.12E$-$3	-3.36E$-$4
$(0, .25, .25)$	.21	3.82E$-$5	3.82E$-$5	-1.45E$-$6	-1.45E$-$6	4.59E$-$8
$(-.1, -.2, .2)$	.19	-4.72E$-$3	-5.44E$-$4	3.51E$-$4	7.03E$-$5	-2.61E$-$5
$(0, 0, 1)$	.50	-6.92E$-$5	-6.92E$-$5	1.46E$-$6	1.46E$-$6	3.36E$-$7
$(.5, 0, .5)$	.56	1.73E$-$2	1.36E$-$2	6.75E$-$4	-1.51E$-$3	-3.80E$-$4

Other boundary integral equations and general comments

The exterior Dirichlet problem (9.1.23) can also be reduced to the solution of the first kind BIE $\mathcal{S}\psi = f$, with there being no difference in the numerical theory from what has already been developed. We omit any numerical example, as it would be no different from what has already been given above.

When solving the interior or exterior Dirichlet problem for a harmonic function u, it is sometimes of interest to calculate the normal derivative of u at points of the boundary. This too can be done by means of solving the BIE $\mathcal{S}\psi = f$. For the interior Dirichlet problem, use the Green's representation formula (9.1.6) to write

$$\int_S \frac{\partial u(Q)}{\partial \mathbf{n}_Q} \frac{dS(Q)}{|P - Q|} = f(P), \quad P \in S \qquad (9.5.237)$$

with

$$f(P) \equiv -2\pi u(P) + \int_S u(Q) \frac{\partial}{\partial \mathbf{n}_Q} \left[\frac{1}{|P - Q|} \right] dS(Q)$$

The function f enters into the calculations in the form of the inner products

$$\left(B_\mu^\kappa, f \right) = \int_U S_\mu^\kappa(Q) f(M(Q)) J_M(Q) \, dS(Q)$$

We can calculate these coefficients with the same type of integration techniques as were discussed following (9.5.210) for the double layer boundary integral operator $\mathcal{K}$. Numerical examples of this means of solving directly for the normal derivative of u are given in Chen [116, §3.3].

General comments. The methods used here to solve $\mathcal{S}\psi = f$ can also be used to solve other BIE of the first kind. Extensions to such equations for the Helmholtz equation and the biharmonic equation are given in Chen [116].

Most practical problems do not involve surfaces S that are smooth, such as was assumed here. But for such smooth boundaries, we recommend highly the use of numerical methods based on spherical polynomial approximations to solve BIE and associated boundary value problems. It has been our experience that such methods converge much faster than boundary element methods.

Discussion of the literature

For boundary value problems in three dimensions, BIE methods have been used in engineering and the physical sciences for over a century, and there is a very large literature. The numerical analysis of such BIE methods, however, has lagged behind their use in applications, and much of this numerical analysis has been done from 1965 onward. When the boundary is smooth, the single and double layer operators are compact operators; equations of the second kind involving such operators can be treated by the methods of Chapters 2 through 6. From a theoretical perspective, the more difficult problems are those associated with a surface that is only piecewise smooth. The survey [50] gives a survey of the subject as of 1990, and much has happened since then.

For a discussion of the use of BIE in engineering applications, including a brief bibliography of such applications, we refer the reader to the survey [50] and to the introductory paragraphs in the Discussion of the literature at the end of Chapter 7. In addition to the references in Chapter 7, we note the work on transmission and screen problems by Costabel and Stephan [135], [138] and Stephan [529]; on elasticity problems, we note the work by Costabel and Stephan [139]; Petersdorf [414]; Rizzo et al. [468]; Rudolphi et al. [470]; and Watson [555]. For BIE methods for solving the Helmholtz equation, see Colton and Kress [127] and Kleinman and Roach [314]. For examples of the use of boundary integral equations in the study of fluid flow, see Hackbusch [251]; Hess [269]; Hess and Smith [270]; Hsin; Kerwin; and Newman [280]; Johnson [296a]; Nabors et al. [393]; Newman [400]; Nowak [408]; Power [431]; Pozrikidis [432]; and Tuck [547]. The paper Johnson [296] has some analytic integration formulas for commonly occurring collocation integrals for polyhedral surfaces.

There is a large literature of books and articles on the theory of boundary integral equations. Among the better-known books, we note Dautray and Lions [159]; Günter [246]; Hackbusch [251, Chap. 8]; Jaswon and Symm [286]; Kellogg [307]; Kral [316]; Kress [325]; Maz'ya [367]; Mikhlin [380]; Mikhlin and Prößdorf [381]; and Vainikko [550]. For papers on various aspects of the theory of boundary integral equations, we note Costabel [129], Costabel and Wendland [141]; Gilbert [216]; and Hsiao and MacCamy [276].

For work dealing with the solvability and regularity for integral equations of the second kind for surfaces with edges and corners, we refer to Elschner [184]; Fabes; Jodeit; and Lewis [194]; Petersdorf and Stephan [418], [419]; Rathsfeld [448]; and Rempel and Schmidt [459]. Some of these obtain their results by utilizing results from the regularity theory for elliptic boundary value problems, for example, as expounded in Dauge [158] and Grisvard [240]. For hypersingular integral operators on surfaces, see Burton and Miller [100]; Giroire and Nedelec [217]; Krishnasamy et al. [330]; and Petersdorf and Stephan [420].

When considering the numerical solution of boundary integral equations, one of the fundamental dividing lines is whether the surface is smooth or is only piecewise smooth. For BIE of the second kind on smooth surfaces, the theory of projection methods is entirely straightforward, as we noted earlier in this chapter. In this case, the main questions of interest are (1) how to efficiently approximate the integrals occurring in the discretized linear system, (2) how to approximate the surface, and (3) how to solve the linear system efficiently. For BIE of the second kind on piecewise smooth surfaces, there are fundamental difficulties in constructing a convergence analysis, just as for the planar case studied in Chapter 8. For Galerkin methods for solving BIE of the second kind on piecewise smooth surfaces, see Costabel [129]; Dahlberg and Verchota [150]; Elschner [185]; Petersdorf and Stephan [418], [419]; and Stephan [531]. We especially note Elschner's paper, as it generalizes to $\mathbf{R}^3$ many of the ideas that are fundamental to the convergence investigations for planar problems with piecewise smooth boundaries.

For general results on Galerkin's method for solving BIE on smooth surfaces, including integral equations of the first kind and hypersingular integral equations, see Costabel [129]; Stephan [529], [530]; Stephan and Suri [534]; Stephan and Wendland [536]; and Wendland [558]–[565]. For the use of approximations based on using spherical polynomials, see [40], [41], and [45].

For boundary element methods for BIE of the first kind, in addition to that cited in the preceding paragraph, see Maischak and Stephan [361]; Nabors et al. [393]; Nedelec [396]; and Richter [465]. For Galerkin methods with spherical polynomial approximants, for BIE of the first kind on smooth surfaces, see Chen [116] and Chen and Atkinson [117].

The theory of collocation methods for solving BIE has lagged behind that for Galerkin methods. Again the case of BIE of the second kind on smooth surfaces is relatively straightforward, with the outstanding questions the same as for the Galerkin methods. For BIE of the second kind on piecewise smooth surfaces, the first important paper was that of Wendland [557]. Although the approximations were only piecewise constant, the methods of analysis would generalize to many other collocation methods. The paper also contains important lemmas that are needed when dealing with double and single layer operators on piecewise smooth surfaces. The next substantial contribution is that of Rathsfeld [447], [449], [451]–[453]. In addition see [43], [46], [49]; Atkinson and Chien [59]; Chien [119]; Lachat and Watson [333]; and Watson [555]. The author has developed a large boundary element package, with a user's guide given in Ref. [52], and directions for obtaining the guide and the package are given in the bibliographic listing. The boundary element examples of this book were all computed with this package.

As was noted in §9.2.2, the calculation of the integrals in the collocation matrix is the most expensive part of most computer codes for boundary element methods. As references for work on this problem, we note Allgower, Georg, and Widmann [5]; Ref. [52]; Chien [119]; Duffy [171]; Fairweather, Rizzo, and Shippy [195]; Georg [213]; Guermond [244]; Johnson and Scott [295]; Lyness [351]; Schwab [496]; Schwab and Wendland [497]; Yang [582]; and Yang and Atkinson [584]. For many years the main source of difficulty was believed to be the singular integrals in the collocation matrix. However, with the use of the Duffy transformation to aid in evaluating the singular integrals, the most time-consuming part of the calculation is that of the numerical integration of nonsingular integrals, as was explained earlier in §9.2.1 following (9.2.105).

For discretizations of BIE of the second kind, one can use the two-grid and multigrid methods of Chapter 6, as we have done in this chapter. People who have contributed to this approach to the problem are Ref. [53]; Brandt and Lubrecht [88]; Hackbusch [251]; Hebeker [260], [261]; Mandel [362]; and Nowak [407], [408]. For other approaches to iteration for such problems and for BIE of the first kind, see Kane, Keyes, and Prasad [299]; Nabors et al. [393]; Petersdorf and Stephan [418], [420]; and Vavasis [552].

In recent years, there has been a growing interest in *fast methods*, which avoid the calculation of the numerical integrals of standard collocation and Galerkin methods. For a boundary integral equation, let its discretization be denoted by $A_n \rho_n = f_n$, with n the discretization parameter and r_n the order of the system. Then the number of matrix coefficients to be calculated in r_n^2, and the cost of a matrix-vector multiplication $A_n g_n$ is about $2r_n^2$ operations. The desire behind all of the multiresolution methods is to calculate $\psi_n \approx A_n g_n$ for any given vector g_n, with $\epsilon_n \equiv \psi_n - A_n g_n$ suitably small. This is to be done in $O(r_n \log^p r_n)$ operations, for some small p, say $p \leq 3$, and we refer to such methods as *fast matrix-vector calculations*. This lessens significantly the cost of both the calculation of A_n and the multiplication $A_n g_n$. Then iteration methods for solving $A_n \rho_n = f_n$ will generally have a similar cost. Hackbusch and Nowak [252], [253] and Nowak [408] use a "panel clustering method," which uses local approximations based on Taylor series. Greengard [237]; Greengard and Rokhlin [239]; Rokhlin [469]; and Nabors et al. [393] use a "fast multipole method," and it uses local approximations based on spherical polynomial expansions. More recently, there has been great interest in using wavelet approximations. For such, see Alpert et al. [7]; Bond and Vavasis [83]; Dahmen et al. [151], [154]; Petersdorf and Schwab [415], [416]; Petersdorf, Schwab, and Schneider [417]; and Rathsfeld [454]. For other discussions of such techniques, see Canning [101], [102] and Golberg [222].

An alternative approach to the setup and solution of the linear system $A_n \rho_n = f_n$ associated with a BIE is to use a parallel computer to carry out the needed operations with a traditional method as described in this chapter. For discussions of such, see Natarajan and Krishnaswamy [395]; Pester and Rjasanow [413]; and Reichel [456]. In Edelman [172] a survey of sources of dense linear systems is summarized. It appears that most such linear systems arise from discretizations of BIE, and therefore, it would appear best to take into account the source of these systems when designing methods for their solution.

For Poisson equations and other nonhomogeneous PDE, see Ref. [44], Golberg [221], and Golberg and Chen [223]. For nonlinear BIE on surfaces, see Ref. [54]; Ganesh et al. [209]; and Hardee [257]. Another quite useful approach to solving BIE is to use the Sinc-function methods of Stenger and others. For an introduction to this work, see Stenger [528].

Appendix

Results from functional analysis

In the text we have used several results from functional analysis. We now give those results, but without proof. For a complete development of these results, see any standard introductory text on functional analysis. Among such, we list Aubin [70], Conway [128], Kantorovich and Akilov [304], Kesavan [313], Kreyszig [329], and Zeidler [588], [589]. For interesting texts that apply functional analysis directly to numerical analysis, see Collatz [125], Cryer [146], and Linz [344].

Theorem A.1 (Geometric series theorem; cf. [129, p. 192]). *Let $\mathcal{X}$ be a Banach space, and let $\mathcal{A}$ be a bounded operator from $\mathcal{X}$ into $\mathcal{X}$, with*

$$\|\mathcal{A}\| < 1$$

Then $I - \mathcal{A} : \mathcal{X} \overset{1-1}{\underset{onto}{\to}} \mathcal{X}$, $(I - \mathcal{A})^{-1}$ is a bounded linear operator, and

$$\|(I - \mathcal{A})^{-1}\| \le \frac{1}{1 - \|\mathcal{A}\|}$$

The series

$$(I - \mathcal{A})^{-1} = \sum_{j=0}^{\infty} A^j$$

is called the *Neumann series*; under the assumption $\|\mathcal{A}\| < 1$, it converges in the space of bounded operators on $\mathcal{X}$ to $\mathcal{X}$. This preceding theorem is also called the *contractive mapping theorem*, since the mapping $x \to y + Ax$ is a contractive mapping when $\|\mathcal{A}\| < 1$ and it is also called the *Banach fixed point theorem*.

516

Theorem A.2 (Open mapping theorem; cf. Conway [129, p. 90]). *Let* $\mathcal{X}$ *and* $\mathcal{Y}$ *be Banach spaces, and let* $\mathcal{A}: \mathcal{X} \overset{1-1}{\underset{onto}{\rightarrow}} \mathcal{Y}$ *be a bounded linear operator. Then* $\mathcal{A}^{-1}: \mathcal{Y} \overset{1-1}{\underset{onto}{\rightarrow}} \mathcal{X}$ *is a bounded linear operator.*

Theorem A.3 (Principle of uniform boundedness; cf. Conway [129, p. 95]). *Let* $\{\mathcal{A}_n\}$ *be a sequence of bounded linear operators from a Banach space* $\mathcal{X}$ *to a normed space* $\mathcal{Y}$. *Further, assume*

$$\lim_{n\to\infty} \mathcal{A}_n x$$

exists in $\mathcal{Y}$, *for every* $x \in \mathcal{X}$. *Then*

$$\sup_n \|\mathcal{A}_n\| < \infty$$

Corollary A.1 (Banach-Steinhaus theorem; cf. Conway [129, p. 96]). *Let* $\mathcal{X}$ *and* $\mathcal{Y}$ *be Banach spaces, and let*

$$\mathcal{A}, \mathcal{A}_n : \mathcal{X} \to \mathcal{Y}$$

be bounded linear operators. Let $\mathcal{E}$ *be a dense subspace of* $\mathcal{X}$. *Then in order that* $\mathcal{A}_n x \to \mathcal{A}x$ *for all* $x \in \mathcal{X}$, *it is necessary and sufficient that*

(a) $\mathcal{A}_n x \to \mathcal{A}x$ *for all* $x \in \mathcal{E}$
(b) $\sup_n \|\mathcal{A}_n\| < \infty$

Theorem A.4 (Riesz representation theorem; cf. Conway [129, p. 13]). *Let* $\mathcal{X}$ *be a Hilbert space, and let* ℓ *be a bounded linear functional of* $\mathcal{X}$. *Then there is a unique* $v \in \mathcal{X}$ *for which*

$$\ell(x) = (x, v), \quad all\ x \in \mathcal{X}$$

In addition, $\|\ell\| = \|v\|_{\mathcal{X}}$.

As discussed in Chapter 7, this theorem provides the basis for existence theory for boundary value problems for elliptic partial differential equations. It also provides the basis for the existence of the adjoint operator A^* for bounded operators A on a Hilbert space $\mathcal{X}$ to $\mathcal{X}$.

Theorem A.5 (Completion theorem; cf. Kantorovich and Akilov [304, p. 53]). *Let* $\mathcal{X}$ *be a normed vector space. Then there is a complete normed space* $\mathcal{Y}$

with the following properties:

(a) *There is a subspace $\hat{\mathcal{X}} \subset \mathcal{Y}$ and a linear function $\mathcal{I} : \mathcal{X} \overset{1-1}{\underset{onto}{\rightarrow}} \hat{\mathcal{X}}$ with both $\mathcal{I}$ and $\mathcal{I}^{-1}$ continuous, and*

$$\|\mathcal{I}x\|_{\mathcal{Y}} = \|x\|_{\mathcal{X}}, \quad all \ x \in \mathcal{X}$$

The subscripts on the norms are to indicate clearly that the norms are to be identified with the given spaces. The function $\mathcal{I}$ is called an isometric isomorphism (cf. Conway [128, p. 66]) of the spaces $\mathcal{X}$ and $\hat{\mathcal{X}}$.

(b) *$\hat{\mathcal{X}}$ is dense in $\mathcal{Y}$.*

The space $\mathcal{Y}$ is called the *completion* of the space $\mathcal{X}$. Generally $\mathcal{X}$ and $\hat{\mathcal{X}}$ are identified as being the same. The space $L^2(D)$ can be defined as the abstract completion of $C(D)$ using this theorem, and Sobolev spaces $H^m(D)$ can be defined similarly.

Theorem A.6 (Extension theorem; cf. Kantorovich and Akilov [304, p. 129]).
Let $\mathcal{X}$ be a normed space, let $\mathcal{Y}$ be its completion, and let $\mathcal{Z}$ be a Banach space. Let $\mathcal{A} : \mathcal{X} \to \mathcal{Z}$ be a bounded linear operator. Then there is a unique bounded linear operator $\hat{\mathcal{A}} : \mathcal{Y} \to \mathcal{Z}$ for which

$$\hat{\mathcal{A}}x = \mathcal{A}x, \quad all \ x \in \mathcal{Y}$$

$$\|\hat{\mathcal{A}}\|_{\mathcal{Y} \to \mathcal{Z}} = \|\mathcal{A}\|_{\mathcal{X} \to \mathcal{Z}}$$

$\hat{\mathcal{A}}$ is called the extension of $\mathcal{A}$ to the space $\mathcal{Y}$.

This theorem can be used to extend differentiation to functions that appear not to be differentiable in any usual sense, obtaining distributional derivatives.

Bibliography

[1] M. Abramowitz and I. Stegun, editors (1965) *Handbook of Mathematical Functions*, Dover Publications, New York.

[2] R. Adams (1975) *Sobolev Spaces*, Academic Press, New York.

[3] J. Albrecht and L. Collatz, editors (1980) *Numerical Treatment of Integral Equations*, Birkhaüser-Verlag, Stuttgart, Germnay.

[4] E. Allgower, K. Böhmer, K. Georg, and R. Miranda (1992) Exploiting symmetry in boundary element methods, *SIAM J. Numer. Anal.* **29**, pp. 534–552.

[5] E. Allgower, K. Georg, and R. Widmann (1991) Volume integrals for boundary element methods, *J. Comp. Appl. Math.* **38**, pp. 17–29.

[6] B. Alpert (1993) A class of bases in L^2 for the sparse representation of integral operators, *SIAM J. Math. Anal.* **24**, pp. 246–262.

[7] B. Alpert, G. Beylkin, R. Coifman, and V. Rokhlin (1993) Wavelets for the fast solution of second-kind integral equations, *SIAM J. Sci. Stat. Comp.* **14**, pp. 159–184.

[8] S. Amini and N. Maines (1995) "Regularisation of strongly singular integrals in boundary integral equations," Tech. Rep. MCS-95-7, Univ. of Salford, Manchester, United Kingdom.

[9] S. Amini and N. Maines (1995) "Qualitative properties of boundary integral operators and their discretizations," Tech. Rep. MCS-95-12, Univ. of Salford, Manchester, United Kingdom.

[10] S. Amini and I. Sloan (1989) Collocation methods for second kind integral equations with non-compact operators, *J. Integral Eqns & Applics*, **2**, pp. 1–30.

[11] R. Anderssen and P. Bloomfield (1974) Numerical differentiation procedures for non-exact data, *Numer. Math.* **22**, pp. 157–182.

[12] R. Anderssen and F. de Hoog (1990) Abel integral equations, in *Numerical Solution of Integral Equations*, ed. by M. Golberg, Plenum Publishing Corp., New York, pp. 373–410.

[13] R. Anderssen, F. de Hoog, and M. Lukas, editors (1980) *The Application and Numerical Solution of Integral Equations*, Sijthof & Noordhoff, Alphen aan den Rijn, The Netherlands.

[14] P. Anselone, ed. (1964) *Nonlinear Integral Equations*, Univ. of Wisconsin Press, Madison.

[15] P. Anselone (1965) Convergence and error for approximate solutions of

519

integral and operator equations, in *Error in Digital Computation*, Vol. II, ed. by L. Rall, John Wiley & Sons, New York, pp. 231–252.

[16] P. Anselone (1971) *Collectively Compact Operator Approximation Theory and Applications to Integral Equations*, Prentice-Hall, Englewood Cliffs, NJ.

[17] P. Anselone and J. Lee (1979) A unified approach to the approximate solution of linear integral equations, *Lecture Notes in Math.* **701**, Springer-Verlag, Berlin, pp. 41–69.

[18] P. Anselone and R. Moore (1964) Approximate solution of integral and operator equations, *J. Math. Anal. Appl.* **9**, pp. 268–277.

[19] P. Anselone and T. Palmer (1968) Spectral analysis of collectively compact, strongly convergent operator sequences, *Pacific J. Math.* **25**, pp. 423–431.

[20] P. Anselone and I. Sloan (1985) Integral equations on the half line, *J. Integral Equations* **9** (Suppl), pp. 3–23.

[21] P. Anselone and I. Sloan (1988) Numerical solution of integral equations on the half line II: The Wiener-Hopf case, *J. Integral Eqns & Applics* **1**, pp. 203–225.

[22] R. Ansorge (1992) Convergence of discretizations of nonlinear problems: A general approach, *Hamburger Beträge zur Angewandten Mathematik*, Reihe A, #50.

[23] T. Apostol (1957) *Mathematical Analysis*, Addison-Wesley Publishing Co., Reading, MA.

[24] D. Arnold (1983) A spline-trigonometric Galerkin method and an exponentially convergent boundary integral method, *Math. Comp.* **41**, pp. 383–397.

[25] D. Arnold and W. Wendland (1983) On the asymptotic convergence of collocation methods, *Numer. Math.* **41**, pp. 349–381.

[26] D. Arnold and W. Wendland (1985) The convergence of spline collocation for strongly elliptic equations on curves, *Numer. Math.* **47**, pp. 317–341.

[27] D. Arthur (1973) The solution of Fredholm integral equations using spline functions, *J. Inst. Math. Appl.* **11**, pp. 121–129.

[28] K. Atkinson (1966) *Extension of the Nyström method for the numerical solution of linear integral equations of the second kind*, Ph.D. dissertation, University of Wisconsin, Madison.

[29] K. Atkinson (1967) The solution of non-unique linear integral equations, *Numer. Math.* **10**, pp. 117–124.

[30] K. Atkinson (1967) The numerical solution of the eigenvalue problem for compact integral operators, *Trans. Am. Math. Soc.* **129**, pp. 458–465.

[31] K. Atkinson (1967) The numerical solution of Fredholm integral equations of the second kind, *SIAM J. Numer. Anal.* **4**, pp. 337–348.

[32] K. Atkinson (1972) The numerical solution of Fredholm integral equations of the second kind with singular kernels, *Numer. Math.* **19**, pp. 248–259.

[33] K. Atkinson (1973) The numerical evaluation of fixed points for completely continuous operators, *SIAM J. Numer. Anal.* **10**, pp. 799–807.

[34] K. Atkinson (1973) Iterative variants of the Nyström method for the numerical solution of integral equations, *Numer. Math.* **22**, pp. 17–31.

[35] K. Atkinson (1974) An existence theorem for Abel integral equations, *SIAM J. Math. Anal.* **5**, pp. 729–736.

[36] K. Atkinson (1975) Convergence rates for approximate eigenvalues of compact integral operators, *SIAM J. Numer. Anal.* **12**, pp. 213–222.

[37] K. Atkinson (1976) An automatic program for Fredholm linear integral equations of the second kind, *ACM Trans. Math. Soft.* **2**, pp. 154–171.

[38] K. Atkinson (1976) Algorithm 503: An automatic program for Fredholm integral equations of the second kind, *ACM Trans. Math. Soft.* **2**, pp. 196–199.

(This gives just the preliminary comments. The entire program listings are given in the *Collected Algorithms of the ACM*. The complete programs are available from the internet home page of the author at the University of Iowa.)

[39] K. Atkinson (1976) *A Survey of Numerical Methods for the Solution of Fredholm Integral Equations of the Second Kind*, Society for Industrial and Applied Mathematics, Philadelphia, PA.

[40] K. Atkinson (1980) The numerical solution of Laplace's equation in three dimensions-II, in *Numerical Treatment of Integral Equations*, ed. by J. Albrecht and L. Collatz, Birkhäuser, Basel, Switzerland, pp. 1–23.

[41] K. Atkinson (1982) The numerical solution of Laplace's equation in three dimensions, *SIAM J. Numer. Anal.* **19**, pp. 263–274.

[42] K. Atkinson (1982) Numerical integration on the sphere, *J. Austral. Math. Soc. (Series B)* **23**, pp. 332–347.

[43] K. Atkinson (1985) Piecewise polynomial collocation for integral equations on surfaces in three dimensions, *J. Integral Equations* **9** (Suppl), pp. 25–48.

[44] K. Atkinson (1985) The numerical evaluation of particular solutions for Poisson's equation, *IMA J. Numer. Anal.*, **5**, pp. 319–338.

[45] K. Atkinson (1985) Algorithm 629: An integral equation program for Laplace's equation in three dimensions, *ACM Trans. Math. Soft.* **11**, pp. 85–96.

[46] K. Atkinson (1985) Solving integral equations on surfaces in space, in *Constructive Methods for the Practical Treatment of Integral Equations*, ed. by G. Hämmerlin and K. Hoffman, Birkhäuser, Basel, Switzerland, pp. 20–43.

[47] K. Atkinson (1988) A discrete Galerkin method for first kind integral equations with a logarithmic kernel, *J. Integral Eqns Appl.* **1**, pp. 343–363.

[48] K. Atkinson (1989) *An Introduction to Numerical Analysis*, 2nd ed., John Wiley & Sons, New York.

[49] K. Atkinson (1989) An empirical study of the numerical solution of integral equations on surfaces in $\mathbf{R}^3$, *Reports on Computational Mathematics* #1, Dept of Mathematics, University of Iowa, Iowa City.

[50] K. Atkinson (1990) A survey of boundary integral equation methods for the numerical solution of Laplace's equation in three dimensions, in *Numerical Solution of Integral Equations*, ed. by M. Golberg, Plenum Publishing Corp., New York, pp. 1–34.

[51] K. Atkinson (1991) A survey of numerical methods for solving nonlinear integral equations, *J. Integral Eqns Appl.* **4**, pp. 15–46.

[52] K. Atkinson (1993) User's guide to a boundary element package for solving integral equations on piecewise smooth surfaces, *Reports on Computational Mathematics* #43, Dept of Mathematics, University of Iowa, Iowa City. (This is available from the internet home page of the author at the University of Iowa.)

[53] K. Atkinson (1994) Two-grid iteration methods for linear integral equations of the second kind on piecewise smooth surfaces in $\mathbf{R}^3$, *SIAM J. Sci. Stat. Comp.* **15**, pp. 1083–1104.

[54] K. Atkinson (1994) The numerical solution of a nonlinear boundary integral equation on smooth surfaces, *IMA J. Numer. Anal.* **14**, pp. 461–483.

[55] K. Atkinson (1997) The numerical solution of boundary integral equations, in *The State of the Art in Numerical Analysis*, ed. by I. Duff and A. Watson, Oxford University Press.

[56] K. Atkinson and A. Bogomolny (1987) The discrete Galerkin method for integral equations, *Math. Comp.* **48**, pp. 595–616 and S11–S15.

[57] K. Atkinson and G. Chandler (1990) BIE methods for solving Laplace's equation with nonlinear boundary conditions: The smooth boundary case, *Math. Comp.* **55**, pp. 455–472.

[58] K. Atkinson and G. Chandler (1995) The collocation method for solving the radiosity equation for unoccluded surfaces, *Reports on Computational Mathematics* #75, Dept of Mathematics, University of Iowa, Iowa City.

[59] K. Atkinson and D. Chien (1995) Piecewise polynomial collocation for boundary integral equations, *SIAM J. Sci. Stat. Comp.* **16**, pp. 651–681.

[60] K. Atkinson and J. Flores (1993) The discrete collocation method for nonlinear integral equations, *IMA J. Numer. Anal.* **13**, pp. 195–213.

[61] K. Atkinson and I. Graham (1990) Iterative variants of the Nyström method for second kind boundary integral equations, *SIAM J. Sci. Stat. Comp.* **13**, pp. 694–722.

[62] K. Atkinson, I. Graham, and I. Sloan (1983) Piecewise continuous collocation for integral equations, *SIAM J. Numer. Anal.* **20**, pp. 172–186.

[63] K. Atkinson and F. de Hoog (1982) Collocation methods for a boundary integral equation on a wedge, in *Treatment of Integral Equations by Numerical Methods*, ed. by C. Baker and G. Miller, Academic Press, New York, pp. 239–252.

[64] K. Atkinson and F. de Hoog (1984) The numerical solution of Laplace's equation on a wedge, *IMA J. Numer. Anal.* **4**, pp. 19–41.

[65] K. Atkinson and W. Han (1993) On the numerical solution of some semilinear elliptic problems, *Reports on Computational Mathematics* #49, Dept. of Mathematics, University of Iowa, Iowa City.

[66] K. Atkinson and F. Potra (1987) Galerkin's method for nonlinear integral equations, *SIAM J. Numer. Anal.* **24**, pp. 1352–1373.

[67] K. Atkinson and F. Potra (1988) The discrete Galerkin method for nonlinear integral equations, *J. Integral. Eqns. Appl.* **1**, pp. 17–54.

[68] K. Atkinson and F. Potra (1989) The discrete Galerkin method for linear integral equations, *IMA J. Numer. Anal.* **9**, pp. 385–403.

[69] K. Atkinson and I. Sloan (1991) The numerical solution of first kind logarithmic-kernel integral equations on smooth open arcs, *Math. Comp.* **56**, pp. 119–139.

[70] J.-P. Aubin (1979) *Applied Functional Analysis*, John Wiley & Sons, New York.

[71] I. Babuška and A. Aziz (1972), Survey lectures on the mathematical foundations of the finite element method, in *The Mathematical Foundations of the Finite Element Method with Applications to Partial Differential Equations*, ed. by A. Aziz, Academic Press, New York, pp. 3–362.

[72] I. Babuška, B. Guo, and E. Stephan (1990) On the exponential convergence of the *h-p* version for boundary element Galerkin methods on polygons, *Math. Meth. Appl. Sci.* **12**, pp. 413–427.

[73] C. Baker (1977) *The Numerical Treatment of Integral Equations*, Oxford Univ. Press, Oxford, England, UK.

[74] C. Baker (1987) The state of the art in the numerical treatment of integral equations, in *The State of the Art in Numerical Analysis*, ed. by A. Iserles and M. Powell, Clarendon Press, Oxford, England, UK, 1987, pp. 473–509.

[75] C. Baker and G. Miller, editors (1982) *Treatment of Integral Equations by Numerical Methods*, Academic Press, New York.

[76] P. Banerjee and J. Watson, editors (1986) *Developments in Boundary Element Methods-4*, Elsevier Science Publishing Co., New York.

[77] J. Benveniste (1967) Projective solutions of the Dirichlet problem for boundaries with angular points, *SIAM J. Appl. Math.* **15**, pp. 558–568.

[78] M. Bernkopf (1966) The development of function spaces with particular reference to their origins in integral equation theory, *Arch. Hist. Exact Sci.* **3**, pp. 1–96.

[79] D. Beskos, editor (1986) *Boundary Element Methods in Mechanics*, Elsevier, Amsterdam.

[80] G. Beylkins, R. Coifman, and V. Rokhlin (1991) Fast wavelet transforms and numerical algorithms I, *Comm. Pure Appl. Math.* **44**, pp. 141–183.

[81] J. Blue (1978) Boundary integral solutions of Laplace's equation, *Bell Sys. Tech. J.* **57**, pp. 2797–2822.

[82] H. Blum (1988) Numerical treatment of corner and crack singularities, in *Finite Element and Boundary Element Techniques from Mathematical and Engineering Points of View*, ed. by E. Stein and W. Wendland, Springer-Verlag, New York, pp. 171–212.

[83] D. Bond and S. Vavasis (1994) Fast wavelet transforms for matrices arising from boundary element methods, *Cornell Theory Center Tech. Rep.* CTC94TR174, Cornell University, Ithaca, NY.

[84] M. Bourland, S. Nicaise, and L. Paquet (1991) An adapted boundary element method for the Dirichlet problem in polygonal domains, *SIAM J. Num. Anal.* **28**, pp. 728–743.

[85] H. Brakhage (1960) Über die numerische Behandlung von Integralgleichungen nach der Quadraturformelmethode, *Numer. Math.* **2**, pp. 183–196.

[86] H. Brakhage (1961) Zur Fehlerabschätzung für die numerische Eigenwertbestimmung bei Integralgleichungen, *Numer. Math.* **3**, pp. 174–179.

[87] H. Brakhage and P. Werner (1965) Über das Dirichletsche Außenraumproblem für die Helmholtzsche Schwingungsgleichung, *Arch. Math.* **16**, pp. 325–329.

[88] A. Brandt and A. Lubrecht (1990) Multilevel matrix solution of integral equations, *J. Comp. Phys.* **90**, pp. 348–370.

[89] C. Brebbia, editor (1984) *Topics in Boundary Element Research, Vol. 1: Basic Principles and Applications*, Springer-Verlag, Berlin.

[90] C. Brebbia, editor (1985) *Topics in Boundary Element Research, Vol. 2: Time Dependent and Vibration Problems*, Springer-Verlag, Berlin.

[91] C. Brebbia, editor (1987) *Topics in Boundary Element Research, Vol. 3: Computational Aspects*, Springer-Verlag, Berlin.

[92] S. Brenner and L. Scott (1994) *The Mathematical Theory of Finite Element Methods*, Springer-Verlag, New York.

[93] G. Bruhn and W. Wendland (1967) Über die näherungsweise Lösung von linearen Funktionalgleichungen, in *Functionalanalysis, Approximations-theorie, Numerische Mathematik*, ed. by L. Collatz, G. Meinardus, and H. Unger, Birkhäuser Verlag, Basel, Switzerland.

[94] H. Brunner (1984) Iterated collocation methods and their discretizations, *SIAM J. Numer. Anal.* **21**, pp. 1132–1145.

[95] H. Brunner (1987) Collocation methods for one-dimensional Fredholm and Volterra integral equations, in *The State of the Art in Numerical Analysis*, ed. by A. Iserles and M. Powell, Clarendon Press, Oxford, England, UK, 1987, pp. 563–600.

[96] H. Brunner and H. de Riele (1986) *The Numerical Solution of Volterra Equations*, Elsevier, The Netherlands.

[97] H. Bückner (1952) *Die Praktische Behandlung von Integralgleichungen*, Springer-Verlag, Berlin.

[98] H. Bückner (1962) Numerical methods for integral equations, in *Survey of Numerical Analysis*, ed. by John Todd, McGraw-Hill, New York, Chapter 12.

[99] K. Bühring (1995) *"Quadrature methods for the Cauchy singular integral equation on curves with corner points and for the hypersingular integral equation on the interval,"* Ph.D. dissertation, Fachbereich Mathematik Und Informatik, Freie Universität, Berlin.

[100] A. Burton and G. Miller (1971) The application of integral equation methods to the numerical solution of some exterior boundary-value problems, *Proc. Roy. Soc. Lond. A* **323**, pp. 201–210.

[101] F. Canning (1990) Sparse matrix approximations to an integral equation of scattering, *Comm. Appl. Num. Methods* **6**, pp. 543–548.

[102] F. Canning, Sparse approximation for solving integral equations with oscillatory kernels, *SIAM J. Sci. Stat. Comp.* (in press).

[103] Y. Cao and Y. Xu (1994) Singularity preserving Galerkin methods for weakly singular Fredholm integral equations, *J. Integral Eqns. & Appl.* **6**, pp. 303–334.

[104] C. Carstensen and E. Stephan, Adaptive boundary element methods for some first kind integral equations, *SIAM J. Numer. Anal.*, (in press).

[105] G. Chandler (1979) *Superconvergence of numerical solutions to second kind integral equations*, Ph.D. dissertation, Australian National University, Canberra.

[106] G. Chandler (1984) Galerkin's method for boundary integral equations on polygonal domains, *J. Austral. Math. Soc., Series B*, **26**, pp. 1–13.

[107] G. Chandler (1986) Superconvergent approximations to the solution of a boundary integral equation on polygonal domains, *SIAM J. Numer. Anal.* **23**, pp. 1214–1229.

[108] G. Chandler (1992) Midpoint collocation for Cauchy singular integral equations, *Numer. Math.* **62**, pp. 483–503.

[109] G. Chandler and I. Graham (1988) Product integration-collocation methods for non-compact integral operator equations, *Math. Comp.* **50**, pp. 125–138.

[110] G. Chandler and I. Sloan (1990) Spline qualocation methods for boundary integral equations, *Numer. Math.* **58**, pp. 537–567.

[111] S. Chandler-Wilde (1992) On asymptotic behavior at infinity and the finite section method for integral equations on the half-line, *J. Integral Eqns. & Appl.* **6**, pp. 37–74.

[112] S. Chandler-Wilde (1994) On the behavior at infinity of solutions of integral equations on the real line, *J. Integral Eqns. & Appl.* **4**, pp. 153–177.

[113] F. Chatelin (1983) *Spectral Approximation of Linear Operators*, Academic Press, New York.

[114] F. Chatelin and R. Lebbar (1981) The iterated projection solution for the Fredholm integral equation of the second kind, *J. Austral. Math. Soc., Series B*, **22**, pp. 439–451.

[115] G. Chen and J. Zhou (1992) *Boundary Element Methods*, Academic Press, London.

[116] Y. Chen (1994) *Galerkin methods for solving single layer integral equations in three dimensions*, Ph.D. dissertation, Univ. of Iowa, Iowa City.

[117] Y. Chen and K. Atkinson (1994) Solving a single layer integral equation on surfaces in $\mathbf{R}^3$, *Reports on Computational Mathematics* #51, Dept. of Mathematics, University of Iowa, Iowa City.

[118] R. Cheng (1993) On using a modified Nyström method to solve the 2-D potential problem, *J. Integral Eqns. Appl.* **5**, pp. 167–193.

[119] D. Chien (1991) *Piecewise polynomial collocation for integral equations on surfaces in three dimensions*, Ph.D. dissertation, Univ. of Iowa, Iowa City.

[120] D. Chien (1992) Piecewise polynomial collocation for integral equations with a smooth kernel on surfaces in three dimensions, *J. Integral Eqns. and Appl.* **5**, pp. 315–344.

[121] D. Chien (1995) Numerical evaluation of surface integrals in three dimensions, *Math. Comp.* **64**, pp. 727–743.

[122] D. Chien and K. Atkinson, The numerical solution of a hypersingular planar boundary integral equation, *IMA J. Num. Anal.*, (in press).

[123] J. Cochran (1972) *Analysis of Linear Integral Operators*, McGraw-Hill, New York.

[124] M. Cohen and J. Wallace (1993) *Radiosity and Realistic Image Synthesis*, Academic Press, New York.

[125] L. Collatz (1966) *Functional Analysis and Numerical Mathematics*, Academic Press, New York.

[126] D. Colton (1988) *Partial Differential Equations: An Introduction*, Random House, New York.

[127] D. Colton and R. Kress (1983) *Integral Equation Methods in Scattering Theory*, John Wiley, New York.

[128] J. Conway (1990) *A Course in Function Analysis*, 2nd ed., Springer-Verlag, Berlin.

[129] M. Costabel (1987) Principles of boundary element methods, *Comp. Physics Rep.* **6**, pp. 243–274.

[130] M. Costabel (1988) Boundary integral operators on Lipschitz domains: Elementary results, *SIAM J. Math. Anal.* **19**, pp. 613–626.

[131] M. Costabel (1990) Boundary integral operators for the heat equation, *Integral Eqns. Operator Theory* **13**, pp. 498–552.

[132] M. Costabel and W. McLean (1992) Spline collocation for strongly elliptic equations on the torus, *Numer. Math.* **62**, pp. 511–538.

[133] M. Costabel and E. Stephan (1983) The normal derivative of the double layer potential on polygons and Galerkin approximation, *Appl. Anal.* **16**, pp. 205–228.

[134] M. Costabel and E. Stephan (1985) Boundary integral equations for mixed boundary value problems in polygonal domains and Galerkin approximation, in *Mathematical Models and Methods in Mechanics*, Banach Center Publications **15**, Warsaw, pp. 175–251.

[135] M. Costabel and E. Stephan (1985) A direct boundary integral equation method for transmission problems, *J. Math. Anal. Appl.* **106**, pp. 367–413.

[136] M. Costabel and E. Stephan (1987) On the convergence of collocation methods for boundary integral equations on polygons, *Math. Comp.* **49**, pp. 461–478.

[137] M. Costabel and E. Stephan (1988) Duality estimates for the numerical solution of integral equations, *Numer. Math.* **54**, pp. 339–353.

[138] M. Costabel and E. Stephan (1988) Strongly elliptic boundary integral equations for electromagnetic transmission problems, *Proc. Roy. Soc. Edinburgh Sect. A* **109**, pp. 271–296.

[139] M. Costabel and E. Stephan (1990) Coupling of finite and boundary element methods for an elastoplastic interface problem, *SIAM J. Numer. Anal.* **27**, pp. 1212–1226.

[140] M. Costabel, E. Stephan, and W. Wendland (1983) On boundary integral equations of the first kind for the bi-Laplacian in a polygonal plane domain, *Annali Scuola Normale Superiore – Pisa, Series IV* **10**, pp. 197–241.

[141] M. Costabel and W. Wendland (1986) Strong ellipticity of boundary integral operators, *J. Reine Angewandte Math.* **372**, pp. 34–63.
[142] R. Courant and D. Hilbert (1953) *Methods of Mathematical Physics*, Vol. I, Wiley Interscience, New York, Chapter 3.
[143] T. Cruse (1988) *Boundary Element Analysis in Computational Fracture Mechanics*, Kluwer Academic Pub., Norwell, MA.
[144] T. Cruse, A. Pitko, and H. Armen, editors (1985) *Advanced Topics in Boundary Element Analysis*, American Society of Mechanical Engineers, New York.
[145] C. Cryer (1970) "The solution of the Dirichlet problem for Laplace's equation when the boundary data is discontinuous and the domain has a boundary which is of bounded rotation by means of the Lebesgue-Stieltjes integral equation for the double layer potential," Tech. Rep. 99, Dept. of Computer Sci., Univ. of Wisconsin.
[146] C. Cryer (1982) *Numerical Functional Analysis*, Clarendon Press, Oxford, England, UK.
[147] P. Cubillos (1980) *On the numerical solution of Fredholm integral equations of the second kind*, Ph.D. dissertation, University of Iowa, Iowa City.
[148] P. Cubillos (1987) Eigenvalue problems for Fredholm integral operators, *IMA J. Numer. Anal.* **7**, pp. 191–204.
[149] J. Cullum (1971) Numerical differentiation and regularization, *SIAM J. Numer. Anal.* **8**, pp. 254–265.
[150] B. Dahlberg and G. Verchota (1990) Galerkin methods for the boundary integral equations of elliptic equations in non-smooth domains, in *Proc. Conf. Boca Raton, Fla. 1988*, Amer. Math. Soc., Providence, RI, pp. 39–60.
[151] W. Dahmen, B. Kleemann, S. Prößdorf, and R. Schneider (1994) A multiscale method for the double layer potential equation on a polyhedron, in *Advances in Computational Mathematics*, ed. by H. Djkshit and C. Micchelli, *World Scientific Pub. Co.*, Singapore, pp. 15–57.
[152] W. Dahmen and A. Kunoth (1992) Multilevel preconditioning, *Numer. Math.* **63**, pp. 315–344.
[153] W. Dahmen and C. Micchelli (1993) Using the refinement equation for evaluating integrals of wavelets, *SIAM J. Numer. Anal.* **30**, pp. 507–537.
[154] W. Dahmen, S. Prößdorf, and R. Schneider (1993) Multiscale methods for pseudodifferential equations, in *Recent Advances in Wavelet Analysis*, ed. by L. Schumaker and G. Webb, Academic Press, New York, pp. 191–235.
[155] W. Dahmen, S. Prößdorf, and R. Schneider (1994) Wavelet approximation methods for pseudodifferential equations I: stability and convergence, *Math. Zeitschrift* **215**, pp. 583–620.
[156] W. Dahmen, S. Prößdorf, and R. Schneider (1993) Wavelet approximation methods for pseudodifferential equations II: matrix compression and fast solution, *Adv. Comp. Math.* **1**, pp. 259–335.
[157] W. Dahmen, S. Prößdorf, and R. Schneider (1994) Multiscale methods for pseudodifferential equations on smooth manifolds, in *Proceedings of the International Conference on Wavelets: Theory, Algorithms, and Applications*, ed. by C. Chui, L. Montefusco, and L. Puccio, Academic Press, New York, pp. 385–424.
[158] M. Dauge (1988) *Elliptic Boundary Value Problems on Corner Domains*, Springer Lecture Notes in Math, Springer-Verlag, Berlin.
[159] R. Dautray and J. Lions, editors (1990) *Integral Equations and Numerical Methods*, in *Mathematical Analysis and Numerical Methods for Science and Technology*, Vol. IV, Springer-Verlag, Berlin.

[160] P. Davis (1959) On the numerical integration of periodic analytic functions, in *Numerical Approximation*, ed. by R. Langer, Univ. of Wisconsin Press, Madison, pp. 45–59.

[161] P. Davis (1963) *Interpolation and Approximation*, Blaisdell, Waltham, MA.

[162] F. de Hoog (1973) *"Product integration techniques for the numerical solution of integral equations,"* Ph.D. dissertation, Australian National University, Canberra.

[163] F. de Hoog and I. Sloan (1987) The finite section approximation for integral equations on the half line, *J. Austral. Math. Soc., Series B*, **28**, pp. 415–434.

[164] F. de Hoog and R. Weiss (1973) Asymptotic expansions for product integration, *Math. Comp.* **27**, pp. 295–306.

[165] L. Delves and J. Walsh, editors (1974) *Numerical Solution of Integral Equations*, Clarendon Press, Oxford, England, UK.

[166] L. Delves and J. Mohamed (1985) *Computational Methods for Integral Equations*, Cambridge Univ. Press, Cambridge.

[167] D. Dellwo (1988) Accelerated refinement with applications to integral equations, *SIAM J. Num. Anal.* **25**, pp. 1327–1339.

[168] D. Dellwo (1989) Accelerated spectral refinement with applications to integral operators, *SIAM J. Numer. Anal.* **26**, pp. 1184–1193.

[169] R. Doucette (1991) *"Boundary integral equation methods for the numerical solution of Laplace's equation with nonlinear boundary conditions on two-dimensional regions with a piecewise smooth boundary,"* Ph.D. dissertation, University of Iowa, Iowa City.

[170] M. Dow and D. Elliott (1979) The numerical solution of singular integral equations over (−1, 1), *SIAM J. Numer. Anal.* **16**, pp. 115–134.

[171] M. Duffy (1982) Quadrature over a pyramid or cube of integrands with a singularity at the vertex, *SIAM J. Numer. Anal.* **19**, pp. 1260–1262.

[172] A. Edelman (1993) Large dense numerical linear algebra in 1993: The parallel computing influence, (in press) *J. Supercomputing Appl.* **7**, pp. 113–128.

[173] P. Eggermont and C. Lubich (1994) Fast numerical solution of singular integral equations, *J. Integral Eqns Appl.* **6**, pp. 335–351.

[174] P. Eggermont and J. Saranen (1990) L^p estimates of boundary integral equations for some nonlinear boundary value problems, *Numer. Math.* **58**, pp. 465–478.

[175] D. Elliott (1959/60) The numerical solution of integral equations using Chebyshev polynomials, *J. Austral. Math. Soc.* **1**, pp. 334–356.

[176] D. Elliott (1963) A Chebyshev series method for the numerical solution of Fredholm integral equations, *Comp. J.* **6**, pp. 102–111.

[177] D. Elliott (1987) "Singular integral equations on the arc (−1, 1): Theory and Approximate Solution, Parts I & II," Tech. Reports #218 and #223, Dept. of Mathematics, University of Tasmania, Hobart, Australia.

[178] D. Elliott (1989) A comprehensive approach to the approximate solution of singular integral equations over the arc (−1, 1), *J. Integral Eqns. & Appl.* **2**, pp. 59–94.

[179] D. Elliott (1989) Projection methods for singular integral equations, *J. Integral Eqns. & Appl.* **2**, pp. 95–106.

[180] D. Elliott (1990) Convergence theorems for singular integral equations, in *Numerical Solution of Integral Equations*, ed. by M. Golberg, Plenum Publishing Corp., New York, pp. 309–361.

[181] D. Elliott and S. Prößdorf (1995) An algorithm for the approximate solution of integral equations of Mellin type, *Numer. Math.* **70**, pp. 427–452.

[182] J. Elschner (1988) On spline approximation for a class of integral equations – I: Galerkin and collocation methods with piecewise polynomials, *Math. Meth. Appl. Sci.* **10**, pp. 543–559.

[183] J. Elschner (1990) On spline approximation for a class of non-compact integral equations, *Math. Nachr.* **146**, pp. 271–321.

[184] J. Elschner (1992) The double layer potential operator over polyhedral domains I: Solvability in weighted Sobolev spaces, *Appl. Anal.* **45**, pp. 117–134.

[185] J. Elschner (1992) The double layer potential operator over polyhedral domains II: Spline Galerkin methods. *Math. Meth. Appl. Sci.* **15**, pp. 23–37.

[186] J. Elschner (1993) The h-p version of spline approximation methods for Mellin convolution equations, *J. Integral Eqns. Appl.* **5**, pp. 47–73.

[187] J. Elschner (1993) On the exponential convergence of spline approximation methods for Wiener-Hopf equations, *Math. Nachr.* **160**, pp. 253–264.

[188] J. Elschner and I. Graham (1995) An optimal order collocation method for first kind boundary integral equations on polygons, *Numer. Math.* **70**, pp. 1–31.

[189] J. Elschner and I. Graham (1995) Quadrature methods for Symm's integral equation on polygons, *J. IMA Numer. Anal.* (in press).

[190] J. Elschner and E. Stephan (1995) A discrete collocation method for Symm's integral equation on curves with corners, *J. Comp. Appl. Math.*, to appear.

[191] J. Elschner, Y. Jeon, I. Sloan, and E. Stephan, The collocation method for mixed boundary value problems on domains with curved polygonal boundaries, *Numer. Math.* (in press).

[192] J. Elschner, S. Prößdorf, and I. Sloan (1996) The qualocation method for Symm's integral equation on a polygon, *Math. Nachr.* **177**, pp. 81–108.

[193] H. Engl (1983) On the convergence of regularization methods for ill-posed linear operator equations, in *Improperly Posed Problems and Their Numerical Treatment*, ed. by G. Hämmerlin and K. Hoffmann, Birkhäuser-Verlag, Basel, Switzerland, pp. 81–96.

[194] E. Fabes, M. Jodeit, and J. Lewis (1977) Double layer potentials for domains with corners and edges, *Indiana Univ. Math. J.* **26**, pp. 95–114.

[195] G. Fairweather, F. Rizzo, and D. Shippy (1979) Computation of double integrals in the boundary integral method, in *Advances in Computer Methods for Partial Differential Equations – III*, ed. by R. Vichnevetsky and R. Stepleman, IMACS Symposium, Rutgers University, New Jersey, pp. 331–334.

[196] G. Farin (1988) *Curves and Surfaces for Computer Aided Geometric Design*, Academic Press, New York.

[197] I. Fenyö and H. Stolle (1981) *Theorie und Praxis der linearen Integralgleichungen – 1*, Birkhäuser-Verlag, Switzerland.

[198] I. Fenyö and H. Stolle (1983) *Theorie und Praxis der linearen Integralgleichungen – 2*, Birkhäuser-Verlag, Switzerland.

[199] I. Fenyö and H. Stolle (1983) *Theorie und Praxis der linearen Integralgleichungen – 3*, Birkhäuser-Verlag, Switzerland.

[200] I. Fenyö and H. Stolle (1984) *Theorie und Praxis der linearen Integralgleichungen – 4*, Birkhäuser-Verlag, Switzerland.

[201] J. Flores (1990) *"Iteration methods for solving integral equations of the second kind,"* Ph.D. dissertation, University of Iowa, Iowa City.

[202] J. Flores (1993) The conjugate gradient method for solving Fredholm integral equations of the second kind, *Intern. J. Comp. Math.* **48**, pp. 77–94.

[203] D. Forsyth and A. Zisserman (1990) Shape from shading in the light of mutual illumination, *Image and Vision Computing* **8**, pp. 42–49.

[204] L. Fox and E. Goodwin (1953) The numerical solution of non-singular linear integral equations, *Phil. Trans. Royal. Soc. London* A245, pp. 510–534.

[205] I. Fredholm (1903) Sur une classe d'équations fonctionelles, *Acta Math.* **27**, pp. 365–390.

[206] R. Freund, F. Golub, and N. Nachtigal (1992) Iterative solution of linear systems, in *Acta Numerica – 1992*, Cambridge University Press, New York.

[207] D. Gaier (1964) *Konstruktive Methoden der Konformen Abbildung*, Springer-Verlag, Berlin.

[208] F. Gakhov (1966) *Boundary Value Problems*, Pergamon Press, Oxford, England, UK.

[209] M. Ganesh, I. Graham, and J. Sivaloganathan (1993) A pseudospectral 3D boundary integral method applied to a nonlinear model problem from finite elasticity, *SIAM J. Numer. Anal.* **31**, pp. 1378–1414.

[210] M. Ganesh, I. Graham, and J. Sivaloganathan (1996) A new spectral boundary integral collocation method for three-dimensional potential problems, *SIAM J. Numer. Anal.*, to appear.

[211] P. Garabedian (1964) *Partial Differential Equations*, John Wiley, New York, pp. 334–388.

[212] W. Gautschi (1962) On inverses of Vandermonde and confluent Vandermonde matrices, *Numer. Math.* **4**, pp. 117–123.

[213] K. Georg (1991) Approximation of integrals for boundary element methods, *SIAM J. Sci. Stat. Comp.* **12**, pp. 443–453.

[214] K. Georg and J. Tausch (1994) Some error estimates for the numerical approximation of surface integrals, *Math. Comp.* **62**, pp. 755–763.

[215] A. Gerasoulis, editor (1984) *Numerical Solution of Singular Integral Equations, IMACS*, Dept. of Comp. Sc:, Rutgers Univ., New Brunswick, NJ.

[216] R. Gilbert (1970) The construction of solutions for boundary value problems by function theoretic methods, *SIAM J. Math. Anal.* **1**, pp. 96–114.

[217] J. Giroire and J. Nedelec (1978) Numerical solution of an exterior Neumann problem using a double layer potential, *Math. Comp.* **32**, pp. 973–990.

[218] M. Golberg, editor (1978) *Solution Methods for Integral Equations*, Plenum Press, New York.

[219] M. Golberg, editor (1990) *Numerical Solution of Integral Equations*, Plenum Press, New York.

[220] M. Golberg (1994) Discrete polynomial-based Galerkin methods for Fredholm integral equations, *J. Integral Eqns. Appl.* **6**, pp. 197–211.

[221] M. Golberg (1995) The numerical evaluation of particular solutions in the BEM – A review, *Boundary Element Comm.* **6**, pp. 99–106.

[222] M. Golberg (1995) A note on the sparse representation of discrete integral operators, *Appl. Math. Comp.* **70**, pp. 97–118.

[223] M. Golberg and C. Chen (1994) On a method of Atkinson for evaluating domain integrals in the boundary element method, *Applied Math. Comp.* **60**, pp. 125–138.

[224] G. Golub and C. Van Loan (1989) *Matrix Computations*, 2nd ed., Johns Hopkins University Press, Baltimore, MD.

[225] I. Graham (1980) *The Numerical Solution of Fredholm Integral Equations of the Second Kind*, Ph.D. dissertation, University of New South Wales, Sydney, Australia.

[226] I. Graham (1981) Collocation methods for two dimensional weakly singular integral equations, *J. Austral. Math. Soc. (Series B)* **22**, pp. 456–473.

[227] I. Graham (1982) Singularity expansions for the solution of second kind

Fredholm integral equations with weakly singular convolution kernels, *J. Integral Eqns.* **4**, pp. 1–30.

[228] I. Graham (1982) Galerkin methods for second kind integral equations with singularities, *Math. Comp.* **39**, pp. 519–533.

[229] I. Graham and K. Atkinson (1993) On the Sloan iteration applied to integral equations of the first kind, *IMA J. Numer. Anal.* **13**, pp. 29–41.

[230] I. Graham and G. Chandler (1988) High order methods for linear functionals of solutions of second kind integral equations, *SIAM J. Numer. Anal.* **25**, pp. 1118–1137.

[231] I. Graham, S. Joe, and I. Sloan (1985) Iterated Galerkin versus iterated collocation for integral equations of the second kind, *IMA J. Numer. Anal.* **5**, pp. 355–369.

[232] I. Graham and W. Mendes (1989) Nyström-product integration for Wiener-Hopf equations with applications to radiative transfer, *IMA J. Numer. Anal.* **9**, pp. 261–284.

[233] I. Graham, L. Qun, and X. Rui-feng (1992) Extrapolation of Nyström solutions of boundary integral equations on non-smooth domains, *J. Comp. Math.* **10**, pp. 231–244.

[234] I. Graham and C. Schneider (1985) Product integration for weakly singular integral equations in m-dimensional space, in *Constructive Methods for the Practical Treatment of Integral Equations*, ed. by G. Hämmerlin and K. Hoffman, Birkhaüser, Basel.

[235] I. Graham and Y. Yan (1991) Piecewise constant collocation for first kind boundary integral equations, *J. Australian Math. Soc. (Series B)* **33**, pp. 39–64.

[236] C. Green (1969) *Integral Equation Methods*, Thomas Nelson Pub, Britain, London.

[237] L. Greengard (1988) *The Rapid Evaluation of Potential Fields in Particle Systems*, The MIT Press, Cambridge, Massachusetts.

[238] L. Greengard and M. Moura (1994) On the numerical evaluation of electrostatic fields in composite materials, *Acta Numerica 1994*, pp. 379–410.

[239] L. Greengard and V. Rokhlin (1987) A fast algorithm for particle simulation, *J. Comp. Phys.* **73**, pp. 325–348.

[240] P. Grisvard (1985) *Elliptic Problems in Nonsmooth Domains,* Pitman Publishing, Marshfield, MA.

[241] C. Groetsch (1984) *The Theory of Tikhonov Regularization for Fredholm Equations of the First Kind*, Pitman Publishing, Marshfield, MA.

[242] C. Groetsch (1993) *Inverse Problems in the Mathematical Sciences*, Friedr. Vieweg & Sohn Verlagsgesellschaft, Braunschweig/Wiesbaden, Germany.

[243] T. Gronwall (1914) On the degree of convergence of Laplace's series, *Trans. Am. Math. Soc.* **15**, pp. 1–30.

[244] J. Guermond (1992) Numerical quadratures for layer potentials over curved domains in $\mathbf{R}^3$, *SIAM J. Numer. Anal.* **29**, pp. 1347–1369.

[245] J. Guermond and S. Fontaine (1991) A discontinuous h-p Galerkin approximation of potential flows, *Rech. Aérosp.* **4**, pp. 37–49.

[246] N. Günter (1967) *Potential Theory*, Ungar Pub., New York.

[247] W. Hackbusch (1981) Die schnelle Auflösung der Fredholmschen Integralgleichungen zweiter Art, *Beiträge Numer. Math.* **9**, pp. 47–62.

[248] W. Hackbusch (1985) *Multi-grid Methods and Applications*, Springer-Verlag, Berlin.

[249] W. Hackbusch (1989) *Integralgleichungen: Theorie und Numerik*, B. G. Teubner, Stuttgart, Germany.

[250] W. Hackbusch, editor (1992) *Numerical Techniques for Boundary Element Methods*, Notes on Fluid Mechanics, Vol. 33, Proceedings of the 7th GAMM-Seminar, Kiel, 1991, Friedr. Vieweg & Sohn Verlag, Braunschweig/Wiesbaden, Germany.

[251] W. Hackbusch (1994) *Integral Equations: Theory and Numerical Treatment*, Birkhäuser Verlag, Basel.

[252] W. Hackbusch and Z. Nowak (1986) A multilevel discretization and solution method for potential flow problems in three dimensions, in *Finite Approximations in Fluid Mechanics*, ed. by E. H. Hirschel, Notes on Numerical Fluid Mechanics **14**, Vieweg, Braunschweig.

[253] W. Hackbusch and Z. Nowak (1989) On the fast matrix multiplication in the boundary element method by panel clustering, *Numer. Math.* **54**, pp. 463–491.

[254] P. Halmos and V. Sunder (1978) *Bounded Integral Operators on L^2 Spaces*, Springer-Verlag, Berlin.

[255] G. Hämmerlin and K.-H. Hoffmann, editors (1985) *Constructive Methods for the Practical Treatment of Integral Equations*, Birkhäuser-Verlag, Basel.

[256] G. Hämmerlin and L. Schumaker (1980) Procedures for kernel approximation and solution of Fredholm integral equations of the second kind, *Numer. Math.* **34**, pp. 125–141.

[257] E. Hardee (1993) *"Projection methods with globally smooth approximations for Hammerstein integral equations on surfaces,"* Ph.D. dissertation, University of Iowa, Iowa City.

[258] R. Harrington (1990) Origin and development of the method of moments for field computation, *IEEE Antennas & Propagation Magazine* (June).

[259] M. Hashimoto (1970) A method of solving large matrix equations reduced from Fredholm integral equations of the second kind, *J. Assoc. Comp. Mach.* **17**, pp. 629–636.

[260] F. Hebeker (1986) Efficient boundary element methods for three-dimensional exterior viscous flows, *Numer. Meth. Part. Diff. Eqns.* **2**, pp. 273–297.

[261] F. Hebeker (1988) On the numerical treatment of viscous flows against bodies with corners and edges by boundary element and multi-grid methods, *Numer. Math.* **52**, pp. 81–99.

[262] S. Heinrich (1984) On the optimal error of degenerate kernel methods, Tech. Rep. N/84/2, Friedrich-Schiller-Universität Jena, Germany.

[263] H. Helfrich (1981) Simultaneous approximation in negative norms of arbitrary order, *RAIRO Numer. Anal.* **15**, pp. 231–235.

[264] P. Hemker and H. Schippers (1981) Multiple grid methods for Fredholm integral equations of the second kind, *Math. Comp.* **36**, pp. 215–232.

[265] P. Henrici (1974) *Applied and Computational Complex Analysis*, Vol. 1, John Wiley & Sons, New York.

[266] P. Henrici (1977) *Applied and Computational Complex Analysis*, Vol. 2, John Wiley & Sons, New York.

[267] P. Henrici (1986) *Applied and Computational Complex Analysis*, Vol. 3, John Wiley & Sons, New York.

[268] J. Hess (1990) Panel methods in computational fluid dynamics, *Ann. Rev. Fluid Mech.* **22**, pp. 255–274.

[269] J. Hess and A. Smith (1967) Calculation of Potential Flows about Arbitrary Bodies, in *Progress in Aeronautical Sciences*, Vol. 8, ed. by D. Küchemann, Pergamon Press, London.

[270] S. Hildebrandt and E. Wienholtz (1964) Constructive proofs of representation theorems in separable Hilbert space, *Comm. Pure Appl. Math.* **17**, pp. 369–373.

[271] E. Hille and J. Tamarkin (1931) On the characteristic values of linear integral equations, *Acta Math.* **57**, pp. 1–76.

[272] H. Hochstadt (1973) *Integral Equations*, John Wiley, New York.

[273] G. Hsiao, P. Kopp, and W. Wendland (1980) A Galerkin collocation method for some integral equations of the first kind, *Computing* **25**, pp. 89–130.

[274] G. Hsiao, P. Kopp, and W. Wendland (1984) Some applications of a Galerkin-collocation method for boundary integral equations of the first kind, *Math. Meth. Appl. Sci.* **6**, pp. 280–325.

[275] G. Hsiao and R. Kress (1985) On an integral equation for the two-dimensional exterior Stokes problem, *Appl. Numer. Math.* **1**, pp. 77–93.

[276] G. Hsiao and R. MacCamy (1973) Solutions of boundary value problems by integral equations of the first kind, *SIAM Rev.* **15**, pp. 687–705.

[277] G. Hsiao and W. Wendland (1977) A finite element method for some integral equations of the first kind, *J. Math. Anal. Appl.* **58**, pp. 449–481.

[278] G. Hsiao and W. Wendland (1981) The Aubin-Nitsche lemma for integral equations, *J. Integral Eqns.* **3**, pp. 299–315.

[279] C. Hsin, J. Kerwin, and J. Newman (1993) A higher order panel method based on B-splines, *Proceedings of the Sixth International Conference on Numerical Ship Hydrodynamics*, Iowa City, Iowa, preprint.

[280] D. Hwang and C. T. Kelley (1992), Convergence of Broyden's method in Banach spaces, *SIAM J. on Optimization* **2**, pp. 505–532.

[281] Y. Ikebe (1972) The Galerkin method for the numerical solution of Fredholm integral equations of the second kind, *SIAM Rev.* **14**, pp. 465–491.

[282] Y. Ikebe, M. Lynn, and W. Timlake (1969) The numerical solution of the integral equation formulation of the single interface Neumann problem, *SIAM J. Numer. Anal.* **6**, pp. 334–346.

[283] D. Ingham and M. Kelmanson (1984) *Boundary Integral Equation Analyses of Singular, Potential, and Biharmonic Problems*, Springer-Verlag, Berlin.

[284] V. Ivanov (1976) *The Theory of Approximate Methods and Their Application to the Numerical Solution of Singular Integral Equations*, Noordhoff International, Jeyden, The Netherlands.

[285] E. Isaacson and H. Keller (1966) *Analysis of Numerical Methods*, John Wiley, New York.

[286] M. Jaswon and G. Symm (1977) *Integral Equation Methods in Potential Theory and Elastostatics*, Academic Press, New York.

[287] Y.-M. Jeon (1992) Numerical analysis of boundary integral equations for the harmonic and biharmonic equations, Ph.D. thesis, University of Iowa, Iowa City.

[288] Y.-M. Jeon (1992) An indirect boundary integral equation for the biharmonic equation, *SIAM J. Numer. Anal.* **31**, pp. 461–476.

[289] Y.-M. Jeon (1993) A Nyström method for boundary integral equations on domains with a piecewise smooth boundary, *J. Integral Eqns. Appl.* **5**, pp. 221–242.

[290] Y.-M. Jeon (1995) A quadrature method for the Cauchy singular integral equations, *J. Integral Eqns. Appl.* **7**, pp. 425–461.

[291] Y.-M. Jeon and K. Atkinson (1993) An automatic program for the planar Laplace equation with a smooth boundary, *Reports on Computational Mathematics* #42, Dept. of Mathematics, University of Iowa.

[292] S. Joe (1985) Discrete collocation methods for second kind Fredholm integral equations, *SIAM J. Numer. Anal.* **22**, pp. 1167–1177.

[293] S. Joe (1987) Discrete Galerkin methods for Fredholm integral equations of the second kind, *IMA J. Numer. Anal.* **7**, pp. 149–164.

[294] C. Johnson (1988) *Numerical Solution of Partial Differential Equations by the Finite Element Method*, Cambridge Univ. Press, New York.

[295] C. Johnson and R. Scott (1989) An analysis of quadrature errors in second kind boundary integral equations, *SIAM J. Numer. Anal.* **26**, pp. 1356–1382.

[296] F. Johnson (1980) "A general panel method for the analysis and design of arbitrary configurations in incompressible flow," NASA Tech. Rep. NASA-CR-3079, National Tech. Inform. Service, Springfield, VA.

[297] K. Jörgens (1982) *Linear Integral Operators*, Pitman Publishing, Marshfield, MA.

[298] H. Kadner (1967) Die numerische Behandlung von Integralgleichungen nach der Kollokationsmethode, *Numer. Math.* **10**, pp. 241–260.

[299] J. Kane, D. Keyes, and K. Prasad (1991) Iterative solution techniques in boundary element analysis, *Intern. J. Numer. Meth. Eng.* **31**, pp. 1511–1536.

[300] H. Kaneko, R. Noren, and Y. Xu (1992) Numerical solutions for weakly singular Hammerstein equations and their superconvergence, *J. Integral Eqns. Appl.* **4**, pp. 391–406.

[301] H. Kaneko (1989) A projection method for solving Fredholm integral equations of the second kind, *Appl. Numer. Math.* **5**, pp. 333–344.

[302] H. Kaneko and Y. Xu (1991) Degenerate kernel method for Hammerstein equations, *Math. Comp.* **56**, pp. 141–148.

[303] L. Kantorovich (1948) Functional analysis and applied mathematics, *Uspehi Mat. Nauk* **3**, pp. 89–185.

[304] L. Kantorovich and G. Akilov (1964) *Functional Analysis in Normed Spaces*, Pergamon, Oxford, England, UK.

[305] L. Kantorovich and V. Krylov (1964) *Approximate Methods of Higher Analysis*, Noordhoff, Groningen, The Netherlands.

[306] R. Kanwal (1971) *Linear Integral Equations*, Academic Press, New York.

[307] O. Kellogg (1929) *Foundations of Potential Theory*, reprinted by Dover Publications, New York.

[308] C. Kelley (1989) A fast two-grid method for matrix H-equations, *Trans. Theory Stat. Physics* **18**, pp. 185–204.

[309] C. Kelley (1995) A fast multilevel algorithm for integral equations, *SIAM J. Numer. Anal.* **32**, pp. 501–513.

[310] C. Kelley and E. Sachs (1985) Broyden's method for approximate solution of nonlinear integral equations, *J. Integral Eqns* **9**, pp. 25–43.

[311] C. Kelley and E. Sachs (1993) Multilevel algorithms for constrained compact fixed point problems, *SIAM J. Sci. Stat. Comp.* **15**, pp. 645–667.

[312] C. Kelley and Z. Xue (1996) GMRES and integral operators, *SIAM J. Sci. Comp.* **17**, pp. 217–226.

[313] S. Kesavan (1989) *Topics in Functional Analysis and Applications*, John Wiley, New York.

[314] R. Kleinman and G. Roach (1974) Boundary integral equations for the three-dimensional Helmholtz equation, *SIAM Rev.* **16**, pp. 214–236.

[315] H. König (1986) *Eigenvalue Distribution of Compact Operators*, Birkhäuser-Verlag, Basel.

[316] J. Kral (1980) *Integral Operators in Potential Theory*, Lecture Notes in Math. #823, Springer-Verlag, Berlin.

[317] J. Kral and W. Wendland (1986) Some examples concerning applicability of the Fredholm-Radon method in potential theory, *Aplikace Matematiky* **31**, pp. 293–308.

[318] J. Kral and W. Wendland (1986) Some examples concerning the applicability

of the Fredholm-Radon method in potential theory, *Aplikace Matematiky* 31, pp. 293–308.

[319] M. Krasnoselskii (1964) *Topological Methods in the Theory of Nonlinear Integral Equations*, Pergamon Press, Oxford, England, UK.

[320] M. Krasnoselskii, G. Vainikko, P. Zabreyko, and Y. Rutitskii (1972) *Approximate Solution of Operator Equations*, Wolters-Noordhoff Pub, The Netherlands.

[321] M. Krasnoselskii and P. Zabreyko (1984) *Geometric Methods of Nonlinear Analysis*, Springer-Verlag, Berlin.

[322] M. Krein (1963) Integral equations on the half-line with kernel depending on the difference of the arguments, *AMS Translations, Series II*, Vol. 22, pp. 163–288.

[323] R. Kress (1972) Zur numerischen Integration periodischer Funktionen nach der Rechteckregel, *Numer. Math.* **20**, pp. 87–92.

[324] R. Kress (1985) Minimizing the condition number of boundary integral operators in acoustic and electromagnetic scattering, *Q. J. Mech. Appl. Math.* **38**, pp. 323–341.

[325] R. Kress (1989) *Linear Integral Equations*, Springer-Verlag, Berlin.

[326] R. Kress (1990) A Nyström method for boundary integral equations in domains with corners, *Numer. Math.* **58**, pp. 145–161.

[327] R. Kress (1995) On the numerical solution of a hypersingular integral equation in scattering theory, *J. Comp. Appl. Maths.* **61**, pp. 345–360.

[328] R. Kress and I. Sloan (1993) On the numerical solution of a logarithmic integral equation of the first kind for the Helmholtz equation, *Numer. Math.* **66**, pp. 199–214.

[329] E. Kreyszig (1978) *Introductory Functional Analysis with Applications*, John Wiley & Sons, New York.

[330] G. Krishnasamy, T. Rudolphi, L. Schmerr, and F. Rizzo (1988) Hypersingular integral formulas and wave scattering with the boundary element method, *Proceedings 1st Japan-US Symposium on Boundary Elements,* Tokyo.

[331] R. Kußmaul (1969) Ein numerisches Verfahren zur Lösung des Neumannschen Außenproblems fur die Helmholtzsche Schwingungsgleichung, *Computing* **4**, pp. 246–273.

[332] R. Kußmaul and P. Werner (1968) Fehlerabschätzungen für ein numerisches Verfahren zur Auflösung linearer Integralgleichungen mit schwachsingulären Kernen, *Computing* **3**, pp. 22–46.

[333] J. Lachat and J. Watson (1976) Effective numerical treatment of boundary integral equations, *Intern. J. for Numer. Meth. Eng.* **10**, pp. 991–1005.

[334] P. Lancaster and K. Šalkauskas (1986) *Curve and Surface Fitting: An Introduction*, Academic Press, New York.

[335] L. Landweber (1951) An iteration formula for Fredholm integral equations of the first kind, *Am. J. of Math.* **73**, pp. 615–624.

[336] P. Lax and A. Milgram (1954) *Parabolic Equations*, Annals of Math. Studies No. 33, pp. 167–190.

[337] B. Lee and M. Trummer (1994) Multigrid conformal mapping via the Szegö kernel, *Elect. Trans. Numer. Anal.* **2**, pp. 22–43.

[338] H. Lee (1991) *The multigrid method for integral equations*, Ph.D. dissertation, University of Iowa, Iowa City.

[339] R. Lehman (1959) Developments at an analytic corner of solutions of elliptic partial differential equations, *J. Math. Mech.* **8**, pp. 727–760.

[340] T.-C. Lin (1982) *"The numerical solution of the Helmholtz equation using integral equations,"* Ph.D. thesis, University of Iowa, Iowa City.

[341] T.-C. Lin (1984) A proof for the Burton and Miller integral equation approach for the Helmholtz equation, *J. Math. Anal. Appl.* **103**, pp. 565–574.

[342] T.-C. Lin (1985) The numerical solution of Helmholtz's equation for the exterior Dirichlet problem in three dimensions, *SIAM J. Numer. Anal.* **22**, pp. 670–686.

[343] T.-C. Lin (1988) Smoothness results of single and double layer solutions of the Helmholtz equation, *J. Integral Eqns. Appl.* **1**, pp. 83–121.

[344] P. Linz (1979) *Theoretical Numerical Analysis*, John Wiley, New York.

[345] P. Linz (1985) *Analytical and Numerical Methods for Volterra Equations*, Society for Industrial and Applied Mathematics, Philadelphia, PA.

[346] P. Linz (1991) Bounds and estimates for condition numbers of integral equations, *SIAM J. Numer. Anal.* **28**, pp. 227–235.

[347] A. Lonseth (1954) Approximate solutions of Fredholm-type integral equations, *Bull. AMS* **60**, pp. 415–430.

[348] W. Lovitt (1923) *Linear Integral Equations*, Dover, New York (reprint).

[349] D. Luenberger (1984) *Linear and Nonlinear Programming*, 2nd ed., Addison-Wesley, Reading, MA.

[350] J. Lyness (1976) Applications of extrapolation techniques to multidimensional quadrature of some integrand functions with a singularity, *J. Comp. Phys.* **20**, pp. 346–364.

[351] J. Lyness (1991) Extrapolation-based boundary element quadrature, *Rend. Sem. Mat. Univ. Pol. Torino, Fasc. Speciale 1991, Numerical Methods*, pp. 189–203.

[352] J. Lyness (1992) On handling singularities in finite elements, in *Numerical Integration*, ed. by T. Espelid and A. Genz, Kluwer Pub., pp. 219–233.

[353] J. Lyness (1994) Quadrature over curved surfaces by extrapolation, *Math. Comp.* **63**, pp. 727–747.

[354] J. Lyness (1994) Finite-part integrals and the Euler-MacLaurin expansion, in *Approximation & Computation*, ed. by R. Zahar, ISNM 119, Birkhäuser Verlag, Basel.

[355] J. Lyness and R. Cools (1994) A survey of numerical cubature over triangles, *Proc. Symp. Appl. Math.* **48**, pp. 127–150.

[356] J. Lyness and E. de Doncker (1993) On quadrature error expansions, Part II: The full corner singularity, *Numer. Math.* **64**, pp. 355–370.

[357] J. Lyness and D. Jespersen (1975) Moderate degree symmetric quadrature rules for the triangle, *J. Inst. Math. Appl.* **15**, pp. 19–32.

[358] J. Lyness and K. Puri (1973) The Euler-MacLaurin expansion for the simplex, *Math. Comp.* **27**, pp. 273–293.

[359] M. Lynn and W. Timlake (1968) The numerical solution of singular integral equations of potential theory, *Numer. Math.* **11**, pp. 77–98.

[360] T. MacRobert (1967) *Spherical Harmonics*, 3rd ed., Pergamon Press, Oxford England, UK.

[361] M. Maischak and E. Stephan (1994) The *hp*-version of the boundary element method in R^3: The basic approximation results, *Math. Mech. Appl. Sci.*, in press.

[362] J. Mandel (1985) On multilevel iterative methods for integral equations of the second kind and related problems, *Numer. Math.* **46**, pp. 147–157.

[363] H. Maniar (1995) "*A three dimensional higher order panel method based on B-splines,*" Ph.D. thesis, MIT, Cambridge, MA.

[364] E. Martensen (1968) *Potentialtheorie*, Teubner-Verlag, Stuttgart, Germany.

[365] J. Marti (1986) *Introduction to Sobolev Spaces and Finite Element Solution of Elliptic Boundary Value Problems*, Academic Press, New York.

536 Bibliography

[366] J. Mason and R. Smith (1982) Boundary integral equation methods for a variety of curved crack problems, in *Treatment of Integral Equations by Numerical Methods*, ed. by C. Baker and G. Miller, Academic Press, New York, pp. 239–252.

[367] V. Maz'ya (1991) Boundary integral equations, in *Analysis IV*, ed. by V. Maz'ya and S. Nikolskii, Springer-Verlag, Berlin, pp. 127–222.

[368] W. McLean (1985) *"Boundary integral methods for the Laplace equation,"* Ph.D. dissertation, Australian National University, Canberra.

[369] W. McLean (1986) A spectral Galerkin method for boundary integral equations, *Math. Comp.* **47**, pp. 597–607.

[370] W. McLean (1989) Asymptotic error expansions for numerical solutions of integral equations, *IMA J. Numer. Anal.* **9**, pp. 373–384.

[371] W. McLean (1991) Variational properties of some simple boundary integral equations, in *Proc. Centre for Maths. & Applics.* **26**, Australian National University, Canberra, pp. 168–178.

[372] W. McLean (1995) Fully-discrete collocation methods for an integral equation of the first kind, *J. Integral Eqns. Appl.* **6**, pp. 537–572.

[373] W. McLean, S. Prößdorf, and W. Wendland (1993) A fully-discrete trigonometric collocation method, *J. Integral Eqns. Appl.* **5**, pp. 103–129.

[374] W. McLean and I. Sloan (1992) A fully-discrete and symmetric boundary element method, *IMA J. Numer. Anal.* **14**, pp. 311–345.

[375] W. McLean and W. Wendland (1989) Trigonometric approximation of solutions of periodic pseudodifferential equations, *Operator Theory* **41**, pp. 359–383.

[376] G. Meinardus (1967) *Approximation of Functions: Theory and Numerical Methods*, Springer-Verlag, New York.

[377] S. Mikhlin (1950) *Singular Integral Equations*, AMS Translation Series I, #24, American Math. Soc., Providence, RI.

[378] S. Mikhlin (1964) *Integral Equations*, 2nd ed., Pergamon Press, Oxford England, UK.

[379] S. Mikhlin (1965) *Multi-dimensional Singular Integrals and Singular Integral Equations*, Pergamon Press, Oxford England, UK.

[380] S. Mikhlin (1970) *Mathematical Physics: An Advanced Course*, North-Holland Publishing, Amsterdam, The Netherlands.

[381] S. Mikhlin and S. Prößdorf (1986) *Singular Integral Operators*, Springer-Verlag, Berlin.

[382] S. Mikhlin and K. Smolitskiy (1967) *Approximate Methods for Solutions of Differential and Integral Equations*, American Elsevier, New York.

[383] C. Miller (1979) *"Numerical solution of two-dimensional potential theory problems using integral equation techniques,"* Ph.D. dissertation, University of Iowa, Iowa City.

[384] R. Miller (1971) *Nonlinear Volterra Integral Equations*, Benjamin/Cummings Publishing, Menlo Park, CA.

[385] R. Milne (1980) *Applied Functional Analysis: An Introductory Treatment*, Pitman Publishing, Marshfield, MA.

[386] G. Miranda (1970) *Partial Differential Equations of Elliptic Type*, 2nd ed., Springer-Verlag, Berlin.

[387] R. Moore (1968) Approximations to nonlinear operator equations and Newton's method, *Numer. Math.* **12**, pp. 23–34.

[388] I. Moret and P. Omari (1989) An iterative variant of the degenerate kernel method for solving Fredholm integral equations, *Math. Nachr.* **143**, pp. 41–54.

[389] M. Mortenson (1985) *Geometric Modelling*, John Wiley & Sons, New York.

[390] N. Muskhelishvili (1953) *Singular Integral Equations*, Noordhoff International, Leyden, The Netherlands.

[391] N. Muskhelishvili (1953) *Some Basic Problems of the Mathematical Theory of Elasticity*, Noordhoff International, Leyden, The Netherlands.

[392] I. Mysovskih (1961) An error estimate for the numerical solution of a linear integral equation, *Dokl. Akad. Nauk SSSR* **140**, pp. 763–765.

[393] K. Nabors, F. Korsmeyer, F. Leighton, and J. White (1994) Preconditioned, adaptive, multipole-accelerated iterative methods for three-dimensional first-kind integral equations of potential theory, *SIAM J. Sci. Comp.* **15**, pp. 713–735.

[394] Z. Nashed, editor (1976) *Generalized Inverses and Applications*, Academic Press, New York, pp. 193–243.

[395] R. Natarajan and D. Krishnaswamy, A case study in parallel scientific computing: The boundary element method on a distributed memory multicomputer, *Proceedings of Supercomputing 95*, Assoc. of Computing Machinery.

[396] J. Nedelec (1976) Curved finite element methods for the solution of singular integral equations on surfaces in $\mathbf{R}^3$, *Comp. Meth. Appl. Mech. Eng.* **8**, pp. 61–80.

[397] J. Nedelec and J. Planchard (1973) Une méthode variationnelle d'éléments finis pour la résolution numérique d'un problème extérieur dans R^3, R.A.I.R.O. 7 R3, pp. 105–129.

[398] C. Neumann (1877) *Untersuchungen über das logarithmische und Newtonsche Potential*, Teubner, Leipzig, Germany.

[399] O. Nevanlinna (1993) *Convergence of Iterations for Linear Equations*, Birkhäuser, Basel, Switzerland.

[400] J. Newman (1992) Panel methods in marine hydrodynamics, *Proc. Conf. Eleventh Australasian Conf. on Fluid Mechanics*, ed. by M. Davis and G. Walker, Vol. I, pp. 123–130, ISBN 0 85901 519 X.

[401] J. Nitsche (1970) Zur Konvergenz von Näherungsverfahren bezüglich verschiedener Normen, *Numer. Math.* **15**, pp. 224–228.

[402] B. Noble (1958) *Methods based on the Wiener-Hopf Technique for the Solution of Partial Differential Equations*, Pergamon Press, Oxford England, UK.

[403] B. Noble (1971) Some applications of the numerical solution of integral equations to boundary value problems, *Springer Lecture Notes in Math.* **228**, pp. 137–154.

[404] B. Noble (1971) Methods for solving integral equations, [bibliography] Tech. Rep. **1176** (author listing), **1177** (subject listing), *U.S. Army Mathematics Research Center*, Madison, Wisconsin.

[405] B. Noble (1973) Error analysis of collocation methods for solving Fredholm integral equations, in *Topics in Numerical Analysis*, ed. by J. H. Miller, Academic Press, New York.

[406] B. Noble (1977) The numerical solution of integral equations, in *The State of the Art in Numerical Analysis*, ed. by D. Jacobs, Academic Press, New York, pp. 915–966.

[407] Z. Nowak (1982) Use of the multigrid method for Laplacian problems in three dimensions, in *Multigrid Methods*, ed. by W. Hackbusch and U. Trottenberg, Springer-Verlag, Berlin.

[408] Z. Nowak (1988) Efficient panel methods for the potential flow problems in the three space dimensions, Rep. #8815, Institut für Informatik und Praktische Mathematik, Christian Albrechts Universität, Kiel, Germany.

[409] E. Nyström (1930) Über die praktische Auflösung von Integralgleichungen mit Anwendungen auf Randwertaufgaben, *Acta Math.* **54**, pp. 185–204.

[410] P. Omari (1989) On the fast convergence of a Galerkin-like method for equations of the second kind, *Math. Zeit.* **201**, pp. 529–539.

[411] J. Osborn (1975) Spectral approximation for compact operators, *Math. Comp.* **29**, pp. 712–725.

[412] W. Patterson (1974) *Iterative Methods for the Solution of a Linear Operator Equation in Hilbert Space – A Survey*, Springer Lecture Notes in Mathematics #394, Springer-Verlag, Berlin.

[413] M. Pester and S. Rjasanow (1993) A parallel version of the preconditioned conjugate gradient method for boundary element equations, Preprint SPC-93-2, Chemnitzer DFG-Forschergruppe "Scientific Parallel Computing," Technical University of Chemnitz-Zwickau, Germany.

[414] T. von Petersdorff (1989) *Elasticity problems in polyhedra – singularities and approximation with boundary elements*, Ph.D. dissertation, Technische Hochschule Darmstadt, Germany.

[415] T. von Petersdorff and C. Schwab, Wavelet approximations for first kind boundary integral equations on polygons, *Numer. Math.* (in press).

[416] T. von Petersdorff and C. Schwab (1996) Fully discrete multiscale Galerkin BEM, in *Multiresolution Analysis and Partial Differential Equations*, ed. by W. Dahmen, P. Kurdila, and P. Oswald, Academic Press, New York.

[417] T. von Petersdorff, C. Schwab, and R. Schneider (1996) Multiwavelets for second kind integral equations, *SIAM J. Num. Anal.*, to appear

[418] T. von Petersdorff and E. Stephan (1990) Decompositions in edge and corner singularities for the solution of the Dirichlet problem of the Laplacian in a polyhedron, *Math. Nachr.* **149**, pp. 71–104.

[419] T. von Petersdorff and E. Stephan (1990) Regularity of mixed boundary value problems in $\mathbf{R}^3$ and boundary element methods on graded meshes, *Math. Meth. Appl. Sci.* **12**, pp. 229–249.

[420] T. von Petersdorff and E. Stephan (1990) On the convergence of the multigrid method for a hypersingular integral equation of the first kind, *Numer. Math.* **57**, pp. 379–391.

[421] T. von Petersdorff and E. Stephan (1992) Multigrid solvers and preconditioners for first kind integral equations, *Numer. Meth. Partial Diff. Eqns.* **8**, pp. 443–450.

[422] W. Petryshyn (1963) On a general iterative method for the approximate solution of linear operator equations, *Math. Comp.* **17**, pp. 1–10.

[423] D. Phillips (1962) A technique for the numerical solution of certain integral equations of the first kind, *J. Assoc. Comp. Mach.* **9**, pp. 84–97.

[424] J. Phillips (1972) The use of collocation as a projection method for solving linear operator equations, *SIAM J. Numer. Anal.* **9**, pp. 14–28.

[425] J. Pitkäranta (1979) On the differential properties of solutions to Fredholm equations with weakly singular kernels, *J. Inst. Math. Appl.* **24**, pp. 109–119.

[426] W. Pogorzelski (1966) *Integral Equations and Their Applications*, Pergamon Press, Oxford, England, UK.

[427] D. Porter and D. Stirling (1990) *Integral Equations: A Practical Treatment, from Spectral Theory to Applications*, Cambridge Univ. Press, New York.

[428] D. Porter and D. Stirling (1993) The re-iterated Galerkin method, *IMA J. Numer. Anal.* **13**, pp. 125–139.

[429] F. Postell and E. Stephan (1990) On the *h*-, *p*-, and *h*-*p* versions of the boundary element method-Numerical results, *Comp. Meth. Appl. Mech. Eng.* **83**, pp. 69–89.

[430] M. Powell (1981) *Approximation Theory and Methods*, Cambridge Univ. Press, New York.

[431] H. Power (1993) The completed double layer boundary integral equation method for two-dimensional Stokes flow, *IMA J. Appl. Math.* **51**, pp. 123–145.

[432] C. Pozrikidis (1992) *Boundary Integral and Singularity Methods for Linearized Viscous Flow,* Cambridge Univ. Press, Cambridge, UK.

[433] P. Prenter (1973) A method of collocation for the numerical solution of integral equations of the second kind, *SIAM J. Numer. Anal.* **10**, pp. 570–581.

[434] P. Prenter (1975) *Splines and Variational Methods*, Wiley-Interscience, New York.

[435] S. Prößdorf (1978) *Some Classes of Singular Integral Equations*, North-Holland Publishing, Amsterdam, The Netherlands.

[436] S. Prößdorf (1991) Linear Integral Equations, in *Analysis IV*, ed. by V. Maz'ya and S. Nikolskii, Springer-Verlag, Berlin, pp. 1–125.

[437] S. Prößdorf and B. Silbermann (1977) *Projektionsverfahren und die nä herungsweise Lö sung singulä rer Gleichungen*, Teubner, Leipzig, Germany.

[438] S. Prößdorf and B. Silbermann (1991) *Numerical Analysis for Integral and Related Operator Equations*, Birkhäuser Verlag, Basel, Switzerland.

[439] S. Prößdorf and I. Sloan (1992) Quadrature method for singular integral equations on closed curves, *Numer. Math.* **61**, pp. 543–559.

[440] J. Radon (1919) Über die Randwertaufgaben beim logarithmischen Potential, *Sitzungsberichte der Akademie der Wissenschafter Wien* **128 Abt. IIa**, pp. 1123–1167.

[441] D. Ragozin (1971) Constructive polynomial approximation on spheres and projective spaces, *Trans. Am. Math. Soc.* **162**, pp. 157–170.

[442] D. Ragozin (1972) Uniform convergence of spherical harmonic expansions, *Math. Annalen* **195**, pp. 87–94.

[443] L. Rall (1955) Error bounds for iterative solution of Fredholm integral equations, *Pacific J. Math.* **5**, pp. 977–986.

[444] L. Rall (1969) *Computational Solution of Nonlinear Operator Equations*, John Wiley, New York.

[445] A. Ramm (1980) *Theory and Applications of Some New Classes of Integral Equations*, Springer-Verlag, New York.

[446] E. Rank (1989) Adaptive *h*-, *p*-, and *hp*-versions for boundary integral element methods, *Intern. J. Numer. Meth. Eng.* **28**, pp. 1335–1349.

[447] A. Rathsfeld (1992) The invertibility of the double layer potential operator in the space of continuous functions defined on a polyhedron: The panel method, *Appl. Anal.* **45**, pp. 135–177.

[448] A. Rathsfeld (1992) On the finite section method for double layer potential equations over the boundary of a polyhedron, *Math. Nachr.* **157**, pp. 7–14.

[449] A. Rathsfeld (1992) Piecewise polynomial collocation for the double layer potential equation over polyhedral boundaries. Part I: The wedge; Part II: The cube, *Preprint No. 8,* Institut für Angewandte Analysis und Stochastik, Berlin.

[450] A. Rathsfeld (1992) Iterative solution of linear systems arising from the Nyström method for the double layer potential equation over curves with corners, *Math. Meth. Appl. Sci.* **15**, pp. 443–455.

[451] A. Rathsfeld (1993) On quadrature methods for the double layer potential equation over the boundary of a polyhedron, *Numer. Math.* **66**, pp. 67–95.

[452] A. Rathsfeld (1994) Piecewise polynomial collocation for the double layer potential equation over polyhedral boundaries, in *Boundary Value Problems*

and Integral Equations in Non-smooth Domains, ed. by M. Costabel, M. Dauge, and S. Nicaise, Lecture Notes in Applied Mathematics 167, Marcel Dekker, Basel, pp. 219–253.

[453] A. Rathsfeld (1995) Nyström's method and iterative solvers for the solution of the double layer potential equation over polyhedral boundaries, *SIAM J. Numer. Anal.* **32**, pp. 924–951.

[454] A. Rathsfeld (1994) A wavelet algorithm for the solution of the double layer potential equation over polygonal boundaries, *J. Integral Eqns. Appl.* **7**, pp. 47–98.

[455] A. Rathsfeld, R. Kieser, and B. Kleemann (1992) On a full discretization scheme for a hypersingular boundary integral equation over smooth curves, *Z. Anal. Anwendungen* **11**, pp. 385–396.

[456] L. Reichel (1987) Parallel iterative methods for the solution of Fredholm integral equations of the second kind, in *Hypercube Multiprocessors*, ed. by M. T. Heath, Society for Industrial and Applied Mathematics, Philadelphia, pp. 520–529.

[457] L. Reichel and Y. Yan (1994) Fast solution of a class of periodic pseudo-differential equations, *J. Integral Eqns. Appl.* **6**, pp. 401–426.

[458] H. Reinhardt (1985) *Analysis of Approximate Methods for Differential and Integral Equations*, Springer-Verlag, Berlin.

[459] S. Rempel and G. Schmidt (1991) Eigenvalues for spherical domains with corners via boundary integral equations, *Integral Eqns. Oper. Theory* **14**, pp. 229–250.

[460] M. Renardy and R. Rogers (1992) *An Introduction to Partial Differential Equations*, Springer-Verlag, New York.

[461] J. Rice (1969) On the degree of convergence of nonlinear spline approximation, in *Approximations with Special Emphasis on Spline Functions*, ed. by I. J. Schoenberg, Academic Press, New York, pp. 349–365.

[462] G. Richter (1976) On weakly singular Fredholm integral equations with displacement kernels, *J. Math. Anal. Appl.* **55**, pp. 32–42.

[463] G. Richter (1977) An integral equation method for the biharmonic equation, in *Advances in Computer Methods for Partial Differential Equations: II*, ed. by R. Vichnevetsky, IMACS, Dept. of Computer Science, Rutgers Univ., New Brunswick.

[464] G. Richter (1978) Superconvergence of piecewise polynomial Galerkin approximations for Fredholm integral equations of the second kind, *Numer. Math.* **31**, pp. 63–70.

[465] G. Richter (1978) Numerical solution of integral equations of the first kind with nonsmooth kernels, *SIAM J. Numer. Anal.* **15**, pp. 511–522.

[466] H. te Riele, editor (1979) *Colloquium Treatment of Integral Equations*, Mathematisch Centrum, Amsterdam, The Netherlands.

[467] T. Rivlin (1990) *Chebyshev Polynomials*, 2nd ed., John Wiley & Sons, New York.

[468] F. Rizzo, D. Shippy, and M. Rezayat (1985) A boundary integral equation method for radiation and scattering of elastic waves in three dimensions, *Intern. J. Numer. Meth. Eng.* **21**, pp. 115–129.

[469] V. Rokhlin (1983) Rapid solution of integral equations of classical potential theory, *J. Comp. Phys.* **60**, pp. 187–207.

[470] T. Rudolphi, G. Krishnasamy, L. Schmerr, and F. Rizzo (1988) On the use of strongly singular integral equations for crack problems, in *Proc. 10th International Conference on Boundary Elements*, Springer-Verlag.

[471] K. Ruotsalainen (1992) On the Galerkin boundary element method for a mixed nonlinear boundary value problem, *Appl. Anal.* **46**, pp. 195–213.

[472] K. Ruotsalainen and J. Saranen (1987) Some boundary element methods using Dirac's distributions as trial functions, *SIAM J. Numer. Anal.* **24**, pp. 816–827.

[473] K. Ruotsalainen and J. Saranen (1989) On the collocation method for a nonlinear boundary integral equation, *J. Comp. Appl. Math.* **28**, pp. 339–348.

[474] K. Ruotsalainen and W. Wendland (1988) On the boundary element method for some nonlinear boundary value problems, *Numer. Math.* **53**, pp. 229–314.

[475] Y. Saad and M. Schultz (1986) GMRES: A generalized minimal residual algorithm for solving nonsymmetric linear systems, *SIAM J. Sci. Stat. Comput.* **7**, pp. 856–869.

[476] J. Saavedra (1989) *Boundary Integral Equations for Nonsimply Connected Regions*, Ph.D. dissertation, University of Iowa, Iowa City.

[477] J. Saranen (1988) The convergence of even degree spline collocation solution for potential problems in the plane, *Numer. Math.* **53**, pp. 499–512.

[478] J. Saranen (1989) Extrapolation methods for spline collocation solutions of pseudodifferential equations on curves, *Numer. Math.* **56**, pp. 385–407.

[479] J. Saranen (1990) Projection methods for a class of Hammerstein equations, *SIAM J. Numer. Anal.* **27**, pp. 1445–1449.

[480] J. Saranen (1991) The modified quadrature method for logarithmic-kernel integral equations on closed curves, *J. Integral Eqns. Appl.* **3**, pp. 575–600.

[481] J. Saranen (1993) A modified discrete spectral collocation method for first kind integral equations with a logarithmic kernel, *J. Integral Eqns. Appl.* **5**, pp. 547–567.

[482] J. Saranen and L. Schroderus (1995) Some discrete methods for boundary integral equations on smooth closed curves, *SIAM J. Numer. Anal.* **32**, pp. 1535–1564.

[483] J. Saranen and I. Sloan (1992) Quadrature methods for logarithmic-kernel integral equations on closed curves, *IMA J. Numer. Anal.* **12**, pp. 167–187.

[484] J. Saranen and G. Vainikko (1996) Trigonometric collocation methods with product-integration for boundary integral equations on closed curves, *SIAM J. Numer. Anal.* **33**, pp. 1577–1596.

[485] J. Saranen and W. Wendland (1985) On the asymptotic convergence of collocation methods with spline functions of even degree, *Math. Comp.* **45**, pp. 91–108.

[486] S. Sauter and C. Schwab (1996) Quadrature for hp-Galerkin BEM in $\mathbf{R}^3$, Res. Rep. #96-02, Seminar für Angew. Math., Eid. Tech. Hoch., Zurich.

[487] F. Javier Sayas (1994) *"Asymptotic expansion of the error of some boundary element methods,"* Ph.D. dissertation, Dept. of Applied Mathematics, University of Zaragoza, Spain.

[488] A. Schatz, V. Thomée, and W. Wendland (1990) *Mathematical Theory of Finite and Boundary Element Methods*, Birkhäuser Verlag, Basel, Switzerland.

[489] H. Schippers (1982) Application of multigrid methods for integral equations to two problems from fluid dynamics, *J. Comp. Physics* **48**, pp. 441–461.

[490] H. Schippers (1985) Multi-grid methods for boundary integral equations, *Numer. Math.* **46**, pp. 351–363.

[491] H. Schippers (1987) Multi-grid methods in boundary element calculations, in *Boundary Elements IX*, ed. by C. Brebbia, G. Kuhn, and W. Wendland, Springer-Verlag, Berlin, pp. 475–492.

[492] W. Schmeidler (1950) *Integralgleichungen mit Anwendungen in Physik und Technik*, Akademische-Verlag, Leipzig, Germany.

[493] C. Schneider (1979) Regularity of the solution to a class of weakly singular Fredholm integral equations of the second kind, *Integral Eqns. Oper. Theory* **2**, pp. 62–68.

[494] C. Schneider (1981) Product integration for weakly singular integral equations, *Math. Comp.* **36**, pp. 207–213.

[495] E. Schock (1985) Arbitrarily slow convergence, uniform convergence, and superconvergence of Galerkin-like methods, *IMA J. Numer. Anal.* **5**, pp. 153–160.

[496] C. Schwab (1994) Variable order composite quadrature of singular and nearly singular integrals, *Computing* **53**, pp. 173–194.

[497] C. Schwab and W. Wendland (1992) On numerical cubatures of singular surface integrals in boundary element methods, *Numer. Math.* **62**, pp. 343–369.

[498] A. Sidi (1989) Comparison of some numerical quadrature formulas for weakly singular periodic Fredholm integral equations, *Computing* **43**, pp. 159–170.

[499] A. Sidi and M. Israeli (1988) Quadrature methods for periodic singular and weakly singular Fredholm integral equations, *J. Sci. Comp.* **3**, pp. 201–231.

[500] F. Sillion and C. Puech (1994) *Radiosity and Global Illumination*, Morgan Kaufmann Pub., San Francisco, CA.

[501] I. Sloan (1976) Convergence of degenerate kernel methods, *J. Austral. Math. Soc. Series B* **19**, pp. 422–431.

[502] I. Sloan (1976) Error analysis for a class of degenerate kernel methods, *Numer. Math.* **25**, pp. 231–238.

[503] I. Sloan (1976) Improvement by iteration for compact operator equations, *Math. Comp.* **30**, pp. 758–764.

[504] I. Sloan (1980) The numerical solution of Fredholm equations of the second kind by polynomial interpolation, *J. Integral Eqns* **2**, pp. 265–279.

[505] I. Sloan (1981) Analysis of general quadrature methods for integral equations of the second kind, *Numer. Math.* **38**, pp. 263–278.

[506] I. Sloan (1988) A quadrature-based approach to improving the collocation method, *Numer. Math.* **54**, pp. 41–56.

[507] I. Sloan (1990) Superconvergence, in *Numerical Solution of Integral Equations*, ed. by M. Golberg, Plenum Press, New York.

[508] I. Sloan (1993) Polynomial interpolation and hyperinterpolation over general regions, School of Mathematics, Univ. of New South Wales, Sydney, Australia.

[509] I. Sloan (1992) Error analysis of boundary integral methods, *Acta Numerica* **1**, pp. 287–339.

[510] I. Sloan (1992) Unconventional methods for boundary integral equations in the plane, in *Numerical Analysis 1991*, ed. by D. F. Griffiths & G. A. Watson, Longman Scientific and Technical Pub., Britain, pp. 194–218.

[511] I. Sloan and K. Atkinson (1995) Semi-discrete Galerkin approximations for the single-layer equation on Lipschitz curves, *Reports on Computational Mathematics* #72, Dept of Mathematics, University of Iowa.

[512] I. Sloan and B. Burn (1979) Collocation with polynomials for integral equations of the second kind, *J. Integral Eqns* **1**, pp. 77–94.

[513] I. Sloan, B. Burn, and N. Datyner (1975) A new approach to the numerical solution of integral equations, *J. Comp. Physics* **18**, pp. 92–105.

[514] I. Sloan and W. Smith (1982) Properties of interpolatory product integration rules, *SIAM J. Numer. Anal.* **19**, pp. 427–442.

[515] I. Sloan and A. Spence (1988) Galerkin method for integral equations of

the first kind with logarithmic kernel: theory, *IMA J. Numer. Anal.* **8**, pp. 105–122.

[516] I. Sloan and A. Spence (1988) Galerkin method for integral equations of the first kind with logarithmic kernel: applications, *IMA J. Numer. Anal.* **8**, pp. 123–140.

[517] I. Sloan and E. Stephan (1992) Collocation with Chebyshev polynomials for Symm's integral equation on an interval, *J. Austral. Math. Soc., Series B*, **34**, pp. 199–211.

[518] V.I. Smirnov (1964) *Integral Equations and Partial Differential Equations: A Course of Higher Mathematics*, Vol. IV, Pergamon Press, Oxford, England, UK, Chapters 1, 4.

[519] W. Smith and I. Sloan (1980) Product integration rules based on the zeros of Jacobi polynomials, *SIAM J. Numer. Anal.* **17**, pp. 1–13.

[520] F. Smithies (1958) *Integral Equations*, Cambridge Univ. Press, New York.

[521] I. Sneddon (1966) *Mixed Boundary Value Problems in Potential Theory*, North-Holland, Amsterdam, The Netherlands.

[522] A. Spence (1974) *"The numerical solution of the integral equation eigenvalue problem,"* Ph.D. dissertation, Oxford University, Oxford, UK.

[523] A. Spence (1975) On the convergence of the Nyström method for the integral equation eigenvalue problem, *Numer. Math.* **25**, pp. 57–66.

[524] A. Spence (1978) Error bounds and estimates for eigenvalues of integral equations, *Numer. Math.* **29**, pp. 133–147.

[525] A. Spence and K. Thomas (1983) On superconvergence properties of Galerkin's method for compact operator equations, *IMA J. Numer. Anal.* **3**, pp. 253–271.

[526] I. Stakgold (1979) *Green's Functions and Boundary Value Problems*, Wiley-Interscience, New York.

[527] H. Starr (1991) *"On the numerical solution of one-dimensional integral and differential equations,"* Ph.D. dissertation, Yale University, New Haven, CT.

[528] F. Stenger (1993) *Numerical Methods Based on Sinc and Analytic Functions*, Springer-Verlag, New York.

[529] E. Stephan (1987) Boundary integral equations for screen problems in $\mathbf{R}^3$, *J. Integral Eqns Oper. Theory* **10**, pp. 236–257.

[530] E. Stephan (1987) Boundary integral equations for mixed boundary value problems in $\mathbf{R}^3$, *Math. Nachr.* **131**, pp. 167–199.

[531] E. Stephan (1988) Improved Galerkin methods for integral equations on polygons and on polyhedral surfaces, in *First Joint Japan/U.S. Symposium on Boundary Element Methods*, Tokyo, pp. 73–80.

[532] E. Stephan (1992) Coupling of finite elements and boundary elements for some nonlinear interface problems, *Comp. Meth. Appl. Mech. Eng.* **101**, pp. 61–72.

[533] E. Stephan (1996) The *hp*-version of the boundary element method for solving 2- and 3-dimensional problems – especially in the presence of singularities and adaptive approaches, *Comp. Meth. Appl. Mech. Eng.* **133**, pp. 183–208.

[534] E. Stephan and M. Suri (1989) On the convergence of the *p*-version of the boundary element Galerkin method, *Math. Comp.* **52**, pp. 32–48.

[535] E. Stephan and M. Suri (1991) The h-p version of the boundary element method on polygonal domains with quasiuniform meshes, *RAIRO Math. Modelling Numer. Anal.* **25**, pp. 783–807.

[536] E. Stephan and W. Wendland (1976) Remarks to Galerkin and least squares methods with finite elements for general elliptic problems, *Manuscripta Geodactica*, **1**, pp. 93–123.

[537] A. Stroud (1971) *Approximate Calculation of Multiple Integrals*, Prentice-Hall, Englewood Cliffs, NJ.

[538] F. Stummel (1970) Diskrete Konvergenz Linearer Operatoren I, *Math. Annalen* **190**, pp. 45–92.

[539] J. Tausch (1995) Perturbation analysis for some linear boundary integral operators, *J. Integral Eqns. Appl.* **7**, pp. 351–370.

[540] J. Tausch (1996) Equivariant preconditioners for boundary element methods, *SIAM J. Sci. Comp.* **17**, pp. 90–99.

[541] K. Thomas (1973) *"The numerical solution of integral equations,"* Ph.D. dissertation, Oxford University, Oxford, UK.

[542] K. Thomas (1975) On the approximate solution of operator equations, *Numer. Math.* **23**, pp. 231–239.

[543] A. Tikhonov and V. Arsenin (1977) *Solutions of Ill-posed Problems*, John Wiley, New York.

[544] F. Tricomi (1957) *Integral Equations*, Wiley-Interscience, New York.

[545] E. Tuck (1993) Some accurate solutions of the lifting surface integral equation, *J. Austral. Math. Soc., Series B* **35**, pp. 127–144.

[546] S. Twomey (1965) The application of numerical filtering to the solution of integral equations encountered in indirect sensing measurements, *J. Franklin Inst.* **279**, pp. 95–109.

[547] G. Vainikko (1967) On the speed of convergence of approximate methods in the eigenvalue problem, *USSR Comp. Math. & Math. Phys.* **7**, pp. 18–32.

[548] G. Vainikko (1969) The compact approximation principle in the theory of approximation methods, *USSR Comp. Math. & Math. Phys.* **9**, pp. 1–32.

[549] G. Vainikko (1976) *Funktionalanalysis der Diskretisierungsmethoden*, Teubner, Leipzig, Germany.

[550] G. Vainikko (1993) *Multidimensional Weakly Singular Integral Equations*, Lecture Notes in Mathematics #1549, Springer-Verlag, Berlin.

[551] G. Vainikko (1996) *"Periodic integral and pseudodifferential equations,"* Res. Rep. C13, Institute of Math., Helsinki Univ. of Tech., Helsinki, Finland.

[552] S. Vavasis (1992) Preconditioning for boundary integral equations, *SIAM J. Matrix Anal.* **13**, pp. 905–925.

[553] E. Venturino (1986) Recent developments in the numerical solution of singular integral equations, *J. Math. Anal. Appl.* **115**, pp. 239–277.

[554] W. Wasow (1957) Asymptotic development of the solution of Dirichlet's problem at analytic corners, *Duke Math. J.* **24**, pp. 47–56.

[555] J. Watson (1979) Advanced implementation of the boundary element method for two- and three-dimensional elastostatics, in *Developments in Boundary Element Methods*, ed. by P. Banerjee and R. Butterfield, Elsevier Appl. Sci. Pub., London.

[556] R. Weiss (1974) On the approximation of fixed points of nonlinear compact operators, *SIAM J. Numer. Anal.* **11**, pp. 550–553.s

[557] W. Wendland (1968) Die Behandlung von Randwertaufgaben im $\mathbf{R}^3$ mit Hilfe von Einfach und Doppelschichtpotentialen, *Numer. Math.* **11**, pp. 380–404.

[558] W. Wendland (1981) On the asymptotic convergence of boundary integral methods, in *Boundary Element Methods*, ed. by C. Brebbia, Springer-Verlag, Berlin, pp. 412–430.

[559] W. Wendland (1981) Asymptotic accuracy and convergence, in *Progress in Boundary Element Methods*, Vol. 1, ed. by C. Brebbia, J. Wiley & Sons, New York, pp. 289–313.

[560] W. Wendland (1983) Boundary element methods and their asymptotic convergence, in *Theoretical Acoustics and Numerical Techniques*, ed. by P. Filippi, Springer-Verlag, Berlin. CISM Courses and Lectures No. 277, International Center for Mechanical Sciences.

[561] W. Wendland (1985) Asymptotic accuracy and convergence for point collocation methods, in *Topics in Boundary Element Research, Vol. II: Time-Dependent and Vibration Problems*, Springer-Verlag, Berlin, pp. 230–257.

[562] W. Wendland (1985) On some mathematical aspects of boundary element methods for elliptic problems, in *The Mathematics of Finite Elements and Applications–V*, ed. by J. R. Whiteman, Academic Press, New York, pp. 193–227.

[563] W. Wendland (1985) "Splines versus trigonometric polynomials: *h*-version *p*-version in 2D boundary integral methods," Tech. Rep. #925, Fachbereich Mathematik, Technische Hochschule Darmstadt.

[564] W. Wendland (1987) Strongly elliptic boundary integral equations, in *The State of the Art in Numerical Analysis*, ed. by A. Iserles and M. Powell, Clarendon Press, Oxford, England, UK, pp. 511–562.

[565] W. Wendland (1990) Boundary element methods for elliptic problems, in *Mathematical Theory of Finite and Boundary Element Methods*, by A. Schatz, V. Thomée, and W. Wendland, Birkhäuser, Boston, pp. 219–276.

[566] W. Wendland, E. Stephan, and G. Hsiao (1979) On the integral equation method for the plane mixed boundary value problem of the Laplacian, *Math. Meth. Appl. Sci.* **1**, pp. 265–321.

[567] R. Whitley (1986) The stability of finite rank methods with applications to integral equations, *SIAM J. Numer. Anal.* **23**, pp. 118–134.

[568] H. Widom (1969) *Lectures on Integral Equations*, Van Nostrand, Reinhold, New York.

[569] H. Wielandt (1956) Error bounds for eigenvalues of symmetric integral equations, *Proc. AMS Symposium Appl. Math.* **6**, pp. 261–282.

[570] L. Wienert (1990) "*Die numerische approximation von randintegraloperatoren für die Helmholtzgleichung im* **R**^{3}," dissertation, der Georg-August-Universität zu Göttingen, Germany.

[571] D. Willis (1986) *Numerical solution of the heat equation by the method of hat potentials*, Ph.D. dissertation, University of Iowa, Iowa City, Iowa.

[572] G. Wing (1991) *A Primer on Integral Equations of the First Kind: The Problem of Deconvolution and Unfolding*, Society for Industrial and Applied Mathematics, Philadelphia, PA.

[573] R. Winther (1980) Some superlinear convergence results for the conjugate gradient method, *SIAM J. Numer. Anal.* **17**, pp. 14–17.

[574] K. Wright (1982) Asymptotic properties of matrices associated with the quadrature method for integral equations, in *Treatment of Integral Equations by Numerical Methods*, ed. by C. T. H. Baker and G. F. Miller, Academic Press, London, pp. 325–336.

[575] Y. Xu and Y. Cao (1994) Singularity preserving Galerkin methods for weakly singular Fredholm integral equations, *J. Integral Eqns. Appl.* **6**, pp. 303–334.

[576] Y. Xu and Y. Zhao (1994) Quadratures for improper integrals and their applications in integral equations, *Proc. Symp. Appl. Math.* **48**, pp. 409–413.

[577] Y. Xu and Y. Zhao (1996) Quadratures for boundary integral equations of the

first kind with logarithmic kernels, *J. Integral Eqns. Appl.*, **8**, pp. 239–268 (in press).

[578] Y. Yan (1990) The collocation method for first kind boundary integral equations on polygonal domains, *Math. Comp.* **54**, pp. 139–154.

[579] Y. Yan (1990) Cosine change of variable for Symm's integral equation on open arcs, *IMA J. Numer. Anal.* **10**, pp. 521–535.

[580] Y. Yan and I. Sloan (1988) On integral equations of the first kind with logarithmic kernels, *J. Integral Eqns. Appl.*, **1**, pp. 517–548.

[581] Y. Yan and I. Sloan (1989) Mesh grading for integral equations of the first kind with logarithmic kernel, *SIAM J. Numer. Anal.* **26**, pp. 574–587.

[582] Y. Yang (1993) *Multidimensional numerical integration and applications to boundary integral equations*, Ph.D. dissertation, University of Iowa, Iowa City, Iowa.

[583] Y. Yang (1995) A discrete collocation method for boundary integral equations, *J. Integral Eqns. & Appl.*, **7**, pp. 233–261.

[584] Y. Yang and K. Atkinson (1993) Numerical integration for multivariable functions with point singularities, *SIAM J. Numer. Anal.* **32**, pp. 969–983.

[585] A. Young (1954) The application of approximate product integration to the numerical solution of integral equations, *Proc. Royal Soc. London* **A224**, pp. 561–573.

[586] P. Zabreyko, A. Koshelev, M. Krasnosel'skii, S. Mikhlin, L. Rakovshchik, and V. Stet'senko (1975) *Integral Equations – A Reference Text*, Noordhoff International, Leyden, The Netherlands.

[587] E. Zeidler (1986) *Nonlinear Functional Analysis and its Applications – Vol. I: Fixed-Point Theorems*, Springer-Verlag, Berlin.

[588] E. Zeidler (1995) *Applied Functional Analysis: Applications to Mathematical Physics*, Springer-Verlag, Berlin.

[589] E. Zeidler (1995) *Applied Functional Analysis: Main Principles and Their Applications*, Springer-Verlag, Berlin.

[590] R. Zheng, N. Phan-Thien, and C. Coleman (1991) A boundary element approach for non-linear boundary-value problems, *Comp. Mech.* **8**, pp. 71–86.

[591] A. Zygmund (1959) *Trigonometric Series*, Vols. I and II, Cambridge Univ. Press, New York.

Index